The Quantum Beat

Springer
New York
Berlin
Heidelberg
Barcelona
Budapest
Hong Kong
London
Milan
Paris
Santa Clara
Singapore
Tokyo

F.G. Major

The Quantum Beat

The Physical Principles of Atomic Clocks

With 230 Illustrations

F.G. Major
284 Michener Court E.
Severna Park, MD 21146
USA

Library of Congress Cataloging-in-Publication Data
Major, Fouad G.
The quantum beat: the physical principles of atomic clocks/Fouad G. Major
p. cm.
Includes index.
ISBN 0-387-98301-5 (alk. paper)
1. Atomic clocks. I. Title.
QB107.M35 1998
522′.5—dc21

Printed on acid-free paper.

Figure 1.4. Courtesy of the British Museum, London.
Figure 8.8. Courtesy of Tekelec Neuchâtel Time, S.A. Avenue du Mail 59, CH 2000 Neuchâtel, Switzerland.
Figure 18.5. Courtesy of NASA-Headquarters, Still Photograph Library, Washington, D.C.
All other illustrations by Monica Williams of Williams Designs, 418 Edgewater Road, Pasadena, MD 21122, U.S.A.

Production managed by Timothy Taylor; manufacturing supervised by Joseph Quatela.
Photocomposed copy prepared using Microsoft Word.
Printed and bound by Maple-Vail Book Manufacturing Group, York, PA.
Printed in the United States of America.

9 8 7 6 5 4 3 2 1

ISBN 0-387-98301-5 Springer-Verlag New York Berlin Heidelberg SPIN 10635140

A new scientific truth does not triumph by convincing its opponents and making them see the light, but rather because its opponents eventually die, and a new generation grows up that is familiar with it.

Max Planck

Preface

The field of frequency and time measurement has undergone revolutionary changes in recent years, with regard both to its precision and its extension to include optical frequencies. The introduction of techniques to cool atoms and ions through interaction with suitable laser beams, coupled with methods of particle suspension in ultra-high vacuum, has proved astonishingly successful in bringing us closer to the ideal first expressed by Dehmelt of making observations on isolated atomic particles at rest in space. This became the author's own motivating principle in the development of a field-confined mercury ion standard for space applications. The recent implementation of a satellite-based global navigation system, the Global Positioning System, is the most visible example of the enormous impact that atomic frequency standards have had on the civilian and the military sectors of society. The crucial elements in this system are the spacecraft atomic clocks that can maintain a submicrosecond synchronization, putting the accuracy of position determination globally in the submeter range!

This work attempts to convey a broad understanding of the physical principles underlying the workings of these quantum-based atomic clocks, with introductory chapters placing them in context with the early development of mechanical clocks and the introduction of electronic time-keeping as embodied in the quartz-controlled clocks. While the book makes no pretense at being a history of atomic clocks, it nevertheless takes a historical perspective in its treatment of the subject.

Intended for nonspecialists with some knowledge of physics or engineering, *The Quantum Beat* covers a wide range of salient topics relevant to atomic clocks, treated in a broad intuitive manner with a minimum of mathematical formalism. Detailed descriptions are given of the design principles of the rubidium, cesium, hydrogen maser, and mercury ion standards; the revolutionary changes that the advent of the laser has made possible, such as laser cooling, optical pumping, the formation of "optical molasses," and the cesium "fountain" standard; and the time-based global navigation systems, Loran-C and the Global Positioning System. Also included are topics that bear on the precision and absolute accuracy of standards, such as noise, resonance line shape, the relativistic Doppler effect as well as more general relativistic notions of time relevant to synchronization of remote clocks, and time reversal symmetry.

I am greatly indebted to the following for the encouragement I derived from their willingness to read and provide valuable suggestions on parts of the manuscript: Professors Norman Ramsey, Claude Cohen-Tannoudji, Gisbert zu Putlitz,

Charles Drake, Hugh Robinson, and especially my friends and former colleagues Herbert Ueberall of the Catholic University of America, and Claude Audoin of the CNRS Laboratoire de l'Horloge Atomique in Orsay.

Severna Park, Maryland, USA F.G. Major
May 27, 1997

Contents

Preface vii

Chapter 1. Celestial and Mechanical Clocks 1

1.1 Cyclic Events in Nature 1
1.2 The Calendar 2
1.3 Solar Eclipses as Time Markers 3
1.4 The Tides 5
1.5 The Sidereal Day 7
1.6 The Precession of the Equinoxes 8
1.7 The Sundial 9
1.8 The Astrolabe 10
1.9 Water Clocks 12
1.10 Tower Clocks 14
1.11 The Pendulum Clock 16
1.12 The Spring–Balance-Wheel Clock 19

Chapter 2. Oscillations and Fourier Analysis 23

2.1 Oscillatory Motion in Matter 23
2.2 Simple Harmonic Motion 24
2.3 Forced Oscillations: Resonance 26
2.4 Waves in Extended Media 29
2.5 Wave Dispersion 32
2.6 Linear and Nonlinear Media 33
2.7 Normal Modes of Vibration 35
2.8 Parametric Excitations 37
2.9 Fourier Analysis 39
2.10 Coupled Oscillations 43

Chapter 3. Oscillators 47

3.1 Feedback in Amplifiers 47
3.2 Conditions for Oscillation 50

3.3 Resonators 51
3.4 The Klystron Microwave Tube 54
3.5 Oscillators at Optical Frequency 56
3.6 Stability of Oscillators: Noise 58

Chapter 4. Quartz Clocks 63

4.1 Historical Antecedents 63
4.2 Properties and Structure of Crystalline Quartz 66
4.3 Modes of Vibration of a Quartz Plate 71
4.4 X-Ray Crystallography 73
4.5 Fabrication of Quartz Resonators 75
4.6 Factors Affecting the Resonance Frequency 76
4.7 The Quartz Resonator as a Circuit Element 78
4.8 Frequency Stability 80
4.9 Frequency/Time Measurement 84
4.10 Quartz Watches 88

Chapter 5. The Language of Electrons, Atoms, and Quanta 89

5.1 Classical Lorentz Theory 89
5.2 Spectrum of Blackbody Radiation 90
5.3 The Quantum of Radiation: The Photon 91
5.4 Bohr's Theory of the Hydrogen Atom 92
5.5 The Schrödinger Wave Equation 94
5.6 Quantum Numbers of Atomic States 96
5.7 The Vector Model 98
5.8 The Shell Structure of Electron States 99
5.9 The Pauli Exclusion Principle 101
5.10 Spectroscopic Notation 103
5.11 The Hyperfine Interaction 104
5.12 Electrons in Solids: The Band Theory 109

Chapter 6. Magnetic Resonance 117

6.1 Introduction 117
6.2 Atomic Magnetism 117
6.3 The Zeeman Effect 118
6.4 Gyroscopic Motion in a Magnetic Field 122
6.5 Inducing Transitions 123
6.6 Motion of Global Moment: The Bloch Theory 127

6.7 Production of Global Polarization 128

Chapter 7. Corrections to Observed Atomic Resonance 141

7.1 Homogeneous and Inhomogeneous Broadening 142
7.2 The Special Theory of Relativity 144
7.3 The Doppler Effect 146
7.4 The Thermal Doppler Line Shape 148
7.5 Sub-Doppler Line Widths: the Dicke Effect 150
7.6 The General Theory of Relativity 153
7.7 Conclusion 159

Chapter 8. The Rubidium Clock 161

8.1 The Reference Hyperfine Transition 161
8.2 The Breit–Rabi Formula 163
8.3 Optical Pumping of Hyperfine Populations 164
8.4 Optical Hyperfine Pumping: Use of an Isotopic Filter 167
8.5 The Use of Buffer Gases 169
8.6 Light Shifts in the Reference Frequency 172
8.7 Rubidium Frequency Control of Quartz Oscillator 173
8.8 Frequency Stability of the Rubidium Standard 176
8.9 The Miniaturization of Atomic Clocks 178

Chapter 9. The Classical Cesium Standard 179

9.1 Definition of the Unit of Time 179
9.2 Implementation of the Definition: The Cesium Standard 180
9.3 The Physical Design 183
9.4 The Ramsey Separated Field 190
9.5 Detection of Transitions 196
9.6 Frequency-Lock of Flywheel Oscillator to Cesium 198
9.7 Corrections to the Observed Cs Frequency 201

Chapter 10. Atomic and Molecular Oscillators: Masers 205

10.1 The Ammonia Maser 205
10.2 Basic Elements of a Beam Maser 206
10.3 Inversion Spectrum in NH_3 207
10.4 The Electrostatic State Selector 210
10.5 Stimulated Radiation in the Cavity 214
10.6 Threshold for Sustained Oscillation 216

10.7 Sources of Frequency Instability 217
10.8 The Rubidium Maser 221

Chapter 11. The Hydrogen Maser **223**

11.1 Introduction 223
11.2 The Hyperfine Structure of H Ground State 225
11.3 Principles of the Hydrogen Maser 228
11.4 Physical Design of the H-Maser 235
11.5 Automatic Cavity Tuning 245
11.6 The Wall Shift in Frequency 247
11.7 The H-Maser Signal Handling 250
11.8 Hydrogen as a Passive Resonator 253

Chapter 12. The Confinement of Ions **255**

12.1 Introduction 255
12.2 State Selection in Ions 256
12.3 The Penning Trap 259
12.4 The Paul High-Frequency Trap 267

Chapter 13. The NASA Mercury Ion Experiment **285**

13.1 Introduction 285
13.2 Ground State Hyperfine Structure of Hg^{199} 286
13.3 Hyperfine Optical Pumping 288
13.4 Detection of Microwave Resonance 292
13.5 Microwave Resonance Line Shape 293
13.6 The Magnetic Field Correction 296
13.7 The Physical Apparatus 297
13.8 Hg^+ Ion Frequency Standard System 301

Chapter 14. Optical Frequency Oscillators: Lasers **307**

14.1 Introduction 307
14.2 The Resonance Line Width of Optical Cavities 307
14.3 Conditions for Sustained Oscillation 312
14.4 The Sustained Output Power 317
14.5 Laser Optical Elements 318
14.6 The Ruby Laser 322
14.7 The Helium–Neon Laser 327
14.8 The Argon Ion Laser 332

14.9 Liquid Dye Lasers 334
14.10 Semiconductor Lasers 339

Chapter 15. Laser Cooling of Atoms and Ions **345**

15.1 Introduction 345
15.2 Light Pressure 346
15.3 Scattering of Light from Small Particles 348
15.4 Scattering of Light by Atoms 350
15.5 Optical Field Gradient Force 352
15.6 Doppler Cooling 353
15.7 Theoretical Limit 356
15.8 Optical "Molasses" 357
15.9 Polarization Gradient Cooling: "The Sisyphus Effect" 358
15.10 Laser Cooling of Trapped Ions 363

Chapter 16. Application of Lasers to Microwave Standards **369**

16.1 Observation of Individual Ions 369
16.2 The Cooling Laser System 373
16.3 Laser Detection of Hyperfine Resonance 376
16.4 Laser-Based Mercury Ion Standards 382
16.5 The Proposed Ytterbium Ion Standard 383
16.6 Beating Liouville's Theorem 384
16.7 The Cesium Fountain Standard 390

Chapter 17. Measurement of Optical Frequency **395**

17.1 Introduction 395
17.2 Definition of the Meter in Terms of the Second 396
17.3 Theoretical Limit to Spectral Purity of Lasers 396
17.4 Stabilization of Lasers Using Atomic/Molecular Resonances 399
17.5 Stabilization of the He–Ne Laser 400
17.6 Stabilization of the CO_2 Laser 406
17.7 Stabilization Using Two-Photon Transitions 407
17.8 Frequency Comparisons in the Optical Range 409
17.9 Measuring Optical Frequencies Relative to a Microwave Standard 414

Chapter 18. Applications: Time-Based Navigation **419**

18.1 Introduction 419

18.2 "Deep" Space Probes 419
18.3 Very Long Baseline Interferometry 420
18.4 The Motion of the Earth 421
18.5 Radio Navigation 422
18.6 Navigation by Satellite 429
18.7 The Global Positioning System (GPS) 432

Chapter 19. Concluding Thoughts 449

19.1 The Synchronization of Clocks 449
19.2 The Direction of Time 451
19.3 Time-Reversal Symmetry in Subatomic Events 453

References 457

Further Reading 461

Index 465

Chapter 1
Celestial and Mechanical Clocks

1.1 Cyclic Events in Nature

From the earliest times in the course of human development, a recurring theme has been the inexorable passage of time, bringing with it ever-changing aspects of Nature and the cycle of life and death. Only in the realm of mythology do immortal gods live outside of time in their eternal incorruptible abodes.

The discernment of an underlying order in the evolution of natural phenomena, and the cyclic repetition of the motions of the sun, moon, and stars, may be taken as a measure of man's intellectual development. The ease with which early man was able to recognize that certain changes in nature were cyclic depended on the length of the cycle. That of the daily rising and setting of the sun is so short that it must have soon been accepted with confidence as being in the natural order of things, that if the sun disappeared below the horizon, there was little doubt that it would reappear to begin another day. It is otherwise with the much longer period of the seasons; there is evidence to suggest that for some primitive peoples, a year was so long and the means of recording the passage of time so imperfect that they were unable to perceive a cyclic pattern at all in the changing seasons. They must have watched the changing elements with perpetual wonder. To them the onset of winter must have been filled with dire foreboding, giving rise, according to Frazer (1922) in his classic *The Golden Bough*, to magical, and later religious, rites to ensure the return of spring.

This introduces a connection between the timing of important cultural events in the life of early man and that of the cyclic events in nature. This overlays an inherent connection on a biological level: The workings of the human body and indeed of all living creatures follow rhythmic patterns that shadow those in nature. The most obvious are the so-called *circadian rhythms* (from the Latin *circa* (about), and *dies* (day)) with an approximate 24-hour repetitive cycle. Much research has been conducted in recent years on the human asleep–awake cycle, with particular interest in the extent to which the cycle is governed by some internal timing mechanism, as opposed to the external environment. The need to adapt one's schedule of activities to be in harmony with nature, a task compounded by the differing cycles of natural events, is evident in the early development of calendars. Of course, the beginnings of agriculture gave a great impetus to this development in order to keep track of the seasons and accurately plan the cultivation of the soil, the planting of seeds, and the eventual harvest. The cyclic succession of the seasons—from the shedding of leaves in the fall to the cold dormancy of winter, the return to life in the spring, and the warm summer that followed—bore witness to some order underlying the vagaries of daily life. For those early societies whose life and livelihood were closely tied to the sea, the periodic rise and fall of the tide reinforced the same perception of unalterable periodic changes underlying unpredictable short-term changes.

A cyclic phenomenon clearly allows time to be quantified; the period of time to complete a cycle, called briefly the *period*, provides a unit in terms of which any given length of time can be expressed as so many of those units. An obvious example is the use of the (solar) day as a unit, defined as the time between the sun passing overhead one day until it comes to the same point the next day. Another common example is the lunar month, which is the time it takes the moon to go from (say) a new moon to the next new moon. As units of time, it is relevant to ask just how constant these units are, and how accurately they can be measured. Such questions are, of course, at the heart of our subject and are taken up in the chapters that follow.

1.2 The Calendar

It is unfortunate for those whose primary interest is in keeping track of the seasons that they do not recur after a whole number of days. As we all know, the year is about one-fourth of a day in excess of 365 days. In terms of the planetary motions of the earth, this is the same as saying that the period of the earth in its orbit around the sun does not contain a whole number of rotations of the earth about its axis. It is this simple fact that throughout history has complicated the lives of those charged with keeping the calendar. Another such astronomical fact that has challenged the keepers of the calendar is that the orbital period of the moon does not contain a whole number of days, nor are there a whole number of periods of the moon in one year. However, it happens that after a period of 8 years the moon does return to approximately the same position relative to the earth and sun, that is, to the same lunar phase. This, according to Frazer, accounts for the period of 8 years figuring in certain traditions among some ancient peoples. It is not surprising that the degree to which primitive societies have succeeded in developing a calendar has become a measure of the state of advancement of these societies.

Perhaps the most celebrated of these is the ancient Mayan calendar, a remarkable achievement, often described with such lavish admiration as to convey a sense that this New World culture has surpassed some unspoken expectations. The Maya had in fact two calendars (Morley, 1946): a sacred calendar and a civil calendar with a complicated way of enumerating the days. The sacred year was not divided into groups of days, such as months, but consisted of 260 days enumerated by a number from 1 to 13 followed by one of twenty names. However, curiously, the sequence did not simply run through the numbers from 1 to 13 for each name before running through the numbers again with the next name, which would be tantamount to using 20 "months" of 13 days each. Instead, the name was incremented along with the number in going from one day to the next. After the number 13 was reached, the number sequence was repeated again, incrementing the name at the same time. It would be as if we wrote for a sequence of days 1 Feb.,2 Mar.,3 Apr., etc. It is almost as if the Maya were generating a cryptic code! The Mayan civil calendar was based on groupings of 20 days each, so that there were 18 such "months" and a closing month of 5 days to yield 365 days in a year. If a particular day was specified simultaneously using designations according to both calendars, that specification was repeated every 52 years; that is, within a 52-year span the designation would be unambiguous. For longer periods the Maya developed an enumeration system with base 20, a vigesimal system, which is distinguished in having introduced the *zero* independently of the Old World discovery of that concept. It will be recalled that the place-value system of representing numbers, and therefore arithmetic as we know it, would be impossible without the zero. It is curious that the characters we use to represent the digits, namely what we call Arabic numerals, are not used in the Arabic language; instead, Indian characters are used, in which zero is simply a dot.

The Maya also kept detailed watch on the phases of the moon, and the enumeration of the lunar months played an important part in their elaborate religious calendar. In common with other societies of antiquity, the Maya were in awe of celestial events, which they saw as ominous manifestations in which the mysteries of the universe and the future of human destiny were to be read.

1.3 Solar Eclipses as Time Markers

The occurrence of astronomical phenomena such as solar and lunar eclipses, meteors, and comets were recorded with awe from the earliest times and became associated with religious observances or superstitious omens. The seemingly eternal constancy of the motions of the heavenly bodies came to define time and regulate the affairs of many societies, not only in a chronological sense, but also in a mystical astrological sense.

Because of the superstitions that attached to these observations, evidence has been found that records of eclipses reach as far back as 2000 B.C. The sifting of ancient records to discriminate between objectively reported events and those reported spuriously either by accident or by design is a task that has occupied specialists for some time. By now, a large body of data has been compiled from which a chronology of sightings has been constructed, scattered throughout history and over the entire globe.

If we recall how eclipses are produced, we will be better able to appreciate that recordings of the time and place of their occurrences give sensitive time markers in establishing a long-term chronology. Figure 1.1 depicts the positions of the sun, moon, and earth (not to scale) momentarily along a straight line when a solar eclipse occurs. The three bodies will pass through the aligned condition only when the moon and earth are simultaneously at particular points in their respective orbits. These orbits lie in fixed planes: the plane of the moon's orbit passes through the earth's center, while that of the earth (called the *ecliptic*) passes through the sun's center. These orbital planes are inclined at a constant angle of about 5 degrees; hence there will not be a solar eclipse observed on the earth every lunar month, as would be the case if the orbital planes coincided. However, it can happen that as the earth travels along its orbit around the sun, it will reach a point where the sun and earth are in line with where the moon is just crossing the earth's orbital plane.

As will be recalled from optics, the shadow produced by the moon on the earth consists of regions called the umbra and penumbra, corresponding respectively to total eclipse, in which the complete disc of the sun is obstructed, and partial eclipse, where the moon obstructs only part of the sun's disc. Although the sun is immensely larger in diameter than the moon, it is so much farther away from the earth that to an observer on the earth, the moon's disc can cover the sun's disk that is, both bodies subtend about equal angles at the earth (about 0.5°). Actually, a partial eclipse of the sun will not cause "darkness to fall upon the land" unless it is very nearly total, that is, with over ninety-five percent of the sun's disc obstructed.

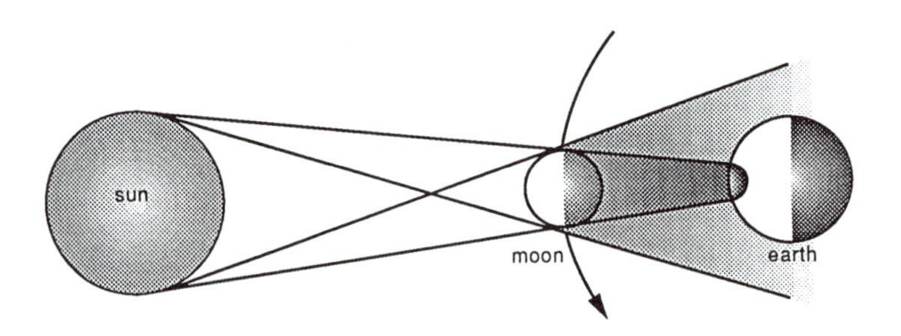

Figure 1.1 The formation of a solar eclipse

The reason that reliable records of total solar eclipses are such useful time-markers is that they occur only when a very special set of astronomical variables such as the diameters and distances of the three bodies and the positions of the earth and moon in their respective orbits fall within very narrow limits. Moreover, these events, particularly total solar eclipses, are so awe-inspiring and so imprinted themselves on the minds of the ancients that it is not to be expected that many went unnoticed. In fact, the problem is to sift out those sightings spuriously injected into records to lend a supernatural weight to some historical event, such as the death of a king! It has been argued that when Joshua refers to the sun as having "stopped in the middle of the sky" he may have witnessed a solar eclipse.

A detailed analysis (Schove and Fletcher, 1984) of the chronology of eclipses reported in the historical records, in which times and places of observation are compared to computed predictions, has revealed the remarkable result that the earth's rotation about its axis has been slowing down over the centuries, as has the moon in its orbit around the earth. It has been estimated on the basis of fossil evidence that over geologic time the length of the day has increased from 390 days per year to the present 365.25. If, following Stephenson (Stephenson and Morrison, 1982), we plot the track of totality for a solar eclipse observed at Athens in A.D. 484 and compute a similar plot based on current astronomical data and a constant rotation of the earth, we would find the path 15° west of where it was recorded, corresponding to a difference in time of one hour. Based on more recent precision measurements, about which more will be said in a later chapter, the slowing of the earth's rotation about its axis amounts to about one part in 43,000,000 per century, or about 4.5° of rotation in 1,500 years.

1.4 The Tides

If the earth were a uniform hard sphere spinning about its axis in the vacuum of space, it would continue spinning at a constant rate indefinitely. In reality, the earth has topographically complicated bodies of water on its surface, and a molten interior; even the "solid" regions are to a degree plastic. Furthermore, it is not perfectly spherical, having an equatorial bulge with a slight north–south asymmetry. The tidal action in the world's oceans—involving as it does the movement of water, which like most liquids has some viscosity (internal resistance to flow)—can create a drag on the earth's rotation, causing it to slow down. In this process the kinetic energy of the rotational motion of the earth is slowly (and irreversibly) converted to random motion on the molecular scale, that is, heat, in the waters of the oceans.

The predominant cause of the tidal action referred to above is the gravitational pull of the moon, with a smaller contribution from the sun. It arises not so much from the moon drawing towards itself the waters of the earth's oceans by its gravitational pull as from the variation in the gravitational field across the earth that the moon superimposes on top of the smaller variation of the sun's pull. The effect of such a variation can be illustrated by considering what has now become familiar: Astronauts aboard a spacecraft orbiting the earth in a circular orbit. Suppose the spacecraft is equipped with "stabilizing booms," that is, two long straight poles fixed to the spacecraft, one pointing toward the earth and the other in the opposite direction, as shown in Figure 1.2.

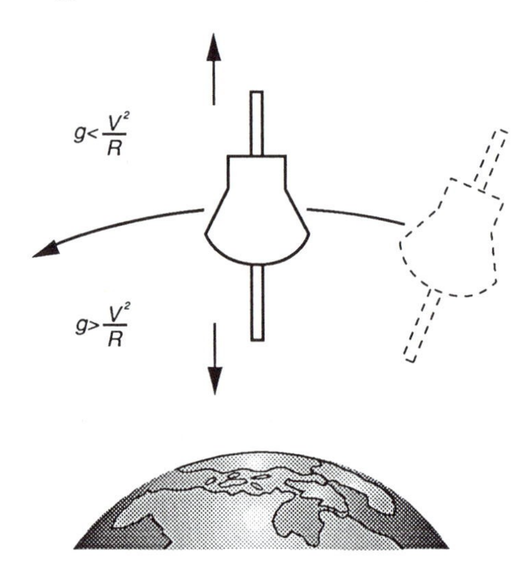

Figure 1.2 Forces acting on an orbiting spacecraft due to variation in the gravitational field across it

As the spacecraft swings quietly along its orbit with all its propulsion systems shut down, the astronauts in the main cabin may float around in a more or less "weightless" state. However, if an astronaut is required to go out to attend to some problem at the ends of the stabilizing booms, then he will notice there that he is no longer weightless. At the end nearest the earth he will have a small but positive weight tending to pull him toward the earth; at the farthest end he will have a small but negative weight, that is, he will have a tendency to be lifted farther out. This means also that the booms themselves experience a stretching force tending to separate the ends. The basic explanation is that for objects in the main cabin there is a balance between the gravitational force of the earth and the dynamical centrifugal force due to the curving trajectory of the spacecraft; a balance that tips in favor of the gravitational force at the end of one boom and the dynamical force at the end of the other. A similar basic argument can be made to explain the fact that tidal motion results in two diametrically opposite bulges in the equilibrium water surface on the earth: one towards the moon and the other away from the moon, as if the system were being stretched along the line joining the centers of the earth and moon. Of course, the actual rise and fall of the tide at any given geographical point is the result of many contributing factors, particularly the topography of the ocean beds, the coast lines, and the resonant response of the tidal motion to the twelve-hour periodic lunar force as the oceans are swept around in the earth's daily rotation.

1.5 The Sidereal Day

In specifying the period of rotation of the earth and its variability, a certain frame of reference is of course implied. In our case, the frame of reference is that defined by "the fixed stars." The period of rotation so defined is called the "sidereal day," as contrasted with the "solar day," which is the period between successive transits of the sun across any given meridian (a great circle with a specified longitude). Since the earth sweeps around the sun in a nearly circular orbit while it is spinning around its axis, the time between successive passes of a given meridian under the sun will differ from what the rotation period would be in the absence of the orbital motion. This is made clear by noting that if the earth had zero spin, the sun would still cross a given meridian every time the earth completed a revolution around the sun. The difference in the values of the solar and sidereal days may be easily approximated if we assume a circular orbit. Referring to Figure 1.3, we note that the sense of rotation (whether clockwise or counterclockwise) is the same for the rotation and revolution of the earth. It follows that as the diagram shows, the sidereal day is shorter than the solar day by the time it takes the earth to turn the angle that the sun's direction has turned in one day by virtue of the orbital motion of the earth. This latter angle is (360°/365.25), and the earth rotates at the rate of 360°/(24×60) degrees per minute; hence the difference in the length of the two days is (360/365.25)×(24×60/360)=3.95 min (sidereal).

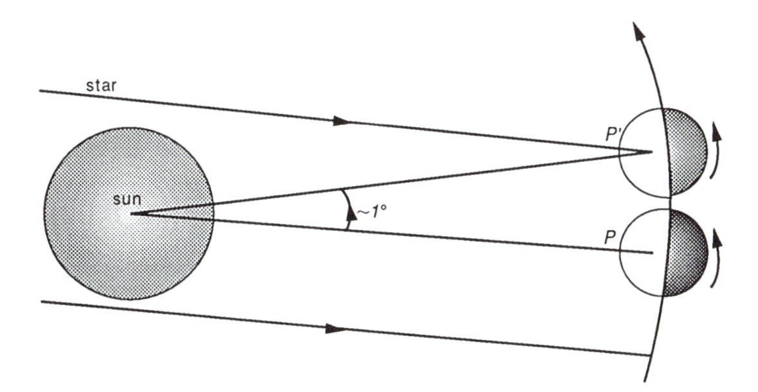

Figure 1.3 The motions of the earth and the difference between the sidereal and solar days

In making the simplifying assumption that the orbit is circular, we have ignored the fact that the orbit in reality is elliptical. According to one of Kepler's laws of planetary motion, the empirical pillars on which Newton's theory stands, the earth moves with a speed such that the area swept out by a radius drawn from the sun to the earth increases at a constant rate. Since in an elliptical orbit the length of this radial arm varies, going from a minimum at the *perihelion* to a maximum at the *aphelion* (a relatively small change for the earth's orbit), this means that the angle swept out by the radial arm in a given time varies from point to point along the orbit. It follows that the length of the solar day varies throughout the year but is always in the neighborhood of four minutes longer than the sidereal day. This variation of the solar day must not, of course, be confused with the seasonal variation of daylight hours, which has to do with the inclination of the earth's axis to the ecliptic plane.

1.6 The Precession of the Equinoxes

To complicate matters further, the nonspherical shape of the earth (which would be symmetric about its axis of rotation if we ignored tidal action) brings into play a torque, originating from the gravitational pull of the moon, tending to turn the axis of the earth towards a direction perpendicular to the plane of the moon's orbit. To those unfamiliar with gyroscopic motion, the effect of this torque is rather remarkable: instead of simply turning the axis directly from the old direction towards the new, it causes the axis to swing around, tracing out the surface of a geometric cone around the new direction as axis. This motion is familiar to anyone who has watched a spinning top; as a result of the torque tending to make it fall, its axis instead swings around in a vertical cone. This motion is called precession of the axis of spin. In the context of planetary motion, and in particular the earth's motion, this motion causes the points along the orbit of the earth where the seasons of the year occur to shift from year to year along the orbit. The reason for this is that the direction of the earth's axis determines the line of intersection of the earth's equatorial plane with the plane of its orbit. The two points around the orbit where this intersection occurs mark the vernal and autumnal *equinoxes*, which are conventionally taken to be the beginning of spring and autumn. Thus as the axis precesses, the equinoxes will also, and for this reason astronomers call this motion the precession of the equinoxes. Although the rate of precession is small, amounting to about one cycle in 26,000 years, nevertheless it says something about the constancy of the solar day, which you will remember varied from point to point along the earth's orbit.

The precession was first detected by one who is arguably one of the greatest astronomers of antiquity, Hipparchus, in the second century B.C. He made careful measurements of star positions, assigning to each star coordinates analogous to longitude and latitude. By comparing his observations with astronomical records dating back over 150 years before his time, he made the incredible discovery that the point in the night sky about which stars appear to rotate (because of the earth's rotation), that is, what is called the *celestial pole*, had definitely shifted. In view of how small the rate of precession is, amounting to no more than one minute of arc per year, this was no mean accomplishment.

1.7 The Sundial

The earliest devices for measuring the elapse of time *within* the span of a day were a natural derivative of the notion of time as being defined by the motions of celestial objects. In order to keep track of the sun's journey across the sky, the shadow clock was devised, which later developed into the sundial. In its most primitive form, it was simply a vertical straight pole, called a *gnomon*, whose shadow is cast upon a horizontal plate marked with lines corresponding to different subdivisions of the day. The principle was applied in many different forms: One of the earliest ancient Egyptian shadow clocks used the shadow of a horizontal bar placed in a north–south direction above a horizontal scale running east–west. These early shadow clocks did not indicate the time in hours, but rather in much larger subdivisions of the day. Through the centuries these clocks evolved into very sophisticated sundials, some of which even were designed to be portable.

The division of the day into 24 hours is traceable back to the ancient Sumerians, who inhabited the land that was known in classical times as Babylonia (Kramer, 1963). Their number system was sexagesimal in character, that is, based on 60, although the separate factors 6 and 10 do occur in combinations such as 6, 10, 60, 600, 3600. They actually had two distinct systems: an everyday mixed system and a pure sexagesimal system used exclusively in mathematical texts. The latter had the elements of a place-value system like our decimal system; however, the Sumerians lacked the concept and notation for zero; furthermore, their way of writing numbers did not indicate the absolute scale; that is, their representation of numbers was unique only to within multiplication by any power of 60. Nevertheless, the impact of the ancient Sumerian culture is evident in the way we subdivide the day into hours, minutes, and seconds; and the circle into degrees, minutes, and seconds of arc.

The Sumerian version of the shadow clock, like those of other ancient cultures, suffered from the same critical flaw: The shadow of a vertical shaft moves at a variable rate over the span of a day, and what is worse, the variability itself changes with the seasons and the latitude where the device is used. It would require a sophisticated knowledge of celestial mechanics to derive corrections to the observed readings for each day of the year and for different latitudes.

A radical improvement in the design of what came to be called sundials was made by Arab astronomers in the Middle Ages. This consisted in mounting the gnomon (which you will recall is the name given to the object producing the shadow) as nearly parallel as possible to the axis of rotation of the earth, that is, pointing toward the celestial pole, which is currently within 1° of the North Star (Polaris). This revolutionary change in design transformed the sundial into a serious instrument for the measurement of time. In order to appreciate the significance of this innovation, let us recall that the apparent daily motion of the sun, and indeed all celestial objects, is due simply to the earth's rotation about its axis; and therefore, to the extent that the earth's rotational motion is uniform, the sun's apparent angular position around that axis will also progress uniformly. It follows that the shadow cast by a shaft parallel to the axis onto a plane perpendicular to it will have an angular position that follows the sun, and it will therefore reproduce the rotation of the earth. Of course, it is only during about half of every rotation that the sun's rays will reach a given point on the earth's surface; however, unlike the shadow clock, the seasonal variation in the relative lengths of daytime and nighttime will in this case have no effect. Since the earth's rotation is very nearly constant, a circle drawn on the plane with the gnomon as center can be divided into 24 equal parts corresponding to the 24 hours of the day. In practical portable sundials, such as might have been used aboard ship at the time of Sir Francis Drake, provision must be made for setting the direction of the gnomon. This they could achieve by finding north with a magnetic compass and the latitude by means of a forerunner of the sextant.

The accuracy one can achieve with this type of sundial, while incomparably greater than the earlier primitive versions, nevertheless is limited by the extreme accuracy with which angular displacements of the shadow would have to be measured. Thus the shadow moves only 0.25° per minute; this implies that an error of 0.1° in angle measurement translates into an error of 24 seconds. This level of accuracy, though unimpressive by more recent standards, coupled with the fact that it provided an absolute reference with which to compare mechanical clocks, ensured the continued use of the sundial, in one form or another, from antiquity until the Enlightenment.

1.8 The Astrolabe

In this context we should include another astronomical instrument of ancient origin, also perfected by medieval Arab astronomers and instrument makers, called the astrolabe, shown in Figure 1.4 (Priestley, 1964). It is a combination of an observational instrument and a computational aid enabling the determination of not only latitude, but also the time of day. It ultimately spread to Western Europe and was there in common use by navigators until the advent of the sextant in the eighteenth century.

Figure 1.4 The astrolabe as depicted in Chaucer's *Treatise on the Astrolabe:* (a) the front side (b) the back side

The astrolabe consisted of a disk on whose rim was engraved a uniform scale with 24 divisions, surmounted by a plate engraved with a projection of the celestial sphere over which arcs of circles were inscribed. Pivoted concentrically were also a metal cutout star pattern called a *rete* and a metal pointer called the *rule*. On the back were concentric scales graduated in degrees, the signs of the zodiac, and the calendar months; another pointer was pivoted at the center. It would be out of place, and probably well beyond the interest of the reader, to devote much space to describing the intricacies of this instrument and how to get the most information out of it. Briefly, it may be said that it is assumed that the calendar month and day are known for the time it is to be used. The altitude of the sun is first observed using the degree scale on the back of the instrument while sighting the sun. From that side also one reads, for that date, the position of the sun along the zodiac. Using this information on the front side of the instrument, the star pattern is turned with respect to the projection of the celestial sphere until the sun's position in the zodiac agrees with its observed altitude. The time is read by appropriately setting one end of the rule and reading the other end on the 24 division scale. It is interesting that Geoffrey Chaucer, of *Canterbury Tales* fame, also wrote a *Treatise on the Astrolabe* in 1391. Since it gave both latitude and time of day, the astrolabe was used by navigators well into the eighteenth century.

1.9 Water Clocks

Among the earliest nonastronomical devices for measuring time was the water clock, of which rudimentary examples have been found among ancient Egyptian artifacts dating back to 2000 B.C. It was essentially a conical stone vessel filled with water that escaped slowly through a small hole at the bottom provided for that purpose. A uniform scale was marked along the side of the vessel to enable the elapsed time to be gauged by how far the water level had fallen. The ancients were led through experience to the need for a conical shape in order to achieve an approximately uniform scale. It would have been fairly obvious that the level in a straight cylinder falls faster when nearly full than when nearly empty. However, that the shape should be conical, rather than spherical or some other shape, is less obvious and must have been arrived at rather through convention than careful observation. The hydrodynamic problem that the design presents is actually a fairly complicated one; a lot depends on whether or not the water passes through a channel-like opening, in which case the viscosity of the water plays an important role, making the rate of flow of water more or less proportional to the pressure, and therefore to the height of the surface above the opening. On the other hand, if the opening is such that the effect of viscosity is negligible, then we have ideal conditions where *Bernoulli's principle* should apply and the *kinetic energy* (that is, the *square* of the velocity) of the escaping water should be determined by the pressure, or depth below the surface.

Since the pressure in the water at the depth of the opening is proportional to the depth, irrespective of the shape of the container, it follows that if the flow rate is proportional to pressure, a constant rate of fall of the water level will be achieved if the area of the surface of the water is proportional to the depth of the opening. This is ideally satisfied by a cylinder whose axial cross section is approximately parabolic.

Rather than attempt to perfect the shape of the container, which we now appreciate is a good deal more complicated than would at first appear, the actual development of water clocks took a much more promising tack in achieving a constant rate of flow by providing, in the words of the plumbing profession, a "constant head," that is, a fixed water level above the hole. This advance is attributed to an Alexandrian by the name of Ktesibios (also given the Latinized spelling Ctesibius), a celebrated inventor of Ptolemaic Alexandria around 250 B.C. (de Camp, 1960). His accomplishments included other mechanical and hydraulic devices, such as a water pump and pipe organ. From his work evolved the Hellenistic type of water clock, called *clepsydra*, that was in common use throughout classical times. Such clocks were commonly used then to allot time to speakers in a debate: When the water ran out, it was time to stop. Successive speakers were assigned the first water, second water, and so on. This may have something to do with the expression "of the first water" as something of the finest quality. The essential design is illustrated schematically in Figure 1.5.

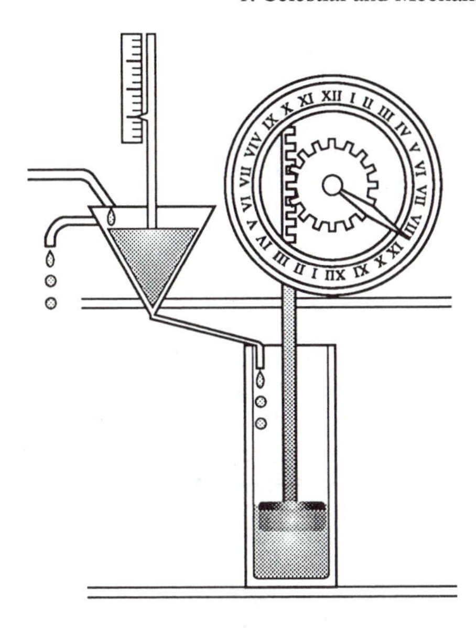

Figure 1.5. Schematic drawing of an ancient Greek water clock

The constant pressure head is achieved simply by allowing an adequate continuous flow from some source into the vessel and preventing the level from rising above a fixed point by having an overflow outlet. The constant flow was collected in a graduated straight cylinder. The design often incorporated various time-display mechanisms that were actuated by the constant rise in the water level. In one instance a float supported a straight ratchet engaging a toothed wheel; to this was attached a pointer to indicate the time on a circular dial. In other designs a pointer was joined to the float by a vertical shaft, enabling the rise in the float to be read on a vertical scale drawn on the surface of a rotatable drum. By varying the scale at different points around the periphery of the drum, it was possible to accommodate the seasonal variations in the length of the hour, which was then defined as a certain fraction of the period from sunrise to sunset.

In China, water clocks are known to have existed at least from the sixth century of the Christian era; but their development took a more elaborate mechanical turn. In place of the time being measured by the continuous motion of a simple float rising in a linear fashion, the Chinese took things to a higher level of sophistication by introducing the idea of using the flow of water to control the rate of turning of a water wheel; not continuously, but in discrete steps, much like the crown wheel in the tower clock escapement mechanism, to be discussed in the next section. The water flowed at a constant rate into successive buckets mounted on short swivel arms between numerous equally spaced spokes of a wheel, free to turn in a vertical plane about a fixed axle. By a clever arrangement of balanced beams, levers, and connecting rods, the rotation of the water wheel was automatically stopped by blocking one of the spokes while a predetermined amount of water flowed into each bucket in succession. When the critical amount of water had been reached, the bucket arm was able to tilt against an accurate counterweight at the other end of a balance beam, in effect "weighing the contents of the bucket" before allowing the wheel to turn until the next spoke was engaged and the wheel stopped again for the next bucket to fill.

The accuracy achieved in a well-constructed clepsydra was comparable to the sundial, but of course their time scale was not absolute, in the sense that they had to be calibrated against a scale based on astronomical observations. A serious limitation of the water clock is its obvious nonportability; it is difficult to imagine how it could be made suitable for use aboard a ship on the high seas.

Another device that we should mention based on the flow of material through a small hole is, of course, the *hourglass*, the universal symbol of the fleeting nature of time. Sand was not the only substance used; the choice was directed towards greater reproducibility. Many granular solids are efficient absorbers of moisture, a property that clearly disqualifies them, since they would have a greater tendency to form clusters. The rate of flow through a constriction clearly depends on the grain or cluster size as well as friction between grains. Hourglasses provided a convenient way of determining a fixed interval of time, and sets of hourglasses were used aboard ships to mark the *watch*, the 4-hour spell of duty.

1.10 Tower Clocks

A great deal has been written about mechanical clocks and clockmaking, a testament to the enduring fascination with which the subject has been regarded through the ages. We will limit our discussion of this subject to a review of their design and performance from the vantage point of present-day horology.

In the analysis of the operation of mechanical clocks it is useful to separate the mechanism into three essential functional parts: first, the energy source, which has usually taken the form of a falling weight or a coiled spring; second, a mechanical system capable of inherently stable periodic motion to serve as regulator; third, a mechanism to derive and display the time in the desired units. The third part consists of gear trains and a dial. Of these the most critical and challenging is the regulating system, and the history of the advancement in mechanical clockmaking is the history of the development of this part of clock design.

The fundamental problems in regulator design reduce to two in number: first, an oscillatory system must be found whose period of oscillation is constant and insensitive to changes in the physical environment and operating conditions; second, a method must be found for the regulator to control the transmission of power from the energy source to the rest of the clockwork with the least possible reaction on the regulator. Some interaction is essential to sustain the oscillation of the regulator. However, this must be small compared to the oscillation energy of the reference system of the regulator. As with any mechanical system, there will always be frictional forces present, which in the absence of an adequate source of excitation energy will cause the oscillator to come to rest. Therefore, a small driving force must act on it in step with the motion to maintain a nearly constant level of excitation. But this requirement runs counter to the function of the regulator as a controller: Rather than controlling, it is being controlled. The ideal situation would be one in which the reference oscillator was free to execute its natural oscillations without any external perturbations acting on it.

To reconcile these opposing requirements and achieve the best possible outcome requires the following: First, the oscillator must have very low inherent friction, enabling it to continue to execute its motion with only a very weak driving force; and second, it must exert its control in a "trigger" fashion. This means that a small force exerted by the controlling element, acting for only a small fraction of the period of oscillation, must control a much larger force transmitted from the energy source to the gear train driving the clockwork. The mechanical means of achieving this is called the *escapement.*

An early version extant in the fourteenth century was widely used in tower clocks for cathedrals, public squares, etc. for almost three centuries. It is called the *verge* and *foliot* escapement and is shown in its basic form in Figure 1.6. A straight horizontal beam, the foliot, with equal weights balanced at its ends, is suspended at its middle. Rigidly attached to the foliot at its point of suspension is a vertical spindle, called the verge, to which are rigidly attached two small flat projections, called *pallets*; these engage at diametrically opposite points a vertical wheel (the scape or crown wheel) with cogs perpendicular to its face. The planes of the pallets are parallel to the axis of the verge, but are typically ninety degrees apart. This balanced foliot–verge system is capable of simple periodic angular motion about the verge as axis. The torsion in the suspension of the foliot provides the necessary restoring torque when the foliot is turned away from its equilibrium position. The action of the pallets as the foliot rotates back and forth is to momentarily block the cogwheel alternately by one pallet, then the other. The rotation of the cogwheel, which derives its torque from the energy source, is thereby regulated. The reaction back on the oscillating foliot occurs at the contact between the pallets and the cogs. As earlier pointed out, this reaction tends to sustain the oscillation of the foliot. If the suspension material is chosen to have good elastic properties with a particularly low internal friction and if the foliot is massive to increase the energy and the period of oscillation, this regulator can be expected to be relatively stable and insensitive to small perturbations, such as air drafts, noise, and vibration.

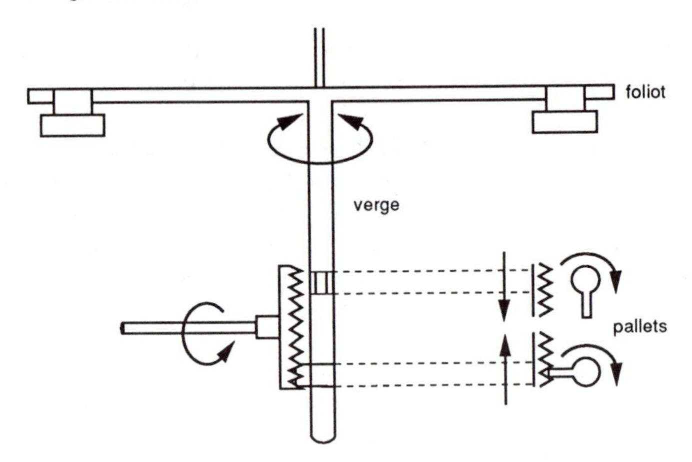

Figure 1.6 The foliot and verge escapement for tower clocks

In judging these mechanical clocks we should separate the principles upon which the design is based from the implementation of those principles, that is, the choice of materials and the level of precision in the manufacture of the clocks. If we consider the operating principles of the clocks we have been describing, we note that the foliot–verge system is really a type of torsion pendulum, and as such is capable of as great a constancy of oscillation period as are later developments, for example the pendulum. Its limitations arise principally from the design of its escapement; the force of reaction is too large and acts for too large a fraction of the period. The choice of material for the suspension is also critical; a fused quartz fiber suspension would have excellent elastic properties. Such quartz suspensions have been widely used in torsion balances since the seventeenth century, because of both their strength and elastic properties; the restoring torque they provide is linearly dependent on the rotation angle.

1.11 The Pendulum Clock

Two important advances were made in the seventeenth century: First came the pendulum as the regulator, and then, of equal importance, came the "deadbeat" anchor escapement. Let us consider these in turn.

The story of Galileo's timing the swings of a chandelier at the cathedral in Pisa using his pulse is well known. The discovery of the "isochronism" of the pendulum, that is, taking an equal time to complete a swing no matter how widely it swings, dates from 1583 when Galileo was a medical student. The story is usually repeated as an example of extraordinary resourcefulness in his desire to study the pendulum. This may be so, but it should be noted that he was at the time also interested in medicine and in particular the pulse rate as an indicator of fever (Drake, 1967). In fact, there is no published account by him at this time suggesting the use of the pendulum as a regulator for mechanical clocks. He did, however, use it to construct an instrument to conveniently measure the pulse rate in patients. It consisted essentially of a pendulum with variable length, which was adjusted to match the pulse rate of the patient. It was calibrated to read directly conditions such as "slow" or "feverish." It was not until a few months before his death, in 1642, that Galileo suggested the application of the pendulum to clocks. He had become blind in 1638 and was no longer able to put his idea into practice. He dictated a design to his son Vincenzo, who made drawings but did not actually complete a working model. The credit for actually incorporating a pendulum into the design of a clock around 1656 goes to Huygens, a name associated in the mind of every physics student with the wave theory of light.

The pendulum is essentially an object pivoted or suspended so that it swings freely. For purposes of analysis, we distinguish between a simple pendulum, which consists of a small object suspended by a thin string of negligible mass, and a compound pendulum, in which the mass distribution along the pendulum is not negligible. The essential characteristic of the pendulum, as Galileo noted, is that for small swings the period of oscillation, that is, the time to complete a swing in one direction and back to the starting point, is the same no matter how wide the swing, provided that it remains small. This property caught Galileo's attention because it appears to run counter to what might superficially be expected: After all, with a large swing, the pendulum bob has farther to travel, and it is indeed remarkable that its speed varies in just such a way that the oscillation period is always the same. Actually, the same could be said of a beam suspended by a material with suitable elastic properties, such as fused quartz. They both display *simple harmonic motion.* But while the latter depends on the property of the suspension material, the pendulum has no such dependence on materials, which are generally subject to variation. However, the period of the pendulum does depend on its *radius of gyration*, which depends on the distribution of mass along the length of the pendulum. Since all materials expand and contract with the rise and fall of temperature, the constancy of the period is limited by any fluctuations in the temperature. We may attempt to overcome this limitation by taking one or all of the following steps: Choose a material that has extraordinarily low thermal expansion, such as the alloy *invar*; regulate the temperature to reduce its fluctuations; and use a composite pendulum incorporating two materials of differing expansion coefficients, such as brass and steel, in such a way that the expansion of one is compensated by the other.

Another important limitation of the pendulum as a reference oscillator is that its period depends on the strength of gravity, which varies from point to point on the earth's surface. This is because the earth is neither spherical nor homogeneous. As far back as 1672, it was established through pendulum measurements that the acceleration due to gravity is different for different geographical locations. This was explained by Newton by assuming a model of the earth as a uniform gravitating plastic body, which, by virtue of its spin, would bulge around the equator into an oblate spheroid. The value of the fractional difference between the earth's radii at the equator and poles was later computed in 1737 by the Frenchman Clairaut to be 1 in 299. The acceleration due to gravity is also dependent on altitude in a way that may be affected by local topography and geology. The differences in the times indicated by a pendulum clock at different geographical locations could be on the order of one minute per day. Another source of fluctuation in the period of a pendulum is air resistance, whose drag on the swinging pendulum depends on the density of the air and thus the atmospheric pressure.

The other major development, which came around 1670, was a much improved escapement: the *anchor* escapement, and later the *deadbeat anchor* introduced by Graham in 1715. Figure 1.7 shows the essential design.

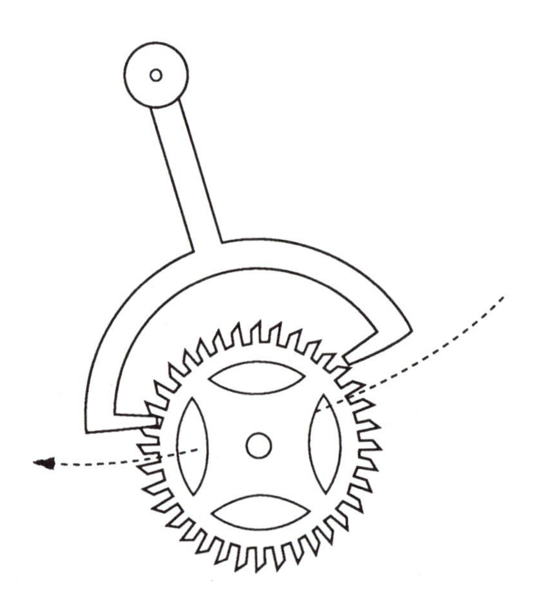

Figure 1.7 The anchor escapement

Unlike the verge–foliot escapement, where the pallets engage diametrically opposite points on the scape wheel, the anchor escapement acts on a sector of a rachet wheel having radial teeth. This geometric difference allows the pallets to be separated by some distance along the rim of the scape wheel, and in consequence a smaller angular movement of the anchor about its axis is needed to engage and disengage the pallets and the wheel. This is advantageous to the performance of any regulator based on a mechanical oscillator, since its oscillation is simple harmonic only in the limit of small oscillations. But an even more important difference to note is that the pallets move at right angles to the direction of motion of the scape wheel teeth. This means that the force of interaction between the pallets and the teeth has little torque around the axis of the pallet mount and is therefore ineffective in disturbing the oscillation of the pendulum. Moreover, the pendulum is entirely free of the scape wheel for part of its oscillation, a first step toward the ideal condition. The deadbeat design is so called because unlike the anchor escapement just described, there is no recoil of the scape wheel and the gear train behind it during a swing, or beat, of the pendulum. This was achieved by a careful contouring of the faces of the pallets and the teeth of the scape wheel. This refinement further improved the isolation of the pendulum and enhanced its performance as a regulator.

1.12 The Spring–Balance-Wheel Clock

For fixed installations, such as in observatories or clock towers, the pendulum-controlled clock became the most widely used, and by the end of the eighteenth century it had reached an accuracy sufficient to the demands of the day. However, there remained one area of need that the pendulum clock could not satisfy: shipboard and, in today's jargon, other mobile environments, where the clock may be subjected to erratic inertial forces. Moreover, any attempt to scale down the size of the clock to make it more portable would aggravate the problems of air resistance, friction, and curvature in the knife edge on which the pendulum is pivoted. A further disqualification for shipboard use is the variability of the period with geographical location, as previously described.

Although brave attempts were made to develop a pendulum clock that would be reliable in the field, it finally became clear that a new approach was required. This came in the form of the balance wheel and the spiral hairspring, which ultimately became universally used in all mechanical watches. The hairspring, or balance spring, was a spiral of fine resilient metal fixed at the outer end to the body of the watch and at the center of the spiral to the arbor of the balance wheel, which is delicately pivoted on jeweled bearings to reduce the rate of wear. The basic advantage of the spiral spring is that it provides a restoring torque on the balance wheel independent of gravity and permits a reduction in the strain (degree of internal deformation) in the material of the spring for a given rotation of the balance wheel. This is important if the restoring torque is to remain proportional to the angle of rotation of the balance wheel and lead to simple harmonic motion.

With almost every important advance in the world of ideas or in the practical world of devices one name has, through common usage, become associated as the one to whom all credit is due. However, we all know that almost always there were other thinkers and inventors who had made critical contributions to those advances. When the contest for recognition is between two equally prominent personalities, the controversy is resolved, if at all, along national lines. In the present instance there is no doubt that the Englishman Robert Hooke, of *Hooke's law* fame, had indeed proposed a spring–balance mechanism sufficiently accurate to determine longitude at sea. It seems that Hooke had ambitions of exploiting his ideas in an entrepreneurial spirit, not commonly avowed by physicists of his day. In any event, Hooke failed to form a syndicate, and he never actually "reduced his ideas to practice," as patent lawyers would say. On the other hand, Huygens, already credited with implementing the use of the pendulum as a regulator, did in fact have a clock constructed that was regulated by a balance wheel.

Much has been written about the British Admiralty's quest in the early eighteenth century to simplify the solution of an age-old problem in navigation: the determination of longitude at sea (Sobel, 1995). Unlike latitude, which can be deduced from straightforward observations, such as the altitude of the sun on the meridian (at noon) or the altitude of the star Polaris, longitude was computed by a rather complicated procedure devised by the astronomer Edmund Halley, better known for his comet. It had been recognized for some time that if a mariner at sea had a precise clock indicating Greenwich Mean Time (GMT), he could determine longitude by using it to find the time of local noon, for example. To spur interest in the development of such a shipboard clock, the British Admiralty established a Board of Longitude in 1714 that offered a reward of £20,000 to anyone who could determine longitude at sea with an error less than thirty miles. At the equator this corresponds to an error of about 1.7 minutes in time. For a voyage lasting one month this implies an error less than about 4 parts in 10^5, beyond the capability of the existing clocks under shipboard conditions. A Yorkshireman named John Harrison, a far more gifted instrument maker than politician, perfected his first chronometer by 1735, an intricate piece of ingenious mechanical design to minimize friction, etc. Sadly, because of the novelty of his ideas and a prejudice in favor of the astronomical technique called *lunars*, the Board of Longitude was not much impressed and denied Harrison the award. Not until 1761, after his chronometers had been generally admitted to have merit, was the Admiralty willing to try one out on a voyage to Jamaica, during which it lost less than 2 minutes at a fairly constant rate. It was Captain Cook, another Yorkshireman, who by carrying Harrison's timepieces on his long voyages demonstrated finally that old Harrison's chronometers had indeed met the Admiralty requirements and had fully deserved the award.

The principal problem with the balance wheel is its susceptibility to thermal changes in dimensions and consequent changes in the period of oscillation. As with the pendulum there are three remedies; of these the most universal is compensation of expansions and contractions due to temperature fluctuations through the use of two metals in the form of a *bimetallic strip*.

The ultimate success of the balance wheel as a regulator in precision mechanical clocks was made possible by further progress in the design of the escapement, culminating in the *détente*, or chronometer spring, escapement.

This brought the performance level to a height unmatched until the arrival of electronic timekeeping. This escapement approximates more closely than any other the ideal of allowing the regulator to oscillate freely except for a very short period of interaction with the scape wheel.

Over the succeeding centuries timepieces were progressively refined and made smaller; from the pocket watch to the dainty ladies' wristwatch. Figure 1.8 illustrates schematically the essential features of the escapement commonly used in high-quality wrist watches. Where it took Harrison literally years to painstakingly construct by hand a single clock, it ultimately became possible to mass produce them, thus making them universally affordable. But it is remarkable that from the point of view of accuracy, no purely mechanical clock has surpassed some of Harrison's later chronometers.

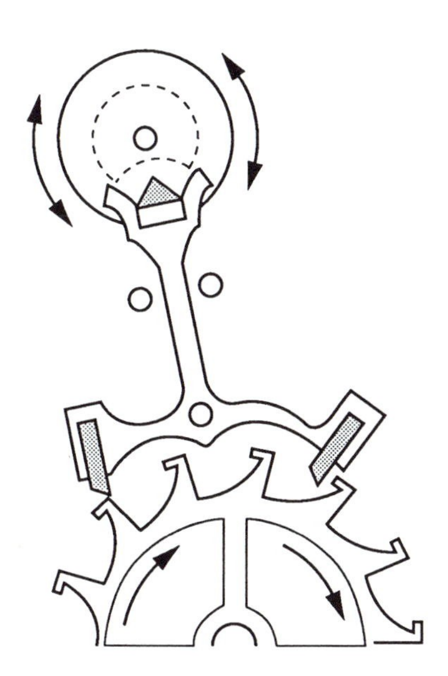

Figure 1.8 A typical escapement design in a high-quality wristwatch

Chapter 2
Oscillations and Fourier Analysis

2.1 Oscillatory Motion in Matter

A universal property of material objects is their ability to vibrate, whether the vibration results in an audible sound, as in the ringing of a bell, or is subtle and inaudible, as the motion in a quartz crystal. It can be a microscopic oscillation on an atomic scale, or as large as an earthquake. Oscillations in any part of an extended object or medium with undefined boundaries almost always propagate as waves.

If any solid object is struck with a sharp blow at some point, vibrations spread throughout the body, and waves are set up in the surrounding medium. If the medium is air, and we are within hearing range, the waves fall on our eardrums and are perceived as a loud sound, whose quality experience teaches us to differentiate according to the kind of object and the way it was struck. Unless the shape of the body and the way it was struck satisfy very particular conditions, the sound produced will be far from a pure tone. The sounds produced by different objects are recognizably different; even if we play the *same* note on different musical instruments, the quality of the sound, or *timbre*, as musicians call it, is different. It is a remarkable fact, first fully appreciated by Alexander Graham Bell, that just from the rapidly fluctuating air pressure of a sound wave falling on our eardrums we are able to construct what we should call an "acoustic image." That is, we are able to sort out and recognize the various sources of sound whose pressure waves have combined to produce a net complex wave pattern falling on the eardrum. To really appreciate how remarkable this facility is, imagine that a microphone is used to convert the complex fluctuations of pressure into an electrical signal that is connected through appropriate circuits to an oscilloscope, and you watched these fluctuations on the screen. Now, without being allowed to hear the sounds, imagine trying to recognize, just from the complex pattern, a friend's voice, or even that it is a human voice at all.

The reason that oscillatory motion is so universally present stems from two fundamental properties of matter. First, objects as we normally find them are in stable equilibrium; that is, any change in their shape brings into play a force to restore the undisturbed shape. Second, all objects have *inertia*; that is, once a body or part of a body has been set in motion, it will tend to continue in that state, unless forces are impressed upon it to change its state; this is the well-known first law of motion of Newton. It follows that when, for example, an external force causes a momentary displacement from equilibrium, the restoring force arising from the body's inherent equilibrium will cause the affected part of the body not only to return to the undisturbed state, but, because of inertia, to overshoot in the other direction. This in turn evokes again a restoring force and an overshoot, and so on.

2.2 Simple Harmonic Motion

The simplest form of oscillatory motion is *simple harmonic motion*, as exemplified by the swinging of a pendulum. This will ensue whenever a physical system is displaced from stable equilibrium by a sufficiently small amount that the restoring force varies nearly linearly with the displacement. Thus a Taylor expansion of the energy U in terms of a small displacement ξ about the point of stable equilibrium yields the following:

$$U = U_0 + a_2\xi^2 + a_3\xi^3 + \ldots \qquad (a_2 > 0), \tag{2.1}$$

and for sufficiently small ξ the restoring force $F=-dU/d\xi$ may be taken as linear in the displacement. It follows that the equation of motion is given by

$$\frac{d^2\xi}{dt^2} + \frac{2a_2}{m}\xi = 0 \qquad (a_2 > 0), \tag{2.2}$$

which has the well-known periodic solution

$$\xi = \xi_0 \cos[\omega t + \phi_0] \tag{2.3}$$

characterized by a unique (angular) *frequency* ω, and initial *amplitude* ξ_0 and *phase* ϕ_0. In a useful graphical representation, the displacement ξ is the projection onto a fixed straight line of a radius vector ξ_0 rotating with constant angular velocity ω; the quantity $(\omega t+\phi_0)$ is then the angular position of the radius vector, giving the phase of the motion. Such a representation is a *phasor diagram*, illustrated in Figure 2.1.

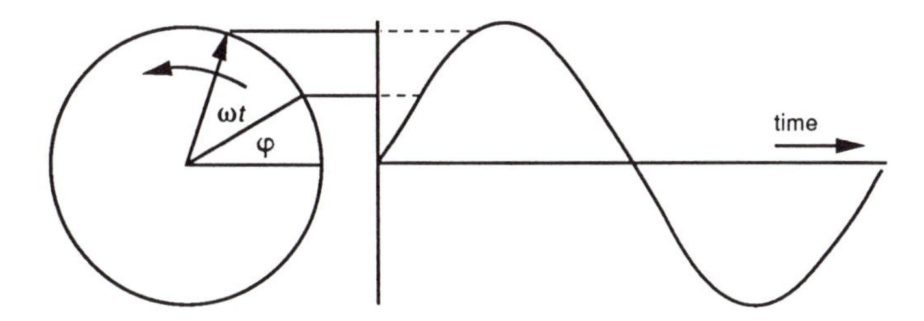

Figure 2.1 Simple harmonic motion as a projection of uniform circular motion: phasor diagram

As a corollary, or simply by rewriting the solution in exponential form, it follows that the motion is the sum of two phasors of equal length rotating in opposite directions, thus

$$\xi = \frac{\xi_0}{2} e^{+i(\omega t + \varphi_0)} + \frac{\xi_0}{2} e^{-i(\omega t + \varphi_0)} . \qquad 2.4$$

In the assumed linear approximation of the equation of motion, if ξ_1 and ξ_2 are two solutions of the equation, then any linear combination $(a\xi_1+b\xi_2)$, where a and b are constants, is also a solution.

If the next higher term in the expansion of U is retained, we are led to a nonlinear (or *anharmonic*) oscillator. The prototypical example is the pendulum when the finite amplitude of oscillation is treated to a higher order of approximation than the simple linear one. Thus the exact equation of motion, expressed in terms of the angular deflection of the pendulum θ, is nonlinear, as follows:

$$l\frac{d^2\theta}{dt^2} + g\sin\theta = 0 . \qquad 2.5$$

If $\theta \ll 1$, we may expand sinθ in powers of θ to obtain a higher-order approximation to the equation of motion than the linear one. Thus

$$l\frac{d^2\theta}{dt^2} + g\left(\theta - \frac{1}{6}\theta^3\right) = 0 . \qquad 2.6$$

Assume now that the amplitude of the motion is θ_0, so that in the linear approximation the solution would be $\theta_0 \cos(\omega_0 t+\phi_0)$, where $\omega_0=\sqrt{(g/l)}$. We can obtain an approximate correction to the frequency by using the method of successive approximation; this we do by assuming the following approximate form for the solution:

$$\theta = \theta_0 \cos\omega t + \varepsilon \cos 3\omega t, \tag{2.7}$$

On substituting this into the equation of motion and setting the coefficients of $\cos\omega t$ and $\cos 3\omega t$ equal to zero, we find the following:

$$\omega = \omega_0\left(1-\frac{\theta_0^2}{16}\right); \ \varepsilon = \frac{1}{3}\left(\frac{\theta_0}{4}\right)^3 \quad (\theta_0 \ll 1), \tag{2.8}$$

which shows that the pendulum has a longer period at finite amplitudes than the limit as the amplitude approaches zero.

In the simple pendulum the suspended mass is constrained to move along the arc of a circle. It was this motion that Galileo thought to have the property of isochronism (or tautochronism), that is, requiring equal time to complete a cycle starting from any point on the arc. In fact, the mass must be constrained along a cycloid, the figure traced out by a point on a circle rolling on a straight line, rather than a circle, in order to have this property. A more famous, related problem, one first suggested and solved by Bernoulli and independently by Newton and Leibnitz, has to do with the curve joining two fixed points along which the time to complete the motion is a *minimum* with respect to a variation in the curve; again the solution is a cycloid.

Attempts in the early development of pendulum clocks to realize in practice the isosynchronism of cycloidal motion were soon abandoned when it became apparent that *other* sources of error were more significant. In any event, in order to maintain a constant clock rate it is necessary only to regulate the amplitude of oscillation.

We should note that the presence of the nonlinear term in the equation of motion puts it in a whole different class of problems: those dealing with nonlinear phenomena. One far-reaching consequence of the nonlinearity is that the solution will now contain, in addition to the oscillatory term at the fundamental frequency ω, higher harmonics starting with 3ω. We will encounter in later chapters electronic devices of great practical importance whose characteristic response to applied electric fields is nonlinear.

2.3 Forced Oscillations: Resonance

Although our main concern will be the resonant response of atomic systems, requiring a quantum description, some of the basic classical concepts provide at least a background of ideas in which some of the terminology has its origins.

Imagine an oscillatory system, such as we have been discussing, having the further complication that its energy is slowly dissipated through some force resisting its motion. This is most simply introduced phenomenologically into the equation of motion as a term proportional to the time derivative of the displacement. The response of such a system to a periodic disturbance is governed by the following equation:

$$\frac{d^2\xi}{dt^2}+\gamma\frac{d\xi}{dt}+\omega_0^2\xi=\alpha_0 e^{ipt}, \qquad 2.9$$

which has the well-known solution

$$\xi=\frac{\alpha_0}{\sqrt{\left(\omega_0^2-p^2\right)^2+\gamma^2 p^2}}e^{i(pt-\phi)}+\xi_0 e^{-\frac{\gamma}{2}t}e^{+i(\omega t+\psi)}, \qquad 2.10$$

where $\phi=\arctan[\gamma p/(\omega_0^2-p^2)]$ and $\omega=\sqrt{\omega_0^2-\gamma^2/4}$. The important feature of this solution is, of course, the resonantly large amplitude of the first term, the *particular integral*, at $\omega=p$; but an equally significant point is that its *phase*, unlike that of the second natural oscillation term, bears a fixed relationship to that of the driving force. This means that if we have a large number of identical oscillators initially oscillating with random phase, and they are then subject to the same driving force, the net global disturbance will simply be the sum of the resonant terms, since the other terms will tend to average out.

2.3.1 Response near Resonance: the Q-Factor

In order to analyze the behavior near resonance of a lightly damped oscillator for which $\gamma<<\omega_0$, let us assume that $p=\omega_0+\Delta$, where $\Delta<<\omega_0$. Then we can write the following for the amplitude and phase of the impressed oscillation:

$$A=\frac{\alpha_0}{2\omega_0}\frac{1}{\sqrt{\Delta^2+\left(\frac{\gamma}{2}\right)^2}}; \quad \phi=\arctan\left(-\frac{\gamma}{2\Delta}\right), \quad \Delta\langle\langle\omega_0, \qquad 2.11$$

which, when plotted as functions of Δ, show for the amplitude the sharply peaked curve characteristic of resonance, falling to $1/\sqrt{2}$ of the maximum at $\Delta=-\gamma/2$ and $\Delta=+\gamma/2$, and for the phase, the sharp variation over that tuning range from $\pi/4$ to $3\pi/4$, passing through the value $\phi=\pi/2$ at exact resonance when $\Delta=0$. A measure of the sharpness of the resonance, a figure of merit called the *Q-factor*, is defined as the ratio between the frequency and the resonance frequency width γ. Thus

$$Q=\frac{\omega_0}{\gamma}. \qquad 2.12$$

An equally useful result is obtained by relating Q to the rate of energy dissipation by the oscillating system. Thus from the equation of motion of the free oscillator we find after multiplying throughout by $d\xi/dt$ the following:

$$\frac{d}{dt}\left[\frac{1}{2}\left(\frac{d\xi}{dt}\right)^2+\frac{1}{2}\omega^2\xi^2\right]=-\gamma\left(\frac{d\xi}{dt}\right)^2, \qquad 2.13$$

from which we obtain by averaging over many cycles (still assuming a weakly damped oscillator) the important result

$$\frac{d\langle U_{tot}\rangle}{dt}=-2\gamma\langle U_k\rangle; \quad \langle U_k\rangle=\frac{1}{2}\langle U_{tot}\rangle, \qquad 2.14$$

From this follows the important result that we shall have many occasions to quote in the future:

$$Q=\omega_0\frac{\langle U\rangle}{\frac{d\langle U\rangle}{dt}}. \qquad 2.15$$

Associated with the rapid change in amplitude is, as we have already indicated, a rapid change in the relative phase between the driving force and the response it causes. This interdependence between the amplitude and phase happens to be of particular importance in the classical model of *optical dispersion* in a medium as a manifestation of the resonant behavior of its constituent atoms to the oscillating electric field in the light wave.

As we shall see in the next chapter, the sharp change in the phase ϕ as a function of frequency near resonance is of critical importance to the *frequency stability* of an oscillator, wherever the resonance is used as the primary frequency-selective element in the system. An important quantity from that point of view is the change in the phase angle produced by a given small detuning of the frequency from exact resonance. Figure 2.2 shows the approximate shapes of typical frequency-response curves. If we make the crude approximation that the phase varies linearly in the immediate vicinity of resonance, then since ϕ varies by π radians as the frequency is tuned from $-\gamma/2$ to $\gamma/2$, it follows that the change in phase $\Delta\phi$ is given approximately by the following:

$$\Delta\phi=\frac{(\omega_0-\omega)}{\gamma}\pi. \qquad 2.16$$

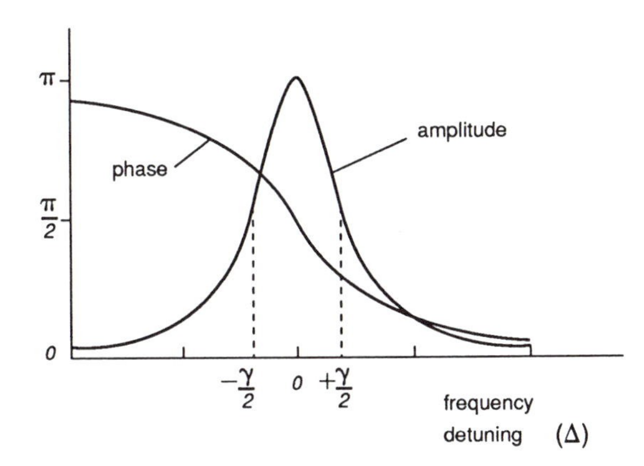

Figure 2.2 Amplitude and phase response curves versus frequency of a damped oscillator

We note that having a very small γ, or equivalently, a very small fractional line width, favors a small change in frequency accompanying any given deviation in phase; and it is the phase that is susceptible to fluctuation in a real system.

2.4 Waves in Extended Media

In a region of space where a momentary disturbance takes place, whether among interacting material particles, as in an acoustic field, or in the field vectors of an electromagnetic field, such a disturbance generally propagates out as a wave. A historic example is the first successful effort to produce and detect electromagnetic waves as predicted by Maxwell's theory. Heinrich Hertz , at the University of Bonn, detected electromagnetic waves radiating from a "disturbance" in the form of a high-voltage spark. One of the physical conditions found in the propagation of a disturbance as a wave is a delay in phase between the oscillation at a given point and that at an adjacent point along the direction of propagation; this is inevitably associated with a finite wave velocity.

The simplest case to analyze is that of transverse waves on a stretched string. It is evident in this case that the net force on a small element of the string depends on the *difference* between the directions of the string at the two ends of the given small segment and therefore depends on the *curvature* of the string. It follows by Newton's second law that the acceleration of this segment is proportional to the curvature; or stated symbolically, we have the well-known form of the (one-dimensional) *wave equation*:

$$T\frac{\partial^2 y}{\partial x^2} - \rho\frac{\partial^2 y}{\partial t^2} = 0, \qquad 2.17$$

where T and ρ are constants, the *tension* and *linear density* of the string. If we rewrite the equation as

$$\frac{\partial^2 y}{\partial x^2} - \frac{1}{V^2}\frac{\partial^2 y}{\partial t^2} = 0, \qquad 2.18$$

we can verify that a general solution, called *D'Alembert's solution,* can be written as follows:

$$y = f_1(x - Vt) + f_2(x + Vt), \qquad 2.19$$

where f_1 and f_2 are any (differentiable) functions, the first of which represents a disturbance traveling with a velocity V in the positive x direction, while the other is one traveling in the opposite direction, *without change of shape*: This is ultimately because V was assumed to be a constant.

In the case of the electromagnetic field, Maxwell's theory, the triumph of nineteenth-century physics, predicts that the electric and magnetic field vectors **E** and **B** propagate in a medium characterized by the electric permittivity ε and magnetic permeability μ according to the following wave equation expressed with reference to a Cartesian system of coordinates x, y, z:

$$\frac{\partial^2 E_x}{\partial x^2} + \frac{\partial^2 E_x}{\partial y^2} + \frac{\partial^2 E_x}{\partial z^2} - \varepsilon\mu\frac{\partial^2 E_x}{\partial t^2} = 0, \qquad 2.20$$

with similar equations for the other components. It follows that for an unbounded uniform medium, the velocity of propagation $V = 1/\sqrt{\mu\varepsilon}$ is a constant, which in a vacuum has a numerical value in the MKS system of units of $2.9979... \times 10^8$ m/s.

The simplest solutions to the wave equation in an unbounded medium have a simple harmonic dependence on the coordinates and time, which in one dimension may be written in the form

$$E_y = E_0 \sin(kz - \omega t + \phi). \qquad 2.21$$

where k is the magnitude of the wave vector, ω is the (angular) frequency, and ϕ is an arbitrary phase.

The surfaces of constant phase, defined by $(kz - \omega t)$=constant, travel with the velocity V given by $V = \omega/k$. If we write, as is conventionally done, $V = c/n$ where c is the velocity of light *in vacuo,* then the quantity n, originally defined for frequencies in the optical range, is the refractive index that appears in Snell's law. This is the velocity of propagation only of the phase of a simple harmonic wave having a single frequency; for any more complicated wave, it becomes necessary to stipulate exactly what it is that the velocity refers to. Clearly, the concept of a wave velocity has meaning only if some identifiable attribute of the wave is indeed traveling with a well-defined velocity. If, for example, the wave has only one large crest like the bow wave of a ship traveling with sufficient speed, then the velocity with which that crest travels can differ from the phase velocity if the particular medium is *dispersive,* that is, if the phase velocity is a function of the frequency. This is readily seen if we recall that such a waveform can be thought of as a Fourier sum of simple harmonic waves, which now are assumed to travel at different velocities. In fact, there is no a priori reason for the wave to preserve its shape as it progresses; if it does not, the whole notion of wave velocity loses meaning. However, under some conditions a *group velocity* can be defined for a *wave packet* given by $V=d\omega/dk$. More will be said about dispersive media in the next section.

It will be useful to review some of the fundamental properties of waves. Without going into great detail in the matter, we will simply state that at a boundary surface, where there is an abrupt change in the nature of the medium, waves will be partially reflected, and partially transmitted with generally a change in direction, that is, *refraction*, governed by Snell's law. The geometric surface joining all points that have the same phase is the *wavefront*, and in an unbounded medium the wavefront will advance at each point along the perpendicular to the surface, called a *ray* at that point.

If there is an obstruction in the medium, that is, a region where, for example, the energy of the wave is strongly absorbed, the waves will "bend around corners": the phenomenon of *diffraction*. This, it may be recalled, was the initial objection to the wave theory of light, an objection soon removed by the argument that the wavelength of light is extremely small compared with the dimensions of ordinary objects, and that diffraction is small under these conditions. The analysis of diffraction problems is based on *Huygens's principle*, as given exact mathematical expression by Kirchhoff, who showed that the solution to the wave equation at a given field point can be expressed as a surface integral of the field and its derivatives on a geometrical surface surrounding the field point. The evaluation of that surface integral is made tractable in the case of optical diffraction around large-scale objects by the smallness of the wavelength, which justifies a number of approximations. If an incident wave is delimited, for example by the aperture of an optical instrument or the antenna of a radio telescope, the field, of course, is nonzero only over the surface of the aperture, and the integral is simply over that surface. Application of the theory to the important case of a circular aperture under conditions referred to as *Frauenhoffer diffraction*, where the diffracted wave is brought to a focus onto a plane, yields the following result for the intensity distribution in the focal plane:

$$I = 4 I_0 \frac{J_1^2(ka \sin\theta)}{k^2 a^2 \sin^2\theta}, \qquad 2.22$$

where a is the radius of the aperture and θ is the inclination of the direction of the field point with respect to the system axis. The Bessel function J_1 $(ka \sin\theta)$ oscillates as the argument increases, implying an intensity pattern that consists of a central disk, called the *Airy disk,* surrounded by concentric bands that quickly fade as we go out from the center. Since the first zero of the Bessel function occurs when its argument is about 3.8, the radius of the Airy disk is therefore given by

$$\sin\theta \approx \theta \approx \frac{3.8}{ka} = 1.2\frac{\lambda}{D}. \qquad 2.23$$

In the approximation where *ray optics* are used, the image in the focal plane would of course have been a geometrical point.

2.5 Wave Dispersion

Another fundamental wave phenomenon is *dispersion,* the same phenomenon that was made manifest by Isaac Newton in his classic experiment on the dispersion of sunlight into its colored constituents using a glass prism. It occurs when the refractive index varies from one frequency to another; this can occur only in a material medium, never in vacuum, at least according to Maxwell's classical theory. The dispersive action of nonmagnetic dielectric materials is wholly due to the frequency dependence of the electric permittivity ε; this ultimately derives from the frequency dependence of the dynamical response of the molecular charges in the medium to the electric field component in the wave. This is a problem in quantum mechanics. However, H.A. Lorentz was able on the basis of his *electron theory* to account, at least qualitatively, for the gross features of the phenomenon. He assumed that the atomic particles exhibited resonant behavior at certain natural frequencies of oscillation and that the damping arises from interparticle collisions interrupting the phase of the particle oscillation.

According to this model, the oscillating electric field in the wave induces an oscillating polarization in each of the atomic particles with a definite phase relationship to the field, leading to a total global polarization, which for field vectors with the time dependence $\exp(-i\omega t)$ adds a resonant term to the permittivity, as follows:

$$\varepsilon=\left[1+\frac{\sigma^2}{\omega_0^2-\omega^2-i\gamma\omega}\right]\varepsilon_0 . \qquad 2.24$$

where σ is a measure of the atomic oscillator strength. It follows that the (complex) refractive index n is given by

$$n=\frac{c}{V}=c\sqrt{\varepsilon_0\mu_0\left[1+\frac{\sigma^2}{\omega_0^2-\omega_0^2-i\gamma\omega}\right]}, \qquad 2.25$$

from which we finally obtain, assuming that σ is small,

$$n=1+\frac{\sigma^2}{2}\frac{\left(\omega_0^2-\omega^2\right)}{\left(\omega_0^2-\omega^2\right)^2+\gamma^2\omega^2}+i\frac{\sigma^2}{2}\frac{\gamma\omega}{(\omega_0^2-\omega^2)^2+\gamma^2\omega^2} . \qquad 2.26$$

Finally, substituting this result in the assumed (complex) form for the plane wave solution,

$$E_x=E_0e^{i(nkz-\omega t)}, \qquad 2.27$$

we see that the real part of n determines the phase velocity and hence the dispersion, while the imaginary part yields an exponential attenuation of the wave amplitude, corresponding to absorption in the medium, provided that γ is a positive number. This shows explicitly how the real and imaginary parts of the atomic response determine the frequency dependence of the real and imaginary parts of the complex *propagation constant* through the medium, that is, of the refractive index and absorption of the wave. The complex propagation constant, as a function of frequency, exhibits a relationship between the real and imaginary parts that is an example of a far more general result that finds expression in what are called the Kramers–Kronig *dispersion relations*. It is far beyond the scope of this book to do more than mention that in a relativistic theory these relations are involved with the question of causality and the impossibility of a signal propagating faster than light.

2.6 Linear and Nonlinear Media

So far we have considered media that are *linear*, which means in the case of acoustic waves that a stress applied at some point produces a proportional strain; and conversely, a displacement from equilibrium brings about a proportional restoring force, resulting in simple harmonic motion. In the case of electromagnetic waves the classical theory leads to strictly linear equations *in vacuo*. A linear medium has an extremely important property: It obeys the *principle of superposition*. This states roughly that if more than one wave acts at a certain point, the resultant wave is simply the (vector) sum of these. At first, this may sound like pure tautology. The real meaning of the statement is that it is valid to talk about several waves being present simultaneously at a certain point as if they were individual entities that preserve their identity at the point where they overlap. A corollary is that in a linear medium, a wave is unchanged after it passes a region of overlap with another wave. According to classical theory, two light beams, no matter how powerful, intersecting in a vacuum will not interact with each other: each emerges from the point of intersection as if the other beam were not there. In the realm of quantum field theory, however, it is another story: The vacuum state is far from "empty"!

However, it is possible to increase the strength of a disturbance in a *material* medium to such a point that the medium is no longer linear, and the principle of superposition no longer valid. Waves would then interact through the medium with each other, generating other waves at higher harmonic frequencies. We have already seen this in the case of the pendulum, where the presence of a nonlinear (third-degree) term in the equation of motion led to the presence of a third harmonic frequency.

In the more important circumstance where the field equations describing propagation through a given medium have a significant quadratic term, as in the frequency mixing devices we shall encounter later, two overlapping waves of frequencies ω_1 and ω_2 would interact, and the total solution would include the following:

$$\alpha\left[E_1(t)+E_2(t)\right]^2 = \alpha E_1^2\cos^2(\omega_1 t)+\alpha E_2^2\cos^2(\omega_2 t) + 2\alpha E_1 E_2\cos(\omega_1 t)\cos(\omega_1 t)+\ldots \qquad 2.28$$

Using the trigonometric identities:

$$\cos^2(\omega t) = \frac{1}{2}\left[\cos(2\omega t)+1\right],$$

$$\cos(\omega_1 t)\cos(\omega_2 t) = \frac{1}{2}\left[\cos(\omega_1+\omega_2)\,t + \cos(\omega_1-\omega_2)\,t\right], \qquad 2.29$$

we see that with the assumed degree of nonlinearity, the second harmonic as well as the sum and difference frequencies appear in the output. By suitable filtering, any one of these frequency components can be isolated. We will have occasion to discuss in a later chapter the use of nonlinear crystal devices to produce intercombination and harmonic frequencies in the radio frequency and optical regions of the spectrum.

2.7 Normal Modes of Vibration

When waves are set up in a medium with a closed boundary surface, there will be reflections at different parts of the boundary, with the possibility of multiple reflections in which reflected waves are themselves reflected from opposing surfaces, all combining to produce a resultant wave pattern. If the medium is linear, the problem of finding the resultant is simply a matter of summing over the individual waves. It is one of the fundamental characteristics of waves that the resultant amplitude at a given point can be large or small depending on the relative phase of the combining waves at that point, producing an *interference pattern.*

Let us consider a homogeneous medium with a pair of parallel planes forming part of its boundary surfaces; the remainder of the boundary is immaterial. Let us assume that a disturbance has been created at some point in this medium, giving rise to a wave that will travel out and be reflected by each of the plane boundary surfaces, return to the opposite surfaces, and be reflected again to pass through the initial point. The total distance traversed in making this round trip will be the same for all initial points and equal to twice the distance between the plane boundary surfaces. If this distance happens to be equal to a whole number of wavelengths of the wave, the waves arriving back at any initial point will, after an even number of reflections, be in phase with the initial disturbance, and the wave amplitude will build up at all points, as long as the external excitation continues. By contrast, if the round trip distance is not a whole number of wavelengths, the reflected waves will not be in phase with the exciting source, nor with waves from prior reflections, and the resultant of many even slightly out of phase waves will be weak and evanescent. Note that it is not necessary that the phase difference be near 180° to lead to cancellation and a weak resultant wave; even a small difference in phase produced in each round trip will accumulate after many successive reflections to result in the presence of waves having a phase ranging from 0° to 360°. In that event, for every wave of a given phase, there will be another wave 180° out of phase with it, leading to cancellation.

The condition for a buildup of the wave can be simply stated as follows:

$$2L = n\lambda_n, \tag{2.30}$$

where n is any positive integer. This allows us to calculate the corresponding frequencies $\nu_n=V/\lambda_n=nV/2L$. Thus if we know the wave velocity V in the given medium and the distance between the reflecting surfaces, we can predict that certain frequencies of excitation will find a strong response, while any other, even neighboring, frequencies will not do so. Since n can be any whole number, there is an infinite number of frequencies forming a *discrete* spectrum, in which the frequencies have separate, isolated values, as opposed to a continuous spectrum in which frequency values can fall arbitrarily close to each other and merge into a continuum. The simplicity of the result, that the frequencies in the spectrum are whole multiples of the fundamental frequency $V/2L$, is due to the simple geometry of two plane reflecting surfaces in a homogeneous medium. However, even for more complicated geometries, part of the spectrum may still be discrete; but the frequencies will not necessarily be at equal increments.

To further elaborate on these basic concepts, let us consider another system, one that better lends itself to graphical illustration: a vibrating string stretched between two fixed points. Note that we can think of the fixed points merely as points where the string joins another string of infinite mass, and therefore we can regard the fixed points as the "boundaries" between two media. It has a discrete spectrum consisting of a *fundamental* frequency $\nu=V/2L$ and integral multiples of it called *harmonics*. In a musical context the harmonics above the first are called overtones, whose excitation determines the quality of the sound. These are the frequencies of the so-called *normal modes* of vibration of the string, shown in Figure 2.3.

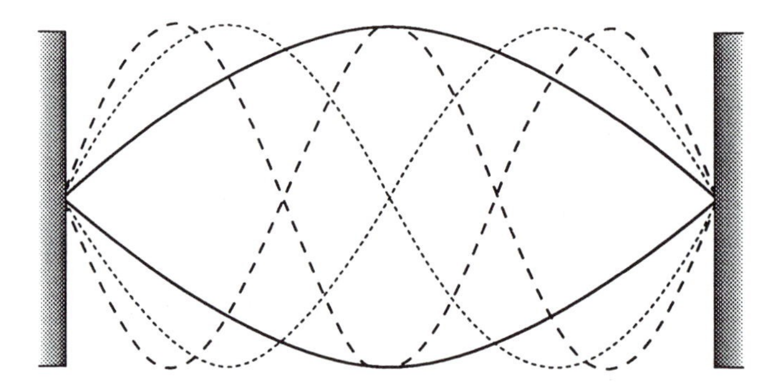

Figure 2.3 The natural modes of vibration of a stretched string

Each can be excited by applying an external periodic force, and the amplitude resulting from such excitation is qualitatively easy to predict: it is essentially zero unless the frequency is in the immediate neighborhood of one of the natural frequencies. At that point, the amplitude would grow indefinitely if it were not for frictional forces, or the onset of some amplitude-dependent mechanism to limit its growth. This phenomenon is of course *resonance*, which provides a method of determining the normal mode frequencies of oscillation of the system. At other frequencies the buildup of excitation is weak because of the mismatch in phase, as already described. Just how complete the cancellation will be depends on the highest number of reflections represented among the waves contributing to the resultant. It may be said approximately that for complete cancellation, the number of waves must be large enough that phase shifts spanning the entire 360° will be present. Now, the increment in phase per round trip is $360(\Delta\nu \cdot 2L/V)$ degrees, where $\Delta\nu$ is a small offset in frequency from one of the discrete frequencies in the spectrum. Thus for cancellation, we require a number N of traversals such that $N \cdot 360(\Delta\nu \cdot 2L/V) = 360$; that is, $\Delta\nu \cdot 2NL/V = 1$. But $2NL/V$ is simply the total time the wave has traveled back and forth, which in reality will be limited by internal frictional loss of energy in the string and imperfect reflections at the end points. Thus if we write $\Delta\tau$ for the mean time it takes the wave to become insignificant, then the smallest $\Delta\nu$ for cancellation is given by $\Delta\nu \cdot \Delta\tau \approx 1$; a smaller frequency offset gives only partial cancellation. This implies that in determining the frequency of resonance there is effectively a spread, or uncertainty, in the result if the measurement occupies a finite interval of time. This result, arrived at in a simpleminded way, hints at a much more general and fundamental result concerning uncertainties in the simultaneous observation of physical quantities: the now famous *Heisenberg Uncertainty Principle.* This principle applies to the simultaneous measurement of such quantities as frequency and time, which are said to be *complementary*, for which a determination of the frequency implies a finite time to accomplish it. Therefore, by its very nature, we cannot specify the frequency of an oscillation at a precise instant in time. To quantify this idea requires a precise definition of "uncertainty" in a physical measurement, which Heisenberg did in the context of quantum theory.

2.8 Parametric Excitations

The most often cited and certainly most dramatic example of the effects of resonance is the collapse of the suspension bridge across the Tacoma Narrows in Washington State, USA. Although its failure was due to violent oscillations, there was no external periodic force acting on it, but rather a buildup of what are called *parametric* oscillations, much like the fluttering of window blinds in a steady wind. Such oscillations are characterized by a buildup resulting from some dynamic parameter varying in a particular way within each cycle.

There is another interesting phenomenon, in which a steady stream of air excites sound vibrations in a stretched string: the aeolian (from the Greek *aiolios*, wind) harp or lyre. This is a stringed instrument consisting of a set of strings of equal length stretched in a frame. When a steady air current passes over the strings, it emits a musical tone. The mechanism by which this occurs is rather subtle, as shown by the observed fact that the pitch of the tone does not seem to depend on the length or tension in the string, which would certainly be the case if it were simply a matter of the resonant frequencies being excited. It is observed, however, that if the resonant frequencies of the strings are made to equal the tone produced by the wind, the sound is greatly reinforced. The pitch of the sound depends on the velocity of the wind and the diameter of the string. According to Rayleigh, the great nineteenth-century physicist, noted for his theory of sound, the sound arises from vortices (eddies) in the air produced by the motion of air across the strings.

The simplest example of parametrically driven oscillations is the "pumping" of a child's swing, in which the child extends and retracts its legs, thereby varying the effective length of the suspension, during each swing. If we assume that a parameter that determines the frequency ω_0, in this case the length of the pendulum, is modulated harmonically at double the oscillation frequency, then the equation of motion will have the following form:

$$\frac{d^2\theta}{dt^2} + \omega_0^2\left[1 + \varepsilon\cos(2\omega_0 t)\right]\theta = 0. \qquad 2.31$$

If we assume $\varepsilon \ll 1$, then we can look for an approximate solution of the following form:

$$\theta = a(t)\cos\omega_0 t + b(t)\sin\omega_0 t. \qquad 2.32$$

where $a(t)$ and $b(t)$ vary negligibly during an oscillation. By substituting this form into the equation of motion, we find by setting the coefficients of $\cos\omega_0 t$ and $\sin\omega_0 t$ equal to zero, and neglecting higher harmonic frequencies, that the amplitudes $a(t)$ and $b(t)$ must satisfy the following equations:

$$\begin{aligned} \frac{da}{dt} + \frac{\varepsilon\omega_0}{4} b = 0, \\ \frac{db}{dt} + \frac{\varepsilon\omega_0}{4} a = 0, \end{aligned} \qquad 2.33$$

from which we obtain finally the possible solution

$$a(t) = a_1 e^{+\frac{\varepsilon\omega_0}{4}t} + a_2 e^{-\frac{\varepsilon\omega_0}{4}t}, \qquad 2.34$$

with a similar result for $b(t)$. The presence of the first term, with the positive exponent, shows that the amplitude will grow exponentially. It is important to note (although our simple discussion does not deal with it) that the excitation of a *parametric resonance* will occur over a precise *range* of frequencies of modulation of the parameter; and further, that if the system is initially undisturbed, so that both θ and $d\theta/dt$ are initially zero, the system will not be excited into oscillation.

2.9 Fourier Analysis

When a system is subjected to a *simple periodic* disturbance, its response, in general, will be an oscillation at the frequency of that disturbance, superimposed on whatever free, natural oscillations were already present. As we have seen in the case of a simple physical system consisting of a vibrating string, a large resonant response is induced by a simple periodic force only at one of its natural frequencies. In general, however, when a violin string is excited into vibration, for example by plucking it, the shape of the string is a complicated function of time. We might imagine a high-speed movie camera recording this complex wave motion frame by frame. Predicting the motion of a system produced by an arbitrary initial displacement from its quiescent state is a fundamental problem of physics. The term "motion" used here is not restricted to movement in space; it could be, for example, the variation of temperature throughout a body as determined by the laws that govern the flow of heat.

Since any given natural frequency can effectively be excited only by an oscillatory force at that frequency, it is reasonable to assume that if the excitation is a complicated function of time, the response at the different natural frequencies somehow is representative of the "amount" of those frequencies in the excitation function. From this it seems plausible that to every given excitation function of time there corresponds a unique set of amplitudes (and phases) of the natural-mode responses. This would imply that any given excitation function of time can be regarded as a sum over a unique set of harmonic oscillations at the natural frequencies. In fact, this is given precise mathematical expression in the *Fourier expansion theorem*, one of the most useful theorems in physics, named for Joseph Fourier, a French mathematician who made a systematic study of what is now called Fourier analysis. It applies equally to the representation of an arbitrary initial *shape* of the string as a sum over a unique set of simple harmonic functions of position, making up the natural modes of vibration. This is of such importance to the understanding of what we shall encounter in succeeding chapters on atomic resonance that we must devote some effort to understanding it. The theorem proves that almost any periodic waveform, of whatever shape, can be expressed as the sum over a series of harmonic functions having amplitudes unique to the waveform, and it gives formulas for computing those amplitudes. In the context of high-fidelity audio systems the term "harmonic distortion" is familiar: It refers to the power in the second and higher harmonics of the given frequency being reproduced. This assumes that a distorted waveform can be unambiguously specified as consisting of a fundamental and harmonic components. The theorem is based on a special property called *orthogonality* of the functions describing the normal modes of vibration. The term means the property of being "perpendicular," as might be applied to two vectors; for functions, the test for this property is that the average of the product of the functions be zero, when taken over the appropriate interval. In that sense they are "uncorrelated." In the case of the normal mode functions of the vibrating string, $\sin(n\pi x/L)$ and $\sin(m\pi x/L)$, where n and m are integers, their product averaged over the interval $0 < x < L$ is zero. Thus

$$\int_0^L \sin\left(n\pi\frac{x}{L}\right)\sin\left(m\pi\frac{x}{L}\right)dx = 0, \quad n \neq m. \qquad 2.35$$

In general, for any given periodic function, that is, one satisfying $f(x)=f(x+2\pi)$, orthogonality allows the amplitudes of the harmonics in the following *Fourier series* expansion of the function to be determined:

$$\begin{aligned} f(x) = a_0 + a_1\sin(x) + a_2\sin(2x) + a_3\sin(3x) + \ldots \\ + b_1\cos(x) + b_2\cos(2x) + b_3\cos(3x) + \cdots. \end{aligned} \qquad 2.36$$

Thus by multiplying both sides of equation 2.36 by $\sin(nx)$ and integrating over the fundamental interval we immediately obtain the amplitude a_n. Thus

$$\int_0^{2\pi} \sin(nx) f(x) dx = \pi a_n , \qquad 2.37$$

with a similar result for the amplitudes b_n by replacing $\sin(nx)$ with $\cos(nx)$. We note that the amplitude is in a sense a measure of the extent to which the given function correlates with the harmonic mode function.

The theorem proves that by including higher and higher harmonics, the exact function can be represented as closely as we please. It follows that the amplitudes must decrease as we go to higher-order harmonics, so that a fair representation may be achieved with a finite number of harmonics. As an example, in Figure 2.4a is shown a periodic sawtooth waveform and beside it, in Figure 2.4b, are shown the amplitudes of the first few harmonics plotted against frequency to display the spectrum of the wave. The effect of a filter that removes all but the first three harmonics is shown in Figure 2.4c. We should note that to represent sharp changes in the waveform requires the inclusion of the higher harmonics in the sum.

It is clear from what has been said that for a plucked string, the extent to which each of the natural frequencies will be excited will depend first on the amplitude of each Fourier component in the initial displacement and second on the degree to which each component is able to build up its amplitude in the presence of losses at the boundaries, etc. Since the initial amplitude of a given Fourier component according to the theorem is computed as an overlap integral between the given harmonic function and the function representing the initial displacement, the excitation of that particular harmonic is favored by having the initial displacement large where the harmonic displacement is large.

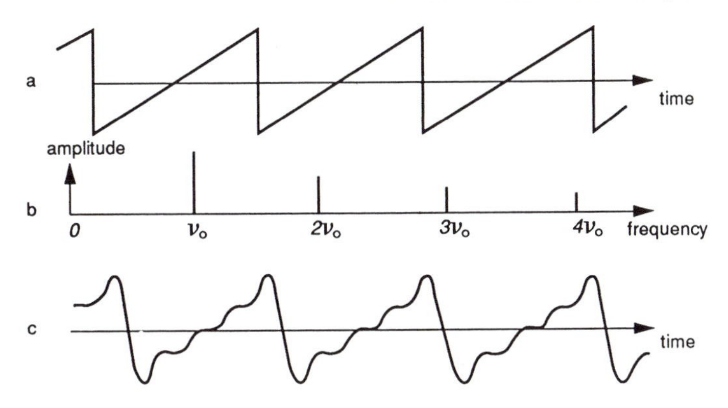

Figure 2.4 (a) A sawtooth waveform, (b) its Fourier spectrum, (c) the sum of the first three harmonics

For nonperiodic functions, there is a corresponding *Fourier integral theorem,* according to which, as a particular example, an even function $f(t)$ of time (that is, one satisfying $f(t)=f(-t)$) can be represented by the following integral:

$$f(t)=\int_0^\infty F(\omega)\cos(\omega t)d\omega \,, \qquad 2.38$$

where $F(\omega)$, now a function of a continuous variable, rather than the discrete mode index number n, gives the amplitude distribution over frequency, that is, the *Fourier spectrum* of the function $f(t)$. $F(\omega)$ has a unique, one-to-one relationship with $f(t)$, which the Fourier theorem proves is a reciprocal one, in the sense that $F(\omega)$ is obtained from $f(t)$ simply by interchanging their roles. The one function is called the *Fourier transform* of the other.

It frequently happens that where we have a complex signal consisting of what may appear as an unintelligible fluctuation in voltage, we are able to present the information in a far more useful way by applying the Fourier integral theorem. To show in a concrete way how this may be accomplished, consider the following hypothetical experiment. An input signal, which could, for example, be a sound wave or a microwave of complex waveform, is connected to an infinite number of ideal resonators tuned to progressively higher frequencies, with only a small increment in frequency between each resonator and its successor. This, it may be recalled, is the way it is thought that the human ear processes incoming sounds and is thereby able to separate the various types of sources that make up the complex waveform it receives. Let it be assumed that the input signal is switched on for a predetermined period, after which it is switched off, and the amplitudes and phases of the oscillations in all the resonators are measured and then plotted against their resonant frequencies. Such a plot is the frequency spectrum of the incoming complex waveform, a waveform that begins as zero, jumps to the signal value when the switch is turned on, and goes back to zero when the switch is turned off. It is assumed that the frequency difference between consecutive resonators is small, so that there will be a very large number of them. The two principles that are the essence of this method of analysis are these: First, the phases and amplitudes of the resonators are unique to the incoming signal, and second, if we simply add simple harmonic oscillations at the frequencies of the resonators with those amplitudes and phases, the sum will reproduce the original signal waveform.

In later chapters we will have frequent occasion to refer to the Fourier spectra of signals. Two examples of Fourier transforms are shown in Figure 2.5. The first is a signal in which a simple oscillation is switched on at some point and thereafter slowly decays. The second represents a signal that really contains just one frequency, but the phase changes at irregular intervals of time.

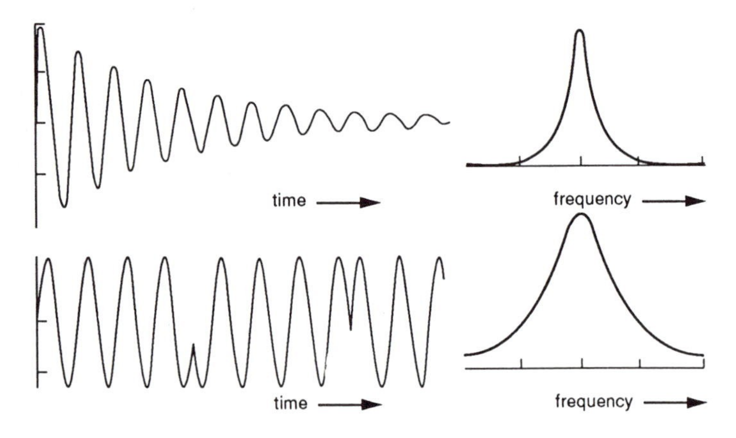

Figure 2.5 Examples of Fourier spectra

In some important cases the phases are either indeterminate or inaccessible; in such cases the power spectrum, showing only the square of the amplitude at each frequency, is nevertheless very useful. The most obvious example is in the analysis of optical radiation, where of necessity we are limited to studying the power spectrum, since no common detector exists that can follow the extremely rapid oscillations in a light wave. Thus when sunlight, for example, is passed through a glass prism to separate the colors of the rainbow, as Newton did in his classic researches on the composition of white light, we are in a sense transforming the fluctuating field components in the incoming electromagnetic wave (the optical signal) into a continuous distribution of intensity over frequency, its Fourier spectrum. In this particular case, as blackbody radiation, the light from the sun will have phases that are random, which makes the availability of a representation in the form of a power spectrum, free of the phases, particularly crucial.

2.10 Coupled Oscillations

An important situation often arises in which one oscillatory system interacts with another. This often occurs where the oscillations of the one are to be synchronized with the other, a process familiar in television receivers. There the local sweep circuits that scan the picture have to be synchronized with the received horizontal and vertical synchronization pulses to obtain a stable picture. This, however, is synchronization under conditions in which the aspect we wish to consider is clearly absent: The oscillating systems do not interact directly. Let us consider, instead, two oscillating systems in which a resonant frequency in one nearly coincides with one in the other system, and assume that there is a weak coupling between them. A somewhat contrived example is shown in Figure 2.6, which depicts two pendulums (or is it pendula) of nearly equal natural oscillation period whose suspension is from a massive body that can slide horizontally without friction. If the coupling body were so massive that it may be regarded as immovable, then the pendulums would be independent of each other. However, for a large but finite mass, any oscillation in one pendulum affects the other. Perhaps the most striking phenomenon is seen in this system if we set one pendulum in motion while the other is left initially undisturbed. If we watch the subsequent motion of the two pendulums, a curious thing happens: The pendulum initially at rest will begin oscillating with increasing amplitude while the amplitude of the other simultaneously decreases. This will continue until the pendulum that was originally set in motion comes to rest, and the two have exchanged the initial state. Then the process reverses, and the two return to the original state. The energy of oscillation would continue to be exchanged back and forth indefinitely if it were not for the inevitable presence of frictional forces at the points of suspension and air resistance, which will cause the energy to be dissipated as heat and the system to come to rest. It is as if the system cannot "make up its mind" which state to be in; its oscillatory state is continually changing.

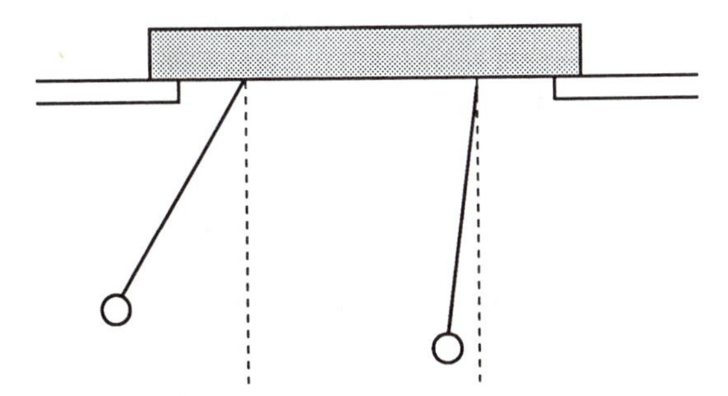

Figure 2.6 Two identical coupled pendulums

An interesting question to ask about the coupled system is whether it can be set oscillating in some mode in which all parts of the system execute oscillation at the same frequency, with a stable amplitude. The attempt to answer this type of question, particularly to more complex systems involving several coupled systems, has led to a sophisticated theory and the concept of *normal* vibrations. To illustrate what is meant by the term, let us go back to the two coupled pendulums. We will state without proof that if this system is initially set in motion, either with the two pendulums in phase or the two exactly 180° out of phase, they will continue to oscillate in those modes with a constant amplitude. These two modes, illustrated in Figure 2.7, are called the normal modes of vibration for this particular system. It is important to note that for these modes to be preserved, the two pendulums must oscillate with a common frequency. This is, in fact, the defining characteristic of the normal modes: In a given mode, all parts of the coupled system must oscillate at one frequency belonging to that mode. The common frequency will, in general, vary from one mode to another. In the case of the two coupled pendulums, the frequencies of the two modes differ to an extent determined by the degree of coupling between them;

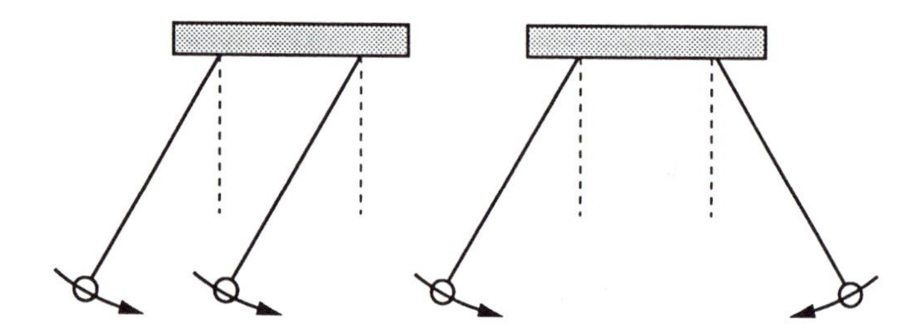

Figure 2.7 The normal modes of oscillation of two identical pendulums

this can be shown to be m/M, where m is the mass of the pendulum bob and M is the coupling mass. In terms of this coupling parameter m/M, the frequencies of the modes are approximately $\nu_a=\nu_o(1+m/M)$ and $\nu_b=\nu_o$, where ν_o is the frequency of free oscillation in the absence of coupling.

It is interesting to view the original bizarre behavior, in which the oscillation went back and forth between the pendulums, in terms of the normal modes. We see that when only one pendulum is set in motion, the system is not in a normal mode but could be looked on as a "mixture" (or more precisely, a *linear superposition*) of the two normal modes; that is, the motion of each pendulum in our particular example is the *sum* of equal amplitudes of the two normal modes. But these modes do not have exactly the same frequency, and their relative phase will continuously increase, passing periodically through times when their phases differ by a whole number of cycles and are in step, and times when they get

180° out of step. When they are in step, they reinforce each other and produce a large amplitude, while the opposite is true when they get out of step and cancel each other. Thus the amplitude of each pendulum alternately rises and falls periodically, a phenomenon called "beats," from the way it is manifested when two musical notes having nearly the same pitch are sounded together. The number of beats per second can be shown to equal the difference between the frequencies of the two normal modes and therefore proportional to the strength of coupling, between the two pendulums; the tighter the coupling the higher the frequency at which the energy is exchanged back and forth between the two pendulums.

Chapter 3
Oscillators

3.1 Feedback in Amplifiers

As already noted, an oscillatory system will, in the absence of a driving force to maintain the oscillation, eventually come to rest. In order to keep a constant level of oscillation, it is necessary to inject energy into the system, an action most efficiently performed by a periodic force at a resonant frequency. It is not necessary that the external source of energy itself be periodic, since the oscillating system can be made to draw energy automatically at its own frequency; it is then called a self-sustained *oscillator*. In essence, this is accomplished by driving the oscillator from a power source from which the transfer of energy to the oscillator is regulated by the oscillator itself. This amounts to using a power amplifier to drive the oscillator with an amplified version of its own oscillation. To sustain the oscillation, the amplified power must be fed back to the oscillating system in the proper phase to reinforce the oscillation already present. This is called *positive feedback* and is generally associated with a rapid buildup of energy, which in every practical situation, however, is always limited by the onset of some degradation of the conditions that led to the buildup, or some imposed limit. Self-sustaining motion (not perpetual motion) occurs in many kinds of systems. For example, in a very broad sense a steam engine is a self-sustaining rotator, in the sense that valves controlling the flow of steam into the cylinders to drive the pistons are in fact acting like power amplifiers, whose output, the force on the pistons, in addition to driving the train, also actuates the valves, providing positive feedback.

Positive feedback is familiar to most people as the cause of loud whistling in a public address system when the amplification is set beyond a certain point for a given disposition of microphone and loudspeakers. To understand the conditions necessary to produce a self-sustaining oscillation, we must recognize that it is not sufficient that there be feedback, it must be sufficient and in the right phase. In a sound system the microphone converts any sound waves impinging on it into a weak fluctuating electric current, or voltage, which we will simply call a signal. This is amplified to a much larger signal, more or less faithfully reproducing the same fluctuations, and applied to the loudspeakers, which convert the current fluctuations back into sound pressure waves in the surrounding air. At some distance this sound wave will reach the microphone, completing a feedback loop. Depending on the directionalities of the microphone and the loudspeakers, and the distance between them, the feedback signal may be stronger or weaker than the original signal at the beginning of the loop. To stipulate the conditions for self-sustaining oscillation, imagine the following experiment: Imagine that a magical shield could be placed between the microphone and the speakers without in any way affecting the acoustics, so that the loop is broken by the shield. Now imagine a frequency synthesizer placed next to the shield, and let a succession of pure tones be sounded at a known, constant intensity. For each of these tones suppose the level of intensity arriving just on the other side of the shield is measured. Then the ratio of the two intensities is called the loop (power) gain. Further assume that we could see and analyze the waveforms of the sound waves on the two sides of the shield and thereby determine their relative phase. We are now ready to state the conditions for self oscillation: The loop gain must reach unity, and the loop phase difference must be zero or a whole number of cycles. The question remains as to why the oscillation takes place at nearly a single frequency.

To see why the oscillation condition is usually limited to a single frequency requires a somewhat more detailed study of feedback amplifiers, a subject of great practical interest and sophistication, not only for oscillators, but especially for the converse problem of maintaining "stability," that is, avoiding a system's breaking into oscillation. Systems involving the feed-back of signals for the purposes of automatic control of devices constitute the whole important subject of servo mechanisms, a field central to the control of the frequency of free-running oscillators by atomic resonances, an essential feature of all atomic clocks.

The fundamental question we can ask about an amplifier is the following: If we apply a time-varying signal at its input terminals, what signal will appear at the output terminals? A perfect amplifier would be by definition one in which the output signal is just a scale factor times the input signal, so that if the input and output signals were plotted as functions of time, they would be indistinguishable, apart from a change of scale. In reality, we can only hope to approach this ideal by careful design. An actual amplifier can fall short of the ideal with respect to two independent requirements: speed and linearity. We have already mentioned linearity in connection with waves in a medium; a similar definition applies to amplifiers. It implies that the gain (or scale factor) should be constant, independent of the amplitude of the signal, and if the input signal is the sum of two signals, the output will be the sum of the same two signals magnified by a certain factor. In a practical device this cannot be realized over an indefinitely large input signal; when driven beyond a certain signal level, the gain starts to decrease, and the output will become distorted. A single frequency input signal will then yield a distorted output with higher harmonic frequencies in its Fourier spectrum. We will therefore assume that the signal amplitude is in the range where the amplifier is linear. There remains the speed; that is, how rapidly the input may fluctuate and yet be faithfully amplified. On the assumption that the amplifier is linear, we may apply Fourier analysis, based on the Fourier expansion theorem, which we have already encountered. Using it we can express any periodic signal having an arbitrary waveform as the sum of a series of simple harmonic oscillations having frequencies that are whole multiples of a fundamental frequency corresponding to the period of the signal. Since the amplifier is linear, we may treat each simple harmonic frequency separately, find its output amplitude and phase, and then sum the outputs for all the harmonics in the Fourier series to get the actual output waveform. In order to be able to carry out this procedure, we need the gain and phase shift for each harmonic frequency in the input. For this reason it is customary to specify the performance of an amplifier by giving its *frequency response curves*, that is, a plot of its gain and phase shift as a function of input frequency over the range for which the gain is significant. Armed with these plots, we can compute the output waveform for any input, no matter how complex.

An ideal amplifier would have a constant gain and phase shift for all input frequencies. In reality, amplifiers will have a maximum frequency beyond which the gain falls gradually to zero, accompanied by a variable phase shift. The aim in amplifier design is, of course, first to have a stable amplifier, one that will not break into oscillation; and second, to have the desired frequency response curves.

For example, a high-fidelity audio-frequency amplifier would be designed to have a constant gain, that is, a flat curve, for frequencies lying in the audible range, typically around 15 Hz to 15 kHz, and falling to zero outside these limits, as shown in Figure 3.1. A radio-frequency receiver, on the other hand, may have

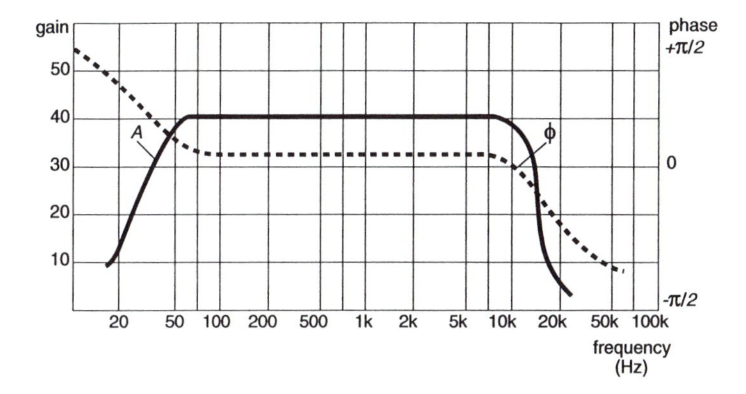

Figure 3.1 The frequency response curves for a typical high-fidelity audio amplifier

a tuned "front-end amplifier" that for station selectivity purposely has a response curve that rises steeply at the tuned-in frequency to a narrow plateau, perhaps 30 kHz wide, and falls as steeply on the other side. The 30 kHz band is to permit the audio-modulated radio signal to be amplified without distortion. Engineers speak of "wide-band" and "narrow-band" amplifiers in referring to the gain and phase plots versus frequency.

3.2 Conditions for Oscillation

If there is positive, or *regenerative*, feedback present, and the fraction of the signal fed back is increased from zero, the system at first will not oscillate, but will remain an amplifier with an enhanced gain and narrower bandwidth. However, as the feedback is increased, a point may be reached when the system will break into oscillation, or in the context of servo systems, will become unstable. If the system is linear, a powerful way to analyze the conditions under which the system becomes unstable, a way that deals simply with amplitude and phase at the same time, is to describe the system in terms of its response to the (complex) exponential form $\exp(i\omega t)$ so that the amplifier gain $A_0(\omega)$ in the absence of feedback, and $\beta(\omega)$, the fraction of the output fed back to the input of the amplifier, are complex functions of the frequency. The product $A_0(\omega)\beta(\omega)$ is then called the (open) *loop gain*, or better, *loop transfer function*. The closed transfer function of the system can be shown to be given by the following:

$$A(\omega)=\frac{A_0(\omega)}{1+A_0(\omega)\beta(\omega)}. \qquad 3.1$$

If we plot the locus of the loop transfer function for different values of the frequency, ranging from $\omega=-\infty$ to $\omega=+\infty$, we obtain what is called then *Nyquist diagram*, as illustrated in Figure 3.2. The condition for stability can now be stated under some very broad restrictions: A system is stable if the locus of $A_0(\omega)\beta(\omega)$ does not encircle the point $(-1, i0)$ as ω varies over its entire range.

This criterion predicts that as the loop gain approaches the point $(-1, i0)$, the gain increases without limit—there would then be an output without an input. What happens at that point, in fact, is that the circuit breaks into oscillation. This will first happen at that frequency ν where the phase change around the feedback loop is zero or a multiple of 360° and the loop gain first reaches one. However, once a buildup of oscillation begins, the amplifier will be driven to voltage levels where it becomes quite nonlinear, and the gain will begin to fall drastically. This leads to distortion in the output waveform from the ideal pure sine wave and sets a limit to the amplitude of oscillation.

We are now in a little better position to understand how positive feedback in a public address system can cause a whistle, rather than a roar. The loop comprising the microphone amplifier, loudspeaker, and the air medium has a loop gain and phase shift that is a function of frequency; there will generally be one frequency at which the phase shift is zero (or a whole number of cycles), and if the gain at that frequency is unity or larger, oscillation will take place at that frequency. If this frequency is in the audible range, a whistle will be heard.

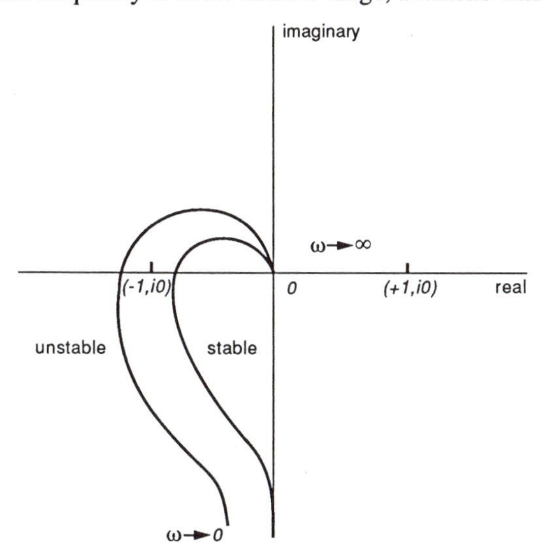

Figure 3.2 The Nyquist stability diagram

So far, we have made it plausible that if the loop gain is at least unity and the phase shift effectively zero at a certain frequency, then if a signal is present in the system at that frequency, that signal will build up rapidly and be sustained. But the question remains, Where did this signal come from in the first place? The answer is that it is in the nature of electrical circuits that there will always be low-level random fluctuations in the current and voltage forming a base on which applied currents and voltages are superimposed. This random fluctuation is electrical *noise*. Even when all extraneous sources of noise, whether "brush noise" from electric motors or noise produced by the mechanical vibration of the circuit itself, called "microphonics," or atmospheric "static," there will always remain two fundamental types of noise: *thermal* (or *Johnson*) noise and *shot* noise, about which more will be said later in this chapter.

3.3 Resonators

The fundamental questions in the design of self-sustained oscillating systems, oscillators for short, that are destined to be used as a reference in the regulation of a clock, are: Precisely what are the factors that determine the frequency, and how can we minimize any instability in that frequency? From what has been said, it is clear that the phase shift and gain around the loop should satisfy the condition for oscillation only at the desired frequency, even in the presence of inevitable fluctuations in the operating conditions. This can be achieved if the conditions for oscillation hold only at precisely the desired frequency, where the gain curve has a very sharp peak rising above unity and the total phase shift at that point passing through zero. The way this can be accomplished is to incorporate into the feedback loop a highly frequency-selective element: a resonator.

The practical form the resonator takes will, of course depend on the desired stability and frequency range. One extremely important example in the radio frequency range is the *quartz crystal* resonator, the subject of the next chapter. A less stable choice in the radio-frequency range would be a simple combination of an inductor and capacitor, which would take the form of a copper coil between the ends of which is connected a parallel plate capacitor, as shown schematically in Figure 3.3a. The analogous mechanical system is a mass connected to a spring, shown in Figure 3.3b, in which the energy oscillates between the kinetic energy of the mass and the elastic energy of the spring. In the case of the inductor and capacitor, the electrical energy oscillates between that of a current in an inductor (with its associated magnetic field) and that of a charge on a capacitor (with its associated electric field). If the inductance of the coil is represented by L and the capacitance by C, then the resonant L–C circuit has a resonance frequency $\nu = 1/\left(2\pi\sqrt{LC}\right)$. Thus for a radio-frequency resonance, a small coil may have typically an inductance of 10μH and the capacitor a capacitance of 10pF. Substitution into the formula yields for the resonant frequency about 500 kHz. Figure 3.4 shows an L–C tuned oscillator in which by proper design, there is a net 360° phase shift in the feedback loop starting (say) at the input to the amplifier, going through the amplifier, and returning a fraction of its amplified output by way of the feedback coupling provided by a resistor and capacitor. For suitable quiescent voltages supplied to the amplifier, it will be, for small signals, quite linear, and the condition for the onset of oscillation can be computed on that basis.

If we wish to construct a resonator with a very much higher resonant frequency by reducing L and C, a point will be reached when it becomes impossible to use "lumped" components, that is, objects that are constructed to have predominantly only inductance or capacitance. A coil designed to have extremely low inductance takes on the aspect of a U-shaped strip of copper, with not only inductance but also a significant capacitance between its ends. Such might be the resonant element in a UHF oscillator operating in the 100 MHz range.

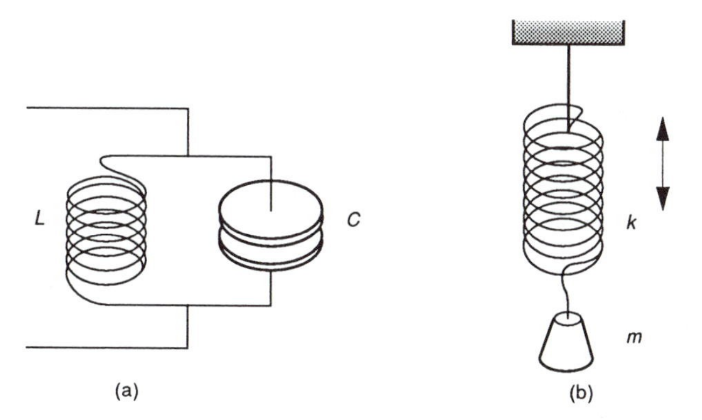

Figure 3.3 (a) Resonant L–C circuit (b) analogous mechanical spring–mass system

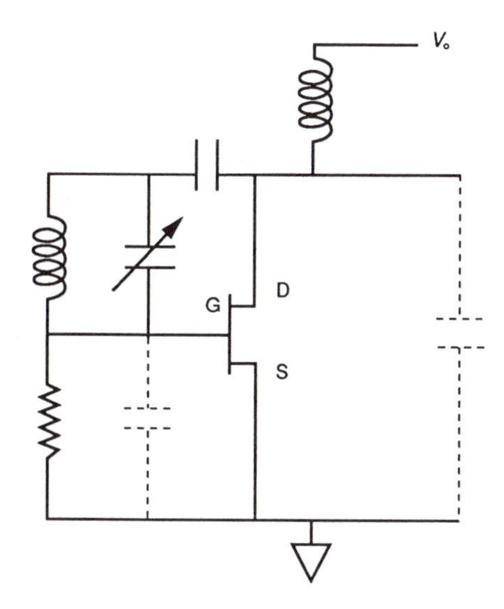

Figure 3.4 An L–C tuned transistor oscillator.

If we consider even higher frequencies, reaching to the microwave region around 1 gigahertz, that is, 1000 MHz, we note that at that frequency an electromagnetic wave, which has a velocity of 3×10^8 meters per second, has a wavelength of 30 cm, which is on the order of the dimensions of ordinary objects. This means that the microwave current in a wire of that length would not be the same at all points, as is taken for granted at lower frequencies.

In treating phenomena at microwave frequencies and above, the focus is on the electromagnetic field in the space around the conductor, which assumes the role of a boundary surface at which the electromagnetic wave is reflected or absorbed. Resonators in the microwave region are generally closed, hollow conductors called *cavities*, which are usually cylindrical, with either a rectangular or circular cross section and provided with a plunger for tuning the resonant frequency. Like the natural modes of vibration of a violin string or the acoustic resonant modes in an organ pipe, these resonators have a series of characteristic field patterns, called modes of vibration, each with a discrete frequency. Surface electric currents do flow on the surface of the cylinder, but the description in terms of the electric and magnetic field patterns is generally more useful.

The different modes of vibration are classified and labeled with three numerical indices to indicate the manner in which the field varies with respect to the three spatial coordinates. In the case of the cylindrical resonator, the three mode indices are related to the number of nodes various components of the field have in the azimuthal direction (around the axis) and in the radial and axial directions. A node in the field is simply where the field passes through a zero value. Since there are two types of fields present, the electric and magnetic fields, a distinction is made between those modes in which the electric field is perpendicular to the cylinder axis (called TE modes) and those in which the magnetic field is perpendicular to the axis (called TM modes). Thus a typical mode designation would be, say, TE_{011} for the mode having no variation around the axis (zero nodes) and going to zero once in the radial direction (at the cylindrical surface) and zero at the plane end caps. Figure 3.5 shows the field pattern in a cylindrical microwave cavity oscillating in one of its modes.

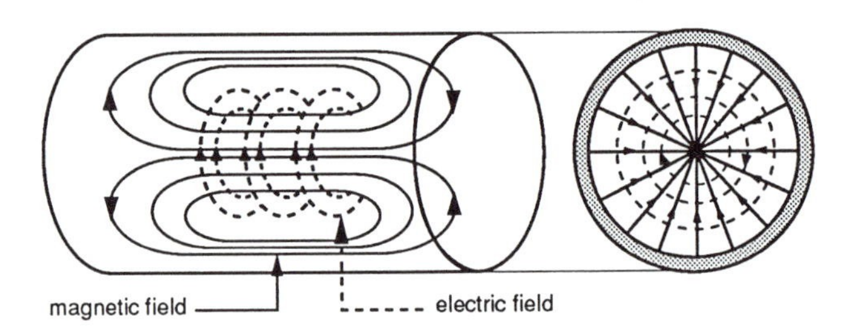

Figure 3.5 The field distribution of the TE_{011} mode in a cylindrical microwave cavity

3.4 The Klystron Microwave Tube

A common oscillator in the microwave region of the spectrum is the *reflex klystron*, a microwave vacuum tube that was developed during the Second World War and was as critical to the development of radar as the triode vacuum tube had been to radio.

Like all electron vacuum tubes it has a heated cathode as a source of electrons, which are formed into a beam drawn towards a positive anode through which it passes to enter the space between two grids that are part of one end of a reentrant microwave cavity, as shown in Figure 3.6. At the opposite end, outside of the cavity, is a negative electrode called the repeller. To show that it is capable of self-sustained oscillation, we have to show that any oscillation that may be present from whatever source will be amplified and fed back with the proper phase to reinforce the oscillation, and make up for any energy losses that may otherwise cause the oscillation to die away. In the case of the klystron, the amplification comes about through the interaction of the electron beam with the electric field between the two grids. If a small oscillation exists in the cavity, the electrons entering the space between the grids in a constant stream will emerge with different velocities at different times. During that part of the cycle when later electrons are given a greater velocity than earlier ones, they will catch up and cause bunching to occur as they travel through the "drift space" to the repeller electrode. As the name suggests, the electron beam is repelled by the negative potential on that electrode, and the beam is folded back on itself, returning in a bunched up form through the space between the two grids. This will cause an increase in the oscillation if the timing of the bunches is such as to reinforce the oscillation originally producing it. Since the cavity is in effect being excited in a pulsed fashion, much like periodically striking a bell, the output may not be spectrally pure unless a very high Q cavity is used as a filter. The timing of the electron bunches is sensitive to the repeller voltage, and therefore it is easy to control the frequency of oscillation; unfortunately, by the same token, any instability in the repeller voltage will translate into frequency instability.

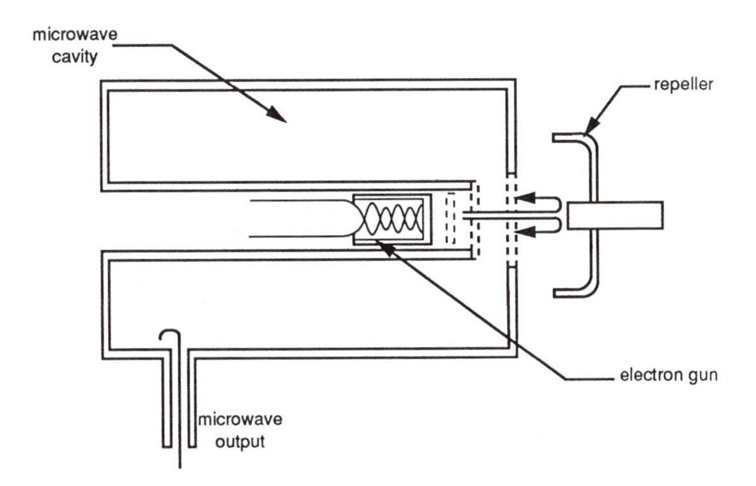

Figure 3.6 The reflex klystron microwave oscillator

3.5 Oscillators at Optical Frequency

If we continue our progression to still higher frequencies, we will reach the infrared region of the electromagnetic spectrum around 10^{13} Hz, or a wavelength in free space of about 30 microns. Beyond that we have the remarkably narrow optical region in the approximate wavelength range of 0.7 microns to 0.3 microns, which perhaps not surprisingly is about where the spectrum of sunlight peaks.

Resonators used in oscillators in this spectral range are still called "cavities," although they are far from complete enclosures. As resonant structures, cavities have to meet standards of precision in their dimensions that are dictated by the wavelength for which they are designed. It follows that resonant cavities for the infrared and optical frequency ranges must be fabricated with the precision one associates with high-quality optical instruments.

The usual form cavities in this region take is the open structure illustrated in Figure 3.7. M and M' are two precisely parallel mirrors, constructed to have extremely high reflectivity at the wavelength for which the cavity is designed. The distance between the mirrors usually contains a very large number of wavelengths of the resonant radiation, that is, it is a very high order axial mode. Unlike the microwave cylindrical cavity, these cavities, because of their open structure, strictly speaking do *not* "support" an infinite discrete set of normal modes; nevertheless, as was theoretically shown prior to any laboratory demonstration, they do support quasi-modes, but only the lowest-order azimuthal and radial ones. The lowest radial mode has the field concentrated along the center line of the mirrors and is therefore efficiently reflected, whereas in the higher radial modes, the field is more spread out radially beyond the edges of the mirrors, and therefore sharply attenuated. The order of the resonant axial mode depends on the precise distance between the mirrors and for plane mirrors is given by $n=2L/\lambda$. If one of the mirrors is moved parallel to itself towards the other mirror, the cavity will pass through an axial mode resonance every half wavelength displacement of the mirror.

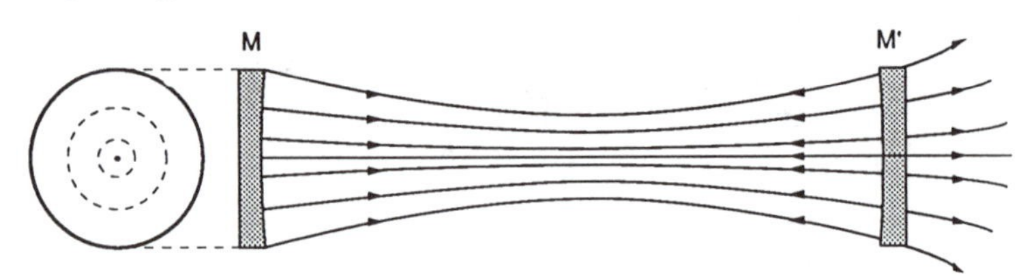

Figure 3.7 An optical cavity

An oscillator operating at an optical frequency using such a cavity would most commonly have the amplification take place in the interior of the cavity itself. Since the amplification is obtained using a process known as *stimulated emission* from suitably prepared atoms or molecules, such an oscillator is called by the acronym LASER (*L*ight *A*mplification by *S*timulated *E*mission *R*adiation), now accepted as a common noun, laser. This designation was a natural derivative of the original term *maser*, a low-noise microwave amplifier also based on stimulated emission of radiation in certain materials.

The most common process of light emission, however, is through *spontaneous* emission; it is the process in all common light sources such as an incandescent solid or gas flame. Whenever the internal motions of constituents of matter are agitated or raised to some excitation level above their *ground state*, the excitation energy is eventually given up either through random collisions with other particles, ultimately being degraded into heat, or in the form of radiation. Unlike the radiation resulting from stimulated emission, whose phase is linked to the radiation stimulating it, in spontaneous emission there is no such phase correlation. Unless very special conditions obtain to allow certain specific modes of the radiation field to build up, spontaneous emission will dominate; on the other hand, if the excited molecules are placed in a suitable optical cavity, the optical field strength can build up efficiently in certain few normal modes of the cavity, and the relative probability of stimulated emission can be greatly enhanced.

Under ordinary circumstances, when light emission is predominantly by spontaneous emission, there is a lack of correlation between the phases of radiation from individual atoms; the result is what is called *incoherent* light. This means essentially that the radiation field is the result of combining a large number of waves with random phases, so that there is no well-defined orderly variation in phase as we go from point to point in the field, or from time to time at any fixed point. A test for coherence is whether it is possible to observe spatial or temporal interference as manifested in *beats*. Spatial beats, called interference fringes, are analogous to moiré patterns, while temporal beats are a periodic rise and fall in the amplitude. Such interference patterns are not observable between different incoherent sources and only under very restricted conditions even from the same source. It is possible to derive a partially coherent wave from an ordinary source, as Young did in his classic two-slit interference experiment to demonstrate the wave nature of light, by using light originating from an extremely small area of a source.

Under conditions obtaining in an optical cavity with carefully aligned highly reflecting mirrors, where longitudinal modes have a high Q, the radiation field emitted by suitably prepared atoms can build up in just those modes to the point where stimulated emission is dominant. Although spontaneous emission is always present to some extent, the stimulated radiation not only has remarkable coherence extending over large distances, but it is also directed in a characteristically sharp beam with a strikingly small divergence angle. We are accustomed to seeing light from ordinary sources spread out in a somewhat diffuse cone even when some provision is made to concentrate the beam; but the sharpness of a laser beam is extraordinary. One of the few examples of a conventional light source producing a remarkably intense beam of relatively small divergence is the searchlight used in World War II to scan the skies for bombers at night, which used a very bright arc source and collimating optics, which directed the light into a well-defined beam in order to increase its range. The explanation of the high directionality of the output light beam of a laser is to be found in the design of the resonant cavity. As already stated, only the lowest radial modes have a sufficient Q to permit oscillation; all others would have a radial distribution of intensity that extends beyond the edges of the mirrors. Nevertheless, even for the lowest radial mode there will be an angular spread of the optical wave through the phenomenon of diffraction by a mirror of finite radius. For a mirror of radius R the divergence is, however, only on the order $\lambda/\pi R$ radians. For example, if λ=0.5μm and the mirror radius is 1cm, we would find for the divergence angle only 9×10^{-4} degrees. In practice, this theoretical limit is rarely achieved; more typically the divergence angle is closer to 0.1 degree, still an extraordinary degree of directionality.

It must not be assumed that an oscillator based on a resonant structure with many modes can or should oscillate in only one mode at a time. It is indeed possible that the gain is sufficiently high that in addition to the mode with the highest Q, others may meet the oscillation criterion, and oscillate at the same time. In fact, it often happens that the total power rather than spectral purity is more important, in which case the oscillator is allowed to oscillate simultaneously in many modes. On the other hand, if spectral purity is the primary objective then steps may be taken to suppress all but the desired mode, by in effect degrading their Q.

3.6 Stability of Oscillators: Noise

We will now attempt to develop the concept of stability in the frequency of oscillators on a more quantitative footing and discuss the factors that may limit it. An ideal oscillator generates a signal that is a pure sinusoidal oscillation with a Fourier spectrum consisting of an infinitely narrow line. We use this ideal as the point of departure and treat any fluctuations in the output of the oscillator as deviations from that ideal. The presumption is that the fluctuations are small, that we are dealing with an approximately harmonic oscillation on which are superposed possibly random fluctuations in amplitude and phase. Of course, that leaves the question open as to what value of frequency the fluctuations should be referred. This can be answered, at least conceptually, by assuming that we have a very large number of identically constructed oscillators, all of which are set oscillating at the same instant t=0. Suppose that after some arbitrary interval of time t we record simultaneously the number of oscillations and phase angles for all the oscillators in the group (or *ensemble*). We can define a reference frequency as the average frequency taken over the ensemble; it would follow that the deviations from that frequency have a zero average.

This assumption deserves examination, since it is clearly possible that there could be sources of instability that tended to produce fluctuations more in one direction than another. On this account we draw a distinction between fluctuations that are reasonably believed to be random and those that are *secular*, a drift in one direction. In practice, it is simpler, and a good deal cheaper, to use one oscillator, rather than a large ensemble, and simply repeat the phase measurement over equal intervals of time as often as desired. Of course, it is assumed in this case first that an acceptable time standard is available, and second, that the fluctuations present are of a nature that it is irrelevant at what point in time a measurement interval begins. Random fluctuations for which that is a valid assumption are said to be *stationary*; it is generally assumed to be the case for all the sources of fluctuation that concern us.

In addition to common man-made, or "artificial," sources of random fluctuation, or *noise*, oscillators are subject to two fundamental types that can be traced to the atomic nature of matter and electrical charge. In a simplified model we may picture a metal conductor as consisting of positive ions arranged in a rigid lattice, embedded in a sea of electrons. At all temperatures above absolute zero (–273°C) the ion lattice and the electrons are in a state of thermal agitation. Imagine a closed geometric surface enclosing a part of the metal; the number of electrons inside that surface will fluctuate as electrons cross the surface in their random motion. This means that part of the metal will have a net electrical charge that fluctuates between positive and negative but, of course, on the average remains neutral. This fluctuation in charge with its concomitant fluctuations in voltage and current is called *Johnson noise*.

Again because of the fact that the charge carriers are individual particles, they do not advance like a continuous band of charge, but rather like a disordered mob, and the number crossing a given surface in unit time, which after all is the electric current, will fluctuate. The precise degree of fluctuation depends on the extent to which there is correlation between the positions of individual electrons, due, for example, to long-range forces of interaction between them. We note that correlation is in principle never totally absent, ultimately because of the quantum effects of an overlap of electron wave functions. Since there is a huge number of electrons in even a moderate current, the current will not, relatively speaking, fluctuate very much. This fluctuation is distinct from thermal noise, and it was called by Schottky, who identified it, *shot noise*. A less sporting metaphor would be raindrops falling on a roof.

As deviations randomly fluctuating in time, noise current has random phase, and what information can be gleaned about the type of noise that might be present can come only from studying the Fourier *power* spectrum. Thus, for example, shot noise has a power spectrum that is flat, that is, the amount of power in a fixed frequency interval is independent of where that interval lies on the frequency scale. On the other hand, the power spectrum of Johnson noise is flat for all frequencies until the size of the quantum $h\nu$ reaches the order of the energy of thermal agitation; except at extremely low temperatures, this means frequencies in the infrared region of the spectrum.

Like many other electronic devices, oscillators exhibit another type of noise: *flicker noise*. The term originated in the age of vacuum electron tubes in reference to the *flicker effect*, which is the fluctuation in the tube current on account of variations in the electron emission from the hot cathode. The term is also applied to noise observed in solid-state devices such as Schottky barrier diodes, where it is attributed to multistep tunneling by charge carriers. As a random time process, it is best characterized by its Fourier spectrum (or to be more precise, the frequency spectrum of the square of the fluctuation); it varies inversely as the frequency and is therefore often referred to as *1/f noise* (*f* for frequency), since its power spectrum has a $1/\nu$ dependence. In the case of oscillators, this type of fluctuation in the frequency is particularly unwelcome, since it increases without limit as the noise frequency approaches zero. This means that slow fluctuations, which correspond to low ν in the Fourier spectrum, are large, and they continue getting larger for longer and longer times of observation. This implies that we will see the frequency wander off without limit if we wait long enough! This incidentally also invalidates the assumption that the noise processes are stationary. It would seem to be a discouraging prospect; it seems to say we can never build a clock that will not eventually drift without limit. But let us not overstate the case; it is probably not a fundamental type of noise in the sense that thermal and shot noise are fundamental. It is possible, as was done in the case of vacuum tubes, to reduce the 1/*f* noise by a proper choice of operating and manufacturing processes. In Figure 3.8 are shown the power spectra for the three types of noise.

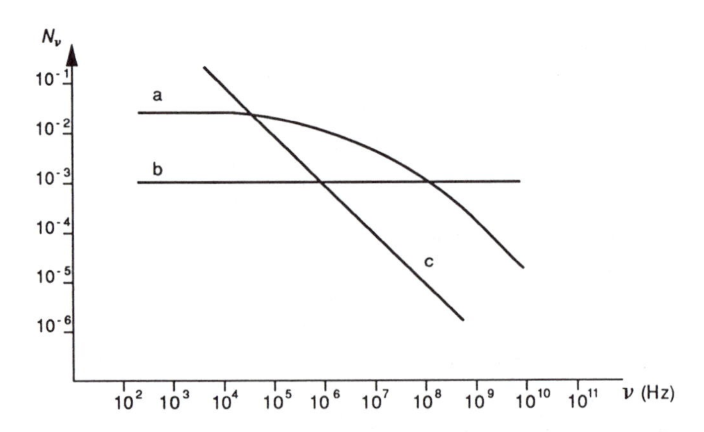

Figure 3.8 The power spectrum of (a) Johnson noise at low temperature, (b) shot noise, and (c) flicker (1/*f*) noise

In practice, the noise present in oscillators may be classified as follows: (a) Those due to the fundamental shot noise and system parameter fluctuations, for example, which modulate the signal itself; and (b) so-called additive noise which, as the name suggests, is noise added to the signal and is therefore independent of the size of the signal. The latter can arise from Johnson noise in the circuits associated with the resonator, or amplifiers, etc.

In precision oscillators the amplitude of oscillation of the resonator is stabilized at low levels; it is expected therefore that additive amplifier noise will be significant for short sampling times but will tend to average out over longer times. In contrast, parameter fluctuations and shot noise lead to a *random walk* in phase.

The random walk problem in statistics is a special case of a classical problem whose solution and its ramifications are associated with such illustrious names as Newton, Poisson, and Gauss. It is simply stated: What is the probability of having a number m of successes in n tries, if the probability of success in one try is given as (say) one in N? Intuitively, we would not be surprised to find that the average number of successes is simply n/N. Of course, we should not expect that every time we make a set of n tries we will have the same number of successes; if we repeat the process over many sets of tries and record each time the number of successes, we should find that number fluctuating equally above and below the average value. A measure of this dispersion in the value of the number of successes is obtained by averaging the square of the deviation of this number from the average value. The result is always a positive number, whose square root gives what is called the *standard deviation*. We state without proof that this quantity, usually denoted by σ, has a value for our problem given by $\sigma^2=np(1-p)$ where $p=1/N$. Now we are ready to pursue the random walk problem, which can arise in various guises, but is usually stated in colorful terms such as: A drunk takes equal steps L feet long, as likely in one direction as in the opposite direction; how far does he get after n steps? Here we can assume that a forward step is counted a success and occurs with a probability of 1/2. Now, if m of the total n steps are forward, then he has advanced mL feet and gone back $(n-m)L$ feet, for a net gain ΔL as follows:

$$\Delta L = mL - (n-m)L = 2\left(m - \frac{n}{2}\right)L \,. \qquad 3.2$$

Now, the average value of m is simply $n/2$, and if we substitute for m this average value in the expression for the net distance, we find that the result is zero, which is not unexpected. Of course, as stated earlier, the actual distance after a particular set of n steps can be anything from zero to nL, with a probability distribution characterized by the standard deviation in ΔL, which we will represent by σ_L. Using the expression we have assumed for the standard deviation in the number of steps forward, in this case $\sigma^2 = n \bullet \frac{1}{2}\left(1-\frac{1}{2}\right)$, we find a formula for the standard deviation in a random walk that will be cited in many contexts in the future:

$$\sigma_L^2 = 4\left\langle\left(m-\frac{n}{2}\right)^2\right\rangle L^2 = 4\sigma^2 L^2 = nL^2. \qquad 3.3$$

where $\langle\ \rangle$ denotes average value.

This indicates that whereas the average of the distance traveled is zero, being equally positive as negative, the more steps that are taken the farther he may be found from his starting point. If these arguments are applied to a process in which each step is repeated at constant intervals of time, then n is simply proportional to the total time, τ, and therefore σ is proportional to $\tau^{1/2}$. It may be hard to draw a comparison between the effect of random noise on the phase of an oscillator and the random walk of a drunkard, but it can be shown that such is the case, and the standard deviation in the phase of an oscillator is proportional to $\tau^{1/2}$, where τ is the length of time the oscillator is free to oscillate before its phase is measured. The corresponding *fractional* standard deviation in frequency is *inversely* proportional to $\tau^{1/2}$.

Chapter 4
Quartz Clocks

The use of the stable vibrations of a quartz crystal to control clocks and watches has become so common in recent years that in this age of digital sophistication, we tend to take for granted the revolutionary advance these quartz-controlled time pieces represent. It is true that through the incomparable skill and ingenuity of Swiss watchmakers, the precision achieved in the fabrication and hence performance of mechanical watches has reached truly admirable heights. However, the microelectronic revolution of the 1960s has made it possible to miniaturize the far superior crystal-controlled clock into something that can be worn elegantly on the wrist.

4.1 Historical Antecedents

4.1.1 Frequency Control of Radio Transmissions

The application of high-frequency quartz resonators to regulate electrical oscillators was originally made to provide a sufficiently stable frequency reference for radio transmitters. Their need for high stability arises principally from two considerations: First, any instability in the frequency of the radio wave, whose frequency or amplitude is modulated to convey the audio signal, would make it difficult, if not impossible, to recover the signal (that is demodulate) at the receiver. To understand this, we need to recall the basic design of radio receivers. When radio waves fall on an antenna, they induce a high frequency current signal, which must be amplified prior to recovering the audio signal superimposed on it in the *detector* stage. It is common practice to circumvent the difficulties of designing a stable multistage radio-frequency amplifier, whose stages would have to be retuned to receive different stations, by using what is called a *superheterodyne* design. This is based on generating from the incoming radio-frequency signal a fixed intermediate frequency (IF) signal impressed with the same audio signal by a *heterodyne* method, that is, by generating a "beat" frequency in close analogy with the beats heard when two close notes are sounded together. This *down-conversion* to a lower fixed frequency can be accomplished using a nonlinear circuit element called a *mixer*. This has two inputs: the incoming radio signal and a pure single frequency signal, generated by a tunable *local oscillator*. The output of the mixer will contain Fourier components not only at the input frequencies, but also at intercombination frequencies, among which is one at the *difference*, or heterodyne, frequency. This signal, whose frequency is intermediate between the incoming radio signal and the ultimate audio output, is filtered and amplified by a multistage amplifier, the IF amplifier, which is narrowly tuned to a *fixed* frequency. As the local oscillator is tuned to different frequencies, different incoming radio frequencies will produce a heterodyne signal within the pass-band of the IF amplifier and ultimately produce an audio output. The audio signal is recovered by passing it through a circuit, the detector stage, which converts the modulations in frequency or amplitude of the IF signal (depending on whether it is an AM or FM signal) into simple modulations of a voltage or current at audio frequencies. Any instability in the radio frequency at the transmitter end, or in the local oscillator at the receiver, will cause a fluctuation in the IF frequency signal and a consequent increase in noise and loss of signal.

The second reason for the need to closely control the frequencies of radio transmitters is that there are so many users of radio communication, especially long-range broadcasts, that it becomes necessary to allocate frequency bands and require broadcasters to adhere to their assigned frequency within very narrow limits. The allocation of broadcast frequencies and the specification of frequency tolerance of transmitters form part of the work of the Comité Consultatif International Radioéléctrique (CCIR). The military services often have even more stringent requirements on stability.

4.1.2 Ultrasonic Transducers for Sonar

Ultrasonic vibrations and waves are those of sound whose frequency is beyond the audible range, which for most people extends to around 15 kHz. The immediate impetus to generate ultrasonic waves came from submarine warfare in World War I, which led both Britain and France to embark on intense research programs to develop underwater acoustic receivers for submarine detection, and ultimately to the idea of sonar. We note that since the velocity of longitudinal sound waves in quartz is around 6000 meters per second, a quartz resonator say 1.5 cm long would have its lowest resonance frequency $V/2L$ at 200 kHz, well beyond the audible range, and hence ultrasonic. It is interesting to note in passing that this ultrasonic wave has a wavelength $\lambda = V/\nu = 3$ cm, the same as the microwaves initially used in radar, and therefore would have the same limits on its ability to "see" detail as those of radar. Of course, radar is impossible underwater because the electrical conductivity of seawater allows only very low frequency radio waves to penetrate it. The successful development of sonar is associated with the names of the Russian Constantin Chilowsky and Paul Langevin, the noted French physicist who is best known for his work in magnetism. Their initial attempts used the *singing condenser* as the ultrasonic source. This is based simply on the electrostatic attraction between oppositely charged plates and is the converse of the *condenser* microphone, in which the sound vibrations in one of a pair of parallel plates connected to opposite poles of a battery produce a corresponding oscillatory current through the battery, thereby converting mechanical vibrations into an electrical signal. The use of piezoelectric sources was considered and rejected a number of times, until finally, in 1917, pure quartz crystals were used both in the source and receiver in a successful demonstration of sonar with a range reaching 6 km.

4.1.3 Discovery of the Piezoelectric Effect

The growing demands for improving the stability of oscillators spurred the search for a resonator having an isolated mode of vibration of the highest possible Q and resonant frequency stability. Quartz, among a select class of crystals, was long known to have excellent elastic properties with very low internal friction. Of equal importance, crystalline quartz also exhibits a phenomenon called *piezoelectricity*, an effect found in some crystals satisfying certain symmetry restrictions in which the application of pressure along particular directions produces electrical polarization; that is, surface charges develop, which are proportional to the pressure. This permits any mechanical vibrations in the crystal to produce an oscillating electric current in an associated electronic circuit. Conversely, if a quartz crystal is placed between a pair of metal plates carrying opposite electric charge, the crystal is stressed as if under pressure. This converse effect provides a way of exciting vibrations in the crystal simply by using an alternating voltage on the metal plates.

These effects are among a number involving solids that were discovered in the 19th century, including magnetostriction, in which a dimension of certain materials is increased by magnetization along that dimension; and pyroelectricity, in which a crystal develops electrical charges through temperature gradients. In 1880 Jacques and Pierre Curie (the latter was to become celebrated as the husband of Marie Sklodowska, the discoverer of radium) first published their studies on piezoelectricity, in which they analyzed the conditions under which the effect can be observed and the restriction on the symmetry of the crystals exhibiting this effect. It was recognized even then that piezoelectric crystals were potential acoustic sources and detectors; however, it was not until the means of producing continuous electric oscillations became available at the beginning of the 20th century through the invention of the vacuum-tube triode amplifier that the use of these crystals as acoustic transducers became a reality. It should be noted that it was only four years prior to the Curies' published studies that Bell invented the telephone, a development that stimulated renewed interest in the science of acoustics, and ultimately the birth of a new science: ultrasonics.

It was out of that wartime stimulus that came the application of quartz resonators in the control of frequency in electrical oscillators. The credit for this development belongs to Cady, who as early as 1917 noted that specimens cut from crystals of quartz and rochelle salt acted in an unusual way when "driven" by electrical oscillations near their natural modes of vibration. He further found in researches after the war, published in 1921, that when placed in the circuit of a vacuum-tube oscillator, these crystals exerted a remarkable stabilizing effect on the frequency of oscillation. As almost always happens, when the course of development of science has reached a certain point, the stage seems to be set for certain discoveries to be made, and many actors are drawn to it. Cady's patents on the piezoelectric resonator did not go unchallenged: Nicholson, of Western Electric Company, had been actively exploiting applications of piezoelectric crystals, for example in microphones, loudspeakers, and phonograph pickups, and he took out patents in 1918 that he claimed justified a judgment against Cady's patents. It was about this time that G.W. Pierce invented an improved crystal-controlled circuit, also unsuccessfully challenged by Nicholson, and which found universal adoption for the frequency control of radio transmitters and receivers.

4.2 Properties and Structure of Crystalline Quartz

It has long been known that crystals such as quartz have excellent elastic properties with extremely low internal friction when deformed, as well as exceptionally high strength and low thermal expansion. In fact, fused quartz fibers have long been used for suspension in torsion balances, by which small torques acting on large suspended masses are measured. A fused quartz fiber has the same breaking strength as a steel wire of the same diameter, but it has a smaller modulus of rigidity under torsional stress, that is, it twists more easily. But the most important property it has, from the point of view of constructing a high-Q resonator, is its nearly perfect elasticity, in the sense that when a stress is removed it returns to its original unstressed form. This implies that whatever work is done in deforming it, that work is stored, without loss, as elastic energy, which will very nearly be totally regained when the stress is relieved. High Q means a low intrinsic rate of loss of vibrational energy, and as we have seen, it is associated with a very sharp resonance spectrum. This quality confers two important advantages on the resonator: First, its relatively undamped oscillation requires a minimal amount of coupling to the amplifier to sustain the oscillations; and second, the sharpness of the resonance is important to minimize the effects of noise and fluctuations in the gain of the amplifier on the oscillation frequency. However, even with these advantages, the stability in frequency will be ultimately limited by noise, both short-term noise, which may be thought of as introducing an uncertainty into the frequency, and long-term drift in the resonance due to structural "aging" in the resonator. Finally, we should recall that provision must be made to limit and stabilize the level of oscillation; this also affects the frequency, since any fluctuation in amplitude shows up as a broadening of the spectrum and uncertainty in the frequency.

Other advantages of quartz are first, that its elastic properties are far less sensitive to environmental conditions, such as temperature and humidity, than they would be for most other solids, and second, the extremely small degree to which its dimensions change with temperature. It is this fact that explains its extraordinary resilience under extreme thermal stress; for example, a quartz rod heated to red heat and plunged into cold water will not crack. Of course, the merits of quartz, or of any material, must ultimately be judged relative to the demands of the application to which it is to be put. We will, in fact, see that in order to realize the best performance in a quartz resonator, the two properties in which it excels over other candidates are the ones that leave room for further improvement: thermal expansion and long-term constancy in elastic properties.

Crystalline quartz, whose chemical composition is silicon dioxide (SiO_2), otherwise known as silica, is a three-dimensional lattice, held together by what chemists call *covalent* bonds, with each Si atom surrounded by four O atoms at the vertices of a regular tetrahedron, and each O atom joined to two Si atoms. The term *bond* refers to an interatomic force that pulls the atoms together up to a certain equilibrium separation and that requires a certain energy to dissolve. These bonds ultimately arise from electrostatic forces between the individual charged particles that make up the atoms in the crystal. However, classical theory fails to predict any stable configuration of charges; only quantum theory admits in a logical way a stable ground state that does not radiate energy. In the case of crystalline quartz, we are dealing *not* with the stability of three particles, as the chemical formula SiO_2 might suggest, but with an entire crystal. That is, the problem does not separate into small aggregates of atoms we call molecules; the crystal is one big molecule. Of course, the regular, ordered arrangement of the atoms in the crystal, and the symmetry it exhibits, will help make the problem more tractable.

The nature and origin of the chemical bond, such as what holds the quartz crystal together, were studied early in the history of quantum theory by Linus Pauling. It would be inappropriate to attempt to delve into that theory here; however, a brief sketch of the structure of quartz is necessary for an appreciation of its remarkable properties. The valence electrons in Si form four covalent bonds, like outstretched arms at the mutually maximum but equal angles in three dimensions, which leads to tetrahedral symmetry, as shown in Figure 4.1.

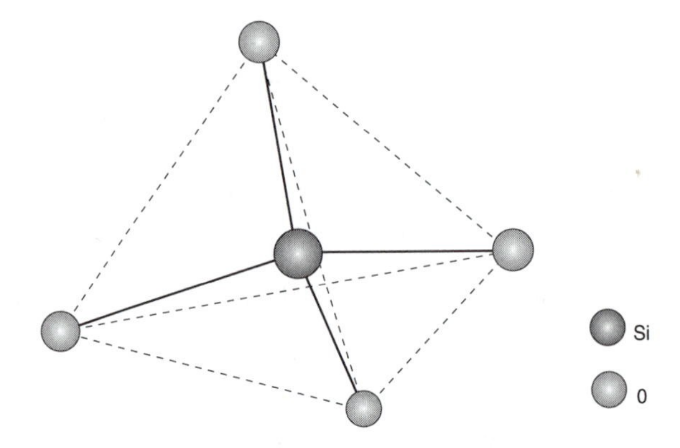

Figure 4.1 Tetrahedral symmetry of the bonds holding the Si atom to the O atoms in crystalline quartz

The O atom, on the other hand, has two unoccupied valence states (orbitals, as the chemists call them), only one of which lines up with a Si atom, to form a single bond. This leaves the O atom with its second vacant orbital, which forms a single bond with another Si atom. This continues indefinitely throughout the solid in one interconnected 3-dimensional lattice, as shown in perspective in Figure 4.2.

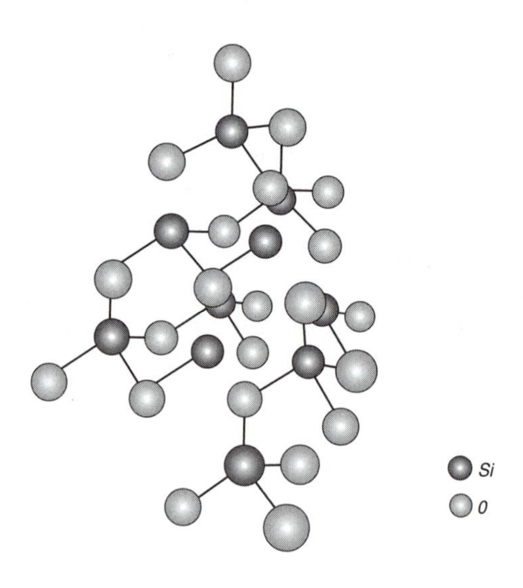

Figure 4.2 The quartz crystal lattice

Thus unlike CO_2, in which the four covalent bonds of C are satisfied by a pair of *double* bonds with two O atoms, thus producing a complete molecule, SiO_2 forms a continuously extended 3-dimensional network. The atoms are strongly bonded, with the crystal having a high melting point at 1710°C.

Like all crystals, that of quartz has its characteristic symmetry properties. From the tetrahedral arrangement of the Si–O bonds, we can see (especially with the aid of a 3-dimensional model) that the crystal has a 3-fold axis of symmetry. What is a little more difficult to see is that there are three 2-fold axes of symmetry perpendicular to the 3-fold axis. Quartz belongs to a crystal symmetry group designated by crystallographers as 32, because of the 3-fold symmetry axis usually designated as the z-axis and the three 2-fold axes, one of which is taken as the x-axis. An axis perpendicular to both the z-axis and the x-axis is defined as the y-axis, as shown in Figure 4.3.

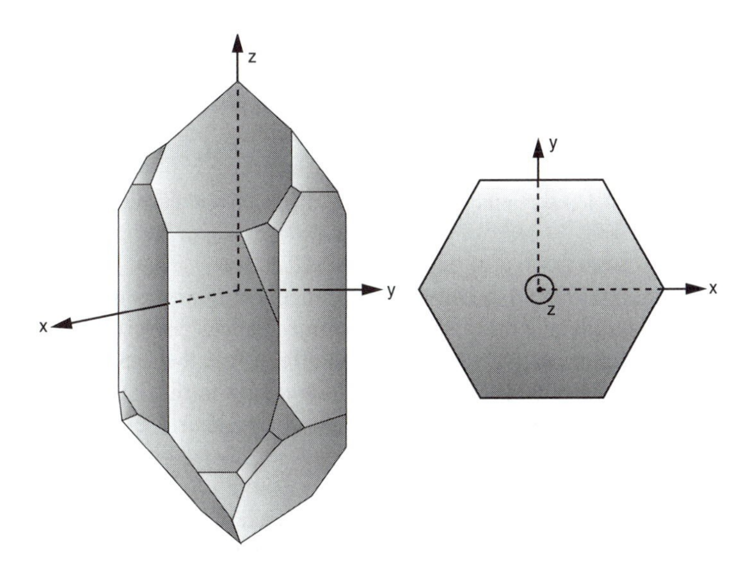

Figure 4.3 The x-, y-, and z-axes in a quartz crystal

However, the important symmetry property quartz has, or we should say does *not* have, is a center of symmetry. This means that if we imagine each atom of Si and O moved to a diametrically opposite position with respect to any fixed point as center, the result would be distinguishable from the initial configuration. Mathematically, it is lack of symmetry under the operation of reversing the signs of all the coordinates of all the atoms in the crystal. The operation of reversing all the signs of the coordinates is equivalent to a rotation through 180 degrees about an axis, followed by taking the mirror image in a plane perpendicular to that axis, as illustrated in Figure 4.4. Almost all common objects have low symmetry and lack a center of symmetry, for example a table or keyboard, although some do, such as a brick or an American football; however, among crystalline substances, a center of symmetry is common.

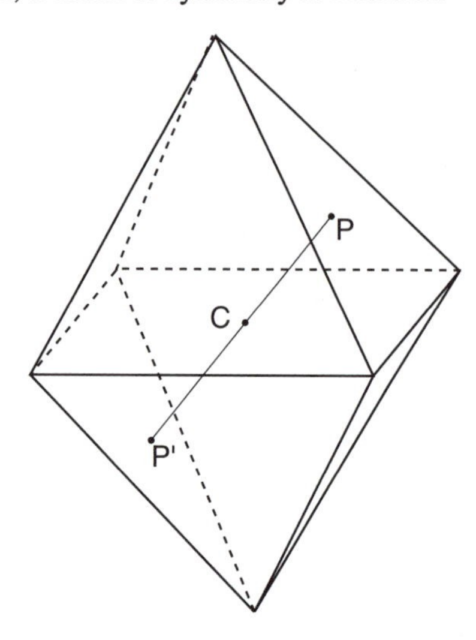

Figure 4.4 The symmetry operation of inversion through a center

Our venture into the crystallographic symmetry of quartz is intended to provide some basic understanding of the electromechanical property that makes it so useful for our purposes: the piezoelectric effect. It is the lack of a center of symmetry that makes it possible for crystalline quartz to display this effect. To justify this statement we can argue that the piezoelectric effect connects a pressure applied to the crystal with an electrical separation of charge, or electrical *polarization* as it is called. The crystal as a whole is, of course, electrically neutral and remains so. However, the balance of charge in the constituent atoms is distorted by the electrons being displaced in a particular direction relative to the positive inner cores of the atoms in the crystal lattice. At least this is an adequate model of what occurs. The result is that unbalanced electrical charges of opposite sign appear on opposite sides of the crystal. Now imagine the same pressure applied to the crystal along the same line after it has undergone the symmetry operation we have just described; the electrons should be displaced in the opposite direction, since pressure is the result of compression and acts symmetrically in both directions. If, in fact, the crystal had a center of symmetry and the crystal were therefore the same after the symmetry operation as before, then unless the electron displacement is zero, we would have a crystal in which the electrons were displaced in opposite directions under the same conditions, which is impossible. It follows that a prerequisite for a crystal to display piezoelectricity is that it should *not* have a center of symmetry.

Actually, we can impose the same restriction on crystals that can display a dependence of their polarization on a quadratic function of an applied electric field, E, since like pressure, E^2 is not changed under a reversal of the coordinates. This is extremely important in selecting crystals suitable for nonlinear optical studies, such as for example producing second-harmonic light waves from intense (laser) light passing through such a crystal. If the optical frequency polarization induced in a crystal by an optical wave of frequency ν has some dependence on E^2 of the incoming wave, so that the time dependence is say $\sin^2(\omega t)$, then a wave at the second harmonic is generated, since $\sin^2(\omega t)=\frac{1}{2}[1-\cos(2\omega t)]$. The first of such work was published in 1961 and consisted in generating a second-harmonic violet beam at $\lambda=0.35\mu m$, using a red beam from a ruby laser at $\lambda=0.69\mu m$ focused on a quartz crystal.

4.3 Modes of Vibration of a Quartz Plate

Because of the crystalline structure, the piezoelectric effect and its converse relate mechanical stress to electrical polarization in ways that depend very much on the direction these quantities have in relation to the crystal axes. Thus an electric field along an x-axis is coupled to a longitudinal strain in the same direction, together with a strain of equal magnitude but opposite in sign along the y-axis. A change in the sign of a strain is simply to replace, for example a compression with a tension and vice versa. Furthermore, an electric field applied along the y-axis causes a shear strain in the x-y plane.

Figure 4.5 illustrates various types of strain, including torsional (twist) and flexural (bending), in addition to the longitudinal (or extensional) and shear we have already encountered.

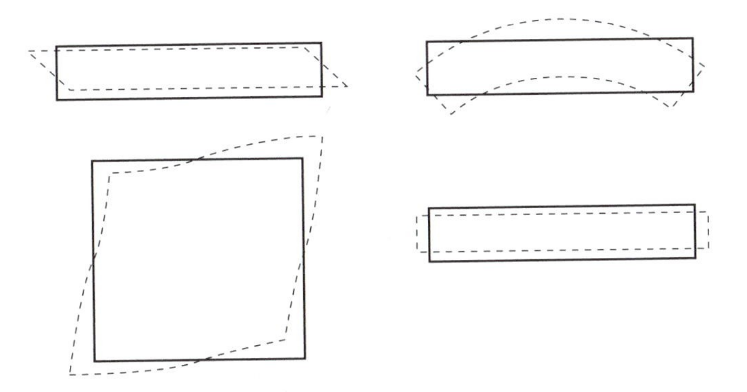

Figure 4.5 The various modes of strain in a quartz crystal plate

To cover a wide frequency range, from say 1 kHz to 100 MHz, quartz bars or plates are used in extensional, shear, or flexural modes of vibration. For the lower end of the frequency spectrum, the flexural mode in a bar can be used up to around 100 kHz, whereas operating in the extension mode extends the range up to 300 kHz. A face shear mode allows the range 300 kHz to 1 MHz to be covered, while a thickness shear mode can extend the range to 30 MHz and beyond. For the higher frequencies, higher odd-harmonic (overtone) resonances, such as the third or fifth overtone, are often used to obviate the need to use fragile crystal plates of extremely small thickness. The best available precision crystals vibrate in the fifth-overtone thickness shear mode at frequencies of 5 MHz or 2.5 MHz.

In order to efficiently and selectively excite these various modes of vibration, the electric field must be applied in the proper direction with respect to the crystal lattice. The choice of field direction bears not only on the efficiency of exciting the desired mode, but also on the dependence of the resonance frequency on the temperature, through the dependence of the elastic properties and dimensions on the temperature. These properties in a crystalline substance, unlike those that lack an ordered structure, are different for different directions relative to the crystal axes. Since the electric field is applied through a pair of parallel plates, the quartz plate or bar must be cut out of the body of the crystal with the desired orientation. Figure 4.6 illustrates three commonly used cuts: the x-cut and two rotated y-cuts. The x-cut has faces cut perpendicular to the x-axis, along which the oscillating electric field is applied. A longitudinal strain is produced in the same direction as the field, resulting in only a longitudinal (extensional) wave propagating in this direction. The x-axis is called a pure mode axis for this type of wave. The two rotated y-cuts are such that the electric field can be applied along directions inclined at the angles of −59° and +31° to the z-axis in the y-z plane. These are pure mode directions for shear waves; that is, a field along these directions excites only shear waves and no other.

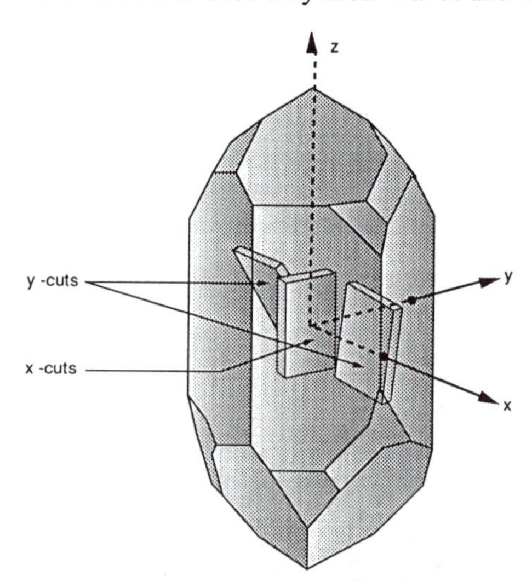

Figure 4.6 The various cuts used in the fabrication of quartz plates

4.4 X-Ray Crystallography

To find the orientation of the crystal axes in a piece of crystalline material relative to some external reference, such as its surface geometry, is a problem for which crystallographers have developed methods involving the angles between the natural facets, which generally characterize crystals, and their optical properties. The most powerful technique, however, involves the use of X-rays: not to produce shadow pictures, of course, but to produce X-ray *diffraction patterns*.

It would take us too far afield to attempt more than a cursory description of this technique, which can be used not only to determine crystal orientation, but more importantly, to analyze the crystal structure. We must begin by stating that X-rays are electromagnetic waves, as are radio waves and light waves, with a wavelength that is typically 1/5000 of a light wave. This fact was exploited in 1912, when at the suggestion of Max von Laue, two young assistants, Friedrich and Knipping, performed an experiment suggested by analogy with a technique well known at the time: the use of a fine grating to analyze an optical spectrum. In this context a *grating* consists of a mirror (or transparent plate) on whose surface a large number of closely spaced grooves are ruled with great precision. Because of its wave nature, light is preferentially reflected from the grating surface along those directions in which the light waves diffracted by the narrow reflecting (or transmitting) stripes reinforce each other. Realizing that the spacing of the fine grooves has to be scaled in proportion to the wavelength, and that therefore it was impossible in practice to simply repeat the optical experiment with X-rays, von Laue suggested that perhaps the regularly spaced atoms in a crystal may act as a 3-dimensional "grating." This indeed proved to be the case, and the classic von Laue X-ray patterns were soon photographed. After this initial success, W.H. Bragg and W.L. Bragg, father and son, established X-ray diffraction as a powerful method of crystal structure analysis. It was W.H. Bragg who gave a simple way of analyzing the way in which the X-rays are scattered by the crystal along many discrete directions.

The analysis proceeds by imagining all the atoms of the crystal assigned to a set of parallel geometrical planes, called *atomic planes*. The regular ordering of the atoms in the crystal ensures that this can always be done, as illustrated for a 2-dimensional lattice in Figure 4.7. Let us imagine that an X-ray wave falling on the atoms produces secondary waves that radiate out in all directions from the atoms, and that these secondary waves have a definite phase relative to the incoming wave. Then it is clear that in any given direction with respect to the atomic planes, the secondary waves will produce a resultant that depends on their relative phases, which differ for different directions. For the secondary waves originating from all atoms in just one atomic plane to be in the same phase requires only that they travel in a direction that obeys the usual law of reflection from a plane mirror coincident with the atomic plane. However, for the secondary waves originating from different parallel atomic planes to be in phase requires that the distance between the planes d satisfy the following condition:

$$2d \sin\theta = n\lambda , \tag{4.1}$$

where θ is the angle between the incoming X-ray and the atomic plane, and n is any whole number. Thus if a collimated beam of X-rays of nearly one wavelength is directed at a single crystal, and a photographic film is placed around it, we would get what is called a Laue pattern of dots of varying densities produced by "reflections" from atomic planes, which of course can have a large number of orientations with different spacing. Such a pattern shows graphically the orientation and symmetry properties of the crystal. However, to determine the crystal orientation requires an instrument called a *goniometer*, which consists essentially of a rotatable crystal mount and a radiation detector on a rotatable arm designed for precise angle measurements. By applying the above formula to the possible sets of reflecting atomic planes, based on the known structure and symmetry, it is possible after a good deal of data analysis to arrive finally at a precise determination of the crystal orientation.

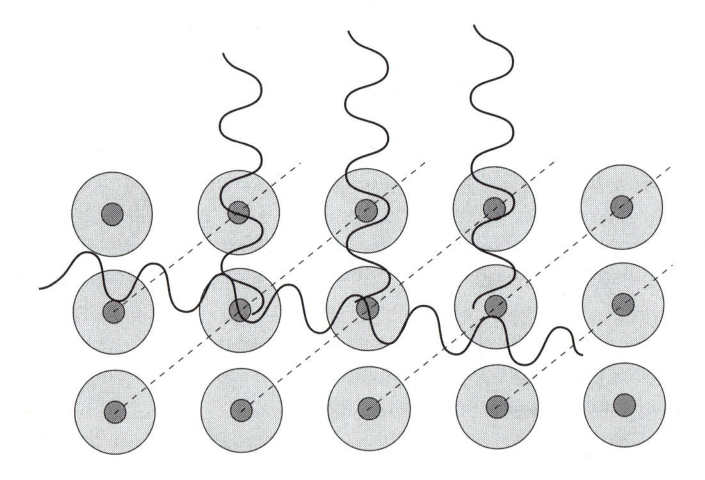

Figure 4.7 Bragg reflection from atomic planes

The application of X-rays to the study of crystals in the context of quartz resonators has also proved extremely useful in another direction: that of making it possible to actually see the distribution of the amplitude of vibration in a crystal plate or bar under actual conditions of excitation. That this is plausible we can see from the formula for an X-ray "reflection" from a set of atomic planes: Even a small relative displacement between the atoms contributing to the reflected beam would change its intensity. This is an invaluable tool in the development of high-Q resonators, because it allows the presence of undesirable lossy modes coexisting with the desired one to be suppressed by various techniques, such as contouring the shape of the crystal.

4.5 Fabrication of Quartz Resonators

The actual performance characteristics of a quartz resonator are determined ultimately by such technical matters as the way the quartz plate is fabricated, how it is mounted, and how its environment is controlled. To prepare the resonator plates, the mother crystal is rough cut into wafers with the desired orientation with respect to the crystal axes. The wafers are then cut into smaller blanks, a little larger than the desired dimensions, and then reduced by several stages of lapping, until the final dimensions are nearly reached. They are then cleaned and etched, the metallic-film electrodes baked on or vacuum deposited, and the plate mounted in a hermetically sealed holder. The blanks are sometimes subjected to heating cycles to accelerate the long-term approach to greater stability, which might appropriately be called aging, although the term usually also includes any long-term drift.

Synthetic quartz has been available for many years on the market; unlike diamond, which can be produced only as tiny grains, quartz and sapphire (crystalline aluminum oxide) can be grown to sizable crystals from the melt in special high-temperature furnaces. By a proper choice of seeds and control of the crystal growth, manufacturers are even able to cut the desired plates with little waste.

A good deal of effort has been devoted in the past to improve the Q-value of resonators made of synthetic quartz. A study of the temperature variation of Q, under conditions where Q is known to be limited by the vibrational energy losses in the quartz rather than some extraneous factors, shows a strong peak in those losses at –223°C. This has been attributed to the presence of *sodium* impurity, whose amount varies from one sample to another. It has been found that the internal energy loss it causes can be reduced significantly by electrolytically removing this impurity ion. This can be accomplished by applying a voltage between electrodes in the molten silica, or more slowly by applying a voltage between the electrodes on the finished plate at an elevated temperature. The realization that the presence of this impurity ion is responsible for some of the internal energy loss has led to ways of improving the purity of synthetic quartz and hence of the Q of quartz resonators. The control of quality in the manufacturing process has been greatly aided by the use of the infrared absorption spectrum as an indicator of acoustic energy loss in the crystal.

4.6 Factors affecting the Resonance Frequency

4.6.1 Aging

The resonance frequency of a quartz resonator typically exhibits a long-term drift extending over perhaps years, a phenomenon called *aging*. The aging rates of high-frequency thickness shear mode resonators designed for precision clocks have reached impressively low values. The actual aging of a particular resonator is determined by the process control used in its manufacture as well as the degree of strain in the mount and contaminant gases in contact with the crystal in its holder. Typical aging rates for solder-sealed metal holders range from as high as a few parts per million per month to as low as one or two parts per million per year for the first two years. The lowest aging rates have been achieved in units in cold-welded metal holders that allow high-temperature bake-out to drive out contaminant gases prior to being sealed. The most important processes that cause aging are briefly (a) thermal stress relaxation, resulting from prior temperature gradients, (b) the adsorption and desorption of gas on the surface of the crystal affecting its mass, and (c) slow structural changes involving imperfections in the crystal.

4.6.2 Surrounding Atmosphere

It is a remarkable fact that the adsorption of what presumably is just a monolayer of gas one molecule thick on the surface of the crystal can produce a very significant change in the natural resonance frequency of the resonator. Here is the basis for a sensitive gas pressure gauge! Of course, it is not that the adsorbed layers of gas are so heavy, but that the precision in frequency measurement has reached such a high level. As is well known from the experience of a half-century when electronics was dependent on vacuum tube technology, the surfaces of all materials, particularly metals, continue to evolve gases into a vacuum and must be "out-gassed" at as high a temperature as possible. The out-gassing is never really complete; gas diffuses out from deeper and deeper layers within the material. The method used in the vacuum tube industry to control the loss of vacuum due to out-gassing, particularly from the incandescent cathode, was to use a "getter," which was typically pure metallic barium evaporated from a small copper or nickel tube containing it by an induction heater. This formed a silvery layer onto the inner surface of the glass envelope of the vacuum tube. Most gases, except the inert gases like helium, react chemically with barium, producing nonvolatile compounds that remain on the surface.

4.6.3 Temperature

The most important physical parameter affecting the frequency of a quartz resonator is the temperature of the crystal. By a proper choice of the orientation of the cut with respect to the crystal axes, the dependence of the frequency of resonance on the temperature can be made minimal at the normal operating temperature of the resonator. The general dependence of frequency on temperature is shown as a set of graphs for various cuts in Figure 4.8. Naturally, the plate orientations in common use have been chosen to have the minimum change in frequency occur around 25°C, as seen in the figure. We also note that the curves marked AT and GT have an inflection point rather than a maximum with respect to $\Delta\nu/\nu$, the fractional change in frequency per degree centigrade; for these the possibility exists of having nearly zero frequency change over a range of temperatures. However, even when the change in frequency is near the minimum, a great deal can be gained by temperature compensation techniques, as well as tight control of the temperature of the crystal unit, by placing it in a constant-temperature oven. In the case of precision quartz-controlled oscillators, the emphasis is usually on controlling the temperature, rather than compensating for it. Not only must the temperature be held constant in time, but also it must be spatially constant across the crystal holder. This usually requires a double oven. The use of the word "oven" suggests operating at high temperature; in fact; the oven is stabilized at a temperature above room temperature, but only moderately so, in order to obviate the need for refrigeration, and make it possible to use natural cooling when the temperature deviates above the set value. Furthermore, precision proportional control is used rather than the familiar on–off control used in home thermostats. Proportional control means that the rate of increase of the amount of power delivered to the oven is proportional to how far the temperature is below the set temperature. To implement this type of control requires some electronics that go well beyond the mercury switch and bimetallic strip used in some on–off controls.

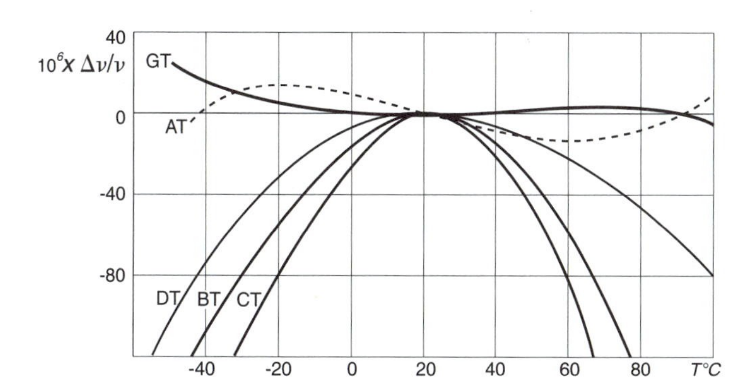

Figure 4.8 The temperature dependence of the frequency of various cuts of quartz (Gerber, 1966)

4.6.4 Excitation Level

Another source of frequency instability, related to the amplifier and the coupling circuits to the crystal, is the power level driving the crystal, which of course determines the level of mechanical vibration of the crystal. While this naturally affects the rate of heat dissipation in the crystal, which could cause some shift in the frequency, there are indications that other possible mechanisms such as changes in the effective elastic properties may play a role. Based on experience, high precision crystal units should be driven at the lowest possible power levels. In order to ensure that the oscillation level is maintained at a constant low level, a special feedback circuit, of a type known from the early days of radio as *automatic gain control* (AGC), is used. Essentially, this is a feedback control loop that performs the following functions: Sample the oscillation level, derive a voltage proportional to its amplitude, compare this voltage to some constant reference voltage to produce an *error signal*, and derive from this a voltage to apply at some point in the oscillator circuit that can affect the oscillation level and bring the error to zero. This type of negative feedback circuit is called a servo control loop, although the term is often associated with the use of an electromechanical device such as a servomotor.

4.7 The Quartz Resonator as a Circuit Element

4.7.1 Equivalent Circuit

To understand the principle of temperature compensation, we must first learn how the actual frequency of oscillation of a crystal is affected by the current–voltage relationships of the circuit elements involved in the way the crystal is mounted and connected to the source driving it. For this we need the *equivalent circuit* of the crystal in its holder, that is, a circuit made up of the basic elements of inductance, capacitance, and resistance that simulates exactly the voltage–current relationship in the vibrating crystal at all frequencies. It cannot be presumed a priori that such an equivalent circuit exists; but recalling that mechanical vibrations universally involve the oscillation of energy from an elastic (potential) form to a kinetic form, it is reasonable to draw a parallel with an electric circuit having inductance and capacitance, as we did earlier in the L–C circuit. In our usual cavalier way we will accept that an analysis of the mechanical vibration of a quartz resonator, taking into account the piezoelectric effect, leads to the equivalent circuit shown in Figure 4.9.

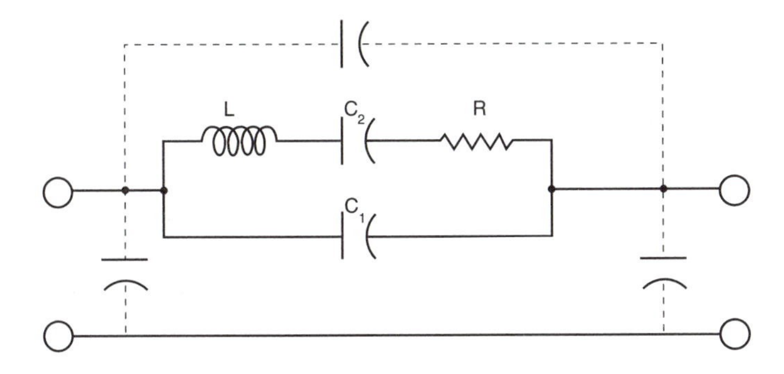

Figure 4.9 The equivalent circuit of a quartz plate

The capacitance C_1 represents that of the crystal between its metal electrodes. The resistance R accounts for the energy losses in the crystal itself as well as its mount. The inductance L and capacitance C_2 represent the inertia and elastic constant that determine the resonance frequency. In general, R is very small, $C_1 >> C_2$, and L is very large, being comparable to the inductance of a coil one meter in diameter and a winding of several thousand turns! The high ratio C_1/C_2 implies very loose coupling between the crystal and the circuit driving it.

The availability of elements such as *thermistors* (temperature sensors made of sintered polycrystalline oxides in bead or rod form) and *varactors* (diodes with voltage-sensitive capacitance) makes temperature compensation possible if we can design a network of these circuit elements that when connected to the resonator circuit tends to cause the opposite change in frequency when the temperature is changed. In fact, with the aid of a computer, circuit designs have been found to give useful compensation.

4.7.2 Frequency Response

As with any other type of feedback oscillator whose frequency is controlled by a high-Q circuit, the quartz crystal is part of a feedback loop of an amplifier with sufficient gain for oscillation. As we emphasized in a previous chapter, the ideal would be to have the frequency-determining high-Q resonator execute its natural oscillation with minimum coupling to the amplifier, whose operating conditions may fluctuate over time. The higher the Q-value, and values reaching 10^6 are not uncommon in precision quartz crystals, the greater the variations in the operating conditions of the amplifier, such as a change in its phase shift, which can be tolerated. The reason is that such phase shifts can be compensated in a high-Q circuit by a very small shift in frequency.

We note parenthetically that although a quartz crystal has only two electrodes, we can think of it as having an input between one electrode and a common ground conductor, and an output between the other electrode and ground, as shown schematically in Figure 4.9. If we imagine an oscillating voltage whose frequency we could vary is applied between one electrode of the crystal and ground, and we measure the relative phase of the voltage appearing on the other electrode, we would find that the phase changed very abruptly as we pass through the resonance frequency, going from –90° through zero at resonance to +90° on the other side of resonance, as shown in Figure 4.10. Almost the entire change in the phase shift occurs within the width of the resonance, which for $Q=10^6$ means a frequency width of 1 Hz in a 1 MHz oscillation frequency.

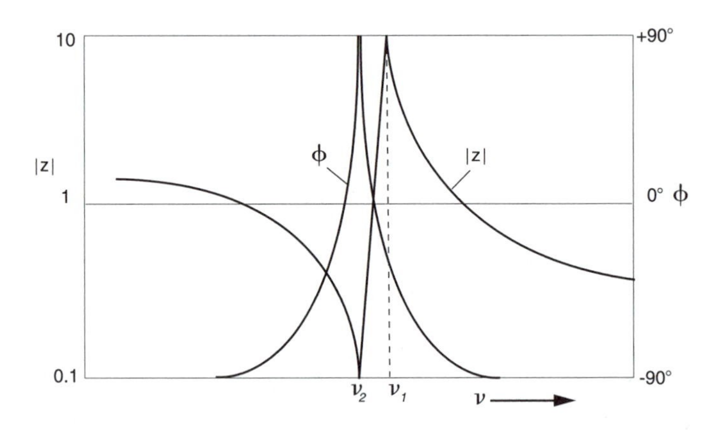

Figure 4.10 The change in impedance and phase as a function of tuning of a quartz crystal

4.8 Frequency Stability

The frequency of any standard is subject in varying degrees to random variation, and the indicated time derived from it will ultimately drift. It is of great importance both to manufacturers and users of standard oscillators to have an agreed-upon fundamental way of specifying the frequency stability, or more accurately, instability of these instruments. In the statistical analysis of their instabilities, it is generally assumed that the fluctuations obey what is called the condition of *stationarity*. This means roughly that the frequency (or phase) as a function of time does not change if the instant from which we start measuring time is displaced. This complicates matters, since we have already described fundamental sources of possible long-term drift in the frequency of a quartz oscillator; in addition, there will be a host of variable environmental and electronic factors that add unpredictably to the instability of the standard, again with the possibility of a steady drift. Such "deterministic" long-term drift would obviously manifest itself, for example, by the given standard appearing to continue to gain or lose time with respect to a primary standard over a protracted length of time. Of course, if we wait long enough, the trend may reverse, so it may be arbitrary to stipulate a span of time beyond which an instability must be separated out in order that the residual obey stationarity. Nevertheless, as a practical matter this must be done, since the duration of a measurement is constrained if nothing else by the lifetime of the person doing the measuring. Thus the statistical development of the theory requires that the time data be numerically fitted by the sum of a random part and a slowly varying deterministic part.

There are two complementary ways of characterizing the random fluctuations in the frequency or phase of an oscillator: the *frequency domain* description in terms of the Fourier power spectrum of the *fluctuations* in the measured quantity, which is obtained using a spectrum analyzer, and the *time domain* description, in which errors in the phase or frequency over different sampling intervals of time are statistically analyzed. The common approach to specifying a random fluctuation in any physical quantity is to repeat its measurement many times under presumed identical conditions, and then to take as a measure of the fluctuation the standard deviation, in the sense of ordinary statistical analysis. The case of instabilities in a time standard is unique in that the physical quantity under study is a time interval and the measurements are repeated sequentially in time, that is, at different values of the quantity being measured. Furthermore, the instability arises from different types of sources, each of which may have a different dependence on time.

To get a complete description of the instabilities in the time domain, therefore, a large number of sets of repeated measurements must be made, one set for each selected time interval in a range of time intervals extending perhaps from 1 second to 10,000 seconds. In the most direct (but not necessarily the most accurate) method of obtaining these data, the time signals from the reference standard are used to gate a frequency counter set to count the output frequency at, say, 5 MHz from the standard under test. The readings of the counter, which can be automatically recorded, give the number of oscillations of the standard contained in each fixed interval. If the frequency counter is zeroed after each interval, and its readings are n_1, n_2, n_3,..., then an accepted measure of the instability, called the *Allan variance*, is defined as follows:

$$\sigma^2 = \left\langle \frac{(n_{i+1} - n_i)^2}{2} \right\rangle_{ave}, \qquad 4.1$$

where the brackets $\langle\ \rangle$ symbolize the average over many equal intervals and i=1, 2, 3,..., N–1, where N is the total number of times the counts are taken for the same time interval. We note that the set of numbers (n_2-n_1), (n_3-n_2),..., (n_N-n_{N-1}) is known in the theory of numerical analysis as the first difference of the set n_1,

$n_2, n_3, \ldots n_N$, and that the variance σ can be zero only if all the n's are equal, that is, if the system under test tracks precisely the standard being used. Without attempting to go any deeper into the matter, we will accept the fact that in defining instability this way, we have a practicable measure that avoids certain difficulties in the statistical analysis of the long-term behavior of time standards.

This definition, however, cannot be implemented with accuracy for very short time intervals; it is supplemented by the frequency domain Fourier spectrum of the fluctuations in frequency (or phase) looked on as functions of time. This presumes that an electronic circuit is used to convert such fluctuations into a proportional, time-varying voltage whose square is a positive definite quantity proportional to the electrical power developed in the circuit. This power is analyzed by an instrument, called for obvious reasons a spectrum analyzer, to give the power per unit frequency interval (hertz) in its Fourier spectrum. This should not be confused with the Fourier spectrum of the oscillatory signal of the oscillator itself; rather, it is the spectrum of frequencies with which the phase of the oscillatory signal varies in time. Had we been dealing with fluctuations of amplitude rather than phase or frequency, it turns out that we would have had a simple relationship between the spectrum of the signal itself and the spectrum of the amplitude looked on as a function of time; in fact, the former is simply the latter displaced along the frequency scale an amount equal to the frequency of the oscillatory signal.

This description of the phase instability in the *frequency domain* based on Fourier analysis provides a useful way of distinguishing the different types of noise that underlie this instability. This is done by specifying the power distribution in the spectrum of the noise; it turns out that the more common types of noise exhibit a simple power-law dependence on frequency. Thus if the spectral power distribution varies as $1/\nu^2$, it is a random walk in frequency; if the dependence is as $1/\nu$, it is *flicker* frequency noise; finally, if the distribution is independent of frequency, that is, the graph is flat, it is called *white* frequency noise. Of course, these distributions are determined not only by the fundamental sources producing them, but also by any frequency dependence in the circuitry. These power laws translate into equally simple dependence of the variance σ on the constant time interval used in its measurement. Thus for the important flicker noise it can be shown that σ is independent of the length of the interval, whereas for white frequency noise σ falls as $1/\tau^{1/2}$. Now, for circuits at ordinary temperatures operating in the radio-frequency range, thermal (Johnson) noise is very nearly "white" (the same power density at all frequencies), so that since this is a universal source of noise, we frequently see a plot of σ versus τ exhibit the $1/\tau^{1/2}$ characteristic of this type of noise, at least up to a certain point, after which flicker noise becomes dominant, and the graph flattens out.

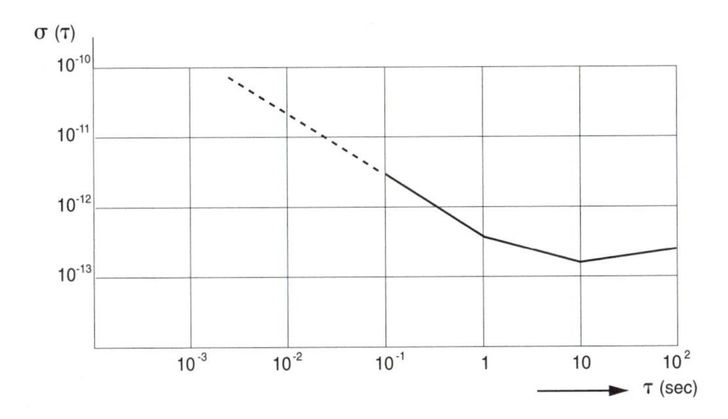

Figure 4.11 Allan variance of frequency versus sampling time for an ultrastable spacecraft quartz oscillator (Norton, 1994)

From the somewhat lengthy discussion we gave earlier of the sources of instability in a quartz-controlled oscillator, it might seem that such oscillators are plagued with errant behavior; that is far from the truth. To help regain our sense of proportion in the matter, let us recall that a precision quartz oscillator, for example a fifth-overtone 5 MHz crystal with Q as high as 1,000,000, may well be stable over a period of an hour to better than one part in 10^{11}. Or to put it another way, while oscillating at the rate of 5,000,000 oscillations per second, it may gain or lose only one oscillation in 5.5 hours! In fact, the quartz oscillator has no equal with respect to short-term frequency stability and spectral purity, which is really a statement about the extremely low noise that accompanies its undistorted output. This unique property makes it invaluable in a number of applications in addition to time measurement. One such application is Doppler radar, in which the shift in the frequency of the echo from a moving target due to the Doppler effect is exploited to discriminate against "ground clutter," since the frequency of the latter is unchanged. The ability to detect small changes in echo frequency and hence in target velocity, as well as the range, depends on the frequency stability of the transmitted microwaves. A viable approach is to start with a precision 100 MHz quartz oscillator and generate the desired microwave frequency as a harmonic using established frequency-multiplying techniques. Since the line width and other frequencies present in the spectrum are spread out by the same multiplying factor, it places such stringent requirements on the short-term frequency stability and spectral purity that quartz controlled oscillators are looked to as superior candidates for such an application.

4.9 Frequency/Time Measurement

4.9.1 Measurement of Time Intervals

We will now take up the question of how to fulfill the normal function expected of a clock, namely to "tell time." But before we embark on what appears as only an innocuous enterprise, let us first consider the use of our oscillator simply to measure the time interval between two events. This will require a calibration of the oscillator against a standard. Depending, of course, on the level of accuracy required in the measurement, it may be necessary to have the calibration traceable back to a national standard of frequency. This standard is no longer based on the so-called "ephemeris second," but has, since the 1967 General Conference on Weights and Measures, been defined in terms of a certain microwave resonance in an isotope of the cesium atom. We shall have a great deal more to say on this subject in succeeding chapters. Let us assume for the present that the frequency of a given crystal-controlled oscillator has been calibrated against some acceptable standard, and we wish to use it to measure the time interval between two events. Unlike a common stopwatch, which is started at the first event and stopped at the second, an oscillator requires time to stabilize; and therefore it must be allowed to continue uninterrupted while its connection to an oscillation counter is made through a gate circuit, which is opened and closed by the two events. Note that we have implicitly limited the accuracy to within a whole number of oscillations, that is, for a 5 MHz oscillator to within 0.2 microsecond; to measure fractions of an oscillation would require special techniques.

4.9.2 Binary Circuits

In this age of digital electronics, circuits such as gates and counters are common as the basic building blocks for more complex digital circuits. These circuits process all signals in binary form; that is, there are only two discrete voltage levels, representing 0 and 1 in the binary system used to represent the numerical magnitudes of signals. In order to use binary circuits to count harmonic oscillations, we must reshape the waveform so that it conforms to the two-level binary format. There are several ways in which this can be done; one possibility is to use a circuit called a Schmidt trigger, whose output is a sudden step in voltage when the input passes smoothly a certain preset trigger level, and will step back to its former level if the input returns to pass through the preset value in the other direction. We could, for example, set the trigger level near the crest of the oscillator waveform, thereby generating a train of narrow rectangular pulses rising from the "low" level to the "high" level of the binary circuits. We can now use standard binary gates and counters to measure a time interval.

Counters are based on a bistable circuit whose history precedes digital computers, integrated circuits, and even solid-state electronics itself. This bistable circuit, once called an Eccles–Jordan pair, is a bistable multivibrator, now generally referred to as a *flip-flop*, which probably first came into wide use with the invention of nuclear radiation counters, such as the Geiger counter, in 1913. It belongs to a distinct class, quite unlike the usual linear amplifiers, in which the voltages and currents at different points in the circuit may assume any of a continuous range of stable values controlled by a continuous input signal. The flip-flop has only two discrete stable states and no others; it passes through intermediate states only in a sudden transit from one state to the other. Such a transition is triggered by a sudden change in voltage appropriately applied to it. The obvious mechanical analogue is a common electrical switch, which has only two stable states: on or off.

The flip-flop can be constructed from two (inverting) single-stage amplifiers with overall regenerative, or positive, feedback; the output of each amplifier is connected to the input of the other in such a way that a downward fall in the current of either amplifier goes precipitously all the way to zero because of the feedback. It is instructive to consider what distinguishes this circuit from the kind of feedback amplifiers we have described in a previous chapter as oscillators. The difference lies essentially in the frequency dependence of the amount of feedback; for a harmonic oscillator the feedback must reach a maximum at only one nonzero frequency, whereas if we are to speak in such terms, we would say that the flip-flop has feedback extending from zero frequency (DC) up to frequencies that determine the speed of switching. A "zero frequency" signal is simply a constant, or DC, voltage, and the feedback in a flip-flop affects constant DC levels on the amplifier, causing it either to shut off or draw the maximum current the circuit will allow. Furthermore, we must remember that a feedback amplifier designed to serve as a pure single-frequency oscillator must be provided with a means of limiting the amplitude of oscillation. Otherwise, it would swing between two extreme states, producing a square wave output. In that case it would be called a free running multivibrator.

However, in keeping with the spirit of the age we will think of gates and other basic circuits as integrated circuit chips specified in terms of the outputs as functions of the inputs, without attempting to analyze their inner workings. Figure 4.12 shows an example of what is called an RS flip-flop using two cross-coupled NOR gates. A NOR gate is a "black box" with two input terminals and one output, typical of logic gates used in digital electronics. In the binary world of computers, the logic values true and false are represented by two voltage levels: high and low, which are typically separated by 5 volts. In the NOR gate the output is high only when neither of the two inputs is high. Like all other embodiments of flip-flops, this has two stable states; one in which the output at Q is high and at Q^* low, and the other in which the outputs are reversed. A momentary high applied to the S input will put the circuit in the former state, independent of what state it was in; and similarly, a momentary high at the R

input results in the reverse output state, again, we should emphasize, independent of its former state. Such a flip-flop can be made into a "scale-of-two," that is, one stage of a binary counter, by providing an input gating circuit that in effect alternates the input pulses between the two input terminals, *R* and *S*, so that on each input pulse the flip-flop changes state. The resulting output at Q*, for example, is then a square wave with alternating high and low voltages, with the transition from (say) low to high occurring every second input pulse. Now, this rising edge of the square wave can be made to generate a positive rectangular pulse by using an appropriate trigger circuit, giving us a train of pulses similar to the original train, except that for every output pulse, two of the original pulses are required. If we connect, for example, 24 such binary units in tandem, then since we are dividing by 2 in each stage, it requires 2^{23} incoming pulses to result in a pulse reaching the last stage, and 2^{24} pulses before all stages return to their original state. Since 2^{24}=16,777,216, if the oscillator frequency is 5 MHz, then the maximum interval that could be measured would be about 16,777,216/5,000,000≈3.3 seconds.

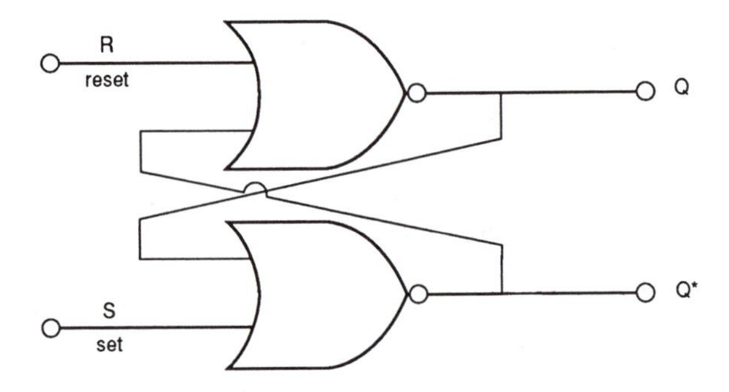

Figure 4.12. An RS flip-flop circuit

The same binary counter stage can be used to provide a "gate" signal, that is, a constant "high" level voltage starting at the first event and ending at the second event, but otherwise at the "low" level. This gate can then be applied to one input of an AND logic gate, while the other is connected to the pulse-shaped output of the oscillator. The output of the AND gate will be a finite pulse train starting at the first event and ending at the second. By counting the number of pulses in this train, we obtain the interval of time between the events.

We should note at this point that the same procedure we have just been describing could be turned around to measure the frequency of the oscillator if the time interval between the two events is independently known with the desired accuracy. In fact, this is the basis of most frequency counters commonly available on the market. In these the time intervals, which can be selected according to the frequency to be measured, are derived from an internal reference.

4.9.3 Frequency Synthesizers

From what has been said we may have created the impression that we are limited to the division of the oscillator frequency by powers of 2, which of course would not give us output displays in seconds, minutes, and hours, as we expect of a clock. This is far from the present reality; not only can we produce outputs at those frequencies coherent with the oscillator, but we can synthesize almost any other conceivable frequency, also coherent with the oscillator. The key to achieving this lies in nonlinear solid-state devices, that is, ones in which the graph of the current through them versus the applied voltage causing it is not a straight line. The simplest example is a crystal diode, the same basic device that was used by amateurs in the early days of radio to build a "crystal set"; it was a fine wire, the "cat-whisker," making a point contact with a silicon crystal.

These nonlinear devices can be used to generate harmonics of the input frequency by introducing sharp distortions in the waveform while preserving its precise period. The resulting Fourier spectrum will have a comb-like appearance consisting of harmonics of the fundamental frequency ranging up to many times that frequency. They can also act as mixers of two incoming frequencies, producing signals at frequencies that are the result of adding and subtracting harmonics of the input frequencies. There is an important distinction that should be drawn here: If, for example, two frequencies, one very much larger than the other, are applied to a *linear* device, then within the device, the oscillation being the sum of the two frequencies will appear as an oscillation whose center is itself oscillating at the lower frequency. However, in a *nonlinear* device the emerging signal will also exhibit a low-frequency modulation of the amplitude of the higher-frequency signal. Such a modulated signal we have just shown to have a Fourier spectrum containing amplitudes at the sum and difference frequencies. Furthermore, the phase information is preserved in these processes, so that the output frequencies are phase correlated to the input frequencies; that is, they are phase coherent.

It follows that since we can add and subtract frequencies, and multiply and divide them by any integer, we can in principle do all the necessary arithmetic to derive any frequency we want. Instruments that implement these techniques in practice, called *frequency synthesizers*, have been developed to a high degree of sophistication, involving crystal filters and phase-lock techniques, in which the phase of one stage is controlled through a feedback loop to lock onto another. Of particular concern is avoiding the introduction of electrical noise and consequent loss of spectral purity when high orders of frequency multiplication are involved. Synthesizers are readily available on the market covering various frequency ranges, in which any of a discrete set of output frequencies can be selected using front panel switches.

4.10 Quartz Watches

We conclude this chapter with a brief description of what has become commonplace since the early 1970s: the quartz watch. Thanks to the revolutionary development of microelectronics, a quartz-controlled clock of miniature dimensions is not only possible, it is commercially available at a cost for the quartz movement of less than U.S.$10! Their time-keeping accuracy is typically a few parts in 10^6, which translates into a few seconds per month!

The immediate predecessor of the quartz watch was an expensive, high-quality electronic model based on an acoustic resonator: the tuning fork. It had an accuracy that surpassed the best mechanical spring–balance movement; however; the improvement was not commensurate with its high cost, and it was soon superseded by the quartz movement.

Figure 4.13 shows schematically the basic elements of a quartz watch. The quartz crystal generally vibrates at a relatively low frequency, and it is incorporated into a miniaturized circuit board on which is mounted an integrated circuit frequency divider and driver for a stepper motor, that is, one that turns a fixed angle (in this case 6°) each time it receives an electrical pulse.

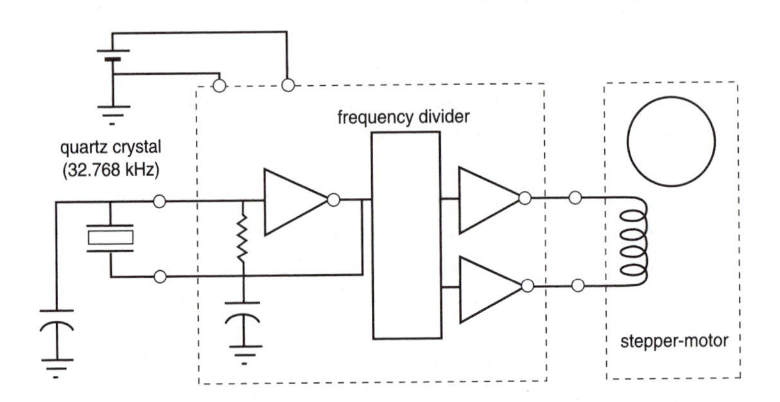

Figure 4.13 Schematic circuit diagram of a typical quartz movement

Chapter 5
The Language of Electrons, Atoms, and Quanta

5.1 Classical Lorentz Theory

When we speak of oscillations at optical frequencies and their amplification, we are indeed a long way from the world of swinging pendulums and oscillating balance wheels. It is true that classical theory based on Newton's laws of motion and Maxwell's theory of electromagnetic radiation are inappropriate to deal with the interaction of radiation with atoms and molecules; for this we need the *quantum theory*. However, from a background of classical theory, certain aspects can be sketched in a semiclassical way, in which quantum ideas are superimposed on a classical base. Historically, this characterized the early development of the theory of radiation and the general features of the theory of *optical dispersion*. In this context "dispersion" refers to the dependence of the refractivity of a medium on the wavelength, which leads to the dispersion of, for example, white light by a glass prism into the colors of the rainbow.

Prior to the advent of quantum theory early in this century, the interaction of radiation with matter was explained on the basis of the *electron theory* of H.A. Lorentz, in which the response of matter to an electromagnetic wave was expressed in terms of "atomic oscillators" pictured as electrons elastically bound to the atomic centers. The interaction of atoms with an electromagnetic wave was imagined as consisting in these electrons being driven into forced oscillation by the oscillating electric field component of the wave. It can be shown, however, that in order to have continuous absorption of energy from the wave (as opposed to a fleeting absorption when the wave first interacts with an electron, setting it in motion), it is necessary to assume that during the interaction with the wave, the driven electron oscillation must in effect experience a resistive force. This clearly cannot be a frictional force in the usual sense; and the *radiation reaction* force, which accounts for the energy radiated by the vibrating electron, proves to be too small to account for the degree of light absorption that can occur. Lorentz attributed the net absorption to the repeated interruption of the electron oscillation by collisions with other atoms, resulting in the randomization of the oscillation phase. In the absence of this, the phase of the periodic electron velocity remains in quadrature with the driving force, and the average work done on the electron averages to zero over a period of oscillation of the field. The result would be that no net absorption of energy takes place. Through those phase-randomizing collisions, there is a continuous transfer of energy to the electrons that appears, through the same collisions, as random kinetic energy of the colliding atoms, that is, heat.

The same model was also used to describe the process of emission of radiation by atomic oscillators, when set into vibration by collisions with other atoms in an electrical discharge, or in a state of thermal agitation, as in a flame. It is well established classically, on the basis of Maxwell's theory, that an oscillating electric charge will radiate electromagnetic waves. In this case, since we have a negative charge (the electron) oscillating with respect to an equal positive charge, the wave that is generated is that of an oscillating electric dipole. This has a characteristic radiation pattern, that is, distribution of intensity in different directions, similar to that from a simple radio transmitter antenna. The frequency of the radiated electromagnetic wave is classically the same as the frequency of oscillation of the supposed atomic oscillator. If through some nonlinearity the atomic oscillator excitation results in some second or higher harmonics, at twice or a higher multiple of the fundamental frequency, the radiation will also contain those harmonic frequencies. It was one of the fatal flaws of classical theory in explaining atomic spectra that the observed frequencies emitted by atoms do not bear a simple harmonic relationship to each other.

5.2 Spectrum of Blackbody Radiation

But the breakdown in the classical theory of radiation, which finally led Planck to postulate the quantum of energy, first came in the explanation of the spectrum of the radiation in thermal equilibrium with matter, the so-called *blackbody radiation*.

This is radiation whose spectrum is characteristic of the equilibrium temperature, and it is independent of the nature of the matter interacting with it. It can be observed only under conditions where the interacting matter can thoroughly absorb and re-emit radiation at all frequencies. In practice this is achieved by studying the radiation inside an enclosure, which is provided with a small hole to allow a sample of the radiation to be analyzed outside the cavity. The observed continuous spectrum, showing the radiated intensity in a small fixed frequency band as a function of the center frequency of that band, is shown in Figure 5.1. Contrary to classical predictions, the graph tends to zero at the upper and lower ends of the frequency scale, with a maximum intensity at some intermediate frequency, which, in accordance with common experience, depends on the temperature: the color varies from red toward the blue as the temperature is raised. This is given precise expression in *Wien's displacement law*: the frequency at which the intensity is maximum shifts to higher values, in direct proportion to an increase in the temperature. Wien derived his law on the basis of classical arguments, prior to Planck's work, and it is borne out by experiment. The spectrum of sunlight is approximately that of a "black body" at a temperature of about 6000°K, with the maximum intensity occurring at around a wavelength of 0.5μm, in the middle of the visible region of the electromagnetic spectrum.

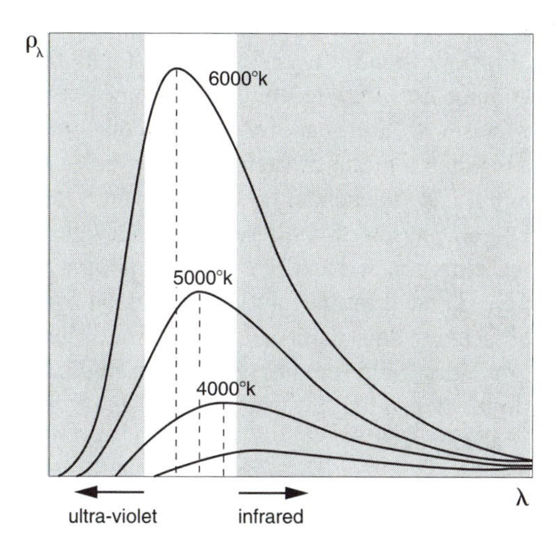

Figure 5.1 The spectrum of blackbody radiation showing the shift in the maximum with temperature

5.3 The Quantum of Radiation: The Photon

After all attempts based on the classical theory of thermal equilibrium and the exchange of energy between radiation and matter failed to explain the observed spectrum, Max Planck in 1901 published a radically new theory, which was able to predict a spectrum in close agreement with experiment. It was based on the postulate that matter contained an immense number of electromagnetic "resonators" that could exchange energy with the radiation field not continuously in arbitrarily small amounts, but only in discrete units he called *quanta*, whose energy is proportional to the frequency: $E=h\nu$, where h is a universal constant of nature, now called Planck's constant, with a numerical value in our system of units of 6.6×10^{-34} joule·second.

A greater understanding of the physical processes that result in the emission of blackbody radiation came with the reinterpretation of the process by Einstein, who introduced the concept of a quantum of electromagnetic radiation, called a *photon*, which in some circumstances manifests a discrete particle nature. On this basis, blackbody radiation results when an equilibrium has been reached between the *photon gas* and the atoms of matter through continual absorption and re-emission of photons by the atoms. When this model was applied to derive Planck's formula, it was found that the well-known *spontaneous* emission process, in which an atom gives up its energy of excitation spontaneously by emitting a photon, alone would not lead to an equilibrium consistent with Planck's formula. Einstein found it necessary to postulate that an atom that has absorbed a photon may not only re-emit it spontaneously, but may also be *stimulated* to re-emit it, with a probability that depends on the number of photons already present. When a group of such atoms or molecules undergo spontaneous emission, they do so independently of each other; there is, therefore, *no* correlation between the phases of their several contributions to the radiation emitted. In contrast, emission induced by existing photons, that is, *stimulated emission*, has a phase dictated by the phase of the existing radiation field, and hence all atoms subjected to this field will have more or less a common phase. This results in the radiation field remaining coherent in phase and increasing or decreasing in amplitude depending on whether the rate of emission is greater or less than the rate of absorption.

5.4 Bohr's Theory of the Hydrogen Atom

The success of the radically new quantum postulate of Planck soon saw the spread of quantum ideas to the hitherto intractable problem of explaining optical emission spectra of atoms. A wealth of accurate experimental data had been accumulated on the wavelengths of the many series of lines that form atomic spectra. Each chemical element emits its own characteristic, and for all but the simplest atoms, complex line spectrum. Intensive efforts had been devoted to finding regularities in these spectra, and a number of empirical rules were enunciated, all of which brought some order to their practical analysis.

The theoretical breakthrough came after the success of the *nuclear model* of the atom, which was postulated by Ernest Rutherford around 1911 to explain unexpectedly large angles of scattering of high-speed α-particles (a product of natural radioactivity of certain elements) by atoms in a gold foil target. The model was strikingly confirmed in subsequent years in his laboratory, a feat he announced as possibly more important than the outcome of what was then called the Great War. For this he received the Nobel Prize and given the title Lord Rutherford of Nelson (his birthplace in New Zealand). Prior to that there was intense speculation as to just how electrons and protons, the elementary particles known at the time, were arranged in atoms. As is now familiar to everyone, the basic arrangement is that almost the entire mass of an atom resides in a small, positively charged nucleus, which is surrounded by a cloud of negatively charged electrons.

Around 1913, the Danish physicist Niels Bohr, by a set of ad hoc quantum notions superimposed onto a classical planetary model of the hydrogen atom, was able to obtain with remarkable accuracy the wavelengths of a series of lines in the spectrum of that atom. The most radical of Bohr's postulates was that there exist certain orbits in which electrons could circulate indefinitely without radiating energy, contrary to the classical prediction that an orbiting charge should lose energy by radiation and eventually spiral into the nucleus. These he called *stationary* orbits, and he postulated that of all the possible orbits that classical theory allows, only those are stationary that satisfy the following condition on their angular momentum L:

$$L = \frac{nh}{2\pi}, \qquad 5.1$$

where n is an integer and h is the same constant Planck had used to define the quantum of energy. For a circular orbit of radius r, $L=mVr$, where m is the mass of the electron and V its velocity. He further postulated that the frequency of radiation emitted by the atoms is not the vibration or rotational frequency of the electron in the classical sense, but is derived from Planck's formula. Thus, when an atom makes a quantum transition from a stationary state of energy E_2 to one having energy E_1, the frequency of the radiation is that of the radiated quantum, that is,

$$\nu = \frac{(E_2 - E_1)}{h}. \qquad 5.2$$

Of course, these radical postulates were not made lightly. The line spectra of atoms show remarkable regularities, with series of lines forming striking patterns, plausibly reminiscent of the classical vibration spectra of complex structures. It would be natural to assume that these vibration spectra should form the basis of an explanation of the spectrum. Unfortunately, of all of the precise experimental data that was available and some empirical formulas that were discovered relating the wavelengths in the spectra, none was consistent with the harmonic relationships characteristic of classical vibration frequencies.

Bohr's ad hoc postulate identifying stationary orbits became a little less so through the work of de Broglie, published in 1924. In this de Broglie argued on the basis of the dual particle–wave nature of light, which was then the subject of much speculation and debate, that material particles have the same duality. The success of Bohr's theory seemed to hint at a wave property of electrons, since at the time the only context in which equations contained integers was in normal modes of vibration, and the interference of waves. On the basis of the special theory of relativity de Broglie was able to find the connection between the particle and wave nature of all matter and radiation, in a theory called *wave mechanics*, the precursor of quantum mechanics. According to de Broglie, a particle of mass m moving with a velocity V has a *wave* associated with it "guiding" its motion, whose wavelength, now called the ***de Broglie wavelength***, is given by $\lambda=h/mV$, where h is, as usual, Planck's constant. If we use this result in Bohr's equation for the stationary orbits, we find $(h/\lambda)r=nh/2\pi$; that is, $2\pi r=n\lambda$. But this is precisely the condition for a resonant mode of vibration of a circular string supporting oscillations with a wavelength λ; any other radius would not have the wave reinforcing itself as it traveled around the circle.

The theory of Bohr, elaborated by Sommerfeld, and now referred to as the "old quantum theory," dealt only with "stationary" quantum states and quantum numbers; it had little to say about nonstationary phenomena such as transitions between states and collisions between particles. This situation changed with the coming of quantum mechanics.

5.5 The Schrödinger Wave Equation

The spirit of de Broglie's description remains in the subsequent quantum theory of Schrödinger. The concept of a wave determining the motion of a particle implies the radical notion that the amplitude of a wave, given as a function of the coordinates and called a *wave function*, is to be used to describe the motion of a particle, rather than regarding a particle as a point mass occupying a certain position in space specified by its coordinates. The physical interpretation of the wave function, conventionally represented by the Greek letter ψ, lends itself to some speculation in the minds of some, hinting at a mysterious wave that guides the motion of matter. However, a probabilistic view prevails in which $|\psi(x,y,z)|^2$ is taken as the space density of the *probability* that the particle is at the coordinates x, y, z, in the sense that $|\psi(x,y,z)|^2 dx\,dy\,dz$ is the probability of the particle being found in a cell of sides dx, dy, and dz centered at the point (x, y, z). Since the particle must be somewhere with a 100% certainty, it follows that the wave function must satisfy the following *normalization* condition:

$$\int_{-\infty}^{+\infty}\int_{-\infty}^{+\infty}\int_{-\infty}^{+\infty}|\psi|^2 dxdydz = 1. \qquad 5.3$$

This, of course, imposes a mathematical restriction on the *wave function*: Its integral must be finite.

In Schrödinger's *wave mechanics*, which is one mathematical representation of quantum mechanics, the equations of motion of classical mechanics are replaced by a differential equation, called the *Schrödinger equation,* to determine the *wave function*. Thus, for example, the equation for a free electron having energy E in a one-dimensional world would be as follows:

$$\frac{d^2\psi}{dx^2}+\frac{8\pi^2 mE}{h^2}\psi=0. \qquad 5.4$$

Of all the mathematical solutions of the Schrödinger equation, those that may be accepted as representing the stationary states of a physical system are defined as those particular solutions, called *eigenfunctions* (German for proper functions), that are finite and satisfy certain conditions at the boundaries of the system. For example, if an electron obeying the above equation is confined between two plane, parallel, "impenetrable walls" forming the boundaries at x=0 and x=L, the stationary solution of the Schrödinger equation describing that electron would be equal to zero at those points and beyond. One can readily verify that the following are solutions:

$$\psi_n = N\sin(k_n x), \qquad 5.5$$

where

$$K_n = n\frac{\pi}{L}, \quad E_n = n^2\frac{h^2}{8L^2 m}, \qquad 5.6$$

and n = 1, 2, 3,.... We note that were it not for the boundary conditions $\Psi(0)=0$ and $\Psi(L)=0$, the equation would have been satisfied by $\sin(kx)$, where k, and therefore E, are continuous variables, and not "quantized" to have the discrete values labeled with the index n: k_n and E_n. The functions $N\sin(k_n x)$ are the stationary wave functions, the eigenfunctions of Schrödinger's equation for the particular system we have assumed. They are analogous to the classical normal modes of vibration of a system.

For the 3-dimensional case of a particle confined in a rectangular box with sides L_1, L_2, L_3, the eigenfunctions have the form

$$\psi_{l,m,n} = \sqrt{\frac{8}{L_1 L_2 L_3}}\sin(k_l x)\sin(k_m y)\sin(k_n z), \qquad 5.7$$

where

$$k_l = \frac{l\pi}{L_1}, k_m = \frac{m\pi}{L_2}, k_n = \frac{n\pi}{L_3}, \qquad 5.8$$

and the quantum energy levels are given by

$$E_{l,m,n} = \frac{h^2}{8\pi^2 m}\left[k_l^2 + k_m^2 + k_n^2 +\right]. \quad 5.9$$

We note that we now have three quantum numbers *l*, *m*, and *n* to distinguish the various possible stationary states, and that these appear in the quantization of the components of the wave vector **k** along the three coordinate axes. If we recall the formula for the de Broglie wavelength, we find that $k=(2\pi/h)mV$; that is, it is the linear momentum that is quantized. The constant factor $\sqrt{8(L_1 L_2 L_3)}$ is introduced to meet the normalization condition.

We note that the stationary states we found for an electron in a box are far from the classical picture of a point mass bouncing back and forth between the boundaries. A particle moving back and forth would be represented as a time-dependent wave function that, at any moment is small everywhere except in the neighborhood of the particle position. Such a wave function, called a *wave packet*, can be synthesized as a sum over the harmonic eigenfunctions, following the spirit of the Fourier expansion theorem. Each eigenfunction corresponds to a different energy, and therefore a different frequency (since Planck's formula $E=h\nu$ still holds), with the result that the wave packet will have a time dependence reflecting the motion of the particle.

5.6 Quantum Numbers of Atomic States

If a particle is subjected to a central force, that is, one directed toward a fixed point, such as the electrostatic Coulomb force that a nucleus exerts on the electrons surrounding it in an atom, three quantum numbers will again be required to specify a stationary state. In this case the spherical symmetry of the equation suggests that the solution is most naturally expressed using the spherical polar coordinates r, θ, ϕ. The quantum numbers conventionally designated as n, l, m play a role in close analogy to the indices used to label the various normal modes of vibrations of a sphere. The values of the quantum numbers are restricted as follows: n=1, 2, 3,...,while $l \leq (n-1)$ and $m=l$, $(l-1)$, $(l-2)$... $-(l-2)$, $-(l-1)$, $-l$. The part of the wave function that is a function of the r-coordinate has a number of nodes (zeros) given by $(n-l-1)$, and the part that is a function of the co-latitude θ has $(l-m)$ nodes off-axis. Following spectroscopic convention, electrons in an atom having l=0, 1, 2, 3,... are called s-, p-, d-, f-electrons, etc., respectively. The quantum numbers l and m, which are associated with the angular part of the wave function, in fact reflect the quantization of the angular momentum and its component along the polar axis, respectively. According to the theory, a system having nominally an orbital momentum quantum number l actually will have orbital angular momentum of $\sqrt{l(l+1)}$ in units of $h/2\pi$, whereas the maximum component along the polar axis is only l. (We will usually omit the unit $h/2\pi$ unless we are doing a numerical calculation.) Thus the theory predicts that the maximum component the angular momentum can have along any given axis is somewhat less than the magnitude of the angular momentum itself. This is a strictly quantum effect, since classically the angular momentum is a vector that can assume any direction, and in particular can point exactly in the direction of any given axis. The effect can be interpreted in terms of vectors by saying it arises from a quantum uncertainty in the angle the angular momentum vector makes with the axis.

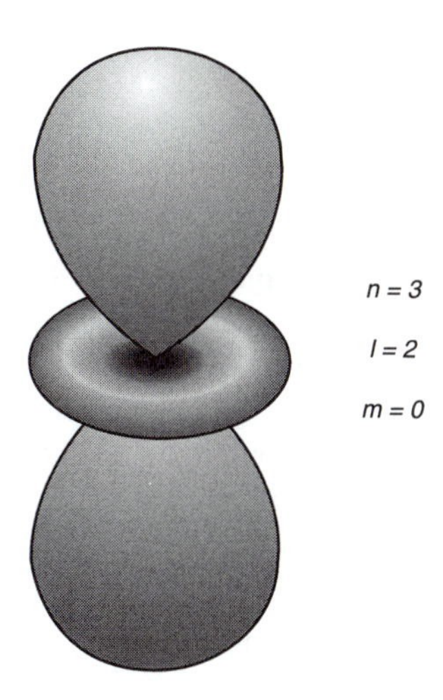

Figure 5.2 Example of the probability distribution for a particle in a central field

The quantization of the component of angular momentum along an axis arises mathematically from the condition that the solution to the Schrödinger equation must be a simple periodic function of the angle ϕ around the axis; that is, it must repeat itself every 360°. This physical requirement imposed on the mathematical solution bears some resemblance to the implied condition for normal modes in the Bohr circular orbits. In any event, it constitutes one of the most radical breaks with classical mechanics: It implies that the angular momentum of a system can only assume certain discrete orientations with respect to a given axis; this is sometimes called *space quantization*, and it is of profound importance in the quantum theory of atoms subjected to an external magnetic field, and the attendant shifts in energy levels: the *Zeeman effect*.

Since an atomic angular momentum will have associated with it a magnetic moment (both due to the orbital motion of the charged electron and its intrinsic spin), the energy shift produced by a magnetic field is expected to depend on the component of the angular momentum along the field and therefore, with the axis chosen along the field direction, on the quantum number m. For this reason m is called the *magnetic* quantum number, and to reiterate, for a state with a *total* angular momentum (including spin) quantum number J (which, as we shall see, may be integral or half-integral) the magnetic quantum number m_J, can have one of the following $(2J+1)$ discrete values: J, $(J-1)$, $(J-2)$,..., $-(J-2)$, $-(J-1)$, $-J$. For example, a particle in an angular momentum state described nominally as a $J=5/2$ state may have as its component along a given axis one of the following values: +5/2, +3/2, +1/2, -1/2, -3/2, -5/2.

5.7 The Vector Model

It should be emphasized that the quantum numbers, while they represent in quantum mechanics the results of measurement of a particular dynamical quantity, such as angular momentum, it is only in systems involving very large quantum numbers that they approximate classical behavior. It happens that we can, according to what is called the *vector model,* retain the concept of angular momentum as a classical vector, provided that we give these vectors properties that are peculiar to quantum mechanics. The uncertainty in pointing the angular momentum exactly along a given direction is one of them. The other concerns combining different angular momenta to obtain a resultant: the result of adding two angular momenta whose quantum numbers are, for example J_1 and J_2, where $J_2 < J_1$, would classically be *any* value between (J_1-J_2) and (J_1+J_2), depending on the angle between the two angular momentum vectors, whereas in quantum theory the resultant is one of a discrete set that starts with (J_1-J_2) and by increments of one unit reaches (J_1+J_2). For example, suppose a system in an angular momentum state with quantum number $J_1=3$ and another with $J_2=1$ interact in such a way that results in stationary states of the combined angular momenta. The quantum numbers belonging to this combined representation would be 2, 3, 4, implying according to the vector model that the angular momentum vectors can make only certain discrete angles with respect to each other, as illustrated in Figure 5.3.

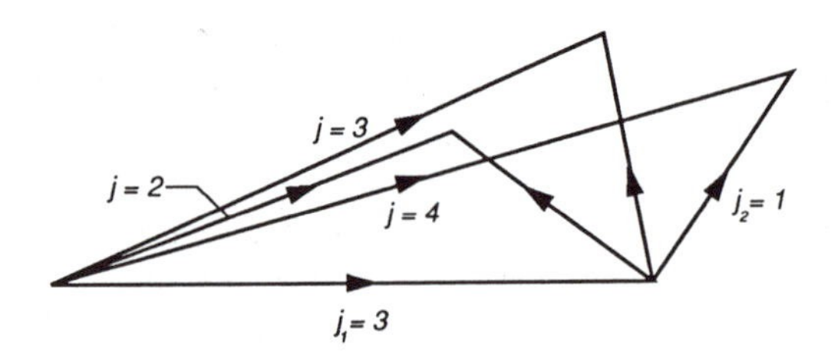

Figure 5.3 Vector diagram for the addition of angular momenta

If the particle is an electron, a complete specification of its quantum state requires not only the dependence of its wave function on the space coordinates, but also the state of another attribute of the electron called the *spin*. This is an *intrinsic* angular momentum of $\frac{1}{2}(h/2\pi)$, part of what it is to be an electron. It was first introduced to explain atomic spectra and later brilliantly shown by Dirac to be a logical necessity, forming an integral part of a relativistic quantum theory. For a free electron, the spin component along any given axis can only be $+1/2$ or $-1/2$, corresponding to only two possible directions of spin.

For electrons in an atom, a total angular momentum larger than 1/2 can result from the spin combining with the orbital angular momentum of its motion around the nucleus, which is conventionally represented by l and is always integral. It can be shown that the magnetic field produced by the orbital motion of the electron can exert a torque on its own spin, a coupling called the *spin–orbit interaction*, which is extremely important in understanding atoms. In the absence of other torques acting separately on the two types of angular momentum, such as a strong external magnetic field, the two will give a resultant angular momentum represented conventionally by j, which is conserved, and quantized both in magnitude and spatial orientation. For example, an electron in an orbit with orbital angular momentum $l=2$ will have a resultant, when combined with its spin of 1/2, equal to either $2+1/2$ or $2-1/2$, that is, 5/2 or 3/2. Recall that these numbers give the maximum component observable along any given axis in units of $h/2\pi$.

5.8 The Shell Structure of Electron States

When there is a large number of electrons in different orbits, the prediction of the possible combined angular momentum states quickly becomes very complicated; not only is there spin–orbit coupling, but also interactions between the spin and orbital magnetic moments of different electrons. Fortunately, it happens that electrons in an atom can be grouped into *shells*, which, as we shall see, can contain only a certain maximum number of electrons. When completely filled, a shell has zero resultant angular momentum; so that only electrons in any incomplete shells need be considered in arriving at the overall atomic angular momentum state.

The reason for venturing a little into the abstruse realm of quantum theory is that it is essential for any basic understanding of atomic and molecular structure and dynamics. We recall that the atoms of the chemical elements have small positive nuclei, where most of the mass resides, surrounded by a cloud of negative electrons that occupy available quantum *states,* each state labeled by a set of three quantum numbers, plus a fourth specifying the spin state. We have already seen that for a given value of the quantum number l there are $(2l+1)$ states with different m; if we include the two possible directions of the spin, this number is doubled. (The presence of spin–orbit coupling requiring a description in terms of the total (spin plus orbital) quantum numbers does not affect the number $2(2l+1)$.) These states correspond to different orientations of the orbital and spin angular momenta with respect to a fixed axis. In the absence of an external field, such as a magnetic field, all directions in space are identical, and the energy of electrons in these states is the same; they are all at one energy level. They are called *degenerate* states. Furthermore, it is found that for a pure Coulomb (inverse square law) electrostatic field, such as we have in the hydrogen atom, the solution to the Schrödinger equation yields possible values of energy that depend only on the quantum number n, and so there is degeneracy with respect to the l quantum number as well. Now, for each value of n, the quantum number l can assume any of $(n-1)$ values, and as we have seen, to each l value there are $2(2l+1)$ degenerate states. The total number of degenerate states having the same n is therefore

$$\sum_{0}^{n-1} 2(2l+1) = 4\frac{n(n-1)}{2} + 2n = 2n^2. \qquad 5.10$$

The common energy of these states can be shown to be

$$E_n = -\left(\frac{2\pi^2 mZ^2 e^4}{h^2}\right)\left(\frac{1}{n^2}\right), \qquad 5.11$$

in agreement with the old quantum theory of Bohr, which was already known to be in remarkable agreement with experiment. There is an infinite number of energy levels corresponding to n ranging from 1 to ∞; Figure 5.4a shows some of the lower states. Unlike the Bohr theory, the electrons are not localized along particular orbits but must be regarded as spread out with a radial density given by $4\pi r^2|\psi|^2$, which is illustrated for several states in Figure 5.4b. We note that the average radius increases with n, the outermost electrons having the highest n and the smallest binding energy.

For atoms having a large nuclear charge, and therefore many electrons, the exact solution of the Schrödinger equation becomes impossible, and approximate numerical methods have been developed. An approximation that has proved very useful is to assume that each individual electron moves in an electrostatic field produced by the nuclear charge and an average *spherically symmetric* distribution of charge due to the other electrons. Of course, after solving the Schrödinger equation using this approximate field and obtaining the charge distribution of each electron from its calculated wave function, the combined charge distribution so derived must agree with the one assumed in the first place. The important point for us is that if the field acting on the electrons can indeed be taken to be spherically symmetric, then the same quantum numbers n, l, m can still be used with the same significance, except that the radial distribution of electrons is no longer purely hydrogen-like, and the energy is no longer a function of the principal quantum number n alone, but depends on l as well. That is, the l-degeneracy is removed. However, the dependence of the energy on l is still generally weaker than that on n, aside from some important exceptions for larger l-values. The m-degeneracy remains, and levels are grouped around the different l-values; these groupings are the *shells* mentioned earlier. It is one of the early triumphs of quantum theory that it was able to predict the number of quantum states in each shell. Thus, for example, states having n=4 and l=2 would be said to belong to the 4d shell, and those having n=5, l=0 are in the 5s shell; the former would number 2(2×2+1)=10 states, and the latter just 2(2×0+1)=2 states.

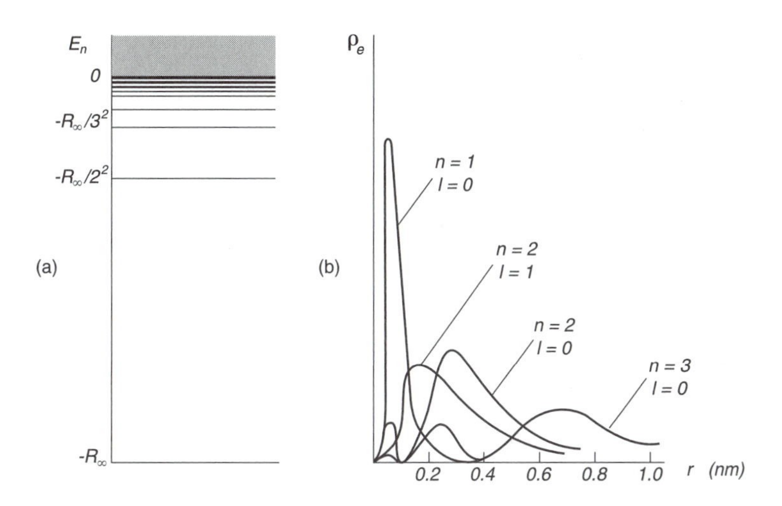

Figure 5.4 (a) The energy levels of the H-atom, and (b) the radial dependence of some of the lower energy wave functions

5.9 The Pauli Exclusion Principle

Starting with a nucleus having a given number of protons, in order to construct a neutral atom in its ground state we must take the same number of electrons and allocate them one by one to progressively higher-energy quantum states beginning with the lowest-energy state first. This atomic building principle is based on the condition that no two electrons can occupy the same quantum state, that is, have the same set of quantum numbers. This is a statement of the *Pauli exclusion principle*, which is at the heart of the quantum explanation of atomic structure and spectra. It can be deduced from a symmetry property of *wave functions* representing a system of electrons and some other elementary particles. Since individual electrons are indistinguishable, in the sense that we cannot know which electron occupies a particular position and spin state, an exchange of the assignment of these between any two electrons in the wave function ψ cannot change the observable $|\psi|^2$. Therefore, an electron exchange must either leave ψ unchanged (symmetric wave function) or at most change its sign (antisymmetric wave function). It happens that photons have the former symmetry, while electrons the latter. For electrons this means that the probability of finding two electrons in identical states is zero, since in that event an exchange of the two electrons must on the one hand leave the wave function unchanged, but on the other its sign must change; this can happen only if it is zero. Once an electron occupies a certain state, that state is said to be filled. This means that in constructing the ground state of an atom, each state must be filled before the next higher energy state is filled. The assignment of electrons to the different possible quantum states is analogous to the assignment of passengers to *single-occupancy* berths on a cruise ship; each berth has a number, and the fare schedule is based mainly on which deck the berth is located, with some differences within a given deck depending on its location. For the electrons in an atom, the "decks" are the shells, and the "fare" is the energy. Unlike a cruise ship, however, the electrons of an atom are in the stable ground state when their total energy ("fare") is a *minimum*.

Since we shall be concerned with crystalline quartz (SiO_2) in the next chapter, let us consider the elements oxygen and silicon as examples. They have (positive) nuclear charges of 8 and 14 respectively, in units of electronic charge. Therefore, oxygen will have the shells 1s, 2s filled and be two short of filling 2p, while silicon will have the 1s, 2s, 2p filled and have two electrons in each of the 3s and 3p outer shells.

It is the outermost electrons in an atom that determine its chemical properties and its interaction with radiation in the optical region of the spectrum. The inner electrons are unable to take part in any small exchange of energy, since all neighboring energy states are filled. Of course, if sufficient energy is involved, as in electron bombardment in an X-ray tube, inner electrons do play a part; but ordinary chemical reactions and optical transitions involve relatively little energy. The Mendeleev periodic system of the chemical elements finds a ready explanation in terms of the filling of shells as the nuclear charge (atomic number, Z) is incremented. Thus the property of having a completely filled outer *p*-shell corresponds to the noble gases and will recur at Z=2 (He), Z=2+8 (Ne), Z=2+8+8 (Ar), etc. Next would be the alkali elements with a single electron outside a closed shell; they are at Z=1 (H), Z=2+1 (Li), Z=2+8+1 (Na), Z=2+8+8+1 (K), etc. Then the alkaline earths, Be, Mg, Ca, Sr,..., with two electrons outside closed shells, and so on. This simple progression is interrupted when we reach a point where it becomes "cheaper" in energy to go to a higher *n*-value than to add to a shell with a high *l*-value. This leads to the so-called *transition elements*, for example, those involved in filling the 3d shell (*after* the 4s shell has been filled), Mn, Fe, Co, Ni.

In all the elements, the inner closed shells and the nucleus form a tightly held inner core, with an unbalanced positive charge equal to the charge of the outer electrons. In the context of chemical bonding, the outer electrons are referred to as the *valence* electrons, of which silicon has four and is therefore tetravalent, and of which oxygen lacks two to complete a shell and is thus divalent. Without going into the subject any more deeply than we absolutely have to, we will simply state that the bonding between atoms to form compounds can be characterized according to the extent that the valence electrons (a) overlap between the atoms (covalent character) or (b) are transposed from one atom to the other, forming positive and negative ions that attract each other (ionic character). Whether the bond between a particular pair of atoms is predominantly covalent or predominantly ionic depends on the relative energy "cost" of the electrons arranging themselves according to the one or the other; recall that stability belongs to the lowest energy. The covalent bond may involve one valence electron, as in the bond between Si and O in quartz, or more than one electron, as typified by the bond between C and O in carbon dioxide (CO_2), in which the carbon atom has a double bond with each oxygen atom.

Now, in a covalent bond, where the dominant feature is the overlap of valence electrons belonging to the two atoms (recall that the electrons are to be viewed as smeared over all space according to the magnitude of their wave function), it is reasonable to expect that the possible distribution of the valence electrons around the inner core will determine the directions along which the bonds occur.

5.10 Spectroscopic Notation

A central problem in the quantum mechanical treatment of atomic observables is to find how the angular momenta of the constituent particles must be coupled in order that the energy and angular momentum are simultaneously in stationary quantum states. Because of the magnetic interactions between the particles, the individual particles will not maintain a constant direction with respect to some fixed axis and can not define a "stationary" quantum state. In a system comprised of many interacting particles, the total angular momentum of the system will always be conserved, remaining constant in magnitude and direction, like an ideal gyroscope. It may also happen that the angular momenta of particles within subsets of the total may be coupled to form conserved parts of the total angular momentum. The magnitude of such conserved angular momenta and their components along an arbitrary axis can serve to describe a stationary quantum state. Thus in one scheme of coupling angular momenta of electrons in a complex atom, called the *Russell–Saunders* coupling, the orbital angular momenta of the electrons are combined, then separately all the spin angular momenta are combined, and finally a resultant of the total orbital and spin angular momenta is obtained.

We recall that in combining angular momentum in quantum theory, we may use the vector model representation, provided that we remember that we are dealing with quantum numbers and that special quantization rules must be observed. Let us consider two examples that will be of considerable interest to us later: the alkali atoms rubidium and cesium. In their ground state, they have only one electron outside closed shells. In the ground state we are considering, this electron has no orbital angular momentum and therefore only the spin angular momentum of 1/2, with two possible components along a given axis, +1/2 or -1/2, and g=2. If this single outer electron occupies the next higher energy state, it would have an orbital angular momentum of one unit, that is, l=1, in addition to its spin. These angular momenta are not individually constant in direction, but the total angular momentum is conserved; according to quantum rules, the total can be only J=1/2 or J=3/2. Because of the relative weakness of the magnetic interactions compared to electrostatic pull of the nucleus, there is a difference in energy between these two states much smaller than would accompany a change in orbit, and this difference is therefore called the *fine-structure* splitting. It is due in this case to the spin–orbit interaction we mentioned earlier in this chapter.

The notation used by spectroscopists to designate these two states in the alkali atoms is $^2P_{1/2}$ and $^2P_{3/2}$. The letter indicates that the orbital angular momentum L=1, the superscript 2 is the value of (2S+1), where S is the spin angular momentum (in this case S=1/2), and finally, the subscripts 1/2 and 3/2 are the two values of total angular momentum J. In this notation the ground state is designated as $^2S_{1/2}$.

5.11 The Hyperfine Interaction

The electron is by no means the only fundamental particle with intrinsic spin and magnetic moment; both the proton and neutron, which are the constituents of atomic nuclei, also have these attributes. These particles have the same magnitude of spin as the electron, but since their charge-to-mass ratio is 2000 times smaller, we would expect, at least classically, that their magnetic moment is also smaller in approximately the same ratio. In fact, as with the magnetic moment of the electron, classical theory is inapplicable, but the classical moments are used as units; for the electron it is the *Bohr magneton*; here it is the *nuclear magneton*. As with the electron, the magnetic moments of the proton and neutron are expressed in terms of g-factors defined as follows: $\mu = g_n I \mu_n$, where μ_n is the classical value of the magnetic moment of a particle with the charge and mass of a proton and an angular momentum of one unit, $h/2\pi$. The measured value for the proton is about g_p=5.586 and for a free neutron g_n=-3.82. Again we see that classical theory is invalid, particularly for the neutron, which, being neutral, should have no magnetic moment at all. The question of what spin and magnetic moment a particular nucleus as a whole exhibits is a complicated one of nuclear structure, involving in general a large number of interacting protons and neutrons. The existence of a nonzero nuclear spin, which like total electronic angular momentum is limited to integral or half-integral values, further complicates the question of the angular momentum states of an atom, since the nuclear magnetic moment associated with it can interact with that of the outer electrons. Since the magnetic moment of the nucleus is so much weaker than the electron moment, it is expected that the different possible orientations of the nucleus will lead only to narrow splitting of the energy states. In recognition of that fact, the interaction between the electron and nucleus is referred to as the *hyperfine* interaction. It is precisely transitions between states separated by this hyperfine interaction that give rise to the sharp resonances used in the atomic standards in the microwave region of the spectrum. The assignment of angular momentum quantum numbers to the quantum states of an atom is very much affected by the addition of the nuclear spin, with important consequences, as we shall see, for any process involving exchange of angular momentum between an atom and radiation, for example.

It is a remarkable fact that in the $^2S_{1/2}$ ground state of the alkali atoms, the electron has zero orbital angular momentum, which classically would be interpreted as a collapsed electron orbit passing right through the nucleus; even the quantum picture is one of an electron spread out in a spherically symmetric way around the nucleus, with a finite probability of being found in the nucleus itself. A thoughtful reaction to this revelation might be, Why is there no nuclear reaction between the electron and the particles that make up the nucleus. The answer is that electron capture by the nucleus can occur in some species of nuclei; but where it is allowed, it is far more likely to involve the innermost electrons in the atom, in a process called K-capture because the innermost shell of an atom is called the K shell. Unlike s-electrons, all others in l=1, 2, 3... orbital angular momentum states have a vanishingly small probability of being in the nucleus. The spherical symmetry of the electron distribution in the $^2S_{1/2}$ state and its finite value in the nucleus have an important bearing on the computation of the interaction energy between the magnetic moment of the nucleus and that of the electron. They mean that we are not dealing with two separated magnetic dipoles, like two little magnets interacting with each other; rather it is a magnetic dipole embedded in a magnetized, spherically symmetric medium, as shown in Figure 5.5.

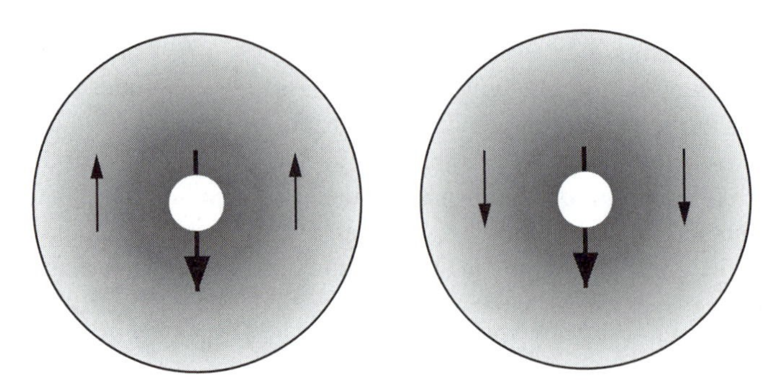

Figure 5.5 The magnetic moment of the nucleus interacts with that of the electron cloud surrounding it

The problem is to compute the amount of energy that would be required to remove the embedded magnet from the center of that magnetized medium. Classically, reversing the relative directions of the magnetization of the magnet and medium merely changes the sign of the energy, the interaction changing from one of attraction to one of repulsion; however, as we have become accustomed by now, this contradicts quantum mechanics. Simply put, the two possible angular momenta given nominally as I+1/2 and I–1/2 cannot be regarded as having the relative directions of the nuclear and electron spins reversed. Figure 5.6 illustrates the addition of an angular momentum of 5/2 with one of 1/2 according to the vector model. We see that since the magnitudes of the vectors have the form $\sqrt{5/2(5/2+1)}$ and $\sqrt{1/2(1/2+1)}$, the vectors for the angular momenta, which are nominally (5/2+1/2) and (5/2–1/2), do not have the 1/2 angular momentum in opposite directions relative to the 5/2.

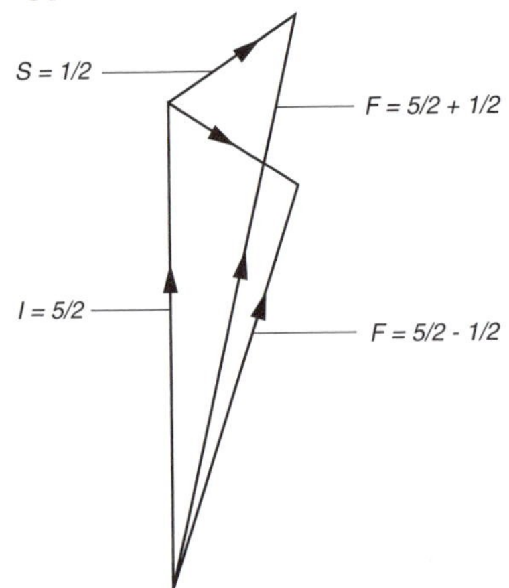

Figure 5.6 The quantum addition of angular momenta 5/2 and 1/2 according to the vector model

The quantum-mechanical solution to the problem of the magnetic interaction between a nuclear moment and an overlapping electron distribution is associated with the name of Fermi, who obtained it as an early application of what was then the new quantum mechanics. The expression he obtained for the energy, in terms of the probability density of the electron at the nucleus and the magnetic moments of the nucleus and electron, is as follows:

$$E=\left(\frac{8\pi}{3I}\right)\mu_e\mu_n|\psi(0)|^2\left[F(F+1)-I(I+1)-J(J+1)\right], \qquad 5.12$$

where $|\psi(0)|^2$ represents the electron density at the nucleus. For zero orbital angular momentum states having the same total electron angular momentum J, we can write for the energy separation between adjacent F values the following:

$$E(F)-E(F-1)=\frac{16\pi}{3}\mu_e\mu_n|\psi(0)|^2\left(\frac{F}{I}\right). \qquad 5.13$$

The application of these formulas to such complex atoms as rubidium and cesium is not expected to yield very accurate results, since many simplifying assumptions have been made; among the more serious are these: A point magnetic dipole was assumed for the nucleus, as was a single electron in an unperturbed state. Even for the hydrogen atom, where these assumptions should be far more tolerable, the drive for accuracy in the theoretical ground state hyperfine separation has led to ever more sophisticated higher-order corrections being computed. As we shall see, thanks to the hydrogen maser this hyperfine separation in hydrogen is undoubtedly the most accurately measured quantity in physics: to better than twelve significant figures! One of the early triumphs in this field was the evidence that there was an "anomaly" in the magnetic moment of the electron; the value deduced experimentally did not agree with the then most advanced relativistic theory of the electron, the Dirac theory, which predicted that the electron g-factor should be exactly 2. In fact, it was found that $g=2(1.00114\ldots)$, a number that has been the subject of precise studies by Dehmelt et al. (Dehmelt, 1981).

In the case of the rubidium atom, there are two naturally occurring *isotopes*, that is, atoms having the same electronic structure (which identifies them as rubidium) and therefore the same nuclear charge, but with a different nuclear mass because of a difference in the number of neutrons (see Figure 5.7). Natural rubidium is about 72% mass 85 with nuclear spin $I=5/2$ and 28% mass 87, which has an extremely weak radioactivity and nuclear spin $I=3/2$. If we follow the quantum rules for combining angular momentum, we will find that the ground state of Rb^{85} splits into energy levels with angular momenta equal to (5/2–1/2) and (5/2+1/2); that is, $F=2$ and $F=3$. Note that we can write symbolically $\mathbf{J}=\mathbf{L}+\mathbf{S}$ and $\mathbf{F}=\mathbf{J}+\mathbf{I}$ to represent the (vector) addition of orbital and spin angular momentum to obtain the total electronic angular momentum, and then the addition of the nuclear moment **I** to get the total conserved angular momentum **F**.

The assignment of angular momentum quantum numbers to the first energy level above the ground state in Rb^{85} is somewhat more complicated, since we have to combine J=3/2 with I=5/2 in addition to the combination of I=5/2 with J=1/2, which leads to the values we have already found for the ground state. In general, we simply write all values between $I+J$ and $I–J$, that is, F=4, 3, 2, 1.

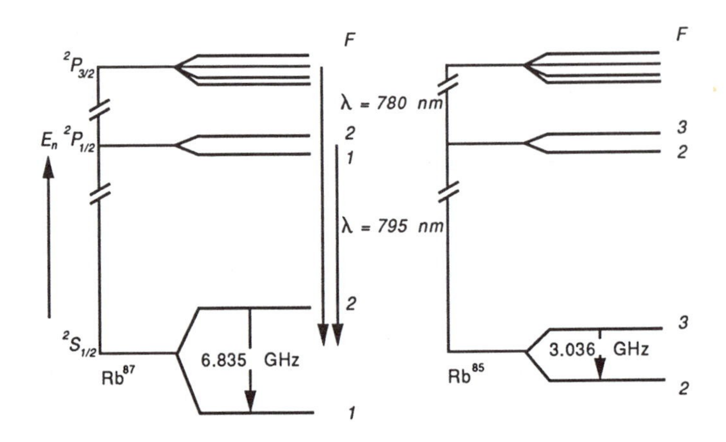

Figure 5.7 Hyperfine structure of low lying states in Rb^{85} and Rb^{87}

Similar arguments may be used to find the angular momenta for the ground state and first excited states of the cesium atom (see Figure 5.8). There is only one stable isotope of cesium, mass 133, with a nuclear spin I=7/2. Hence in the electronic ground state, which has J=1/2, the possible total angular momenta are F=4 and F=3. For the first excited electronic state, which has two electronic angular momentum states, J=1/2 and J=3/2, the coupling with the nuclear spin leads to F=4, 3 for the first J value and F=5, 4, 3, 2 for the other J value.

As already indicated, the magnetic interaction of the nuclear moments with the electrons is expected to be very small compared to the other interactions that determine the quantum energy levels of an atom. Nevertheless, it is precisely the magnetic hyperfine separations in the ground states of rubidium and cesium that have come to be distinguished as fiducial quantities, the latter defining the unit of time.

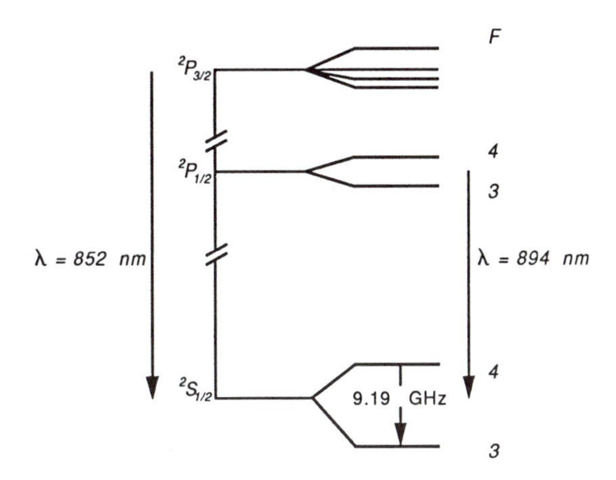

Figure 5.8 Hyperfine structure of low-lying states in Cs^{133}

5.12 Electrons in Solids: The Band Theory

5.12.1 Origin of Energy Bands

In order to understand the principles on which the operation of semiconductor lasers is based, we must review briefly the concepts underlying the theory of electrical conduction in crystalline solids. Apart from some special cases such as the interior of a battery, electrical conduction is a manifestation of the flow of electrons. The conditions, therefore, that determine to what extent a given substance can conduct electricity have to do with the extent to which electrons are able to move freely under the action of an applied electric field.

A crystalline solid is composed of atoms (or ions) arranged in a 3-dimensional array that repeats in a regular pattern. The motion of the electrons and their quantum states are no longer determined just by the electrostatic forces within each atom individually, but rather, particularly the outer valence electrons, by the interaction with all the atoms or ions in the crystal. Instead of the atomic structure problem, where electrons are more or less attracted to a central nucleus, we now have a regular 3-dimensional array of attracting centers. To see what the quantum states of the electrons should be for such an array, let us start with just two centers initially far apart being brought together to their actual separation in the crystal. Since the two-center system is symmetric with respect to an interchange of the positions of the centers, in quantum theory it follows that the wave function representing the two-atom system must be either symmetric (unchanged) or antisymmetric (only change sign) when the electron coordinates with respect to the two centers are exchanged. Initially, when the atoms are very far apart, the energy levels computed on the basis of the two symmetries are equal, and therefore the levels are the same as in the isolated atom, except that to each energy level belong *two* possible quantum states. However, when the atoms approach each other, the energies are no longer the same for the two symmetries, and the levels are split into two close levels. If now a third atom is brought into position from a large distance, it would lead to a 3-fold exchange symmetry and a consequent splitting into three levels. By extension, if N atoms are brought into position to form a crystal, the levels are split into N levels, the widest splitting coming from nearest neighbors. Since the atomic separation determines the maximum splitting, and N for even the smallest visible piece of the crystal is extremely large, on the order of 10^{19} atoms, the result is effectively a continuous band rather than a discrete multiplet. On the basis of this band structure we can now broadly draw the essential distinctions between a conductor, an insulator, and a semiconductor.

5.12.2 Conductors and Insulators

In the lowest-energy state of the system, the electrons fill all the available states, from the lowest up to the energy band that arises from electron states in the outermost shell of the isolated atom. If the last band containing electrons is only partially filled, then there will be within that band a continuum of higher-energy states available to the electrons to go into as a result of gaining kinetic energy from an external electric field, and the crystal is a conductor. For that reason the partially filled band is called the conduction band. For example, an isolated sodium atom has one electron in its outermost 3s shell, which can accommodate, according to the Pauli principle, two electrons. The band that results from this state can therefore accommodate $2N$ electrons, whereas N sodium atoms have only half this number. Therefore, sodium is a good electrical conductor; in fact, the crystal is metallic and like all metals is a good conductor. On the other hand, a crystal is an insulator if all the bands up to a certain uppermost one, called the valence band, are completely filled in the sense of the Pauli principle, and the next higher empty band is so high in energy that no electrons can reach it by thermal agitation. In this case there are no electrons in a position to go into contiguous vacant states in response to an applied electric field, and no change in electron velocity can occur. Hence no current is produced, and the crystal is an insulator.

Finally, we have what are called semiconductors, such as pure silicon, germanium, and gallium arsenide. In these the valence band is filled like an insulator, and the band above it would be empty were it not for the circumstance that it is so close in energy to the top of the valence band that at ordinary temperatures there are appreciable numbers of electrons in it due to thermal agitation. Thus because of the thermal distribution of energy among the electrons, a semiconductor has electrons in a band that would otherwise be empty at absolute zero temperature. The vacancies left behind in the valence band by the electrons that are thermally raised to the conduction band are called *holes* and act like positive electrons. This can be made plausible by thinking of the analogy of a row of seats in a theater all occupied except one; if the person next to the vacant seat gets up and sits in it, the effect is the movement of the vacancy one seat in the opposite direction to that of the person. Clearly, the number of holes left in the valence band must equal the number of electrons in the conduction band. This number depends on the temperature according to the quantum analogue to the Maxwell–Boltzmann distribution, the *Fermi distribution*, which applies to thermal equilibrium of electrons in any system. If the probability of an electron occupying a state of energy in the interval dE centered on the value E is defined as $F(E)dE$, then the distribution function $F(E)$ is a function of temperature of the form

$$F(E)=\frac{1}{\exp\left(\frac{E-E_F}{kT}\right)+1}, \quad 5.14$$

where E_F is a parameter called the *Fermi energy*. Noting that at absolute zero E_F marks the energy at which $F(E)$ abruptly changes from one to zero, we see that the Fermi energy can be described as the highest level reached if all the electrons are distributed one to each of the lowest available states. From Figure 5.9 we see that in order that the number of electrons raised to the conduction band be equal to the number of holes left behind in the valence band, the Fermi energy must be assumed to be midway in the gap between the two bands. The importance of the Fermi level for us is that when a junction is formed between two types of semiconductor, the energy levels on the two sides of the junction must adjust themselves in such a way that the Fermi levels are brought into coincidence.

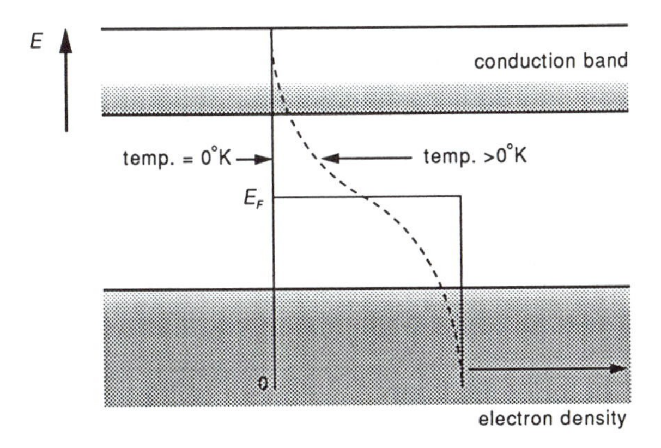

Figure 5.9 The Fermi distribution of electrons in a semiconductor

5.12.3 p-Type and n-Type Semiconductors

So far we have been considering ideally pure semiconductor crystals, the so-called *intrinsic* semiconductors, with impurities well below a few parts in a million. In fact, what made transistors possible and the solid-state revolution in electronics that they brought with them, are the technological advances in purifying and controlling the purity of these materials. By adding minute controlled amounts of "impurities" to the melt during the growth of the semiconductor crystals, a process called *doping*, the electrical conductivity of these semiconductors can be radically altered in useful ways. The result of doping is what is called an *extrinsic* semiconductor, with the number of electrons exceeding that of holes (n-type), or with a preponderance of holes over electrons (p-type).

To understand better the effects of doping, we note first that elements such as silicon and germanium have a valence of four, and they crystallize in the diamond structure in which each valence electron is shared in a covalent bond with one electron from each of four nearest neighbor atoms. These covalent bonds account for all the valence electrons, and therefore at $T \rightarrow 0$ the valence band is completely filled, while the band above it, the conduction band, is empty. Suppose now that as a result of doping, some of the lattice sites in the crystal are occupied not by an atom of the host element, but by an impurity atom with a valence of five, such as arsenic. Four of these five valence electrons will be taken up in forming the four covalent bonds, leaving the fifth electron moving in the field of the remaining ion. This electron and the other such electrons belonging to impurity atoms are more weakly bound to the ions in the crystal environment than they would be in free space and therefore are in discrete states very close to the continuum of free electron states, that is, the conduction band. These discrete states are called *donor* states, because at temperatures above zero they give up electrons to the conduction band, making the crystal n-type with a high conductivity due predominantly to electron flow. The presence of the additional donor electrons puts the Fermi level closer to the conduction band.

Suppose now that the silicon or germanium crystal is doped with an impurity having a valence of three, such as aluminum or gallium. Then where an impurity atom occupies a lattice site there will be one too few electrons to satisfy the four covalent bonds. In this case, an electron from the top of the valence band supplies the missing electron to form a negative ion and leave a hole in the valence band, which, acting like a positive electron, will have weakly bound discrete states, like the mirror image of an electron in the field of a positive ion. These states will be for negative electrons slightly above the top of the valence band, and they are called *acceptor* levels, because they receive electrons from the valence band, leaving holes there to act like positive charge carriers. The resulting semiconductor is called p-type, since the predominant charge carriers responsible for conduction are positive. With fewer electrons in the valence band, the Fermi distribution must be moved lower, with E_F closer to the top of the valence band, in order again to conform with the requirement on the electron number. Figure 5.10 shows schematically the relative positions of the boundaries of the two energy bands, the impurity levels, and the Fermi levels.

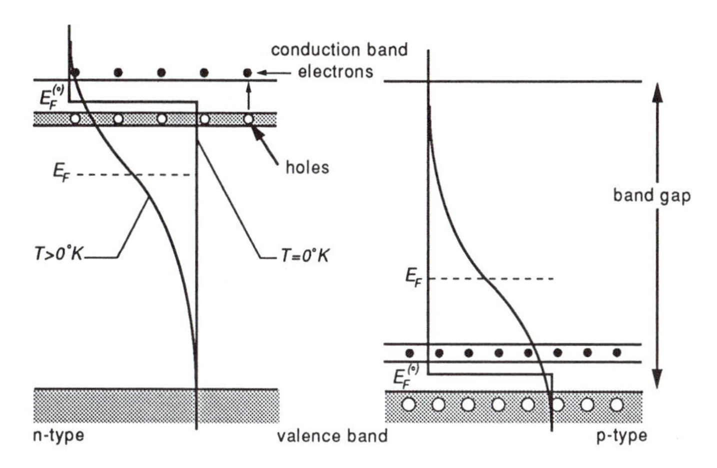

Figure 5.10 Energy bands and impurity levels in a doped semiconductor

5.12.4 Energy–Momentum Relationship

So far we have dealt only with the possible energy states of electrons in the crystal; but a complete dynamical description must include their momentum. This is necessary if we are to deal with the electron transitions accompanying the absorption or emission of radiation. In the case of radiative processes in atoms, conservation laws lead to certain selection rules determining which transitions are allowed and which are forbidden. Here the conservation of linear momentum between the electron making a transition and the photon absorbed or emitted will impose conditions on the crystalline properties that we must now address.

The problem of the motion of electrons acted on by a spatially periodic force such as they experience on an atomic scale from the atoms or ions in the crystal lattice is a quantum-theoretical problem. Their behavior is dominated by their wave nature, and rather than speak of the momentum of an electron, it is more useful to use the de Broglie wave vector $\mathbf{k}=m\mathbf{V}/(h/2\pi)$, whose magnitude is defined as $k=2\pi/\lambda$. The classical (nonrelativistic) relationship between kinetic energy $E=\frac{1}{2}mV^2$ and the wave vector for a free particle is as follows:

$$E=\frac{1}{2m}\left(\frac{kh}{2\pi}\right)^2. \qquad 5.15$$

However, motion in a periodic crystalline field is totally different; in fact, even the most essential attribute of a material particle, namely its mass, is no longer a constant. The change in kinetic energy that a force imparts to an electron, that is, its "inertia," depends on its quantum state, and the concept of an "effective mass" is introduced to frame the problem where possible in Newtonian terms. The way in which the *E–k* relationship for a free particle is modified in an ideal crystal with a lattice spacing of *a* between atoms is shown schematically in Figure 5.11. We notice the band structure and the appearance of "forbidden" gaps around the points $k=n\pi/a$, where n is a whole number. These can be given an electron wave interpretation as the inability of the electron wave to propagate through the crystal with these wave numbers because of coherent reflections from the lattice sites causing destructive interference.

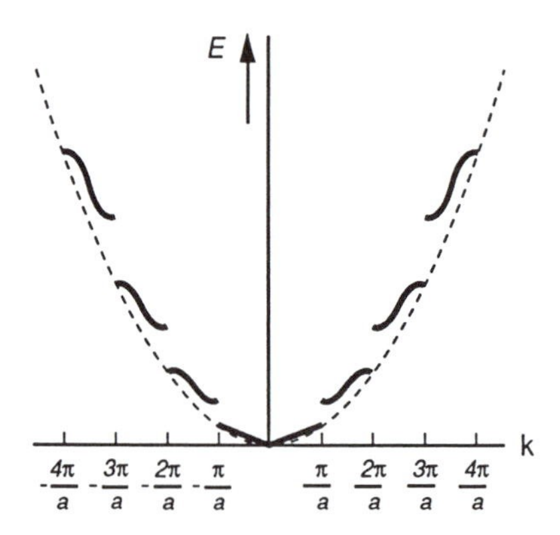

Figure 5.11 The *E–k* graph for an electron in a one-dimensional periodic field: a simple model of a crystal

In a real crystal the detailed *E–k* relationship is in general much more complicated. Figure 5.12 compares graphically the features of that relationship that are of particular relevance to us for two semiconductors: silicon and gallium arsenide. Note that the curves are for specified directions of the electron wave vector with respect to the crystal axes, since most physical properties, including electronic properties, are different in different directions in a crystal. Of particular importance is the fact that the upper boundary of the valence band for GaAs has a maximum at the same value of *k* as a minimum in the lower boundary of the conduction band, whereas for Si this is not the case. Semiconductors that are like GaAs in this respect are said to have a *direct* band-gap, while the others have an *indirect* one. We shall see later that in order that electrons may undergo radiative transitions between bands, involving the emission or absorption of photons, and do it with high probability, it is crucial that the semiconductor be a direct one.

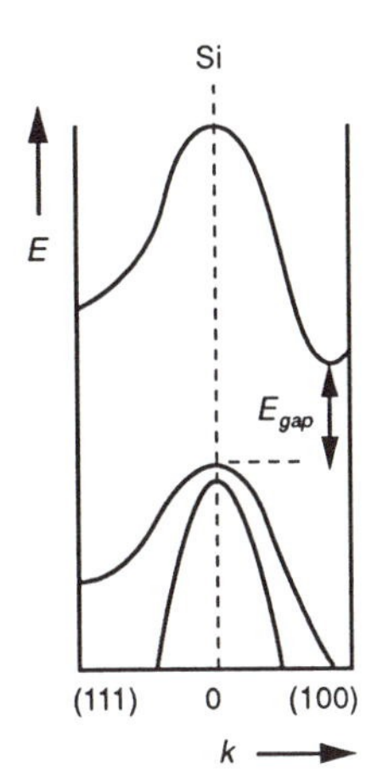

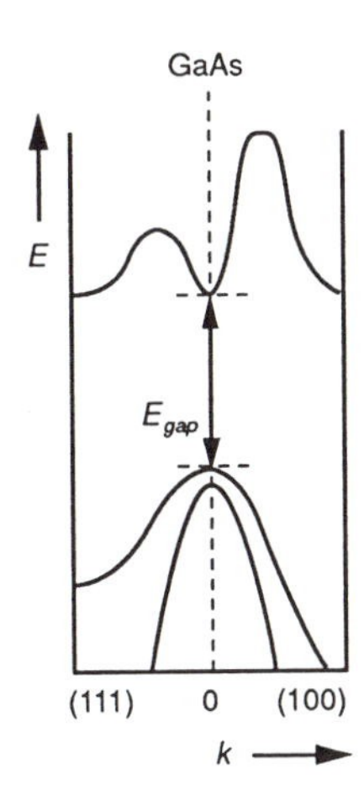

Figure 5.12 The energy–momentum graphs for silicon and gallium arsenide crystals. The indices (100) and (111) specify directions with respect to the crystal axes

Chapter 6
Magnetic Resonance

6.1 Introduction

In the evolution of clocks through the ages, there has been a progression from the use of periodic systems on a large scale with relatively slow movement to increasingly smaller, delicately operated devices running at very much higher frequencies. The large, elaborately built water clocks of China and the pendulum clocks of a later age were by the very nature of their mechanical design vastly more susceptible to environmental sources of error than the balance wheel and ultimately the quartz-controlled clock. The next step in this progression is no less revolutionary than the one from a pendulum to the quartz oscillator; it is clocks based on atomic resonators.

While the quartz resonator involves the vibrations of a single crystal, which is a kind of macromolecule, an atomic or molecular resonator involves the resonant interaction of individual atoms or molecules with electromagnetic oscillations in a microwave or light field. When the resonance occurs in the particle motion in a magnetic field, we have *magnetic resonance*, a technique that was originally applied to the measurement of the magnetic properties of atoms and their nuclei. However, as a laboratory tool, it has found important analytical applications in chemistry, and more recently in the form of (nuclear) magnetic resonance imaging as a powerful, nonintrusive diagnostic tool in medicine.

Although pure magnetic resonance in an external field is not used per se as a reference in clocks, nevertheless, as we shall see, the magnetic interactions within atoms, as well as their interaction with an external magnetic field, are very much involved in atomic resonators. It is for that reason that we will devote this chapter to magnetic resonance and the techniques used to observe it.

6.2 Atomic Magnetism

The outward magnetic properties of matter are the average manifestations of the magnetism of the constituent fundamental particles that make up the atoms of matter. As mentioned in a previous chapter, one of these fundamental particles, the electron, must be attributed with a certain intrinsic angular momentum, namely a spin of ½($h/2\pi$). Although the inescapable image of a tiny electrically charged sphere spinning like a top is not in keeping with the quantum picture, nevertheless, the classical prediction that a rotating charge should produce a magnetic field is qualitatively correct; an electron does act like a small magnet, or more formally, it has a magnetic *dipole moment.* However, the strength of that little magnet is not what a classical (nonquantum) calculation would predict for a spinning particle having the charge and mass of the electron; in fact, it has almost exactly twice that strength. If we represent the classical magnetic moment of a body having the mass and charge of an electron and revolving with angular momentum $h/2\pi$ by μ_B, called the *Bohr magneton*, then we can write for the magnetic moment of the electron $\mu = g\mu_B/2$, where g is a numerical constant yet to be determined. We note that a *classical* particle with spin $\frac{1}{2}h/2\pi$ would have, by the definition of μ_B, a magnetic moment of $\mu_B/2$, and therefore $g=1$. In the case of the electron, however, with the same spin $\frac{1}{2}h/2\pi$ the magnetic moment is μ_B rather than $\frac{1}{2}\mu_B$, that is, for the electron $g \approx 2$, showing just how far classical predictions are invalid in this connection. On the other hand, an electron moving in a closed orbit produces a magnetic field like that of a classical particle executing that motion, for which $g=1$.

In all but the simplest atoms, we have generally a large number of electrons in the outer structure as well as many nucleons (protons and neutrons) in the nucleus, all interacting—the electrons with other electrons predominantly through their electrical repulsion, the electrons and the nucleus through their electrical attraction, and the nucleons with other nucleons through nuclear forces. While the nuclear electrostatic attraction is responsible for the gross energy-level structure, it is the electrical electron–electron interaction and the *spin–orbit interaction* that can involve the orbital and spin motions of the same electron, which is responsible for the detailed energy structure and the stationary angular momentum states of the atom. The magnetic dipole–dipole interactions between particles are much weaker; however, because of their involvement in the hyperfine structure, they play a major role in the present context of atomic clocks.

As we saw in the last chapter, it is in the outermost incomplete electron shell in an atom that we find the electrons whose angular momenta may combine to yield a finite overall resultant; and it is this that can endow an atom with a permanent magnetic moment, making it a so-called *paramagnetic* atom.

6.3 The Zeeman Effect

We must now inquire into the stationary energy states of an atom having a permanent magnetic moment by virtue of its being placed in a static uniform magnetic field. Classically, there is no doubt that its interaction with the field gives it potential energy, since the torque that acts on it can clearly be made to do work in turning towards the direction of the field. By computing the work done in rotating through an arbitrary angle θ, we can show that the potential energy can be written as follows:

$$E_m = -\mu B_0 \cos\theta \,, \tag{6.1}$$

where μ is the magnetic moment; that is, the energy is proportional to the component of the dipole moment vector in the direction of the field. If that direction is taken to be the axis of quantization, then we see that the energy depends on the component of the magnetic moment, and therefore angular momentum, along that axis.

It happens that as long as the applied magnetic field is a weak perturbation compared with the electron interactions *within* the atom, (in the case of the spin–orbit interaction this will remain true even for relatively intense fields), the stationary quantum states are still correctly described with the same set of quantum numbers we have previously introduced; the only change is that states with different components of angular momentum along the field axis will have different energies. That is, the $(2J+1)$ substates with m_J=$-J$, $-(J-1)$, $-(J-2)\ldots+(J-2)$, $+(J-1)$, $+J$ that overlap in energy (and are called *degenerate*) in the absence of a magnetic field will now be separated in energy by an amount that varies with the intensity of the magnetic field. The extent of this splitting also depends on the ratio between the magnetic moment and angular momentum, called the *gyromagnetic ratio* γ, for the particular quantum state. This ratio can be written in terms of an effective g-factor, which for an atomic state is referred to as the *Landé factor*. Since in general, for an atom, the total angular momentum may be the resultant of both spin and orbital types, for which, as we saw, the g-factors are different, the Landé factor will depend on the angular momentum quantum numbers of the given state. In terms of the Landé factor $g(L, S, J)$, the energies of the substates having different components m_J of the total electronic angular momentum J are as follows:

$$E_m = -m_J g(L,\ S,\ J)\mu_B B_0 \,, \tag{6.2}$$

justifying the designation *magnetic* quantum number for m_J. We note that since the energy is proportional to m_J, which is incremented in equal unit steps, the effect of the magnetic field on the energy levels of an atom is to split them into $2J+1$ equally spaced magnetic sublevels.

It follows that in the presence of a magnetic field, what was one energy level becomes a complex of several levels, from which the atom may make optical transitions to other similar complexes at lower levels, as shown in Figure 6.1 for an alkali atom such as rubidium. Under high spectral resolution, the lines in the spectrum of the emitted light will for most atomic species be seen to be split into several closely spaced lines, a phenomenon first noticed by Zeeman around 1896 as a broadening of the lines in the spectrum of light from a sodium flame when placed between the poles of a magnet. The effect was ultimately resolved as a splitting of spectral lines, now called the *Zeeman effect.* Lorentz applied his now classical theory of the electron to explain the effect, with only partial success; the use of the unique classical value of g=1 for all electron states will always lead to a splitting into just *three* lines, corresponding to a change in m_j=+1, −1, or 0. In reality, the effect in the spectra of many atoms exhibits a far more complex pattern, inexplicable according to classical theory; hence such cases were dubbed *anomalous.* Attempts to explain this "anomalous" behavior ultimately led to the discovery of electron spin and the assignment of the non-classical value of 2 for its g-factor.

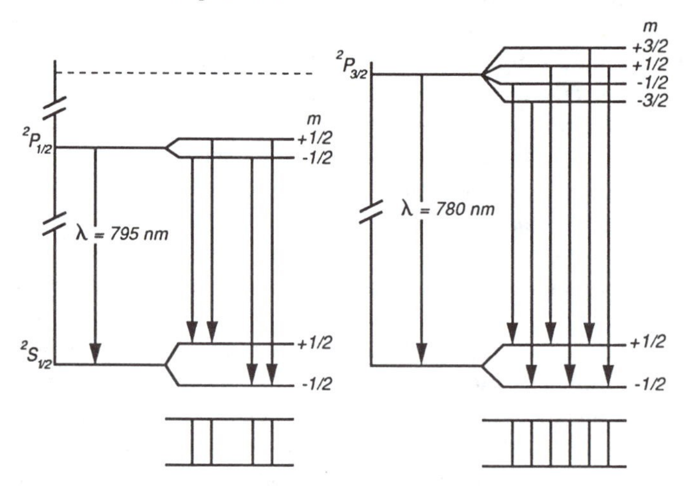

Figure 6.1 The Zeeman effect in the rubidium atom

Of all the initial and final magnetic substates between which we might consider possible transitions to occur, involving the emission or absorption of light, only those will occur with any significant probability that satisfy certain conditions on their quantum numbers, called *selection rules.* The selection rules depend on the mode of vibration within the atom or molecule giving rise to the emission of radiation. We shall limit ourselves to what is called *electric dipole* radiation, which may be pictured as being produced by a linear oscillation of the negative electronic charge in the atom relative to the positive nucleus. An atom can make a transition from a state with energy E_1 and angular momentum quantum numbers (L_1, S_1, J_1, m_1) to another energy state E_2 with quantum numbers (L_2, S_2, J_2, m_2) by radiating one quantum of radiation of frequency $\nu=(E_1-E_2)/h$, provided that the following selection rules are obeyed:

$$L_1 - L_2 = \pm 1;\ S_1 - S_2 = 0;\ J_1 - J_2 = 0\,,\ \pm 1;\ m_1 - m_2 = 0\,,\ \pm 1. \qquad 6.3$$

In complex atoms it often happens that transitions occur between states that do not conform to these selection rules; this arises because in such complex structures the assignment of quantum numbers may be an approximation. These selection rules are arrived at from a computation of the transition rate, or more precisely, the probability that a given atom will undergo an electric dipole transition in unit time, which is a function of the quantum numbers of the initial and final states. It is found that the probability of such a transition taking place is zero unless these selection rules are obeyed. The physical basis for the condition on the orbital quantum number L is rather subtle; it has to do with a symmetry property of the initial and final atomic states of the atom. The condition on the spin angular momentum S states that the process giving rise to this type of radiation cannot affect the total spin. The conditions on J and m have to do with the conservation of angular momentum in the atom–photon system; the radiated photon carries away one unit ($h/2\pi$) of angular momentum, and this combined (vectorially) with the final J value must give a resultant equal to the initial J value. Similarly, the component of angular momentum along any given axis must be conserved, and again the final combined value must equal the initial value. It is this last selection rule that is of special interest for the Zeeman effect: $m_1 - m_2 = 0, \pm 1$; it severely limits the number of possible transitions between the two states.

Transitions in which $m_1 - m_2 = 0$ produce radiation in a pattern similar to that emitted by a simple radio broadcasting antenna, consisting of a straight vertical rod carrying a high-frequency current. On the other hand, the radiation pattern of transitions in which $m_1 - m_2 = \pm 1$ resembles that emitted by a circular loop antenna, in which a high-frequency current is induced. Figure 6.2 illustrates these radiation patterns using the common practice of representing intensity versus angle in a polar diagram.

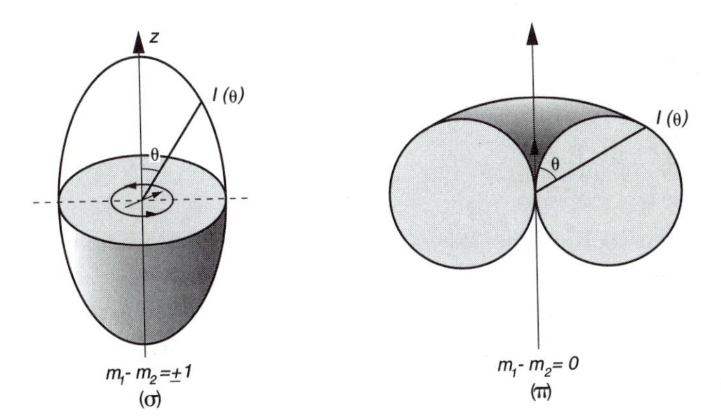

Figure 6.2 Radiation patterns of σ- and π-radiation

6.4 Gyroscopic Motion in a Magnetic Field

We shall now take up the subject of magnetic resonance on systems of free paramagnetic atoms. Some of the fundamental ideas are also applicable to nuclear magnetic resonance in condensed forms of matter. Let us assume that we have paramagnetic atoms placed in a uniform static magnetic field, such as might be produced between the poles of a suitable magnet. We will assume at first that the atoms are free to move without colliding with each other or other particles.

It is fortunate that we can obtain a sufficiently valid description of the motion of such free atoms in a magnetic field using the vector model, and we will be able later to draw a correct correspondence with the proper quantum treatment. After the magnetic field has been established, we can picture the atomic dipole moments, like little compass needles, experiencing a torque tending to turn them towards the direction of the field.

If it were not for the angular momentum associated with the magnetic moments of the atoms, they would simply swing back and forth, again like the needle of a magnetic compass. However, such a torque acting on a spinning body will produce a gyroscopic motion; thus the atoms will precess around the magnetic field direction in such a manner that their angular momentum will sweep out a cone with the field as axis, as shown in Figure 6.3. In this case the torque is proportional to the field strength and the magnetic moment, and hence the angular momentum; but by Newton's laws of motion, the amount of torque needed to produce a certain rate of precession is also proportional to the angular momentum. It follows that the rate of precession depends only on the field strength, and not the amount of angular momentum. This argument is rendered considerably more lucid stated mathematically; thus using the conventional symbols, let $\mathbf{B}_0$ represent the static uniform magnetic field, μ the magnetic dipole moment, and $\mathbf{J}$ the associated angular momentum. Now the torque, which we will represent by Γ, acting on the magnetic dipole in our magnetic field is given in vector notation by $\Gamma=\mu\times\mathbf{B}_0$, and by Newton's laws $d\mathbf{J}/dt=\Gamma$; hence we can write the following:

$$\frac{d\mathbf{J}}{dt}=\mu\times\mathbf{B}_0. \qquad 6.4$$

But μ is proportional to $\mathbf{J}$; hence we can write $\mu=-\gamma\mathbf{J}$, where we have introduced a minus sign because the electron is negatively charged and μ and $\mathbf{J}$ are therefore in opposite directions. The quantity γ is called the *gyromagnetic ratio.* Recalling that the vector product obeys $-\gamma\mathbf{J}\times\mathbf{B}_0=\gamma\mathbf{B}_0\times\mathbf{J}$, we can rewrite the equation as follows:

$$\frac{d\mathbf{J}}{dt} = \gamma \mathbf{B}_0 \times \mathbf{J} . \qquad 6.5$$

Now, $d\mathbf{J}/dt$ for an angular momentum precessing uniformly around a fixed axis at a constant angular velocity ω can be shown to be simply $\omega \times \mathbf{J}$, and hence we see that such a constant precession satisfies Newton's equation of motion, provided that we put $\omega = \gamma \mathbf{B}_0$. Thus we are led to the conclusion that the effect of a magnetic field is simply to cause all systems with the same γ value to precess with the same angular velocity about the field axis, independent of their detailed structure or their initial orientation relative to the field direction. Hence for a system of identical particles, the motion in the magnetic field is indistinguishable from what it would be in the absence of the field, but referred to a set of coordinate axes rotating uniformly with an angular velocity of $-\gamma \mathbf{B}_0$. This result is contained in *Larmor's theorem* of classical theory.

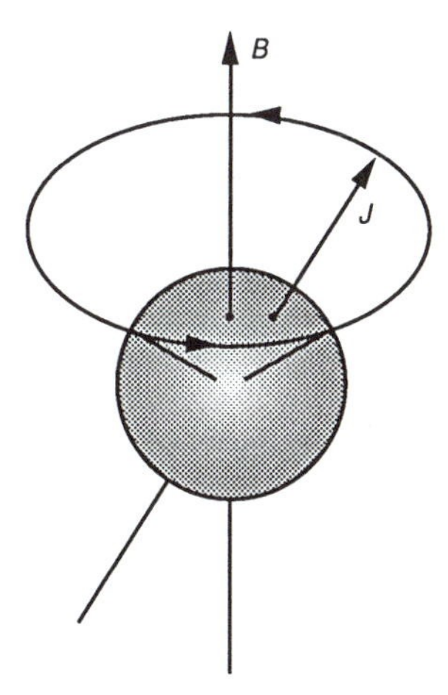

Figure 6.3 A precessing atomic moment according to the vector model

6.5 Inducing Transitions

Let us now take up the phenomenon of magnetic resonance in the simple case of a free paramagnetic atom in a uniform magnetic field. We will assume that J=1/2 so that there are two possible directions for the angular momentum with components along the field direction given by m_J=+1/2 and m_J=−1/2. Let us assume that initially the atom was somehow put into the state with m_J=+1/2; this would require a deliberate physical selection of that state, since in a "natural state" there would be no more reason for an isolated atom to be in one state or the other. We would like to show that if a weak oscillating magnetic field is suitably applied to our atom, we could cause the spin to flip to the opposite direction, that is, into the $m_J=-1/2$ state. To do this we shall use classical theory, confident that results based on the precession of the atom and Larmor's theorem are also valid in quantum theory.

We must digress a little, however, to think about ways in which an oscillating field might be applied to the atom to induce the supposed transition. This devolves around the question of the polarization of the oscillating field. If a vibration takes place along a straight line, or a magnetic field (vector) oscillates along a fixed direction, it has *linear* polarization. If, on the other hand, the magnetic field keeps a constant magnitude but its direction rotates at a constant angular velocity, we say it is *circularly* polarized. Clearly, there are two possible senses (clockwise or counterclockwise) for the rotation, which can be unambiguously stated only with reference to a specified direction along the axis of rotation. According to convention, circular polarization is called right-handed or left-handed according as the field vector rotates clockwise or counterclockwise as seen by someone looking in the direction *opposed* to the direction in which the wave is traveling, as shown in Figure 6.4a. It is interesting to note that an electromagnetic wave not only carries linear momentum, about which we shall have a great deal more to say in a later chapter, but, when circularly polarized, *angular momentum* as well. In terms of the quantum of radiation, the *photon*, this angular momentum is an intrinsic property of each photon amounting to $h/2\pi$. A material particle having this spin would, according to quantum theory, have three possible components along any given axis; the photon is unique in that it has only *two*, corresponding to the types of circular polarization, right-handed ($-h/2\pi$) and left-handed ($+h/2\pi$).

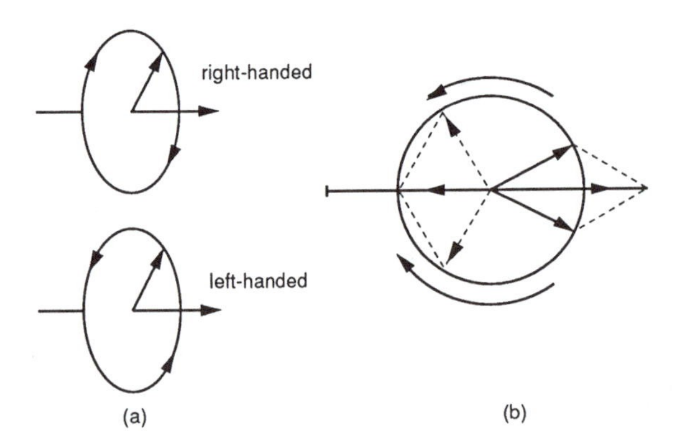

Figure 6.4 (a) Circular polarizations (b) the sum of counter-rotating circular polarizations

Now we can prove a useful relationship between these three types of polarization: If we add two equal but oppositely rotating circularly polarized fields of the same frequency and *coherent in phase*, the result is a linearly polarized field. Figure 6.4b should help convince us of the truth of that statement. The components along the vertical axis will always be equal and opposite and therefore cancel each other, leaving only the horizontal components to add to double the amplitude. Conversely, a linearly polarized oscillation can always be resolved into two equal counter-rotating circularly polarized components.

We are now ready to consider the possibility of inducing a transition from $m_J=+1/2$ to $m_J=-1/2$. Since the two spin states correspond to more or less opposite directions with respect to the field, to bring about such a transition clearly needs a torque to act on the spin; a weak magnetic field at right angles to the main uniform field would create such a torque. Obviously, a static perpendicular field would merely combine with the main field to give a slightly tilted resultant about which the spin would precess, without causing any spin flip. To be precise, the stipulation that the perpendicular field be "static" not to cause transitions means that it must be established slowly compared with the rate of precession; otherwise, a sudden change in the field *could* cause a transition from one spin state to the other, although obviously not in a resonant manner.

What is needed is a small field whose direction precesses in the same direction and at the same frequency as the spin, since then the field and spin would keep their relative directions constant as they both precessed around the main field. But from our digression above this is simply a circularly polarized field rotating in the same sense as the precession, and it can be generated as one of the two counter-rotating components of a field having linear polarization. The other component rotating in the opposite sense would have only a secondary effect on the spin, which we will ignore. Let us assume then that an oscillating magnetic field is applied perpendicular to the main field, resulting in a component of magnitude B_1 rotating in the same sense as our spin with a frequency ω. It simplifies the analysis considerably to imagine turning with the rotating field vector around the axis; that is, refer the motion to a rotating coordinate system, with respect to which the $\mathbf{B}_1$ field is stationary. According to Larmor's theorem, the motions with respect to such a system are indistinguishable from one subject to a magnetic field B_r given by $B_r=-\omega/\gamma$, so that the total axial field in the rotating coordinate system is $B_0-\omega/\gamma$. On adding this vectorially to the transverse field B_1 we get the result,

$$B_{\text{eff}} = \sqrt{\left(B_0 - \frac{\omega}{\gamma}\right)^2 + B_1^2}\,, \qquad 6.6$$

as illustrated in Figure 6.5. The spin will precess about this resultant as axis with an angular velocity $\omega_{\text{eff}}=\gamma B_{\text{eff}}$. If the frequency ω is chosen to equal γB_0; then the spin will precess around the direction of B_1, continuously going from $m_J=+1/2$ to

m_J=−1/2 and back. In the laboratory frame of reference the spin vector sweeps out a cone whose apex angle increases until the cone becomes the median plane at 90°, then continues as a cone in the opposite direction.

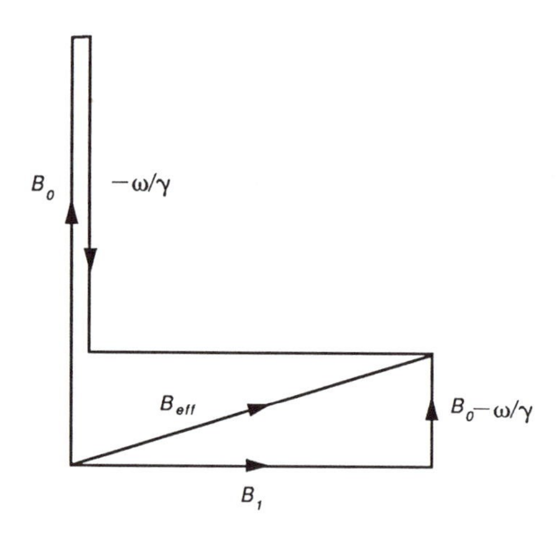

Figure 6.5 The resultant magnetic field in the rotating frame

In a quantum sense the transverse field rotating at the frequency γB_0 causes the initial wave function representing the state with m_J=+1/2 to evolve into one consisting of a *linear superposition* of the two spin states with m_J=+1/2 and m_J=−1/2. This is a characteristically quantum feature in which, in a sense, the description of a system can include simultaneously more than one possible state. It is analogous to the coupled pendulums going back and forth between two modes of oscillation: During the transition it is in neither one nor the other. It is not expected that the motion would stop after the first transition from +1/2 to -1/2 had been completed. However, if a mechanism is present, such as randomizing collisions with other particles, that enables the system to "relax" into thermal equilibrium, then a stationary state is possible. The frequency given by $\omega=\gamma B_0$ is called the magnetic resonance frequency; an offset from this frequency would cause the spin direction to sweep out a cone (in the rotating frame) whose axis is tilted from the perpendicular direction, and the spin does not quite reach the negative field direction.

It remains to make the correspondence between the classical and quantum descriptions correctly. To find the quantum theory probabilities of the spin being observed in one or the other of the two states, we can require that the average component of spin along the direction of the main field be the same as the classically computed component. If we call the angle between the spin direction and the main field at a given time θ, then its component along the field is $(1/2)\cos\theta$; on the other hand, if $P(1/2)$ and $P(-1/2)$ are the probabilities that the spin is in the +1/2 and −1/2 states, respectively, then we must have

$$(+1/2)P(+1/2)+(-1/2)P(-1/2)=(1/2)\cos\theta . \tag{6.7}$$

Since the spin is certain to be in one state or the other, we must also have

$$P(+1/2)+P(-1/2)=1. \tag{6.8}$$

Hence finally we have for the probability of finding the spin in the −1/2 state the following:

$$P(-1/2)=(1/2)(1-\cos\theta). \tag{6.9}$$

At resonance, θ oscillates between 0 and 180 degrees, so that the probability of finding the spin in the −1/2 state oscillates between 0 and 1. Off resonance θ does not reach 180 degrees, so that P(−1/2) never reaches the value 1.

6.6 Motion of Global Moment: The Bloch Theory

We should notice a fundamentally important feature of the argument we have just developed for the probability of inducing transitions, a feature inherent in all quantum transition probability calculations. There is a symmetry between the initial and final states, in that it is immaterial which state we assume as the initial state; in either case the probability of a transition to the other state is the same. Stated more broadly, if a process has a high probability of proceeding one way, it will have an equally high probability of proceeding the opposite way. This imposes an important condition on our ability to observe any large-scale manifestation of transitions occurring in atoms in a large group, as in a volume of gas. Thus if the atoms in a group are assumed to be equally likely to be in one magnetic state as the other, then transitions induced in the manner we have described will not alter the number in each state but merely exchange states of particular atoms, to which we are insensible. It follows that in order to detect the transitions on a group of atoms, they must initially be prepared with a preponderance of atoms in one state or the other. Such a group of atoms is said to be "polarized," with a net global magnetic moment that can easily be calculated if we know the probabilities $P(1/2)$ and $P(-1/2)$; it is given by

$$M = N\mu[P(+1/2)-P(-1/2)], \tag{6.10}$$

where N is the number of atoms in the group. This is reasonable, since $NP(1/2)$ and $NP(-1/2)$ are respectively the average numbers of atoms with their magnetic moments pointing with and against a fixed direction, such as defined by a magnetic field.

In the presence of a resonant oscillating magnetic field, because of the coherent response of the individual atoms, the global polarization vector for a large group of atoms will execute a motion not unlike a single atom. A theory due to F. Bloch, originally developed to explain the dynamic behavior of magnetic moments undergoing magnetic resonance in solids, can, by drawing the correct correspondence with quantum theory, adequately describe our system. In it an equation of motion is set down for the mean global magnetic moment, which is basically similar to the one given earlier for a free atom whose solution yields gyroscopic motion. The theory is described as *phenomenological* in that it is formulated in terms of global average parameters that are directly *observable*; it is characterized by the inclusion of *relaxation* terms that account for the time decay of components of the global vector due to random perturbations. Two characteristic *relaxation times*, denoted by T_1 and T_2, are necessary. The term "relaxation" derives from the original application of the theory to *nuclear* magnetic resonance, in which nuclear magnetic polarization is achieved by the system reaching (or "relaxing" to) thermal equilibrium in a strong external magnetic field. The theory defines the first, T_1, called the *longitudinal* relaxation time, as the mean decay time of the global polarization vector, without regard to any particular mechanism, such as collisions, causing the decay. The other time, T_2, called the *transverse* relaxation time, is the mean decay time of any precessing global polarization of the atoms, such as may be induced by an oscillatory field resonant with the precession frequency. Under the action of such a field, the magnetic moments of individual atoms tend to move in concert, driven by the common magnetic field, producing a net global precessing moment. Any mechanism that causes random fluctuations in the *phase* of the response of the atoms or depletes the number of atoms contributing to that response will reduce T_2.

It can be shown that Bloch's theory, introduced in the context of a vector model description of precessing magnetic moments, can be equally useful in a quantum description, in which the reorientation of a magnetic moment really amounts to transitions between (the magnetic) quantum states. In fact, if the system has only two states between which transitions may be induced, its behavior corresponds to a spin 1/2 particle with its two possible orientations with respect to a given axis.

6.7 Production of Global Polarization

The various techniques for observing magnetic resonance in gases or condensed forms of matter are distinguished essentially by the means used to achieve a polarization in order to render the resonance transitions observable. Other factors such as the frequency range and sample density further differentiate the various techniques.

6.7.1 Thermal Relaxation in Strong Magnetic Fields

The method of producing a polarization common to all magnetic-resonance studies on condensed matter, that is, solids or liquids, whether nuclear or electron resonance, is through the use of very intense magnetic fields. As already pointed out, the different magnetic substates, having different orientations with respect to the external magnetic field, will have different energies, which is called the Zeeman effect. For atoms in a state of thermal agitation and exchanging energy through their mutual interaction, these differences in energy lead to different probabilities for the atoms to be in those substates. All that is necessary is that the interaction between the atoms be able to cause transitions between the different magnetic substates, and hence thermal equilibrium to extend to these states. The higher the energy of a substate, the smaller will be the probability of an atom reaching that state through exchanges of energy with other atoms. A fundamental result in statistical mechanics, due to Boltzmann, is that in thermal equilibrium at (absolute) temperature T, the ratio of the number of atoms having energy E_1 to that having energy E_2 is given by

$$\frac{N(E_1)}{N(E_2)} = \exp\left[-\frac{(E_1 - E_2)}{kT}\right], \qquad 6.11$$

where k is Boltzmann's constant, which has the value $1.38\text{x}10^{-23}$ in the MKS system of units. If we substitute the numerical values for the magnetic energies typical for the field produced by a large laboratory electromagnet, we find that $(E_1-E_2)/kT<<1$ at ordinary temperatures. This shows that the populations of the magnetic states are very nearly equal; that is, the polarization is exceedingly small. The use of high fields to attain polarization of a sample is therefore useful only for condensed forms of matter, in which a sample of reasonable size can contain a sufficient number of atoms to detect the resonance; even then, the highest possible fields, and therefore resonant frequencies, are used to improve the sensitivity. The strength of the field produced by a conventional electromagnet using copper windings is ultimately limited by the amount of electrical power required to maintain the current against the electrical resistance of the windings. The use of superconducting magnets solves the power problem but requires extremely low temperatures to be maintained, incurring a different kind of problem and expense.

To understand the common method used to detect the occurrence of transitions, we must direct our attention to the effect of interactions between the spins of neighboring atoms and possible small differences in the intensity of the main magnetic field at the site of different atoms. We recall that interaction is necessary to achieve thermal equilibrium and hence a certain degree of polarization; but it can also disturb the precessional motion and thereby change its phase, or even cause spin flips, at randomly distributed intervals. Also, in practice the magnetic field may differ slightly from point to point, and therefore the oscillating transverse field, applied to induce transitions, cannot be on exact resonance with all the atoms. The effect of this last circumstance is easy to predict; it simply means that in a solid sample in which the particle positions are fixed, different atoms are brought into resonance as the frequency of the oscillating field is swept, giving what is called an *inhomogeneously* broadened spectral line. One effect of real transitions caused by the spin interaction is only slightly more difficult to predict; it tends to counter the action of the oscillating field by redistributing the populations of the states in the direction of restoring thermal equilibrium. Clearly, if the oscillating field is to have a detectable effect on the polarization, which constitutes our "signal," it must favorably compete with this thermalizing effect. Again a broadening of the resonance frequency occurs; this time because the oscillating field is only allowed a finite time to act coherently on the precessing spin. As we saw in a previous chapter, this means that the frequency of the oscillating field can differ slightly from the precession frequency and still cause transitions; the longer the coherent interaction, the closer the frequencies must be to avoid their getting out of phase during that time. Finally, the interaction between the spins may affect each other's phase and frequency of precession and hence again broaden the resonance frequency response.

6.7.2 Deflection of Atomic Beams

The second important technique for achieving a polarization of atomic and nuclear moments applies to materials in a gaseous form or that can be suitably vaporized. Such materials can, by a suitable nozzle and series of apertures, be formed into a fine jet, called an atomic or molecular *beam*, which in passing between the pole pieces of a special magnet will fan out into a number of separate components each according to its magnetic state. A particular component in the chosen magnetic state can then be isolated using beam stops.

It is interesting that the origins of atomic beams go back to around 1911, early in the history of vacuum technology, when an account was published by Dunoyer of an experiment using the apparatus shown schematically in Figure 6.6, in which sodium vapor issued from a small opening in a heated reservoir of sodium metal at one end of a glass vacuum system equipped with apertures and a cooled surface at the opposite end. From the distribution of the sodium deposited on the cold surface, it was clear that under sufficiently high vacuum conditions the sodium atoms traveled in straight lines like rays of light, in the form of a beam. The production of an atomic beam is based on the fact that particles in a rarefied gas actually travel relatively great distances between collisions with other particles; just how far they travel depends, of course, on the pressure, or the number of atoms occupying a given volume. The distance of free travel by any particular particle varies randomly, but there is an average called "the mean free path"; for example, at a pressure of 10^{-2} Pa (1 mm Hg=133.3 Pa), air molecules have a mean free path of about 45 cm. This pressure is attainable with a good mechanical rotary vacuum pump; pressures below 10^{-4} Pa are obtainable with a molecular diffusion pump. An atomic beam apparatus then is characteristically a vacuum chamber where the pressure is maintained below 10^{-4} Pa, with a smaller chamber within it containing the material and serving as the beam source, which may be a small high-temperature oven, if the material has to be vaporized by heating it. The oven is usually provided with a narrow slit or a bundle of microchannels through which the atoms or molecules stream out to be further limited by a coaxial aperture to select a well-defined beam.

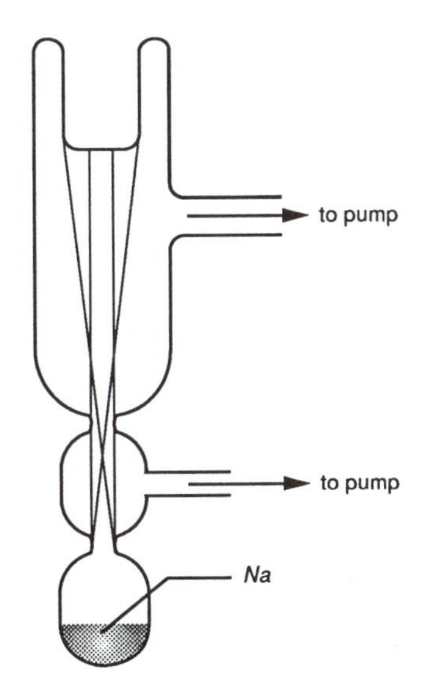

Figure 6.6 Apparatus used by Dunoyer to study the formation of atomic beams

An early exploitation of atomic beams was by optical spectroscopists to reduce Doppler broadening of spectral lines. Since the atoms in a beam have been selected to have velocity directions only in a narrow cone, the absorption or emission of a light wave perpendicular to the beam axis will not, to a first approximation, have any Doppler shift in frequency.

However, a more significant application of atomic beams was the classic Stern–Gerlach experiment. In this the particles in the beam are made to pass at right angles to the field of a special magnet, whose pole shapes are designed to produce a steep gradient in the field strength, as shown in Figure 6.7. This gradient in the field strength translates into a gradient in the magnetic potential energy of the different magnetic substates; this is analogous to rocks placed on the sides of hills of different slopes. The result is that particles in different magnetic substates are deflected by different angles from the initial direction of the beam. Obviously, in order to achieve good separation, the beam must have a very narrow divergence. This method of state separation was first demonstrated by Stern and Gerlach around 1923 in an experiment that ranks as a milestone in the history of physics. As a result of passing through their magnet, the silver beam was split into two distinct beams as quantum theory predicted. The ground state of a silver atom is designated as $^2S_{1/2}$, with J=1/2, and only two possible orientations with respect to the field axis, corresponding to m_J=+1/2 and m_J=−1/2. Classically, of course, the atomic magnetic moment could assume any orientation with respect to the field, and the beam would simply have been smeared out; the fact that it was split into just two components showed for the first time the quantum phenomenon of space quantization: that an angular momentum can be observed only with discrete directions relative to a given axis.

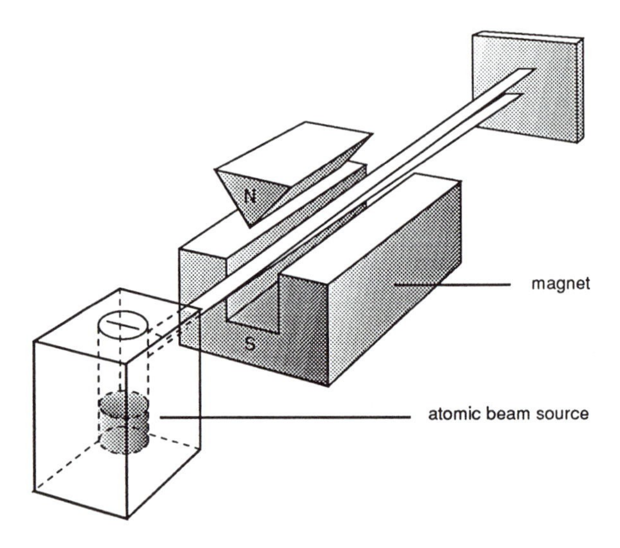

Figure 6.7 The Stern–Gerlach apparatus to show space quantization in a beam of silver atoms

This was followed by the introduction of magnetic resonance on atomic beams in Rabi's laboratory just prior to the Second World War. However, it was shortly after the war that Ramsey introduced the technique of applying the resonant field at two separated points along the beam, a development that made possible the ultimate adoption of the Cs clock as the primary standard.

In a Rabi-type magnetic resonance beam apparatus, the beam that emerges from the high-gradient magnet will have the atoms in the various Zeeman substates fanning out in slightly different directions; and by suitable apertures, particles predominantly in certain substates are selected, in effect producing a polarized beam. The beam then enters the resonance transition region, comprising an extended uniform magnetic field and high-frequency current loops, to produce the oscillating magnetic field. Finally, in order to detect whether transitions have occurred in the transition region, the beam is made to pass through another high-gradient magnet, acting this time as an "analyzer." There are two possible ways in which to analyze the beam: The analyzer magnet and the detector that follows it may be disposed either such that (1) particles that do make a transition are detected (flop-in type), or (2) particles that do *not* make a transition are detected (flop-out type).

The essential difference between the study of atoms in a rarefied beam and a condensed form of matter shows itself fundamentally in the factors that determine resonance line-width. In a beam, line-broadening interaction between atoms occurs only during collisions, which can be rare at low pressure; but the duration of the resonant interaction of the atoms with the oscillating field is limited to the transit time of the particle through that field region. There are only two factors that determine the transit time: the speed of the particles and the length of the uniform field magnet. To reduce the former requires very sophisticated laser techniques (which will be touched on in a later chapter), and there is a practical limit to the extent we can increase the latter. In a later chapter devoted to the primary Cs beam standard we will describe the classic separated transition field design due to Ramsey, which was crucial to achieving narrow resonances in an atomic beam. We will also encounter, in connection with the hydrogen maser, a type of magnetic state selector due to Paul that is radically different from the Stern–Gerlach magnet: It is one that selectively *focuses* rather than simply deflects atoms according to their magnetic substate.

6.7.3 Optical Pumping

The last type of magnetic resonance technique we shall treat is one based on the transfer of polarization from an optical beam to atoms in a process called *resonance fluorescence*. It is a technique that originated in the laboratory of the French spectroscopist Kastler, a technique that has proved extremely fruitful in its applications, not only to atomic timekeeping, but also to magnetometry. To understand the principles underlying this technique, we must digress briefly to consider the interaction of polarized light with atoms.

If an atom experiences a collision with a high-speed electron, as in a neon tube, or with another atom, as in a flame, it may be excited to one of the quantum states above its ground state. However, only if it happens to find itself in a special *metastable* state will it remain excited for long; within possibly less than a fraction of a microsecond it will spontaneously radiate as it makes transitions towards lower-energy states in a cascade fashion. The radiative process is described as spontaneous, since it occurs even in an isolated atom, although according to quantum theory, even in "vacuum" the electromagnetic field is not absolutely zero but has *zero point* oscillations that induce an atom to make a transition to a lower energy state. In doing so, of course, it emits a photon, and the resulting cascade produces an emission spectrum consisting of discrete lines characteristic of the particular atom; hence the use of *spectrum analysis* as an analytical tool in chemistry. As we learned earlier, if the environment of the atoms is such that there is a strong buildup of only a few modes of vibration of the light waves as in a laser, then stimulated emission, which is negligible under "normal" conditions, can become significant; but as we have seen, that requires very special conditions to obtain.

Of particular interest to us in laying the background for an understanding of the optical method of observing magnetic resonance is the polarization of the light emitted in different directions. We have already encountered the term *polarization* in connection with the field necessary to induce magnetic resonance; it is one that is used in several contexts in physics: We have already used it once to mean an unequal population of spin directions relative to a magnetic field. The term also applies to the displacement of the positive charge relative to the negative in individual atoms, as when an electric field is applied to a dielectric material, such as glass. The term is most familiar in the optical context through the common use of polarization filters as sunglasses. It refers to the degree to which the electric (and associated magnetic) field components in a light beam are coherent *in direction*. The question of polarization arises only because in free space, the fields in a light beam lie in a plane perpendicular to the direction of the beam, and may have any angle around the beam as axis; that is, the waves associated with photons are *transverse* waves. An unpolarized beam, such as one from an ordinary light bulb, has the electric (and magnetic) fields oscillating in random directions in any transverse plane, so that if the fields are resolved into any pair of perpendicular components, the intensities of these components would be equal. There are only two pure polarization states a photon can have: right-handed and left-handed circular polarizations. We have already mentioned these states of polarization in connection with the oscillating field required to observe magnetic resonance. Similarly, a photon in a left-handed polarization state is one observed to have the electric (and magnetic) field directions rotate in a counterclockwise direction if we face opposite the direction of travel of the photon; and the reverse is true for right-handed polarization. The other common polarization is called linear or plane polarization, in which the field oscillates along a unique direction, which lies in a plane drawn through the beam axis. It may be regarded as a (linear) superposition of equal and coherent right- and left-handed circular polarizations. This should be apparent from the result obtained in an earlier chapter that simple harmonic motion along a straight line can be regarded as the projection of uniform circular motion. It does not mean that we can take any two oppositely polarized light beams and produce a linearly polarized beam by combining them; it will work only if the two waves are coherent, that is, have a well-defined common phase.

The two circular polarizations correspond to the photon intrinsic angular momentum of $h/2\pi$ pointing with and against its direction of travel, respectively. The experimental confirmation of this was established early in the history of quantum theory by observing the mechanical torque produced on a delicately suspended quartz plate when circularly polarized light is allowed to fall on it. It should be apparent now why a transition in which $m_1-m_2=\pm1$ involves the emission of circularly polarized photons: the law of conservation of angular momentum requires it.

So far, we have dealt only with the two types of atom–photon interactions: spontaneous and stimulated emission of photons with the simultaneous transition of the atom from a higher-energy state to a lower one. There remains the reverse process to stimulated emission, namely absorption, in which a photon disappears and the atom makes a transition from a lower state to a higher one. Since energy must be conserved in these processes, an atom will make a "real" transition to an upper state only if the photon energy $h\nu$ satisfies $h\nu=E_1-E_2$, which is a kind of resonance condition on the frequency of the light. The resonant nature of this condition is amply demonstrated by the fact that the probability of such a transition is the same sharp function of frequency as in the reverse emission process. This sharp function, the spectral line shape, is, as we have already pointed out, fundamentally broadened by the finite radiative lifetimes of the quantum states between which the transition occurs. This inherent spectral line width is, as previously noted, called the natural line width.

We have on a number of occasions used the phrase "probability of a transition"; this requires closer examination, since as we saw in the case of a precessing magnetic moment subjected to a resonant magnetic field, the response of the magnetic moment was not simply a one-time transition from one state to the other. The magnetic moment alternated between the two states at a rate dependent on the strength of the resonant transition-inducing field. In spite of the fact that in that case we were dealing with what is called magnetic rather than electric dipole transitions, as we have here, nevertheless the same behavior would result under similar conditions. These are that the atom be subjected to a single-frequency coherent light beam of such intensity that the atom can alternate between the two states in a time short compared to the radiative lifetimes of those states. This would never have been contemplated as a practical possibility in the optical range until lasers became available; and now the manipulation of optical transitions has become as common as those performed by Ramsey and others in magnetic resonance, where strong coherent resonant fields were readily available decades before lasers. The question remains, How does an atom respond when excited by a relatively weak field that is not of a single frequency but has a relatively broad spectrum? In this limit, it can be shown that the atom has a finite probability, less than 1, of making a transition from its initial state, which increases in proportion to the length of time the atom is subjected to the exciting field. Of course, in quantum theory, if a measurement is made to determine which state it is in, it will be found to be in one or the other; it cannot be found somewhere in between! If the measurement is repeated on a large number of atoms under identical conditions, then the fraction that are observed to have made the transition will increase in proportion to the time they had been subjected to the resonance field.

The constant probability per unit time for absorption of a photon, which applies under "broad excitation," is fundamentally the same as that for stimulated emission, a reflection of a more profound principle called *detailed balancing* on a microscopic scale. Even the probability of spontaneous emission bears a close relationship to stimulated emission, their ratio being independent of the particular properties of any atom. However, their ratio does have a strong dependence on the frequency of the photon involved, according to the following formula attributed to Einstein:

$$A_{nm}/B_{nm}=8\pi h\nu^3/c^3, \qquad 6.12$$

where A_{nm} and B_{nm} are respectively the so-called *Einstein A- and B-coefficients* for spontaneous and stimulated emission, defined such that $(A_{nm}+\rho_\nu B_{nm})$ is the total rate of downward transitions, where ρ_ν is the number of photons of frequency ν already present, per unit volume, stimulating transitions. The strong ν^3 dependence of the ratio of spontaneous to stimulated emission explains why spontaneous emission is negligibly weak at or below the microwave region of the spectrum, becoming dominant at optical frequencies, unless special conditions are created to enhance the density of radiation as in a laser. A consequence of equation 6.12 is that the selection rules governing the quantum numbers of the initial and final states for electric dipole transitions apply equally to absorption. This means, for example, that if an atom is subjected to a light beam with the resonant frequency and pure circular polarization, then transitions occur only if the magnetic quantum number m obeys the selection rule $m_1-m_2=\pm1$. If the light has a pure linear polarization, then $m_1-m_2=0$ must be obeyed.

In the technique developed by Kastler for magnetic resonance by optical means, these selection rules are exploited to produce a magnetic polarization in certain species of atoms, particularly the alkalis: sodium, rubidium, cesium. This is done by a process that is called *optical pumping* of the populations of magnetic substates. It works as follows: Suppose a group of free sodium atoms are somehow confined in such a way that they remain unperturbed in their magnetic substates and are illuminated by a resonant parallel light beam after it has been circularly polarized. Prior to the laser age, such a resonant beam would have been obtained from a sodium vapor lamp, designed to provide the highest possible spectral intensity at the resonance fluorescence wavelength. The selection rules will now apply to the magnetic quantum numbers referred to the light beam as axis for space quantization: Absorption is allowed only if $m_1-m_2=+1$, but not $m_1-m_2=-1$ or 0. Then we see from Figure 6.8 that atoms in the $m=+1/2$ state have nowhere to go and will remain in that state unless perturbed by collisions. However, atoms in the $m=-1/2$ state can make a transition to the $m=+1/2$ substate in the electronic excited state, from which it can return to either substate in the ground state by spontaneous emission. The net effect of this "pumping cycle" is that atoms in the ground state are transferred from the $m=-1/2$ to the $m_2=+1/2$ state, eventually all ending up ideally in that state. But in this state the outer electron has its spin pointing in the direction of the light beam; thus a spin polarization has been achieved by purely optical means, without the need for strong magnetic fields. Furthermore, the same process can be used to monitor the degree of polarization in the group of atoms, since the rate at which photons are absorbed and re-emitted depends on the number of atoms in the $m_2=-1/2$ absorbing substate. Ideally, if 100% polarization is achieved, none of the atoms in the group are in an absorbing substate; interaction with the beam effectively ceases, and the sample becomes quite transparent. If now a resonant high-frequency magnetic field induces transitions between the two magnetic substates, resulting in an increase in the number of atoms in the absorbing $m_2=-1/2$ substate, the transmitted intensity of the pumping light will decrease, providing a way to monitor the occurrence of transitions and the frequency of the inducing field at which magnetic resonance occurs.

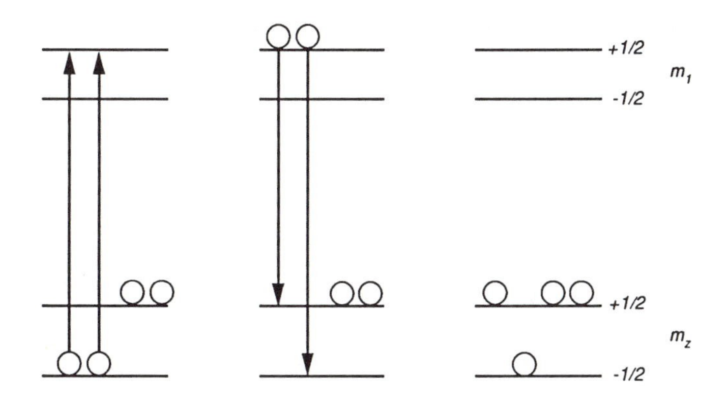

Figure 6.8 Kastler optical pumping of magnetic state populations

The beauty of this technique is that no large electromagnets are involved; indeed, the spins can in principle be polarized along the light beam axis, in zero magnetic field. More than any other technique it is capable of very large degrees of polarization; not only electron spin polarization, but also *nuclear* polarization. This has been demonstrated, for example, in those isotopes of noble gas atoms such as He^3 and Xe^{129}, which have a nuclear moment, in electronic states with zero angular momentum. This has made possible the application of NMRI to the diagnosis of the lung using inhaled optically pumped He^3 gas!

The successful implementation of the optical pumping technique critically depends on the ability first, to have a group of atoms whose spins are more or less free of disorienting collisions, and second, to realize a light source with the desired polarization and spectral properties. The first requirement is far more difficult than might at first appear; collisions with surfaces of containers are found to disorient electron spin directions, and the thermal velocities of atoms are generally so high that the orientation of an atomic spin would be randomized in a very short time. The development of special surface coatings to reduce the randomizing effect has met with enormous success in the case of atomic hydrogen and to a lesser extent with the heavy alkali atoms, which will be discussed at greater length in later chapters.

The first experiments on magnetic resonance using optical pumping were done in Kastler's laboratory on atoms in an atomic beam, in which, as we have already mentioned, the number density of particles is small, and consequently, considerable time elapses between disorienting collisions, thus fulfilling one of the essential requirements of particle containment. However, this is achieved at the cost of having a severely limited number of atoms contributing to the resonance detection signal and aggravated problems in the optics arising from the fact that the atoms are spread over some distance. It is, of course, to the credit of the early experimenters that in spite of these difficulties, they were able to successfully demonstrate a new technique; but in that form it was of limited value for practical applications. It was the introduction of the alkali vapor diffusion cell by Dehmelt that gave the technique its practical importance. This is based on the known inertness of the noble gases—helium, neon, argon, krypton, xenon—due to their closed electron shell structure in the ground state. In that state they are spherical and "rigid" like billiard balls, since it takes a considerable amount of energy to raise an electron to the next available state; moreover, they have no resultant spin or magnetic moment. Therefore, it was reasoned, in a collision with an alkali atom (in the ground spherical state) there should be no magnetic interaction to cause the spin to flip, and yet the colliding atoms could undergo strong momentum-deflecting collisions. To exploit this fact, the atoms under study are contained in a glass cell with a sufficient amount of an inert gas, acting as a *buffer*, to lengthen their diffusion time to the walls, thereby enabling the optical pumping process to develop a significant degree of polarization of the spins.

The second important requirement is an adequate light source with the appropriate polarization. With the development of stable tunable lasers, this requirement has been met with spectacular success. However, if, as was the case at the time the technique was introduced, the source is of necessity a spectral lamp containing the same element, the emitted photons are unpolarized, that is, equally likely to be right-handed as left-handed, with random phase. It happens that in our everyday experience we are more likely to have encountered materials whose properties are naturally described in terms of linearly polarized photons, rather than circularly polarized ones, for example the doubly refracting crystals of calcite, so familiar in science museum gift shops as a curiosity. They are doubly refracting, causing two displaced light waves to be generated in the crystal from the original light falling on it. These are the *ordinary* and *extraordinary* waves, which are linearly polarized at right angles to each other. In fact, this double refraction property is exhibited to varying degrees in many crystals, except those having a cubic symmetry in the unit cell, the building block of the crystal.

Clearly, special polarization-selective optics must be used to alter the state of polarization of the lamp output. Thanks to the genius of Land, of Polaroid camera fame, linear polarizers for optical and near infrared radiation are readily available in plastic sheet form. These polarizers exploit a phenomenon in certain crystals called linear dichroism, which means that one of the two linearly polarized components into which the light is split is selectively more strongly absorbed than the other, leaving the transmitted beam partially polarized. Circular polarizers are also available, which use circular dichroism in certain substances; however, they are unnecessary, since, once we have a beam that has a high degree of linear polarization, we can always convert it to circular polarization by passing it through what is called a *quarter-wave plate*. This is merely a plate cut from a doubly refracting crystal, such as calcite, whose doubly refracting property is a manifestation of a difference in the velocity of light for two perpendicular polarization directions. By setting the polarization direction of the linearly polarized beam at 45° to the two polarization directions in the crystal, called the "slow" and "fast" axes, and by making the light pass exactly the right distance through the crystal, a phase difference of one-quarter of a cycle, or 90°, is introduced between two coherent perpendicular polarization components of equal amplitude. That the emerging beam in this case will be circularly polarized is most easily seen by recalling our discussion on simple harmonic motion and phase in an earlier chapter; a point moving uniformly around a circle will have its shadow (projection) on a diameter follow simple harmonic motion, and the projection on a perpendicular diameter will also be simple harmonic motion but with a phase shift of 90° relative to the first. The converse is obviously also true: If we take the resultant of two perpendicular vibrations of equal amplitude but that differ in phase by 90°, we will get uniform circular motion.

Of equal importance is the task of realizing a light source reasonably free of spectral broadening, and also bright, in the technical sense; that is, with suitable optics it can be made to produce a beam of high intensity and small divergence. The latter immediately suggests a laser source; no other type of source remotely approaches the brightness of a laser source. However, the requirement that it be tunable to the resonance frequency (or wavelength) of a given atom imposes a heavy penalty in complexity. It happens that the two most important alkali atoms from the point of view of applications, rubidium and cesium, have resonance wavelengths that can fall within the tuning range of specially designed solid-state gallium–aluminum–arsenide (GaAlAs) diode lasers. However, their output wavelength is sensitive to temperature and the "injection" current through them, and they must therefore be stabilized using, for example, a precision optical resonator as reference, or a reference absorption cell containing the vapor of the desired element. For economic reasons this may not be an acceptable degree of complexity. Fortunately, a highly satisfactory "conventional" type of source exists using an electrodeless high-frequency electrical discharge in alkali vapor. A small spherical bulb provided with a short hollow stem to hold a small quantity of the alkali metal is evacuated to remove the oxygen, which reacts vigorously with the alkalis, and back-filled with a noble gas to a pressure of a few millimeters of mercury. The lamp is excited by placing the bulb in a radio-frequency field produced by a few turns of copper wire forming the inductance of a tuned L-C circuit driven by a radio-frequency power amplifier. The spectrum of the light output of these lamps when properly operated peaks at the very optical resonance wavelength needed for optical pumping. As we will see later, stability in the intensity and spectral line shape of the lamp output is critically important in the rubidium atomic clock.

Chapter 7
Corrections to Observed Atomic Resonance

All the atomic standards we shall be dealing with are based, in one form or another, on the resonant excitation of a large group of atoms or ions, by which they make transitions from one quantum state to another. From the observed resonance spectrum we must arrive at the intrinsic, or proper, frequency of the atoms, as it would be observed if they were at rest and free from any outside perturbation. Such perturbations will alter and broaden the resonance spectrum and put a limit on the degree of precision with which the intrinsic atomic frequency can be deduced.

It might be thought that a detailed knowledge of the frequency response curve, no matter how broad, should be sufficient for a theoretical analysis to obtain the true resonant frequency. In fact, this is not so; there will inevitably be noise present in any resonance detection system, and the observed response curve will always suffer from a degree of uncertainty. The sharper the response curve, the less important becomes the noise in finding the resonance frequency. A quantitative expression of this fact obviously depends on the detailed shape of the resonance curve; for a Lorentzian line shape we find the following:

$$\frac{\varepsilon}{\Delta\nu} = \frac{A_n}{\sqrt{3}A_0}, \qquad 7.1$$

where ε is the error in finding the line center, $\Delta\nu$ is the line width, and A_0, A_n are the *amplitudes* of the signal at resonance and the noise, respectively. The linear dependence on the ratio of amplitudes comes from the usual practice of defining, in effect, the position of the resonance line in terms of the position of the nearly linear portion of the resonance curve at the inflection point; as a general rule, the $\sqrt{3}$ is ignored.

An understanding of the effects of the physical environment on the resonance line shape and position is crucial in finding ways to minimize these effects in practice, and in correcting for any displacement in frequency they may cause. Ultimately, the stability and reproducibility of the standards depend on how successfully this is accomplished. Such theories have also been developed from the inverse point of view: namely, for what they can reveal about the mechanisms that broaden and/or shift the resonance frequency. The incredibly high degree of spectral resolution that has been reached has raised the level of significance of a number of subtle effects, some involving quantum effects, others Einstein's *general theory of relativity,* about which more will be said later in this chapter.

7.1 Homogeneous and Inhomogeneous Broadening

For resonances observed on a large group of atoms it is useful to distinguish between line-broadening mechanisms according to whether *all* atoms have the *same* broadened spectrum, or the spectrum of the whole group is broadened because *each* atom has a slightly different frequency and the global spectrum merely reflects the distribution of frequencies among the particles. The former is called *homogeneous* broadening, as exemplified by broadening, common to all atoms, due to a finite radiative lifetime, while the latter is *inhomogeneous* broadening, as exemplified by a group in which each atom has a slightly different frequency because of its differing environment.

7.1.1 Homogeneous Broadening

The most common source of homogeneous broadening is the finite time of coherent interaction of the atom with the exciting field. This can be due to the finite radiative lifetime of a quantum state involved in the transition or to phase-randomizing collisions, as was postulated by Lorentz to explain optical dispersion in his *electron theory.* Unfortunately, as Lorentz realized, collisions could not solely explain the width of optical resonance lines; we now know that the radiative lifetimes in optical transitions are usually so short compared to average times between collisions, except at extreme pressures, that the observed broadening is evidence of the "natural" radiative lifetime, and not collisions. The situation is quite different, however, in the radio-frequency and microwave regions of the spectrum, where radiative lifetimes are extremely long. In this case collisions play a dominant role, and by making collisions rare, as in the ion standards, extremely narrow resonances are possible.

The spectral line shape that results from a finite radiative lifetime or collisions that only interrupt the phase is the same *Lorentzian* function $L(\nu)$ we introduced in connection with the response of a damped simple harmonic oscillator, namely

$$L(\nu)=\frac{1}{2\pi}\frac{\Delta\nu}{(\nu_0-\nu)^2+\left(\frac{\Delta\nu}{2}\right)^2}, \qquad 7.2$$

where the factor $1/(2\pi)$ is included so that $\int L(\nu)\, d\nu=1$. In terms of the mean time between phase-randomizing collisions $\Delta\tau$, the expression for $\Delta\nu$ agrees with the approximate result we previously derived for the coherent buildup of oscillation in a resonant structure, namely $\Delta\tau\, \Delta\nu \approx 1$. A more rigorous treatment yields $2\pi\Delta\tau\Delta\nu=1$, which, as already indicated, has a far more general application to the simultaneous measurement of frequency and time. The connection between phase-randomizing collisions and the damping of an oscillator can be shown to arise from the energy dissipation that the continual interruption of phase produces even when each collision is perfectly elastic. The net effect is as though a resistive force were present; in fact, it was shown by Lorentz that $\gamma=2/\Delta\tau$. In the absence of collisions, an undamped oscillator does not, on the average, continuously absorb energy from a driving field, nor does it dissipate energy if left alone. As already mentioned, in the optical part of the spectrum, it is the radiative lifetime of quantum states that usually sets the line width, the so-called *natural line width,* typically in the megahertz range.

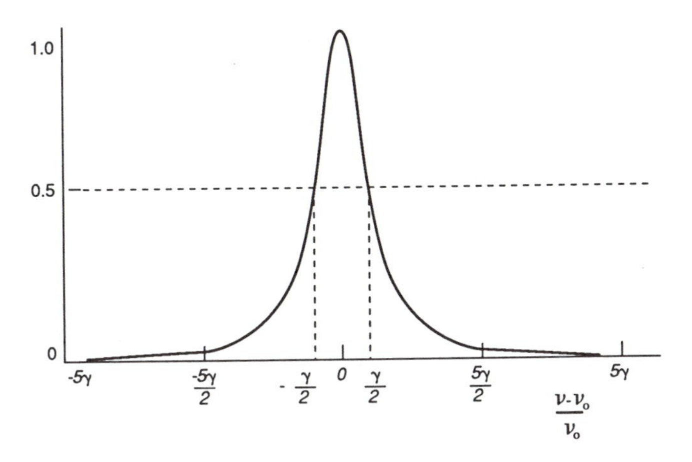

Figure 7.1 The Lorentzian resonance line shape

7.1.2 Inhomogeneous Broadening

The inhomogeneous class of broadening is present whenever different atoms or ions have slightly different resonance frequencies by virtue of, for example, nonuniformities in the distribution of some field that acts on them and displaces the energies of their quantum states. This is of particular concern in nuclear magnetic resonance on solid substances, where an important source of spectral line broadening is the inhomogeneity in the applied magnetic field intensity; in fact, in the original use of the term it was understood to refer to this particular case. The circumstance that makes spatial variations in the applied magnetic field particularly objectionable in solids is that each nucleus is constrained to vibrate with a small amplitude about a fixed lattice site, where the magnetic field may differ from other lattice sites. However, our concern will be with quasi-free individual particles, which far from being constrained, are more or less free to move with their thermal velocity and only rarely collide with other objects. Under these conditions, the most important source of inhomogeneous broadening is the Doppler shift in the frequency of a moving source, a subject that requires us to think about the description of physical phenomena in terms of coordinate frames of reference in relative motion.

7.2 The Special Theory of Relativity

Radically new ideas concerning the relativity of physical laws are derived from the concept of time as a coordinate in a four-dimensional continuum, in terms of which we must describe events in our universe. This is the basic framework of a revolutionary relativistic theory first published by Einstein in 1905 as *the special theory of relativity* (special in the sense that it was restricted in its application). The origins of the theory are found in the attempt to reconcile the way the laws of classical mechanics and electromagnetism are "seen" by observers in different states of motion. The equations of the classical theory of electromagnetism, the beautiful theory of Maxwell, are spectacularly successful in unifying the fields of optics, electricity, and magnetism and are one of the great triumphs of the 19th century. Unfortunately, under the classical coordinate transformation ($x \rightarrow x+v_x t$, etc.) corresponding to going from the coordinate axes of one observer to another in relative motion, they do not retain the same *form*. This would imply that the state of motion can distinguish different observers by the way they see the fundamental laws of nature operate. This would in turn imply, for example, that a particular observer could be singled out as being "at absolute rest," a possibility whose denial defines the *principle of relativity*. Now, the equations of classical mechanics (Newton's laws of motion), on the other hand can easily be shown to preserve their form under such a transformation of coordinates. Faith in Newtonian mechanics ran so deep that at first it was assumed that either Maxwell's theory was at fault or that for electromagnetic phenomena perhaps not all observers are equal; perhaps there *is* a special frame of reference. In fact, it had long been thought that light consists of waves in an all-pervading medium called the *ether*. If so, then for example, since the earth is in constant motion, presumably there should be experienced on the earth's surface an *ether drift*. Such a drift would cause light waves to appear to travel at different speeds depending the direction of propagation, just as the velocity of waves on a river relative to the shore would depend on their direction with respect to the flow. It was to test whether there was any detectable ether drift that the famous Michelson–Morley experiment was designed to do; none was found. Efforts to modify Maxwell's theory to explain this result were overtaken by a radical approach sought by Einstein to modify the coordinate transformations themselves. This was prompted by the work of Minkowski and Lorentz, who found and interpreted transformation equations (now known as the *Lorentz transformation*) under which Maxwell's equations do keep their mathematical form. We owe it to Minkowski for the interpretation of this transformation as an angular shift in the orientation of coordinate axes in four-dimensional space. Einstein's contribution was to take the Lorentz transformation as the correct one and to modify Newton's equations of motion to preserve their form under this transformation. The ramifications of this theory go to all our fundamental concepts of space, time, energy, mass, etc. For us the most relevant result is the transformation of the time variable t. If one coordinate system has a constant velocity V along the x-axis of another system, the space and time coordinates in the two systems are related through the *Lorentz transformation* as follows:

$$x' = \frac{x - Vt}{\sqrt{1 - V^2/c^2}}, \quad y' = y, \quad z' = z, \quad t' = \frac{t - Vx/c^2}{\sqrt{1 - V^2/c^2}}. \qquad 7.3$$

This shows explicitly through the presence of $\sqrt{1 - v^2/c^2}$ in the denominators the physically radical departure from classical physics (and common experience) that the time scale itself varies according to the state of motion of the observer: Clocks will literally run at different rates! The clock reading t' will be ahead of the clock reading t in the system in which the spatial coordinate x is fixed. The latter clock therefore runs slower; hence the effect is called *time dilation*. It would not be in keeping with the spirit of the theory of relativity to say that the observed resonance frequency of a moving atom "only appears" to be lower; we must accept the fact that the time scale itself is not absolute, and a clock or atomic oscillation that defines time in a coordinate frame moving with respect to the observer simply runs slower. This radical break with the classical concept of time was not accepted lightly; it naturally stimulated strenuous efforts in the early establishment of the theory to find direct experimental evidence in the laboratory to support it.

Since the velocity of light is so large (2.99797×10^8 meters/sec) compared to velocities ordinarily encountered in the laboratory, the detection, let alone the measurement, of the dilation of the time scale is very difficult. For all ordinary velocities, $V/c \ll 1$, and the departure from the classical $t'=t$ is extremely small; it is crucial that this be so, of course, since we know that Newtonian mechanics cannot be far from the truth. Nevertheless, the Lorentz transformation does represent a radical break from the classical concepts of space and time, but by now it has become such an integral part of modern physical theory, which has been validated experimentally at so many points, that the invariance of the velocity of light has been taken as a matter of definition: The standard meter is now defined as the distance traveled by light *in vacuo* in a certain (very small) fraction of a standard second. It is no longer in principle meaningful to measure the velocity of light, except as a determination of the meter. In spite of the fact that relativistic effects are extremely small, except for extreme velocities, in the case of atomic clocks, the precision has reached such a level that corrections for such effects are not negligible.

7.3 The Doppler Effect

The term Doppler is of course familiar to everyone in the context of checking the speed of vehicles on our highways. It is, in general, the variation in the observed frequency of any wave whenever the observer and source of the wave are in relative motion. According to a principle enunciated by Christian Doppler in 1842, which applies equally to sound waves and light waves, frequencies observed with respect to reference frames in relative motion are shifted by what is now called the *Doppler effect*. It is a particularly important universal effect in the context of high-resolution spectroscopy because of the ever-present thermal agitation of atoms and molecules.

The frequency shift predicted classically is easily derived: Assume first that we use a frame of reference in which the source of the wave is stationary and the observer is moving relative to this frame in the direction of the source with a velocity V. Let us follow the events after the instant that, say, a crest of the wave passes the observer. If the velocity of the wave is c, then the time that must elapse before the next crest passes the observer must be determined by the condition $Vt+ct=\lambda$, since the distance to the next crest is λ. Solving this for the time we get $t=\lambda/(V+c)$. If t is the time between successive crests passing the observer, then $1/t$ is the number of times a crest will pass in unit time, which is of course the observed frequency of the wave. If we substitute $\nu=1/t$ and $\lambda=c/\nu_0$ in our result for t, we find $\nu=(1+V/c)\nu_0$, showing that the frequency is increased fractionally by V/c. If the observer had been assumed to be moving away from the source, we would obviously have found $\nu=(1-V/c)\nu_0$. In general, if the relative velocity vector makes an angle θ with the direction of propagation of the wave, we can write the following for the change in the observed frequency:

$$\nu-\nu_0=\frac{Vk\cos\theta}{2\pi}, \tag{7.4}$$

where $k=2\pi/\lambda$ is the wave number.

If we use a frame of reference in which the observer is stationary but the source of the waves has a velocity V in the direction of the observer, then if we consider events at the source subsequent to the emission of a crest of the wave, we see that the distance between crests will be shortened by the distance the source travels in one period of oscillation, or symbolically, $\lambda=\lambda_0-V/\nu$. This can be written $\nu=\nu_0/(1-V/c)$.

The Doppler effect is manifested in any type of wave motion. However, in anticipation of the fact that we are concerned here only with light waves, we have used the conventional symbol for the velocity of light, c. We notice that we have obtained different results depending only on whether we chose a frame of reference in which the source is at rest or the observer is at rest. If we had been considering only waves on the surface of water, the difference in the two results would not have been unexpected, since the water itself uniquely defines a frame of reference, and having the observer move in the water is not necessarily the same as having the source move. However in the case of light, the principle of relativity, one of the pillars of Einstein's theory, denies the existence of any "absolute" frame of reference, and the two cases dealt with above must yield the exact same result. This is true in our classical derivation only if we neglect terms of order $(V/c)^2$ and higher.

In the context of atomic resonance standards, even at relatively low atomic velocities an important correction to the resonance frequency is the *relativistic* Doppler effect, by which is meant the second-order term in an expansion of the exact Doppler formula in powers of V/c. The derivation of the exact Doppler formula begins with the recognition that for a plane monochromatic wave, represented as

$$E_y = E_{y0} \cos(kz - \omega t), \tag{7.5}$$

the phase $(kz-\omega t)$ is an invariant (scalar) with respect to a coordinate transformation from one inertial system to another in uniform relative motion. Since (x, y, z, ict) is a (contravariant) 4-vector in a four-dimensional space, it follows that $(k_x, k_y, k_z, i\omega/c)$ is a covariant 4-vector, and thus from the Lorentz transformation equations for the components of a 4-vector, we have the following:

$$i\frac{\omega'}{c} = \frac{i\frac{\omega}{c} + i\frac{V\cos\theta}{c}k}{\sqrt{1-\frac{V^2}{c^2}}}, \tag{7.6}$$

which simplifies to the desired relativistic Doppler formula

$$\omega' = \frac{1+\frac{V}{c}\cos\theta}{\sqrt{1-\frac{V^2}{c^2}}}\omega. \tag{7.7}$$

Since we are concerned only with cases in which $V/c \ll 1$, we can expand in powers of V/c retaining only the second-order term, the so-called "relativistic Doppler effect":

$$\omega' = \omega\left(1 + \frac{V}{c}\cos\theta + \frac{1}{2}\frac{V^2}{c^2} + \cdots\right). \tag{7.8}$$

There was intense interest during the early establishment of the theory in putting this result to the test in the laboratory. The most convincing early experiments were those of Ives, published in 1938. The success of these experiments was largely due to the method developed of bringing out any second-order departure from the ever-present (even classically) linear Doppler effect. This he did by observing a particular line in the spectrum of light emitted by high-speed hydrogen atoms in such a way that he could simultaneously register on a photographic plate the spectrum as seen directly from the atoms and as reflected by a plane mirror to effectively reverse the observed velocity of the atoms. On the same photographic plate the spectrum of slow hydrogen atoms was also registered, providing a fiducial wavelength to compare with the two Doppler-shifted spectral lines on either side of it. Contrary to classical expectations, which are that the Doppler shift simply reverses sign with the velocity and that the two lines from the fast atoms must therefore be symmetrically situated about the center line, he found that the Doppler-shifted lines are both displaced slightly towards the red (lower frequencies) relative to the unshifted line.

7.4 The Thermal Doppler Line Shape

The Doppler broadening of spectral lines is familiar to spectroscopists working in the optical region of the spectrum because it is generally the limiting factor in the attempt to achieve high spectral resolution, and it is universally present. Under conditions where the wavelength of the wave is very much smaller than the average distance a particle travels between collisions, the Doppler shift in the resonance frequency of each atom will result in a spectral profile for the whole ensemble that simply reflects the distribution among the atoms of the frequency shifts associated with their individual thermal velocities. Such conditions commonly exist for light waves, since their wavelength is only on the order of 0.5 μm, compared to mean free paths 100 times longer, at pressures below say 100 pa. The exact line profile when collisions are not negligible is far more complicated; we will not concern ourselves with that, but in the next section we will consider the opposite extreme, where the wavelength is large compared with the average distance an atom is free to travel.

For atoms in thermal equilibrium at absolute temperature T, the components of the velocity of atoms along a given direction, taken to be the z-axis, are distributed among the atoms in accordance with the *Maxwell–Boltzmann* distribution, in which the number of atoms having a z-component of velocity in an infinitesimal range between V_z and (V_z+dV_z) is given by $f(V_z)dV_z$, where $f(V_z)$ is the following function:

$$f(V_z) = N\sqrt{\frac{M}{2\pi kT}}\exp\left(-\frac{MV_z^2}{2kT}\right), \tag{7.9}$$

where M is the atomic mass and k is the Boltzmann constant. Now suppose a monochromatic light beam of frequency ν is directed along the z-axis through an ensemble of atoms whose resonance frequency would be ν_0, measured in their rest frame of reference. We recall that an atom having a velocity component V_z will see (to first-order of approximation in V_z/c) a Doppler shifted frequency $\nu(1-V_z/c)$. Therefore, the light will be in resonance with such an atom not when $\nu=\nu_0$, but rather when $\nu=\nu_0/(1-V_z/c)$, or to the same first-order approximation we have been assuming: $\nu_0 \approx \nu\ (1+V_z/c)$. Therefore, the atoms, regarded as a whole, will behave as one entity with a broadened resonance line shape, obtained by rewriting the velocity distribution function as a *frequency* distribution function by using the fact that the number of atoms having a z-component of velocity in the range dV_z will equal those having a displaced frequency in the interval $d\nu=(\nu_0/c)dV_z$. Now, if we let $g(\nu)$ represent the frequency distribution function, then

$$g(\nu)d\nu = \sqrt{\frac{\alpha}{\pi}}\exp\left[-\alpha\left(\frac{\nu-\nu_0}{\nu_0}\right)^2\right]\frac{d\nu}{\nu_0}; \quad \alpha = \frac{Mc^2}{2kT}. \tag{7.10}$$

Because this function has the form $\exp(-x^2)$ it is called a *Gaussian line shape*, and plotted as a function of frequency, it has the well-known bell shape, shown in Figure 7.2 for Rb vapor at a temperature of 300°K.

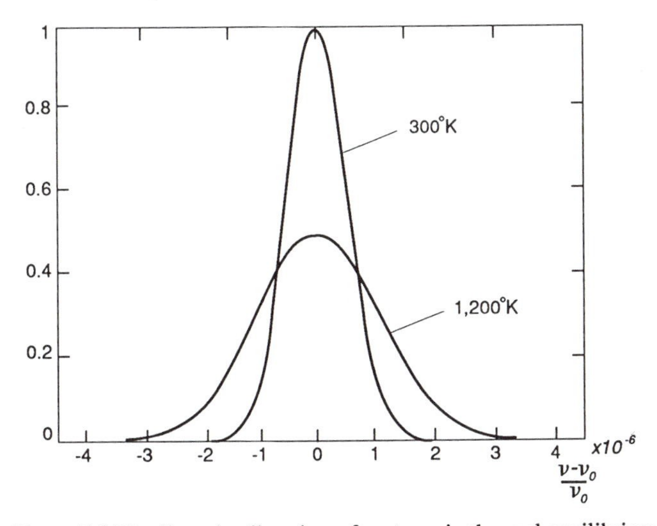

Figure 7.2 The Gaussian line shape for atoms in thermal equilibrium

7.5 Sub-Doppler Line Widths: the DickeEffect

Extremely narrow resonances, far below the Doppler width derived above, are actually observed at *microwave* frequencies on atoms diffusing in a buffer gas, in spite of their thermal agitation. This is due to what has been called the *Dicke effect* (Dicke, 1953). If we substitute in the formula for the first-order Doppler shift, namely $\nu-\nu_0=(V/c)\nu_0$, the numerical values for a Rb atom diffusing through an inert buffer gas with an average thermal velocity of about 10^4 meters per second, we find a Doppler frequency shift in its microwave resonance at 6.8 GHz of about 200 kHz, or 10,000 times the frequency width of the resonance actually observed. Clearly, the conditions for "normal" Doppler broadening are not met.

The first theoretical analysis of the narrowing of Doppler widths through collisions in an inert gas was published by Dicke in 1953. To understand the conditions under which the Dicke effect is expected to be important, let us reexamine our assumptions in arriving at the formulas for the Doppler shifts in frequency. It was assumed that the observer and source continue indefinitely in their state of relative motion, with the observer crossing many undulations of the wave, that is, many wavelengths. To bring out the effects of not fulfilling this condition, consider the contrived example of an observer who is constrained to oscillate back and forth in simple harmonic motion with finite amplitude. The question we have to address is: How does the magnetic field component, for example of the microwave, vary with time as seen by our peripatetic observer; from this we can arrive at the spectrum seen by him using Fourier analysis. Since the relative velocity of the observer is assumed to oscillate with a simple frequency, it follows that the Doppler effect will cause the observer to see a wave whose frequency oscillates about a fixed value. But this is nothing more than a frequency modulated (FM) wave, whose theory is familiar from its common use in radio broadcasting to provide static-free reception of high quality sound. There are three quantities aside from its amplitude that characterize a frequency modulated wave: first, its mean frequency; second, the frequency of modulation; and third, the maximum deviation of the frequency from its unmodulated value, that is, the depth of modulation. We will not reproduce here a derivation of the Fourier spectrum of such a wave, but merely state some of the salient results, some of which may not be altogether intuitive. The most striking is that the spectrum is discrete; it consists of a central line at the unmodulated frequency and *sidebands* consisting of equally spaced lines extending with diminished amplitude to infinity on both sides of the central undisplaced line, as shown in Figure 7.3.

The constant spacing of the lines is just the modulation frequency, so that each line is simply a multiple (harmonic) of the modulation frequency away from the central line. It might be thought that since the frequency "passes" through all values between the limits of modulation, that therefore the spectrum ought to contain all these frequencies; in fact, it does not. Curiously, the sideband amplitudes are not zero even for frequencies that extend beyond the "instantaneous" values the frequency passes through as it swings between its limits. However, if the modulation is infinitely slow, then the sidebands approach each other and will finally merge into a continuum. The amplitude distribution of this continuous spectrum reflects the relative amount of time the frequency spends at different values between the modulation limits.

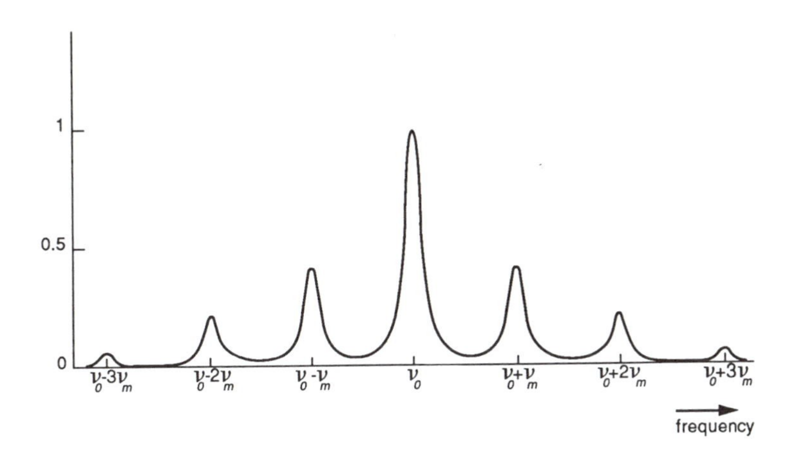

Figure 7.3 The Fourier spectrum of a frequency modulated wave

The way that the amplitudes of the sidebands fall away as we go away from the central line depends on what is called the *modulation index*, which is defined as the ratio of the maximum frequency deviation to the modulation frequency. If the deviation is small in relation to the modulation frequency, that is, if the modulation index is small, then the sidebands will be weak and the central line will predominate.

Let us then compute the modulation index for our oscillating observer. From the Doppler formula, the maximum deviation is the following:

$$\Delta\nu = \frac{V_{\max}}{c}\nu_0 = \frac{2\pi\nu_m a}{c}\nu_0, \tag{7.11}$$

where ν_m is the frequency of the observer's to-and-fro motion, and a is his maximum distance traveled. It follows that the modulation index, which by definition is $\Delta\nu/\nu_m$, is given by the following:

$$\frac{\Delta\nu}{\nu_m} = \frac{2\pi a}{c}\nu_0 = 2\pi\frac{a}{\lambda_0}, \tag{7.12}$$

where λ_0 is the wavelength of the unmodulated wave. This last result contains the essential key to understanding the narrowing of the Doppler effect through collisions, because it tells us that as long as the observer, that is, the atom under study, moves only distances that are small compared to the wavelength, the modulation index will be small, and it will see mainly the undisplaced central frequency, with weak sidebands having an amplitude distribution and spacing determined by the parameters of its particular motion.

Quantitatively, the amplitude of the sideband at the frequency $(\nu \pm n\nu_m)$ is proportional to $J_n(2\pi a/\lambda_0)$, where as usual, J_n represents a Bessel function of order n. If the particle is constrained to oscillate with an amplitude below one wavelength, that is, if $a/\lambda_0 < 1$, then all the amplitudes rapidly approach zero for increasing n above zero, as can be seen from Figure 7.4. In this case the power resides principally in the undisplaced center frequency, which is itself free of the (first-order) Doppler effect.

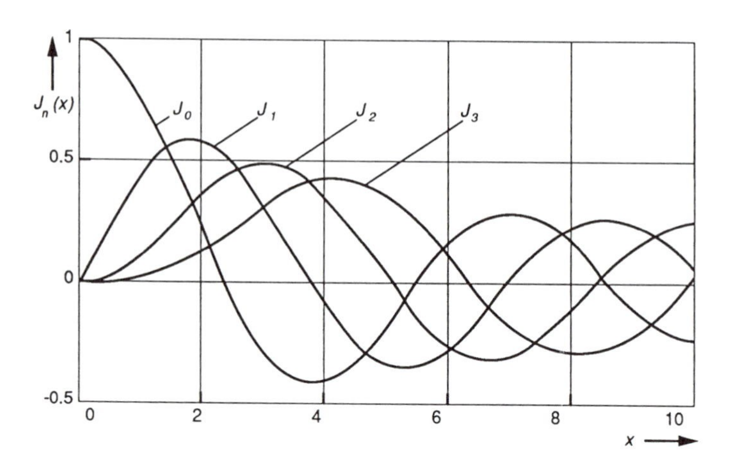

Figure 7.4 The Bessel function $J_n(x)$ for $n = 0, 1, 2, 3$

Although following Dicke, we chose a very special kind of confinement for our observer in the cause of mathematical lucidity, he went on to show that under broad conditions, a rigorous quantum treatment leads to essentially the same qualitative result; namely, whatever the detailed motion of the observer, as long as the motion does not continue uninterrupted for distances much greater than the wavelength, the Doppler spectrum has a sharp central line superimposed on a base that reflects the detailed motion of the observer. In the case of a Rb atom diffusing through a noble gas buffer, a given atom makes frequent random collisions with the gas atoms, collisions that are far more effective in deflecting a Rb atom in its path than in disturbing its internal quantum states. As a consequence, the atom executes a 3-dimensional "random walk" with a net average progress in any direction a slow function of time, a form of statistical confinement, we might say.

Figure 7.5 shows the average spectrum seen by particles of a gas in thermal equilibrium irradiated by a wave of a single frequency, whose spectrum seen by a stationary particle would be just the single central line. The base of the line has the shape expected of particles freely crossing many wavelengths of the wave, that is, the "normal" Doppler line shape. This Doppler base broadens out with increase in temperature, since the thermal velocities of the particles increase, but the central line remains unchanged.

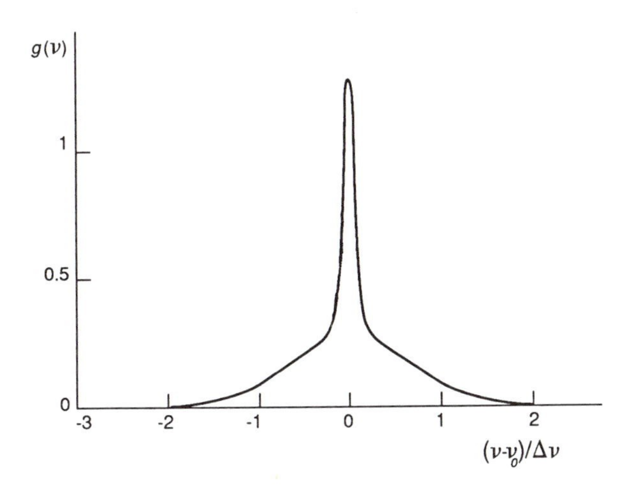

Figure 7.5 The Dicke effect; resonance line shape of an atom diffusing in a buffer gas

7.6 The General Theory of Relativity

In Einstein's *general theory of relativity*, published in 1916, gravitation is explained as a manifestation of the altered geometry of space–time produced by the presence of matter. Under ordinary conditions and over the scale of time and space that are within the realm of human experience (apart from astronomy), the geometry of space is to all intents and purposes that of Euclid, the geometry every schoolboy (and schoolgirl) struggles with.

Abandonment of an absolute and universal time raises a fundamental question: If the time indicated by a clock depends on the position and state of motion of that clock relative to the observer, then the very idea of simultaneity of events occurring "at the same time" at separated points requires careful reexamination.

Thus if two identical clocks are set to display the same time at the same point in space, and then one is removed to another point, the time indicated by the latter could depend on its speed at all points along its path to the new position. If that is the case, then clearly a meaningful way must be defined for comparing remote clocks; that is, a way to synchronize them.

7.6.1 "The Clock Paradox"

An even more drastic situation can be contemplated, in which not only does the time scale depend on the state of relative motion between different clocks, but it does so in a way that even if the clock returns to the original point in space occupied by the first clock, it does not return to synchronism with it. It is this circumstance, which runs counter to common intuition, that spawned the once controversial "clock paradox." (See Figure 7.6.) This "paradox," which has long since been resolved, is the following: Imagine that identical twins decide that one of them will "slip the surly bonds of Earth" and journey at high speed in a spacecraft to a distant point in space and then return to rejoin his twin brother, who has remained on Earth, many years later. If, as relativity theory predicts, the returning twin finds that his Earthbound brother has noticeably aged more than he has, we are led to what appears to be a paradox; that is, it appears we can be led to contradictory conclusions even starting from equivalent premises. (If this were actually the case, then of course it would be a fatal flaw in the theory.) Thus it might be argued that from the point of view of the astronaut-twin, the Earthbound twin recedes at high speed and then returns along a trajectory that is the astronaut's trajectory inverted with respect to their common starting point. The apparent symmetry between the trajectories as seen by the two twins would seem to predict that the astronaut would find that his Earthbound twin had aged less than he! This contradicts the earlier conclusion. If the symmetry assumed between the experiences of the two twins really exists, then there is only one logical conclusion: The twins must have aged precisely the same amount when they are reunited. This conclusion would, in fact, invalidate a fundamental logical consequence of the underlying postulates of the theory of relativity. The argument has been made successfully, however, that the circumstances of the two twins are not symmetrical: One twin actually has to fire up his rocket engines, while the other does not, and this, after protracted analysis and debate, finally was accepted as providing the basis for resolving the "paradox."

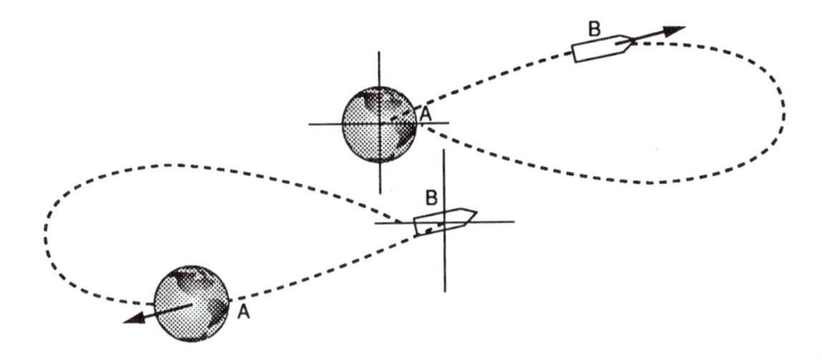

Figure 7.6 The "Clock Paradox": the views of twins A and B

7.6.2 The Synchronization of Remote Clocks: The Formal Definition

If we include the possibility of large masses of matter being present, then Einstein's theory of gravitation leads to even more drastic conclusions about the synchronization of remote clocks. If the distribution of matter is static, then the synchronization of clocks can be achieved by taking advantage of the basic tenet of the theory of relativity: The velocity of light in free space is a fundamental constant of nature and is independent of the state of relative motion of the source or observer. We can in principle synchronize clocks in the neighborhood of some given clock simply by correcting for the transit time of a light signal sent from the given clock to each of the other clocks. This can be done after the fact by a light signal being sent out from the given clock and immediately returned to it from the other clocks. From the elapsed time between the departure of the signal and its return, as measured by the given clock, the correction to be applied to the time of the other clocks is obtained simply by dividing the elapsed time by two.

However, in general, the distribution of matter may not be static, in which case it is impossible to synchronize clocks using light signals as described above. The reason is that if this synchronizing procedure is used to synchronize a remote clock through the intermediary of other clocks placed along a path leading from the given clock to the remote one, the result may depend on the path taken to the remote clock. This would mean, for example, that if the synchronization procedure is carried out over a succession of clocks along a *closed* path ending up at the starting point, it would lead to a different setting than that of the initial clock with which it was intended to be synchronized!

7.6.3 The Sagnac Effect

We should mention at this point an interesting relativistic effect on time measurement associated with the rotation of the earth. Since the coordinate system fixed in the earth is noninertial because its rotation with respect to "the fixed stars" constitutes an accelerated motion (not of speed, but direction), again the theory of *general relativity* is involved. According to the theory, if we imagine we have two identical, precise clocks at some point on the earth's equator, and one remains *fixed* while the other is taken even *slowly* (with respect to the earth) along the equator all the way around until it reaches its starting point, then the times indicated on the two clocks will not agree. The difference $\Delta\tau$ can be shown to be given by the following:

$$\Delta\tau = \pm\frac{2\Omega}{c^2}S, \qquad 7.13$$

where Ω is the angular velocity of the earth (7.3×10^{-5} rad/sec) and S is the area ($\pi R_E^{\,2}=1.3\times10^{14}\mathrm{m}^2$) enclosed by the path of the moving clock. The formula yields a significant time difference of about $\pm1/5$ microsecond, depending on the direction the moving clock takes around the equator. This effect is often referred to as the *Sagnac effect*, after the Frenchman G. Sagnac, who in 1911 detected by optical interference a difference in the time taken by a light wave to complete a round trip in the two possible directions around the mirror arrangement shown in Figure 7.7, when the latter is made to rotate.

Since the velocity of light in free space is constant, this is usually interpreted as a difference in the effective optical path length due to the finite movement of the mirrors in the time it takes the light wave to go from one mirror to the next. In the reference frame of the mirrors themselves, however, it must be interpreted as *time itself* flowing at a changing rate as we go around from one mirror to the next.

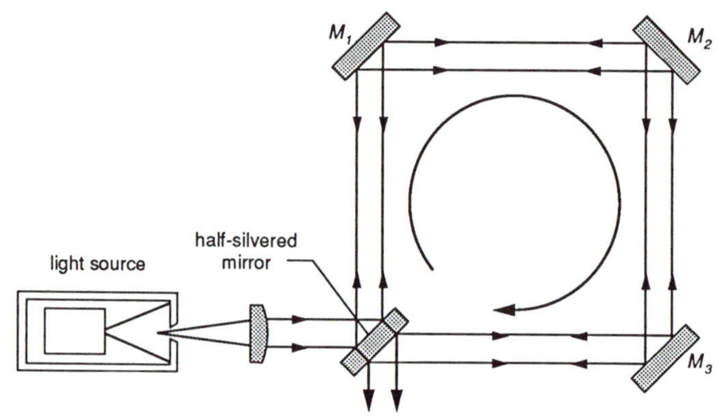

Figure 7.7 The mirror arrangement used by Sagnac to study the propagation of light in a rotating system

7.6.4 The Pound–Rebka Experiment

Einstein's theory predicts that in a static gravitational field such as that of the earth (if we neglect its relatively slow rotation) the specific *proper* time scale that attaches to a particular point in the field differs from the coordinate time scale, which belongs to a general frame of reference defined far from the field region. This means that two *identical* oscillators placed at points in a gravitational field that differ in the value of the gravitational potential Φ will oscillate at different frequencies. To a first approximation, the theory predicts a difference in frequency given as follows:

$$\nu_1 - \nu_2 = \frac{(\Phi_1 - \Phi_2)}{c^2}\nu_2. \qquad 7.14$$

Thus if one oscillator is placed at height L above another oscillator at the surface of the earth, we should expect a difference in frequency between them amounting approximately to

$$\nu_1 - \nu_2 = \frac{gL}{c^2}\nu_2; \quad (L \ll R_E). \qquad 7.15$$

Since the gravitational potential is negative in the neighborhood of a gravitational mass, the proper frequency of an oscillator near such a mass is lower than at a point infinitely far from it; hence the name *gravitational red shift*. The effect is small: Even for the gravitational field of the sun, the fractional shift is only about 2×10^{-6}. To observe shifts in the lines of the solar spectrum is unfortunately less than convincing as a test of the theory, since there are known to exist severe differences of environment between the sun and earth, in addition to the gravitational field.

In the earth's gravitational field the effect is far from insignificant when comparing clock rates aboard high-orbit satellites with ground-based stations. For example, a clock aboard a satellite in a circular orbit of radius (say) 26,000 km, typical of the GPS satellites, would run *faster* than a ground-based clock by the fractional amount given by the following:

$$\frac{\nu - \nu_0}{\nu_0} = \frac{GM_E}{c^2 R}, \qquad 7.16$$

where G is the gravitational constant, M_E is the mass of the earth, and R is the radius of the satellite orbit. If we substitute numerical values, we find a fractional difference of 1.7×10^{-10}, a large number in the context of atomic timekeeping!

For a terrestrial experiment the effect is, of course, very much smaller; for L=30 m we get a fractional difference of only 3×10^{-15}! Fortunately, an experimental breakthrough in γ-ray spectroscopy in 1958 by Mössbauer made it possible to reach the incredibly high level of spectral resolution that a terrestrial red shift experiment requires. In the γ-ray region of the spectrum there are nuclear transitions with long radiative lifetimes and narrow natural line widths. However, the photon momentum is sufficiently high that when it is emitted by a nucleus, part of the transition energy is taken up by the recoil of the nucleus. In fact, the consequent displacement in the photon energy makes it no longer able to be absorbed efficiently by another identical nucleus. What Mössbauer discovered was that if the nuclei are constrained within a suitable crystal lattice in the right temperature range, the recoil is effectively taken up by the entire mass of the crystal, leading to essentially *recoilless* emission and absorption, and hence a degree of spectral resolution unheard of at the time. The γ-ray spectral resolution is so high that even the Doppler effect caused by a slow linear movement is sufficient to provide a sufficient sweep of its energy.

Pound and Rebka, in a classic experiment, exploited this new development in a terrestrial experiment to measure any frequency shift that might be exhibited by photons emitted at one point in a gravitational field and absorbed at another by idential nuclei. Initially, it proved difficult to reach conclusive results; little progress was possible until it was realized that temperature differences between the emitter and absorber can lead to significant second-order Doppler shifts, and that the temperatures must be stabilized and taken into account. Fluctuations as small as $\pm$ 1°C were computed to cause frequency shifts nearly as large as the gravitational shift. They used the Mössbauer effect in the resonance absorption of 14.4 keV γ-rays of Fe^{57}; the sensitivity of their apparatus allowed them to observe the effect by placing the Fe^{57} source in the tower of the physics laboratory at Harvard University at a height of only about 20 m above the resonant absorber.

As with any other experiment designed to test a theory, the salient questions are; First, just how crucial is a positive result to the theory, and second, to what extent does a positive result preclude other theories? It is the uncertainty in answering the latter question that dictates a certain restraint in stating what the experiment actually proves. This is far from a simple matter: A careful analysis is required to strip away all the assumptions that are not in fact proved by the test. Of the elegant mathematical structure that is Einstein's theory of general relativity, this particular test probably only proves that the *equivalence principle*, which states, in effect, that a gravitational field is indistinguishable from an appropriate coordinate transformation, is valid for photons. Falling under gravity is indistinguishable from motion with respect to a frame of reference accelerating upwards. Hence the photons develop a Doppler shift with respect to the accelerating frame of reference given by $(V/c)\nu$, where $V=gL/c$, and we have the same formula as before.

7.7 Conclusion

The are a number of other mechanisms that can affect the spectral profile of an atomic resonance in the microwave region of the spectrum, but none as universal as the Doppler effect. The atomic resonance used as standard in each of the different types of atomic clocks will have its own heirarchy of important factors affecting its width and position on the frequency scale. Thus we shall see in a later chapter that for the rubidium standard it is collisions with the buffer gas atoms and *light shifts* produced by the light used to observe the resonance that are dominant; for the hydrogen maser it is the *wall shift*, and so on. We will discuss these and other cases at greater length in the chapters dealing with specific standards.

However, one subtle effect deserves mentioning, not so much for any practical reason, but because of its fundamental nature. It is the frequency shift due to the *recoil* of an atom as it absorbs or emits a resonant photon. As we have already mentioned in connection with the *Mössbauer effect* and will again encounter when we come to discuss the *laser cooling* of atoms, photons carry momentum, and to conserve linear momentum the atom must recoil. Since the momentum of a single optical photon is exceedingly small, the kinetic energy an atom gains through the recoil is also minuscule and is expected to be near the outer limits of what is observable.

Something similar might be said of some of the "relativistic effects" we have discussed: It is only under extreme conditions of high relative velocity (approaching the speed of light) and extraordinary concentrations of matter, as in neutron stars for example, that such effects become no longer just small corrections. Even at the level of submicrosecond precision and clocks as far removed from each other as intercontinental distances on Earth, the relativistic corrections are still only on the threshold of detection. Thus for Earthbound practical applications, relativistic effects are usually negligible. In fact, one of the most important applications of stable and precise clocks is in terrestrial navigation based on time signals received from radio broadcast stations, either fixed ground-based stations or in orbiting satellites, having precisely known geodetic positions. Special radio receivers equipped with dedicated computers are able to receive these signals and, from the relative differences in the transit times for the signals to reach the field position, compute the coordinates of that field point. Naturally, if future applications call for increasing the accuracy of positioning field points, then not only will more stable clocks be required, but also a point will be reached where relativistic corrections will become important. But they will remain small corrections, rather than gross errors, and the theoretical difficulties in synchronizing remote clocks will not be of practical concern in the foreseeable future.

Chapter 8
The Rubidium Clock

8.1 The Reference Hyperfine Transition

Of the atomic clocks, or more appropriately, frequency/time standards, since their accuracy and sophistication, not to mention their cost, places them far above any other man-made keepers of time, the rubidium clock has the distinction of being the most compact, and therefore the most portable. Rugged versions of the rubidium standard have long been developed for shipboard use as well as for tactical military and missile-borne applications.

The rubidium standard is based on the resonance at microwave frequency of the free rubidium atom between a pair of its quantum states whose separation in energy is due to the electron–nuclear hyperfine interaction. Its compactness is a result of confining the rubidium vapor in a small absorption cell filled with a noble gas to act as a buffer, as mentioned in the last chapter. While this method of confining the atoms of rubidium in order to lengthen their free interaction time with the applied resonant field has been very successful, there are unfortunately residual effects on the frequency of resonance due to the collisions with the buffer gas and to the pumping light itself that disqualify it as an absolute standard. Nevertheless, its general adoption for a variety of applications attests to its usefulness as a secondary standard.

In order to appreciate how the reference transition was chosen from among the many possible resonances observable in the ground state of Rb, we must examine the way in which the energies of the various magnetic substates may depend on the environment, since the resonance frequencies are directly determined by the difference in energy between quantum states. The most important environmental factor is the magnetic field; as earthbound beings immersed in the earth's magnetic field and surrounded by man-made magnetic fields from machinery, etc., it would require elaborate special shielding or active field-cancellation to reduce all static and time-varying magnetic fields. Fortunately, a particular choice of a resonant transition exists, whose frequency is very much less sensitive to the magnetic field than all the others. To see this we must consider how the energy of the atom in the various hyperfine states varies with magnetic field intensity. We recall that Rb^{87} has a nuclear spin I=3/2 and electron angular momentum in the ground state $J=1/2$, leading to total angular momentum (hyperfine) states with $F=2$ and $F=1$. Each of these states comprises a set of substates labeled by their magnetic quantum number $m_F=2, 1, 0, -1, -2$, and $m_F=1, 0, -1$, which give the projections (or components) of the angular momentum, and hence magnetic moment, along an assumed magnetic field direction. In the presence of a magnetic field, this can be only an approximate way to describe the states, since the angular momentum F obtained by adding (vectorially) the nuclear and electron spins will no longer be strictly constant in magnitude and direction, due to the torques exerted on the particles by the magnetic field. In trying to compare the relative strength of the "coupling" between the spins with their tendency to precess independently around the magnetic field direction, the appropriate measure is the amount of energy that would be required on the one hand to turn one spin relative to the other as compared with turning it with respect to the field.

The coupling energy of the spins is the difference in energy between the $F=2$ and $F=1$ hyperfine states in zero magnetic field, and it is precisely the transition between these states that gives rise to the sharp microwave resonance used as the frequency reference. Now, the energy of coupling of the spins to the magnetic field is simply the Zeeman energy, which we have already encountered; it is given by $E_m=-\mu_{\|}B$, where $\mu_{\|}$ is the component of the magnetic moment of the electron along the magnetic field direction. We are now ready to express the condition on the strength of the magnetic field for the representation in terms of F and m_F to be a good approximation; we must have $E_{F=2}-E_{F=1}>>\mu_{\|}B$.

For magnetic fields weak enough to satisfy this condition, the combination of two spins acts as one, with a single angular momentum F precessing around the magnetic field; however, the magnetic moment associated with this is only greater than the electron moment by the small contribution from the nucleus. Thus it is approximately as if we had a single particle with the ratio of magnetic to angular momentum smaller than a free electron in the ratio $(1/2):F$, and a correspondingly slower precession around the magnetic field. Thus in the limit of a vanishingly small magnetic field, the energy of the magnetic substates is simply $E_m=(m/F)\mu B$; that is, the plots of E_m versus B start from $B=0$ with a different slope for each m, as shown in Figure 8.1.

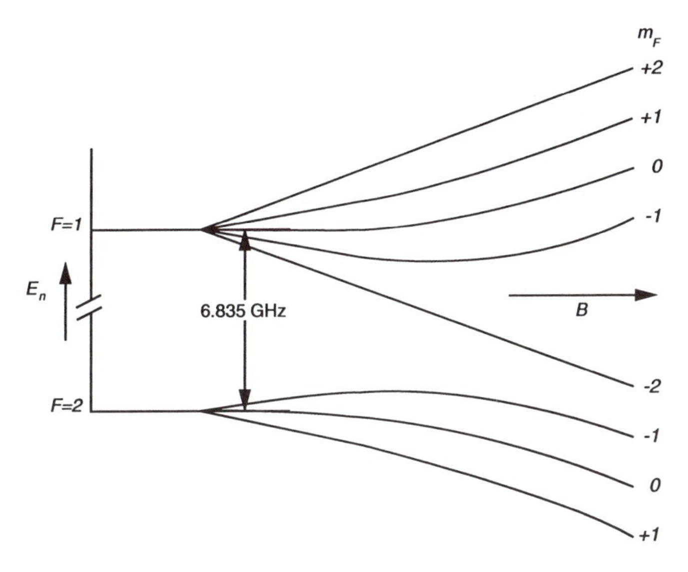

Figure 8.1 The energies of the magnetic hyperfine substates of the ground state in Rb^{87} as a function of applied magnetic field intensity

8.2 The Breit–Rabi Formula

The behavior of E_m as the magnetic field intensity is increased from zero requires an exact quantum treatment; the result is referred to as the *Breit–Rabi formula*, which can be written as follows:

$$E(m_F) = -\frac{E_{hfs}}{2(2I+1)} - \frac{\mu_I B_0}{I} m_F \pm \frac{E_{hfs}}{2}\sqrt{1 + \frac{4m_F}{2I+1}x + x^2}\ , \qquad 8.1$$

where the plus and minus signs refer to the upper and lower hyperfine state respectively, and $x \approx g_J \mu_B B_0 / E_{hfs}$ is the ratio of the Zeeman to the hyperfine energy splitting, and B_0 is the strength of the magnetic field.

There are two features of this solution of great importance to the operation of the rubidium standard: First, the plots of the energies of atoms in levels (F=2, m_F=0) and (F=1, m_F=0) versus the magnetic field start in a horizontal direction before they start curving gently, which means that to a first approximation in x the energy does not change if the magnetic field departs slightly from zero; second, the difference in energy between consecutive Zeeman sublevels near B=0 is proportional to the magnetic field.

Since, as we have already stated, it is difficult in practice to totally shield out the perturbations of a magnetic field or even variations in its intensity over all points in the rubidium absorption cell, the field-insensitive transition between the two m_F=0 levels belonging to F=1 and F=2 is used as the standard. Even the presence of a small inhomogeneity in the magnetic field, which would result in a given rubidium atom in motion "seeing" a variable magnetic field, would cause little spectral broadening of the resonance between these states. The same cannot be said obviously of the "field-dependent" transitions between other substates. Thus the resonance between the m_F=0 states is much sharper than the other Zeeman transitions, a fact of obvious importance for a frequency standard.

The field independence of the energy of the m_F=0 states is true only to a first-order approximation in the immediate vicinity of zero field; beyond that we must use the exact Breit–Rabi formula to calculate the frequency of the transition between these states. A second order approximation in x for that frequency derived from that formula yields

$$\nu = \nu_0 + \frac{(g_J\mu_B - g_I\mu_n)^2}{2h^2\nu_0} B^2, \qquad 8.2$$

where g_J and g_I are respectively the g-factors of the atomic electron and nucleus, numbers that are measures of the strengths of their magnetic moments in the given states. They specify the moments in terms of the fundamental units, the Bohr magneton μ_B and the analogous nuclear magneton μ_n, defined as the classical magnetic moment of a particle having the charge and mass of a proton. This provides a convenient way to make a fine adjustment to the frequency of the standard: Simply vary the current in a magnetic field-producing coil provided for the purpose. As a secondary standard, it is necessary to set the field at such a value that the time scale generated agrees with the atomic time scale, defined in terms of the primary standard. Even after the initial calibration against the primary standard, a readjustment may be necessary after some length of time because of possible long-term drift in the resonance frequency. Furthermore, to estimate the size of the field correction, which is proportional to B^2, we have ready at hand the "field-dependent" Zeeman transitions m_2–m_1=±1, whose frequency gives directly the magnitude of the magnetic field.

8.3 Optical Pumping of Hyperfine Populations

Recall that in our discussion of magnetic resonance in chapter 6 we argued that in order to be able to observe a magnetic transition between two states, there must be a difference in the populations in those states. This is ultimately because the inherent probabilities (per atom) per unit time for absorption and stimulated emission of a quantum of radiation are identical, and unlike transitions in the optical region of the spectrum, the probability for spontaneous transitions is extremely small; the result is that no net global exchange of energy is observed unless the number of atoms in the lower state differs from the number in the upper state. Here we wish to observe the transition between the (F=1, m_F=0) and (F=2, m_F=0) states near zero magnetic field intensity. There are a number of possible ways of achieving this using optical resonance; the choice generally adopted reflects the inevitable concern for commercial viability, which is ultimately a question of performance versus cost. Before describing in detail the method that has been widely exploited commercially, we can get a broader perspective by first considering alternative, but more complex, approaches. The first among them is simply to carry out the usual Kastler optical pumping with circularly polarized light directed along the magnetic field axis, thereby ideally putting all the atoms in one of the extreme m_F-states, that is, m_F=+2 or −2, depending on the sense of the circular polarization. We are assuming here, as we did in the last chapter, that the spectrum of the pumping light is limited to resonance with the transition to the upper $P_{1/2}$ electronic state but otherwise is broad enough in frequency to satisfy energy conservation for all hyperfine transitions that satisfy the angular momentum selection rules. Once we have a large proportion of the atoms in, say, the m_F=+2 state, we can apply a high-frequency magnetic field resonant with transitions among the magnetic substates in the F=2 hyperfine state, in the manner described in the last chapter. The desired population difference between the m_F=0 substates can be achieved under suitable conditions, the most important of which are first, that the atoms be sufficiently free of perturbations; second, that the high-frequency field be strong and uniform; and last, that the static magnetic field be sufficiently uniform. The global effect on the atoms can be pictured classically as a magnetized gyroscope whose axis precesses around the static magnetic field, sweeping out ever wider cones. If the high-frequency magnetic field is on for exactly the short interval it takes the angle of the cone to reach 90°, a so-called 90° pulse, then the axis of the gyroscope precesses in the plane perpendicular to the static field, and the projection of its angular momentum along the field is zero. In quantum terms this is described as having put the atoms in a (linear) superposition of substates with different m_F in which the desired m_F=0 substate has the largest amplitude.

Having a significant fraction of the atoms in the (F=2, m_F=0) substate, a fraction far greater than would ideally be present in the (F=1, m_F=0) substate, meets the first requirement for observing transitions between them. A second and equally critical requirement is the ability to detect the occurrence of transitions; this can be met in principle by the inverse process of applying a 90° pulse of the opposite phase, or a 270° pulse of the same phase, to complete a full circle, bringing the global moment back into alignment with the static magnetic field. If nothing perturbs the atoms in the interval between the two pulses, the atoms would ideally return to their original state, namely, the nonabsorbing m_F=2 into which they were pumped, and the amount of pumping light scattered or absorbed would be the same as it was. However, if in the interval between pulses a resonant microwave field causes transitions to the (F=1, m_F=0) hyperfine state, then they do not all return to their nonabsorbing state, and the amount of pumping light absorbed/scattered will be increased. This change in the interaction between the atoms and the pumping light can be used to monitor the transitions and their resonant dependence on the frequency of the microwave field.

Another technique, which avoids the practical complexity of pulsed operation, is a variant of Kastler optical pumping, in which the circularly polarized beam is directed not along the magnetic field direction, but perpendicular to it, and is therefore called *transverse pumping*. In order to simplify the explanation of this technique, let us assume that the transverse light beam consists of a regular succession of powerful flashes and that the magnetic field is extremely weak. Under these conditions each flash produces a global magnetic polarization in the direction of the beam, that is, perpendicular to the field, which then causes it to precess continuously around the field axis passing periodically through its original direction. Now, if the interval between flashes is adjusted so that each flash coincides in time with the passing of the polarization vector through its original direction, then the polarization is reinforced and will build up to a significant degree. Again, having the polarized atoms precessing predominantly perpendicular to the field implies a preponderance of population in the m_F=0 state, essentially the same state as was produced by a 90° pulse. In the actual implementation of the transverse pumping technique it is not necessary to pulse the light source. Instead, a high-speed electro-optic modulator can be used to impose a harmonic oscillation in the transmitted intensity at the frequency of precession of the atoms in the given magnetic field. If during the pumping process the desired microwave transition is resonantly induced between the hyperfine states having m_F=0, then the distribution of populations of atoms in the different m_F states changes in the direction of increasing those in the absorbing F=1 substates. It will be recalled that the optical pumping process leads to a preponderance of atoms in substates that by reason of the selection rules are unable to absorb light from the pumping beam. Hence by imposing a different distribution with the resonant microwave transitions, the amount of pumping light scattered by the atoms will increase, providing a way of monitoring the resonance. As with the previous technique then, the desired resonance is observed by monitoring the transmitted light intensity as the microwave frequency is swept through resonance; a dip in the intensity of the pumping light transmitted through the absorption cell signals a resonance.

It is interesting to note in passing that the original use of a circularly polarized light beam perpendicular to the magnetic field was first introduced by Hans Dehmelt (Dehmelt,1957) as the inverse process to the foregoing: It was to modulate the intensity of the beam by interaction with a precessing global polarization induced by a resonant high-frequency magnetic field, such as would be used in the 90° pulse, acting on polarization produced by an axial beam. This modulation occurs at the precession frequency and is a direct measure of the (static) magnetic field. Since frequency is measurable with high precision, this has been exploited commercially as an atomic magnetometer of great sensitivity and precision.

8.4 Optical Hyperfine Pumping: Use of an Isotopic Filter

We will now direct our attention to the principle of operation actually implemented in commercial Rb standards; it is called *hyperfine pumping*. Instead of relying on quantum selection rules governing transitions between states of different angular momentum, it is really based on selection of transitions according to the conservation of energy. We focus on the spectrum of the pumping light rather than its polarization; transitions will occur only if the energy and therefore wavelength of the photons in the beam equals the energy difference between the initial and final states. The method therefore relies on having a pumping light source whose spectrum overlaps only one of the two hyperfine components in the resonance optical spectrum of Rb, components that arise from transitions whose initial states are either the F=1 or F=2 hyperfine state. Having a spectrum overlapping only one component, the light from the source can be absorbed by atoms in only one of these states. However, once in the optically excited state, the atoms will spontaneously re-emit photons to both hyperfine states of the electronic ground state, independently of how they came to be in the excited state, and therefore ideally they would all be pumped into the nonabsorbing hyperfine state.

As mentioned in connection with light sources for Kastler pumping, laser sources properly stabilized would be ideal were it not for the added complexity and susceptibility to mechanical shock and vibration. It might be thought that we should be able simply to filter out one of the hyperfine components of the optical resonance line in the spectrum of a Rb vapor lamp; unfortunately, the difference in wavelength between the two components is so small that it would be difficult, if not impossible, using the sharpest type of optical filter available, the interference filter, to separate them without a great loss of intensity.

The original experiments on hyperfine pumping predate lasers, and a suitable light source was achieved through a fortuitous coincidence in the hyperfine structure of the optical resonance spectra of the two isotopes of rubidium, Rb^{85} and Rb^{87}. The difference in nuclear structure and mass of the two isotopes leads to a slight relative shift in their spectra, called, not surprisingly, the *isotope shift*. It happens that one of the two hyperfine components in the Rb^{85} spectrum nearly coincides with the corresponding component in the Rb^{87} spectrum, while the others are well separated, as shown in Figure 8.2.Thus starting with a rubidium vapor lamp filled with enriched Rb^{87} isotope, whose output contains both hyperfine lines, we can partially remove one of them by passing the light through a cell containing enriched Rb^{85} vapor, which will absorb out of the beam (and re-emit in all directions) the line coincident with the Rb^{85} hyperfine component just mentioned. The match in wavelengths can be improved by a process that, however, we shall see is detrimental to the long-term stability of the standard: It is the so-called pressure broadening and shift in spectral lines caused by collisions between the Rb atoms themselves and with others, principally the atoms of the noble gas introduced as a buffer. The direction of the shift, whether to higher or lower wavelengths, depends on which noble gas is used, as does its sensitivity to temperature fluctuations. By using a mixture of two noble gases, which alone would produce opposing shifts, it is possible to choose their proportion in the Rb^{85} vapor absorption cell so as to reduce the temperature dependence as well as improve the wavelength match.

The earliest successful observations of resonance between hyperfine states in the alkali atoms by optical means dates from 1958. Of these, the one of particular interest, because of its adoption for further commercial development, was published by Bender, Beaty, and Chi (Bender et al.,1958), in which the use of the isotope filter in Rb was introduced. Their experimental arrangement is shown in Figure 8.3. The optical hyperfine pumping source was a Rb spectral lamp whose strongest emission occurs at the two "resonance" lines in the red part of the spectrum at wavelengths λ=780 nm and 795 nm (1 nm=1 nanometer=10^{-9} meter). These correspond, in terms of the states between which the transitions occur, to the strong emission lines in a sodium vapor lamp, giving it its familiar yellow color. They arise from radiative transitions between the first two excited electronic states and the ground state, forming a fine structure "doublet." The two hyperfine states with F=1 and F=2, into which the first excited electronic state is split, are much closer in energy than the two corresponding hyperfine states in the ground state, so that under the usual degree of resolution the spectrum appears to have each member of the doublet split into two hyperfine components rather than four. An ordinary Rb vapor lamp, containing a natural mixture of the two isotopes, will therefore emit a spectrum in which each line of the doublet consists of four hyperfine components, two from each isotope. The hyperfine separation in Rb^{85} is about half that in Rb^{87}, with the lower components of the two isotopes much closer in frequency than the upper, as Figure 8.2 attempts to make clear. In the actual experiment, the light from an enriched Rb^{87} spectral lamp is passed through a filter cell containing enriched Rb^{85} vapor and typically 10^4 Pa of argon buffer gas, whose presence broadens the spectral lines and somewhat shifts their centers in a direction to enhance the differential filtering of the two hyperfine lines of Rb^{87}. Ideally, if the filter cell contained only Rb^{85}, it would be almost transparent to the upper component in the output spectrum of the Rb^{87} lamp, while strongly scattering the other component, so that the transmitted light satisfies the basic inequality of intensities required for hyperfine pumping.

The repeated cycle of absorption and re-emission of this light by the Rb^{87} in the resonance absorption cell will pump the atoms into the upper hyperfine state of the electronic ground state, thereby reducing the number left in the lower absorbing state. This has the effect of decreasing the amount of pumping light scattered. If an applied microwave magnetic field causes transitions between the hyperfine states, tending to equalize their populations and thus increasing the number in the lower absorbing state, then more of the pumping light is scattered out of the beam, and the transmitted intensity drops, signaling the occurrence of resonance.

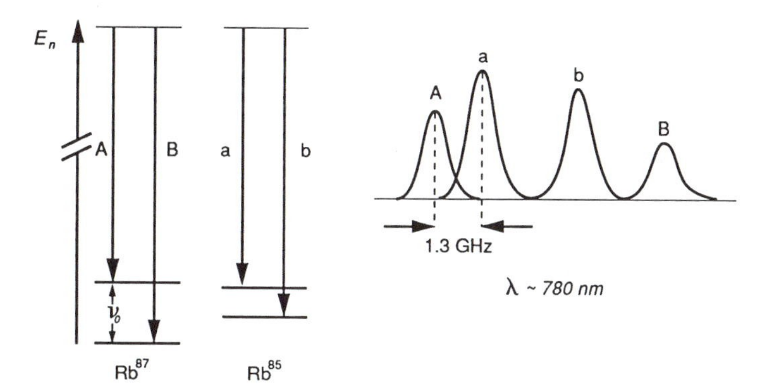

Figure 8.2 The hyperfine structure of the dominant emission lines in Rb^{85} and Rb^{87}

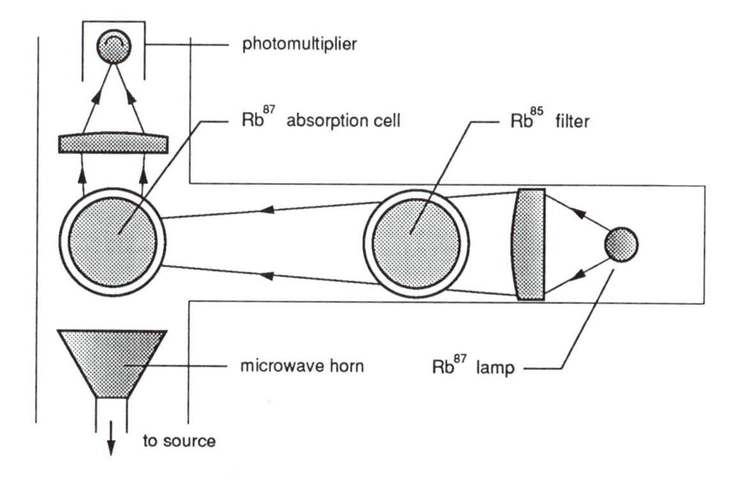

Figure 8.3 The experiment of Bender, Beaty, and Chi on optical hyperfine pumping of Rb^{87}

8.5 The Use of Buffer Gases

In those early experiments, the absorption and filter cells were made of relatively large (≈500 ml) Pyrex bulbs, which were cleaned and baked according to standard vacuum practice, and a small quantity of pure metallic Rb distilled into each bulb. The element Rb, like the other alkalis, is very chemically reactive with air and water; it must be handled either in an inert atmosphere or under vacuum. The cells are back-filled with pure gas before being sealed from the vacuum system.

Experiments carried out with different species of gases at different pressures showed that the hyperfine frequency is shifted in proportion to the pressure, being raised by the light gases—hydrogen, helium, neon, and nitrogen—while for the larger atoms—argon, krypton, xenon, and methane—the frequency is lowered. The linear pressure dependence is to be expected, at a fixed temperature, if the shifts are due to pairs of atoms colliding. There is also an important temperature dependence, however, which is more complicated to predict. An important practical application of these findings is to the reduction of the sensitivity of the frequency to temperature, since some gases cause an increase in frequency with temperature, while others have the opposite effect. It has been found that a mixture of about 12% Ne and 88% Ar gives a temperature dependence of about −10 hertz per degree C at a pressure of about 5.320×10^3 Pa (1 mm Hg≈133 Pa). The sharpest resonance seen for the hyperfine transition at 6.8347...GHz (1 GHz=10^9 Hz) was only about 20 Hz! This is a Q-value of about 300 million; compare this to the best quartz crystal currently available, which may reach a Q-value of 1 million.

The effects of a buffer gas on spectral line shapes and center positions are the large-scale average manifestation of interaction between the Rb atoms and the noble gas atoms; during the collision we have in effect a transient Rb–noble gas "molecule." Under the rarefied conditions obtaining here there would be little error in assuming what may be called the *binary collision approximation*. In this, it is assumed that all the particles have negligible interaction except very briefly during relatively infrequent encounters when the particles come within typical molecular dimensions of each other. This radically simplifies the problem of predicting the effect of the presence of the buffer gas by permitting the problem to be separated into two tractable parts: first, the collision of just two particles under general conditions, and second, the statistical problem of finding the observable averages over a large number of such collisions. Another circumstance that allows an important simplification in the analysis of the collision process is that the relative velocity of the Rb and noble gas atoms at the temperatures under consideration is very much smaller than the speed with which the electrons in the outer structure of the atoms can adjust to the changing internuclear distance. This means that the colliding pair can be thought of as quasi-static at different distances apart, and the energy of the quantum electron state for the two-atom system can be regarded as potential energy in computing the change in kinetic energy of the two particles. The potential energy for the Rb–noble gas collisions, in common with most binary atomic collisions and in a broader sense all matter, corresponds to an attractive force as the particles first approach each other; but as their electronic structures start to interpenetrate, the force turns repulsive and they fly apart. It turns out that during the initial attractive part of their trajectory, the distortion of the electronic state of the Rb atom is accompanied by a reduction in the electron–nuclear interaction, that is, a "red" shift (to lower frequency) in the hyperfine frequency separation, while the repulsive part has the opposite effect. In the heavier noble gases the attractive force is of longer range and leads to a net red shift, while for the lighter gases, He and Ne, it is a blue shift (to higher frequency) because the short-range repulsive force dominates. The length of time a typical collision lasts is extremely small, as can easily be verified: The average relative velocity of the particles due to thermal agitation is on the order of 10^4 meters per second, while the range of interatomic force is typically 10^{-8} meter, so that the time is $t=d/V=10^{-12}$ second. On the other hand, the average time between collisions at the typical gas pressure of 1000 Pa is about 10^{-7} second, or about 100,000 times the duration of a collision. This confirms that the impact approximation is indeed valid for the assumed conditions. Of course, if very much higher buffer gas pressures are used, then the approximation would become invalid and the prediction of the pressure shifts would be very much more difficult. The temperature dependence of the frequency shifts has to do with degree of mutual penetration of the colliding particles; the higher their thermal kinetic energy, the more violent the collisions.

The susceptibility of the resonance frequency, which is to be our standard, to temperature, pressure, and nature of background gases detracts from its accuracy and reproducibility. However, the presence of the buffer gas not only performs the essential function of lengthening the interaction time between the atoms and the resonant field, but also limits broadening of the resonant response by another phenomenon, namely the Doppler effect.

The frequency width at half maximum of the optical resonance line in rubidium is about 700 MHz, which is about 1/10 the hyperfine splitting of the ground state. On the other hand, we can easily verify that for the microwave transition the conditions for the Dicke effect are well satisfied; thus the wavelength of microwave resonance at around 6.8 GHz is $\lambda=c/\nu$, or $\lambda\approx4.3$ cm, and the average distance between collisions at pressures on the order of 10^4 Pa is no more than 0.01 cm, or about 1/400 of the wavelength. Finally, on the subject of the Doppler effect we should note that if the resonant microwave field is applied in the form of an advancing wave, as was done in the experiments of Bender, Beaty, and Chi, the observed resonance will exhibit a small Doppler shift due to a general drift of Rb atoms across the absorption cell. This arises from the fact that Rb reacts chemically with the glass surfaces of the cell, and atoms continually diffuse from their source, a droplet of liquid Rb, towards the walls of the cell. A good deal of effort has been devoted in the past to finding a suitably inert coating for the inner surfaces of the cell, not only to reduce this chemical reaction, but indeed to dispose of the need for a buffer gas altogether. Special aluminosilicate glazes have been developed by the lamp industry to coat the inner surfaces of sodium lamps used for street lighting, lamps that operate at much higher temperatures than the Rb cell, in order to prevent the sodium vapor from chemically attacking the glass and eventually turning it black. The ideal of a cell with surfaces totally unreactive with Rb, which may therefore be "dry filled," that is, not requiring a liquid droplet to maintain the vapor density, has never been achieved. Nevertheless, wax coatings made of high molecular weight paraffins were shown by H. Robinson et al. around 1957 to be highly successful in preventing a randomization of Rb spin direction when the atom collides with the coated surface. Their use instead of a buffer gas to increase the free interaction time between the Rb atoms and the resonant field offers, however, no particular advantage, since like the buffer gases, these coatings also cause frequency shifts. In any event, whether the glass surfaces are coated or not, there will be long-term chemical reaction with the Rb, resulting in the slow evolution of gases, which will cause the resonance frequency to drift. The use of sapphire and other exotic materials to solve this problem continues to be investigated.

8.6 Light Shifts in the Reference Frequency

An equally serious but far more subtle phenomenon that affects the resonant microwave frequency is associated with the pumping light itself; this complex effect is labeled simply the *light shifts*. It was anticipated theoretically by Cohen-Tannoudji and Barrat in 1961 and was soon observed in Kastler's laboratory in the radio-frequency spectrum of Hg^{199}. Its detection in studies of microwave resonance in Rb and Cs was first published in the same year by Arditi and Carver. Although the shift is exceedingly small—its discovery in itself representing no mean accomplishment—yet in the context of a frequency standard it is significant. In addition to the obvious dependence of the shift on the intensity of the light, it also depends on the detailed spectrum of the light, particularly its position in relation to the absorption spectrum of the Rb atoms.

In trying to understand the physical origin of light shifts we must distinguish between two types of quantum transitions involved in the optical pumping cycle: real transitions and the so-called *virtual* transitions. A transition is virtual if the probability of the atom being in the final state returns to zero when the external perturbation is removed rather than stay there, as in a real transition. In the case of Rb subjected to the optical pumping cycle, real transitions would take the atoms up to the first excited electronic state, where they would stay were it not for another mechanism, spontaneous emission, by which they re-emit photons and return to the ground state. At the same time there are virtual transitions in which the electric field of the light wave distorts the electron cloud, and as long as that field is there the atoms are in a quantum state that in terms of the undisturbed Rb stationary states can only be described as a *linear superposition* of them. Once the perturbing electric field is removed, the atom returns to its initial state. This distortion in the electron distribution manifests itself in a shift in the position of lines in the optical spectrum called the (quadratic) *AC Stark effect*, and it depends in general on E^2, where E is the amplitude of the electric field in the light wave. Because of the quadratic dependence on E, the effect does not average to zero for an oscillating optical field that swings symmetrically through positive and negative values about zero. Unfortunately, the energy shift of the quantum states is not the same for the initial and final hyperfine states of the microwave transition we are interested in, resulting in a light shift in the frequency of that transition. Since this source of change in our standard frequency is affected by the many complex factors that determine the detailed spectrum of the light source and absorbing Rb atoms, it seriously detracts from the quality of the Rb standard.

8.7 Rubidium Frequency Control of Quartz Oscillator

We will now take up the subject of the electronic configuration of clocks controlled by the Rb resonance. There are two ways in which the microwave resonance can be used: first, in a passive mode as a resonator or frequency discriminator, and second, in an active mode as an oscillator (maser) generating a signal at the standard frequency. The passive Rb standard is the one that has been commercially developed and is in general use; we will therefore treat it in this chapter, leaving the Rb maser for a later chapter.

As with other standards using an atomic resonator in a passive mode, notably the cesium standard, which we shall study in the next chapter, the resonant response of the atoms to an external microwave field must be monitored. Specifically, it must be possible to discern whether the frequency of the field is below, above, or precisely at the center of the resonance curve. This may be accomplished by a slow periodic modulation of the frequency (or phase) of the microwave field over a small portion of the resonance line profile, as shown in Figure 8.4. We recall that if the modulation is slow enough, it is legitimate to think of the frequency as assuming continuously all frequencies between the limits of the modulation, so that the optical signal will vary in step with the modulation on the side of the resonance with a positive slope and will vary in the opposite direction on the side with the negative slope. If the modulation occurs symmetrically about the center of the resonance, then since the optical signal falls whether the field frequency swings in the positive direction or the negative, the optical signal will oscillate at twice the frequency of modulation. If the modulation is exactly centered on the peak of the resonance, then the optical signal will have no Fourier component at the modulation frequency, but only one at double that frequency.

For the ultimate purpose of controlling the frequency of the microwave field so that it remains locked to the peak of the resonance, we need to derive from the optical signals described above a voltage that can serve as an error signal indicating whether the frequency of the field is too high or too low. This requires a circuit that can selectively amplify signals at the modulation frequency and be sensitive to the relative phase of these signals with respect to the modulating signal. Such a phase-sensitive amplifier is called a *lock-in amplifier*, which in a sense correlates an incoming signal with a reference frequency signal; if the two are in phase, a positive output voltage is given: On the other hand, if they are of opposite phase, we get a negative output voltage. Moreover, if they are not exactly the same frequency, the output will oscillate as the two signals go in and out of phase; in this case the average over a sufficiently long time will be zero. With such a lock-in amplifier we can obtain just the desired error signal for our control circuit; we simply use the signal producing the frequency modulation of the microwave field as the reference, and the optical signal as our input to the lock-in amplifier. Its output as the center frequency of the field is very slowly scanned across the resonance frequency will resemble the plot shown in Figure 8.5.

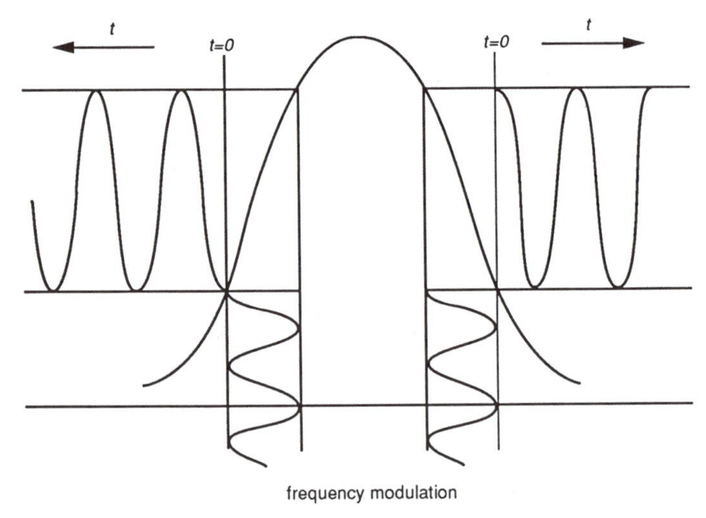

Figure 8.4 The reversal of detector output phase for microwave frequency above and below resonance

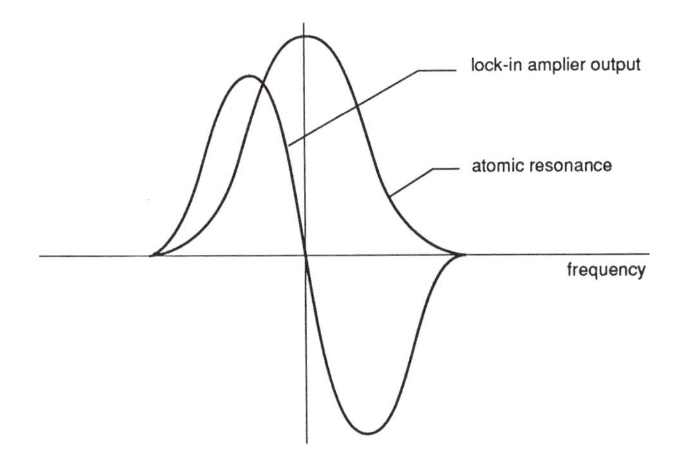

Figure 8.5 The output of the lock-in amplifier as the frequency is swept through resonance

A graph of that form is usually called a dispersion curve, and in the sense used in calculus, it is the derivative function of the bell-shaped absorption curve. We see that the output is indeed negative when the frequency is too high, positive when too low, and zero when at the peak. This is precisely what we need as an error signal in a feedback control loop that seeks to make the error zero by controlling the center frequency of the field.

The field is derived ultimately from a high-quality quartz-controlled oscillator operating typically at 5 MHz (see Figure 8.6). The desired microwave frequency is produced by a frequency synthesizer, which, starting with the 5 MHz oscillation as reference, generates signals at multiples and submultiples of that frequency and then by deriving other signals at the sum or difference of various harmonics, ultimately yields an output signal whose frequency can be preset in fine increments on a front panel keyboard.

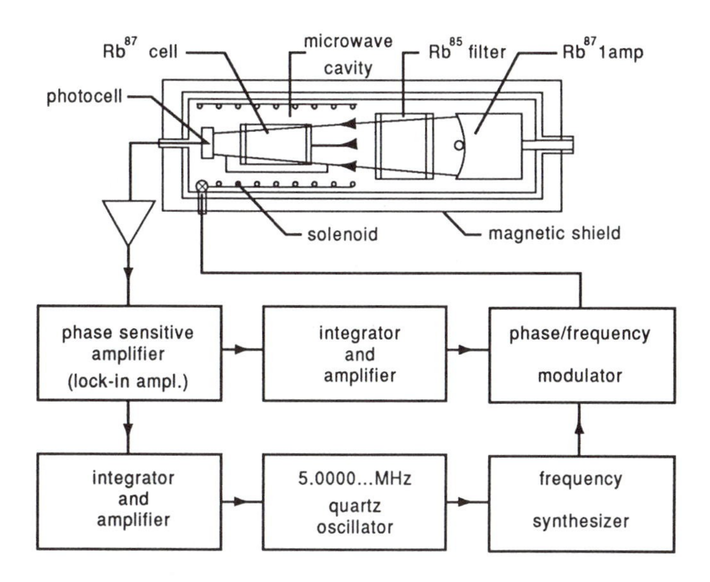

Figure 8.6 Block diagram of basic rubidium frequency standard design

These arithmetical operations on the frequencies of signals are realized through the use of nonlinear solid-state devices, which can act as harmonic generators and frequency mixers. Throughout these operations phase relationships are preserved so that the output is coherent with the stable 5 MHz reference signal. Coherence here simply means that the outputs of two identical synthesizers, sharing the same reference signal but set at different frequencies, can produce a stable "beat," that is, a pure signal whose frequency is the difference between those of the two outputs. Clearly, if the frequency of the reference is changed, the frequency of the output of the synthesizer will change accordingly. Therefore, a voltage-sensitive element in the quartz crystal oscillator circuit is provided to control its frequency and hence that of the microwave field applied to the atoms. If that frequency is too low, the resulting positive error signal must cause the oscillator frequency to rise steadily in order to reduce the error. This requires the voltage appearing on the frequency-controlling element, which is derived from the output of the lock-in amplifier, to increase steadily in magnitude in the proper direction. The circuit that converts a steady (DC) voltage into a linearly increasing one is an *integrator*, which must be included in the feedback loop for stable operation. As the error signal approaches zero at the peak of the resonance, the output of the integrator tends to become constant just at the value to keep the error at zero. Needless to say, a high degree of stability in the operating voltages of the integrator is critical; any drift in DC levels would cause frequency offsets from the atomic resonance, degrading the performance of the standard. Just how closely the frequency is held to the true center of the atomic resonance depends on a large number of factors, some of which are of a fundamental nature, while others are a matter of the performance characteristics of particular devices and the circuits around them.

8.8 Frequency Stability of the Rubidium Standard

We have already mentioned a number of physical phenomena that affect the frequency of the atomic resonance; these are known systematic sources of error, as distinguished from uncontrollable random fluctuations. Among the latter are the various types of electrical noise discussed in a previous chapter. Because of the high order of multiplication of frequency in going from that of the quartz oscillator at 5 MHz to the microwave frequency at 6,800 MHz, any residual fluctuations in the phase of the quartz oscillator are greatly magnified. Therefore, attaining the highest phase stability in the microwave field and hence the sharpest spectrum puts a great burden on the spectral purity and low noise of the quartz oscillator. It is the availability of high-performance quartz oscillators, with extremely high Q-values and low output noise, that has contributed immensely to the success of these atomic standards.

In discussing the accuracy and stability of any type of standard, questions must be addressed that would not arise for ordinary instruments. When we use an ordinary voltmeter, for example, we assume that its calibration and accuracy are traceable ultimately to some acceptable standard. But if the standard be in doubt, what then? This question is really relevant only to the cesium standard, which has been elevated to the status of primary time standard. Nevertheless, as an atomic resonance-based system, the rubidium clock qualifies, for many applications, as a secondary standard and as such, absolute accuracy is not expected of it; its frequency must be set by reference to a primary standard. But how is one to know whether the primary standard is drifting? This question lies at the heart of what is expected of a standard: Standards are not supposed to drift! The pragmatic answer is to have a large collection of embodiments of the standard all purporting to display a unit of time in accordance with its atomic definition. To the extent that there is agreement among the members of this collection, we can have confidence in their accuracy and stability.

The accepted method of specifying the stability of frequency standards, useful particularly for relatively long-term performance, is, as we saw in a previous chapter, in terms of the *Allan variance* of phase or frequency plotted as a function of the sampling time over which that quantity is measured. We recall that this analysis presumes that the condition of stationarity is satisfied, and therefore any long-term drift in the data must first be separated out. We also noted previously that some of the fundamental types of noise can be accurately modeled as having a Fourier spectrum that has a simple power-law dependence on frequency. These power laws translate into equally simple dependence of the Allan variance $\sigma(\tau)$ on the time interval τ used in its measurement. Thus for the important flicker noise it can be shown that σ is independent of the length of the interval, whereas for white frequency noise σ falls as $1/\tau^{1/2}$. Now, for circuits at ordinary temperatures operating in the radio-frequency range, thermal (Johnson) noise is very nearly "white" (the same power density at all frequencies), so that since this is a universal source of noise, we frequently see a plot of σ versus τ exhibit the $1/\tau^{1/2}$ characteristic of this type of noise, at least up to a certain point, after which flicker noise becomes dominant and the graph flattens out.

Figure 8.7 shows plots of the Allan variance for a typical Rb standard together with, for the sake of comparison, several other types of atomic standards we shall be dealing with in succeeding chapters. It is not unusual for a Rb standard to have σ bottom out at around 10^{-12} for time intervals longer than 100 seconds, after systematic drifts have been separated out. This represents an error of about 30 millionths of a second in a year!

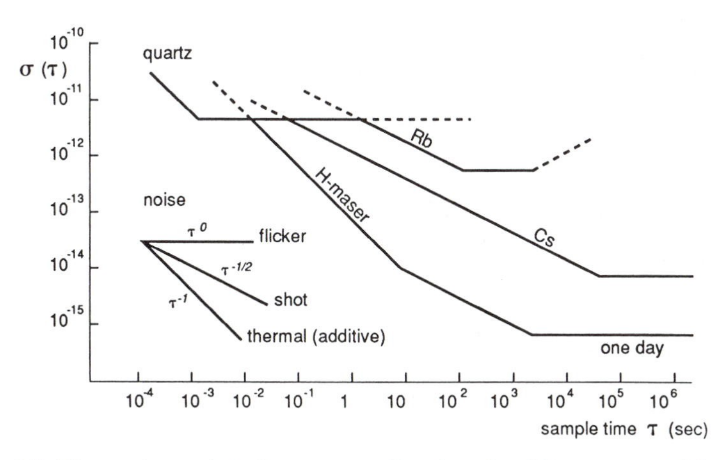

Figure 8.7 Allan variance plotted versus sampling time for different types of frequency standards

8.9 The Miniaturization of Atomic Clocks

The optically pumped alkali vapor gas-diffusion cell resonator lends itself admirably to miniaturization, particularly since the development of semiconductor lasers that are tunable and can be stabilized on the rubidium or cesium optical resonance wavelengths while operating at room temperature. This development obviated the need for the UHF driven lamp and isotope filter of the conventional rubidium standard. Coupled with microelectronic integrated circuitry this has enabled atomic clocks to be built no larger than a walnut!

The fundamental consequences of size reduction are an increased frequency of wall collisions or the need for a higher collision frequency with the molecules of the buffer gas. In either case the undesirable shifts in the reference transition frequency are aggravated, and long-term stability is expected to suffer. The other important consideration is the size of a resonant microwave cavity. For Rb^{87} and Cs^{133} the reference microwave wavelengths are about 4.4 cm and 3.3 cm respectively. Since Rb has fewer magnetic substates than Cs, a larger fraction of Rb atoms can contribute to the signal arising from transitions between one particular pair. In either case the cavity must be "loaded" with a low-loss dielectric material to lower its resonant frequency as its size is made smaller. It has been shown (I. Liberman, 1992), for example, that a miniature standard based on a cesium gas cell not exceeding 4 mm in diameter and 18 mm long, operating under relatively high vapor pressure, so that Cs–Cs collisions are dominant in limiting the free lifetime of the Cs states, would have theoretically a short-term stability $\sigma(\tau)$ on the order of $5\times10^{-12}\tau^{-1/2}$ where τ is in seconds.

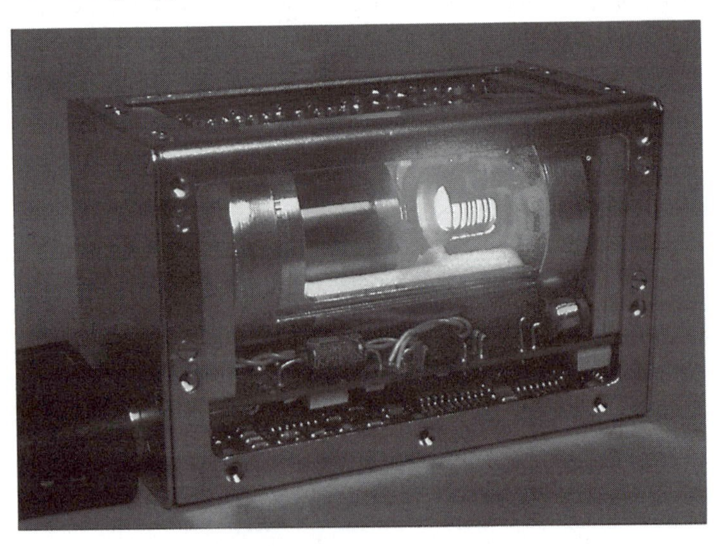

Figure 8.8 Small "industrial" Rb oscillator developed by the Neuchâtel observatory (Rochat, 1994)

Chapter 9
The Classical Cesium Standard

9.1 Definition of the Unit of Time

We will now take up the type of atomic clock that has been elevated to the status of the primary standard of time, displacing the historical role of astronomical observations in the definition of the unit of time, the second. In 1967 the 13th General Conference on Weights and Measures, attended by delegates from about 40 countries, signatories of the Treaty of the Meter, adopted a new definition of the international unit of time. At that conference there was overwhelming support to the idea that the time had come to replace the existing definition, based on the earth's orbital motion around the sun, by an atomic definition. The wording of the new definition is as follows: "The second is the duration of 9,192,631,770 periods of the radiation corresponding to the transition between the two hyperfine levels of the fundamental state of the atom of cesium-133." The ten-digit number assigned in the definition was chosen to agree with the then existing definition of the second, known as the "ephemeris second," which had been adopted in 1956. This latter definition was based on the length of the so-called tropical year, that is, the length of time for the earth to complete its orbit around the sun and return to a point where its axis again makes the same angle with respect to the earth–sun direction; it is the repetition period of the seasons. The obvious drawback to this definition is the practical one of not being available except through the intermediary of stable clocks that must be checked after the fact. But more importantly, a decade after its adoption it had become evident that the accuracy of atomic clocks, which had to be used to implement the ephemeris time, had reached the point where they had become *de facto* standards against which astronomical observations were compared.

9.2 Implementation of the Definition: The Cesium Standard

This new definition is based on the same type of microwave resonance as in the Rb standard, but because of some advantages in detail, the resonance chosen is in the heavier alkali atom cesium. We should point out, however, that the labels Rb standard and Cs standard in common use do not refer merely to the species of atom used, but rather imply certain ways in which they attempt to extend the interaction time between the undisturbed atom and the resonant microwave field. In the common Rb standard we recall that a noble gas is used as a buffer to prevent the free flight of the Rb atoms to the walls of the cells, where their coherent response to the field would be interrupted, and the resonance thereby broadened. By contrast, in the cesium standard the atoms move freely as a beam in a chamber from which the air has been pumped out, that is, in a vacuum. (The use of the word "beam" is more than metaphorical; after all, a light beam can be looked on as a stream of photons.) There is, of course, no fundamental reason that precludes observing the Cs resonance in a diffusion cell by optical methods, or of observing the Rb resonance in an atomic beam machine; in fact, both possibilities have been explored in the past.

In the early development of these devices, they differed not only in the "containment" of the atoms, but also in the way the microwave resonance was made observable: The Rb clock detected resonance by optical hyperfine pumping using a "conventional" uhf-excited vapor lamp as a source, and the Cs standard used magnetic deflection as in the Stern–Gerlach experiment. We will describe in this chapter what might be justly called the classical Cs beam standard using magnetic deflection and reserve to a later chapter the laser-based systems.

Observing atoms in free flight ensures that they suffer only the desired interaction with the resonant field, and not with background particles or optical pumping radiation, both of which, we have seen, produce shifts in the resonance frequency. It is precisely this freedom from unpredictable frequency shifts that made the Cs standard uniquely suitable as a primary standard. Ideally, such a standard must make possible the faithful observation of the sharpest possible resonance with the highest possible signal-to-noise ratio, on a system insensitive to operating conditions. In fact, we can quantify this statement by recalling the result cited in Chapter 7 that for any resonator acting as a frequency reference, the uncertainty in finding the center frequency is $\Delta\nu/(S/N)$, where $\Delta\nu$ is the frequency width of the resonance, and S/N is the signal-to-noise ratio. A figure of merit that increases with decreasing uncertainty can therefore be defined as $F=(S/N)/\Delta\nu$. In the case of the Cs standard, S/N is ultimately limited by shot noise due to the atomic nature of Cs and $S/N=\sqrt{n}$, where n is the number of atoms contributing to the resonance signal.

In the Cs standard the atoms undergoing the resonant transitions move *in vacuo* with thermal velocities, acted on only by a weak uniform magnetic field and the probing resonant microwave field. To deduce the "true" transition frequency of Cs at rest in zero magnetic field, free of interaction with a microwave field generator, involves deterministic or *systematic* corrections based on well-established theory. Thus as long as we believe that a cesium atom is a cesium atom no matter what its provenance, we have a universally reproducible standard. Of course, we can always speculate as to whether it is possible that the fundamental properties of atoms may be slowly evolving relative to a time scale established by other dynamical processes in the universe; however, this is of no practical concern.

The two hyperfine states between which the resonant frequency of transition defines the standard second are indicated in Figure 9.1, which shows the energies of all the hyperfine substates plotted as a function of the intensity of an external magnetic field. The cesium atom has only one stable isotope, mass 133, with nuclear spin $I=(7/2)h/2\pi$, which coupled with the outer electron spin $J=(\frac{1}{2})h/2\pi$ yields according to quantum rules the following total angular momentum: $F=4$ or $F=3$ in units of $h/2\pi$. We saw in an earlier chapter that for magnetic field intensities near zero, the atoms in these two hyperfine states act like bar magnets, which however obey space quantization rules; that is, they can be observed to have only (in this case) integral values (in units of $h/2\pi$) for their components along the field axis. Thus an atom in the $F=4$ state is further characterized by the magnetic quantum number m_F, giving the discrete com-ponents of the angular momentum, which can have only the integral values +4, +3, +2, +1, 0, −1, −2, −3, −4; and similarly for the $F=3$ state. Near zero magnetic field intensity, the energies of the states with different m_F increase initially in a linear fashion with the field, with a gradient proportional to m_F, as would a bar magnet, to give us straight-line graphs. Then as the magnetic field intensity is increased, the energies of all but two of these states no longer increase in proportion to the intensity of the field; instead, the graphs start curving until for large field intensity they become grouped in two nearly parallel sets, as shown in Figure 9.1. The total angular momentum vector is no longer constant in time (because of the torque exerted by the field), and a different set of quantum numbers is required to specify the substates. In the limit, for very intense magnetic field, the electronic moment and the nuclear moment separately have constant components along the field. The appropriate quantum description is in terms of quantum numbers giving the integral or half-integral components of each separately along the field axis. In the present case we have for $I=7/2$ the following 8 possible components, with $m_I= +7/2, +5/2, +3/2, +\frac{1}{2}, -\frac{1}{2}, -3/2, -5/2, -7/2$, and for $J=1/2$ only two possible components, $m_J=+\frac{1}{2}$, and $m_J=-\frac{1}{2}$. There are $8\times2=16$ possible combinations of m_I and m_J, the same number as we have in terms of F and m_F, where we had 9 substates with $F=4$ and 7 substates with $F=3$: 9+7=16. This is as it should be, since increasing the field strength alone cannot generate new quantum states; it can only change their energy.

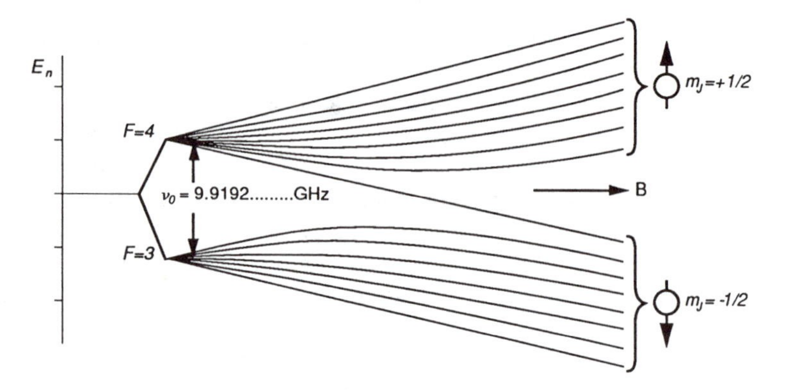

Figure 9.1 The energy of magnetic hyperfine states in cesium 133 as a function of an applied magnetic field

The way the energy of the substates varies with the intensity of the magnetic field is of particular interest for us, since that energy constitutes the potential energy whose gradient determines the force with which the magnetic field acts on an atom to accelerate it. It follows that atoms in the group of substates whose energy increases with magnetic field will experience a force in the direction of decreasing field intensity, while conversely, atoms in the other group of substates will tend to move in the direction of increasing field intensity. Thus atoms acted on by nonuniform magnetic fields will not only execute the usual precessional motion but also experience a body force affecting the motion of their center of mass. There is a further essential point that must be made before we describe the beam machine in more detail: Atoms remain in the same quantum state as long as they move in a smoothly varying magnetic field without going through zero value, ensuring at all times that the time-varying field they see has negligible amplitude in the Fourier spectrum at the precession frequency. These facts are exploited in the atomic beam machines to deflect the atoms selectively according to their quantum state.

We have already been introduced to the idea of atomic beams, their formation and use in the study of magnetic resonance in free atoms and molecules. We have noted the culmination of that technique in the introduction by Ramsey (Ramsey, 1949) of the two separated field regions to induce transitions, which ultimately led to the adoption of the Cs standard as the primary one. The essential elements of an early Cs beam machine using magnetic state selection are shown schematically in Figure 9.2. Atoms from the source enter the strong magnetic field of the A-magnet, where because of a steep transverse gradient, atoms in the two groups having opposite energy-field dependence are deflected in opposite directions. By suitable beam stops, the atoms in the F=3 group, including m_F=0, can be removed, leaving only those in the other group with

F=4, among which are atoms in the desired m_F=0 substate. These atoms leave the intense field of the A-magnet with greater number in the (F=4, m_F=0) state than in the (F=3, m_F=0) state, and *remain* in their respective quantum states as they continue to the much weaker, uniform C-field. If the oscillatory field applied there is *off* resonance with the desired quantum transition, they will again be deflected by the B-magnet in the same direction as in the A-magnet and away from the detector. On the other hand, if the oscillatory field in the C-region is on resonance, most of the atoms in the (F=4, m_F=0) state will make the transition to the (F=3, m_F=0) state, which are deflected in the opposite direction by the B-magnet, towards the detector. This mode of design is called "flop-in," since only atoms that have made the desired transition are detected, to distinguish it from designs in which resonance leads to atoms being deflected away from the detector. A number of different configurations are possible; the choice is ultimately determined by considerations of signal-to-noise ratio.

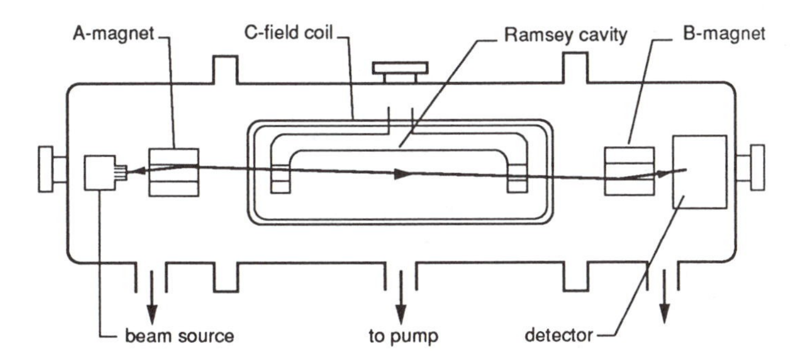

Figure 9.2 The elements of a Cs beam resonance apparatus, the precursor of Cs beam standards

9.3 The Physical Design

9.3.1 The Vacuum System

The entire space through which the atoms pass must be under high vacuum, and therefore a vacuum shell encloses that space, and suitable vacuum pumps and vacuum monitoring instrumentation must be provided. It happens that Cs has, for a metal, a relatively low melting point at 28.5°C and has an equilibrium vapor pressure as high as 10^{-3} Pa at 24°C. This dictates that a means must be provided to remove background Cs vapor, since that vapor density is comparable to that in the beam. In laboratory installations this usually takes the form of "cold traps," liquid containers forming part of the vacuum shell that are cooled by filling them with liquid nitrogen at −196°C. Alternatively, in compact systems designed to be more or less portable, a "getter" may be used; this is a material onto whose surface the Cs either physically attaches in a process called adsorption, or with which it chemically combines, thereby removing it from the volume. In common vacuum practice molecular adsorbents such as carbon, or *zeolites*, which are alkali–metal aluminosilicates, are used. For a chemically reactive element such as Cs, any number of substances will serve as getters; a secondary criterion must be used to make a selection, such as low vapor pressure, temperature stability, and cost. The element that occupies a unique position as a getter is titanium, either as a film deposited by evaporation from a titanium filament or as plates forming the negative electrodes in an electrical discharge. In the latter case the getter action is achieved by having the particles to be pumped impinge on the titanium surface as high-speed ions. The ions are formed in an electrical discharge made possible under very high vacuum conditions by the entrapment of electrons using a special electrode configuration in a strong magnetic field. This class of ion pump, illustrated in Figure 9.3, is effective in pumping all gases, including the noble gases.

It has, since its introduction in the 1950s by Varian Associates, revolutionized vacuum technology, making it possible to reach the vacuum of outer space. It is universally used to maintain the requisite high vacuum in portable Cs systems. The need to operate under high vacuum in a system whose length essentially determines the accuracy largely dictates the physical size and aspect of the Cs standard, and in particular implies that the highest accuracy can be reached only in a fixed laboratory installation.

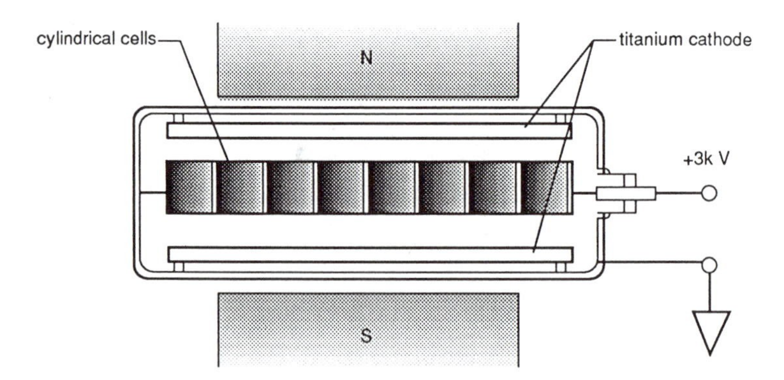

Figure 9.3 The field geometry of the titanium ion pump

9.3.2 The Atomic Beam Source

The source of the Cs beam is a small constant temperature enclosure, the oven, in which the vapor density of the atoms is raised by heating a small quantity of the silvery metal. The Cs vapor from the oven passes through an *effuser*, consisting usually of a bundle of capillary tubes forming a multichannel nozzle that in effect allows only atoms whose velocity lies in a narrow ribbon (or sometimes cone) to leave the source. Figure 9.4 illustrates the basic elements of such an oven. The operating temperature is such that the vapor density is below the point where collisions between atoms can occur with significant probability within the effuser. Under this condition the movement of the atoms through it is described as *thermal effusion*, to distinguish it from the case in which the vapor density is very much higher, in which case the flow is called *hydrodynamic*, as in a gas jet. The most critical part of the source is obviously the effuser, and a great deal of care in the design and operation of the source must be taken to ensure that no buildup of Cs occurs in the effuser, causing fluctuations in the beam intensity.

In spite of all efforts in the design of the effuser to project a sharply narrow beam, it is inevitable in practice that a not inconsiderable amount of the cesium is sprayed out and lands uselessly on the first beam-defining aperture. This, of course, limits the useful life of the charge in the oven. To overcome this limitation, *refluxing* ovens have been designed, in which the emergent cesium outside the useful beam is recovered and "recycled."

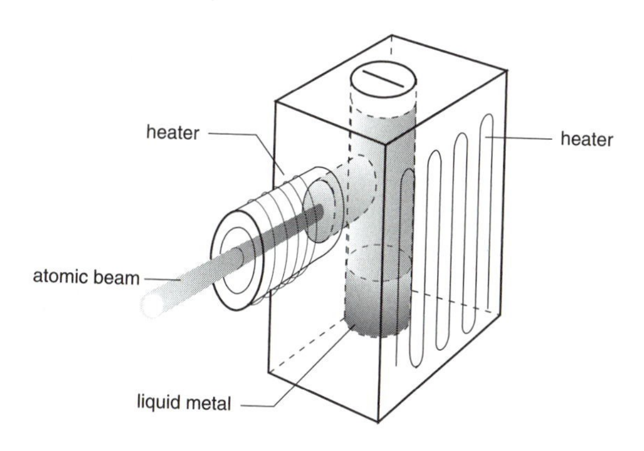

Figure 9.4 The basic design of an atomic beam oven

9.3.3 The Polarizing and Analyzing Magnets

The powerful A-magnet has pole faces specially contoured in order to produce a steeply varying intensity from one point to another. The purpose of this magnet, we recall, is to act on the magnetic moments of the atoms to deflect them and thereby spatially separate them according to their magnetic quantum state. There are essentially two different types of state-selecting magnets: focusing and nonfocusing. The original Stern–Gerlach magnet and its later variants are nonfocusing 2-pole magnets with pole faces contoured to produce the steepest possible descent in the field intensity as we go from one pole to the other.

Atoms in quantum states whose energy increases with magnetic field intensity would experience a force towards a lower field intensity, and conversely, those atoms whose energy decreases with field intensity will experience a force in the opposite direction. Clearly, the force is not the same at all points in the field of the magnet, and consequently, since not all atoms enter the field along precisely the same trajectory, the field distribution is expected to affect the beam "profile," that is, the distribution of atoms over a cross section of the beam. To produce a field gradient more nearly constant over the cross section of an atomic beam than could be achieved by the Stern–Gerlach magnet, a variant shown in Figure 9.5 was developed.

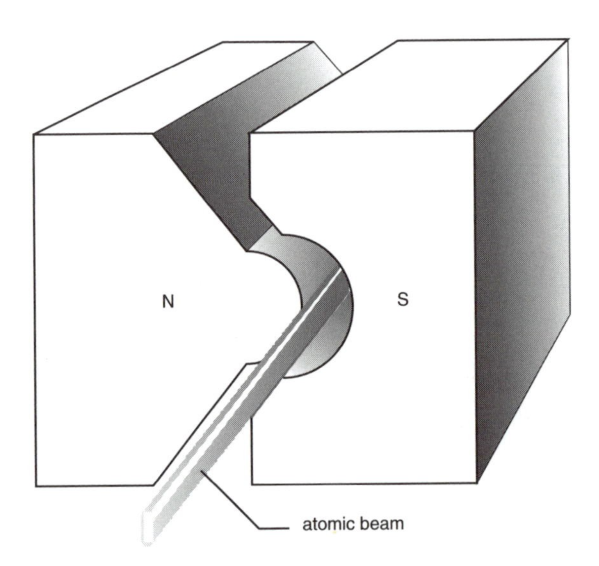

Figure 9.5 A constant gradient state selecting magnet

There are two types of focusing magnets: the quadrupole with a 2-fold axis of symmetry, and the hexapole magnet with a 3-fold axis of symmetry. First let us dispose of the simpler quadrupole type of magnet, which has been exploited far more in the focusing of ion beams than neutral atomic beams. In the neighborhood of the magnet axis it can be shown that the field components are well approximated by $H_x=kx$, $H_y=-ky$, where x, y are coordinates referred to Cartesian axes X,Y chosen to bisect the north and south poles of the magnet. The resultant field is therefore $H=k(H_x^2+H_y^2)^{½}=k(x^2+y^2)^{½}=kr$, where k is a measure of the overall strength of the magnet and r is the radial distance from the axis to the field point. The motion of Cs atoms in such a field is complicated by the fact that their magnetic energy, which acts as potential energy analogous to the potential energy of an object moving under gravity, is not simply proportional to the magnetic field intensity, as it would be for a bar magnet, but rather is a nonlinear function of the field, given by the Breit–Rabi formula we have already encountered. It is as if we were dealing with the motion of a magnet whose strength varied from point to point according to the strength of the magnetic field it is passing through. We recall that at sufficiently high field intensities the plots of the energy versus field strength do tend to become linear, and moreover, in that limit the electronic and nuclear moments separately maintain a constant (quantized) angle with the direction of the magnetic field. In this high field limit, it follows that the force experienced by the atoms is simply proportional to the gradient in the magnetic field intensity, which is constant and in the radial direction. Under the assumed conditions, then, the atoms issuing from the source in the quantum states whose energy increases with field intensity would converge towards the axis, while the others would diverge away from it. The particle trajectory in a radial plane is similar to that of a particle falling under gravity.

In the same limit of high field intensity, the focusing properties of the important hexapole magnet, shown in Figure 9.6 are equally simple to predict. In this case the field in the neighborhood of the axis is approximated by $H_x=k(x^2-y^2)$ and $H_y=-2kxy$, which lead to a resultant field $H=k(x^2+y^2)=kr^2$ and a force that is radial and converging or diverging according to the same condition on atomic state cited above. In this case we see that the gradient of the field, and hence the force, is proportional to the distance from the axis, analogous to the force of an elastic spring. In fact, the radial motion will be a simple harmonic oscillation for atoms in one group of hyperfine states, and rapidly (exponentially) diverging from the axis for the other group. More will be said about the hexapole magnet in connection with the hydrogen maser in a later chapter.

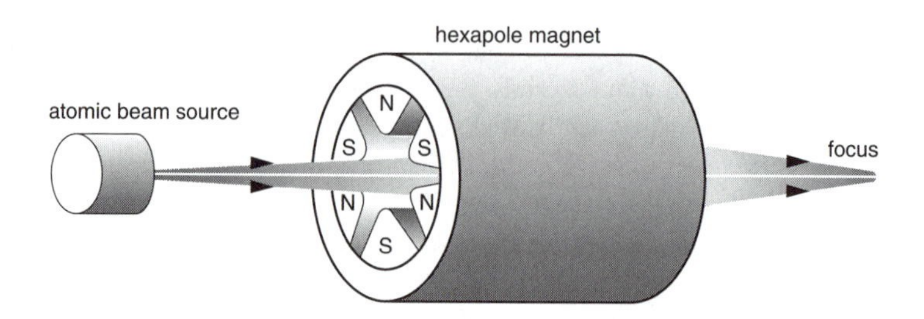

Figure 9.6 The hexapole atomic beam focusing magnet

Since the field intensity tends to zero on the axis for both types of focusing magnets, beam stops must be used to eliminate atoms that would otherwise go through without state selection. Unfortunately, since the beam-forming effuser of the source commonly produces a beam profile that peaks on the axis, such a beam stop would seriously diminish the utilization efficiency of the Cs. A possible solution would be an off-axis ring-shaped source.

We should note one very important limitation of both focusing magnets: unlike the 2-pole magnet, there is no choice as to which states are focused and converge towards the axis and which diverge away from it. In our case, atoms in the F=4 state will always converge, and those in the other F=3 state will diverge from the axis. This makes it impossible to have both the A- and B-magnets focusing in a "flop-in" design.

9.3.4 The Uniform C-Field

After leaving the intense field of the state-selecting A-magnet, the atoms must pass through a gradually decreasing intensity to the uniform C-field, without changing their quantum states. This requires that the time-varying field seen by a moving atom have negligible amplitude in the Fourier spectrum at the transition frequencies between the magnetic substates. Failure to meet this requirement leads to undesirable transitions between the magnetic substates, given the name *Majorana transitions*. Such transitions would cause *relaxation* between the desired m_F=0 substates, defeating the state-selecting function of the magnet. The same situation is encountered in the subsequent transition from the C-field to the powerful B-magnet.

In the elongated C-field region, transitions between the two hyperfine states are resonantly induced by an oscillatory magnetic field. In this region the magnetic field must be relatively weak to take advantage of the first-order insensitivity of the energy of the m_F=0 substates to magnetic field intensity in the neighborhood of zero field. On the other hand, the finite line width of the desired resonance between these substates, dictates that the field be intense enough to produce a sufficient separation among the m-substates, m_F=0, 1, 2, 3, etc., so that the resonant field does not also cause field-dependent $\Delta m_F=\pm1$ transitions; otherwise, field variations would further broaden the transition frequency. Needless to say, the C-field must be as uniform and stable as possible, and therefore magnetic shielding from extraneous magnetic fields is necessary. This is accomplished by enclosing the region with one or more thicknesses of high-permeability magnetic alloys such as mu-metal or supermalloy. A highly uniform magnetic field (the C-field) is produced typically by current flowing in a pair of rectangular coils placed symmetrically parallel to the beam; their separation is chosen to produce a constant field of the highest possible uniformity over the section of the atomic beam where the transitions are induced. Alternatively, an electromagnet with precisely machined plane parallel pole faces and materials selected for homogeneity and freedom from magnetic "hard spots" or other occlusions can be used.

9.3.5 The Transition Field

As already indicated, the one refinement of the Cs beam resonance apparatus that put it in the class of a primary standard is the successive oscillatory field geometry introduced by Ramsey for probing the atomic resonance. To appreciate this, we must go back and examine the problems attendant upon the attempt to observe a microwave resonance in atoms traveling in a beam with thermal velocities on the order of 250 meters per second (about 600 mph). These problems arise from the fact that the length of time the atoms interact with the resonant field is determined by the length of the field region. We recall that the frequency width of the resonance is increased as this time is made shorter; hence the length should be made as great as possible. In fact, the length must be on the order of one or two meters to yield resonance line widths small enough to be interesting for a frequency standard. If L is the length of the transition region and V the average thermal velocity of the atoms, then the average transit time is L/V, and the resonance line width is about $\Delta\nu\approx\frac{1}{2}(L/V)=V/(2L)$. Hence for L=1 m and V=250 m/s we find $\Delta\nu\approx125$ Hz. This is a fundamental width, which can be derived simply from the Fourier spectrum of a pure oscillation of finite duration, as seen by any given atom. This oscillation starts from zero and rises to a constant amplitude for a finite period L/V while the atom is in the transition region, then falls again to zero; its Fourier spectrum is illustrated in Figure 9.7.

A serious consequence of the need to have an extended interaction region is the Doppler shift arising from the directed motion of the atoms though the oscillatory field. It may be thought that this may be overcome simply by using a stationary wave pattern in the interaction region. However, aside from the practical difficulty of ensuring a strictly stationary field, even if there were no net displacement of the resonance frequency, there would nevertheless be a broadening of the spectrum. This may be seen from the following argument: Since the wavelength of the microwave field is only about 3 cm, the atoms would pass through a field whose amplitude and phase vary periodically along their path; that is, they see a modulated field, the frequency of the modulation depending on their velocity. Such a modulation has a Fourier spectrum consisting of two equal sidebands separated from the center frequency by the Doppler frequency $(V/c)\nu_0$. We can reach this same conclusion by thinking of the standing wave as a superposition of two equal waves traveling in opposite directions, as when a wave on the surface of water is reflected back on itself by a straight wall; each wave would have a Doppler shifted frequency in the direction opposite to the other. Since the atoms do not all have the same velocity but are distributed continuously over a wide range of velocity, characteristic of the temperature of the oven, the spectrum would consist of a line broadened out by the Doppler effect.

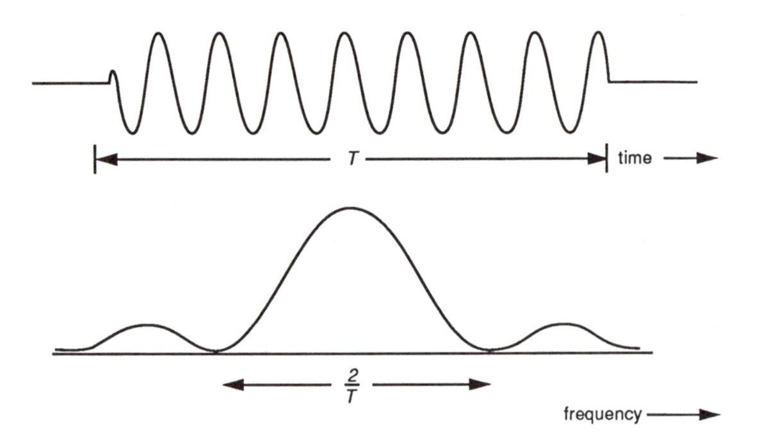

Figure 9.7 The oscillatory field seen by the atoms and its Fourier (power) spectrum

9.4 The Ramsey Separated Field

These problems are alleviated, following Ramsey, by applying the resonant oscillatory field coherently (that is, with a definite phase relationship) in two separated narrow regions, one at the entrance and the other at the exit to the extended C-field transition region. Although the actual length of time a given atom interacts with the oscillatory field is thereby drastically reduced, it can be shown that since the fields in the two regions are in phase, the frequency width of the net response of the atoms traversing the whole transition region is determined by the much longer time the atoms spend in the intervening space. In order to ensure that the fields in the two regions maintain a constant phase relationship, a common microwave source is used, and the fields are symmetrically located at the ends of a single resonant microwave cavity, as shown in Figure 9.8. Thus the field in each narrow region can be limited to one with a single phase and nearly constant amplitude.

The resonant cavity that is the microwave analogue of an echo chamber is usually a section of rectangular wave-guide with 90-degree bends at its ends where apertures are provided for the Cs beam to pass through the standing microwave field pattern. As in all resonant structures, the resonant modes, with their characteristic frequencies and field patterns, are determined by a formula relating the resonant frequencies to the dimensions of the cavity, a formula that contains three integers, the mode indices. A particular mode that might be used is designated as TE_{10n}, which represents what is called a _T_ransverse _E_lectric mode, that is, one in which the electric component of the electromagnetic wave is everywhere perpendicular to the length of the cavity. The indices 1, 0, *n* give the number of times the electric field passes through a maximum amplitude of oscillation as we go in the directions of the three principal dimensions of the wave-guide. Thus if the cross section of the cavity is a rectangle with sides *A* and *B*, where $A>B$, and the length is *C*, then the indices indicate that in this particular mode the field rises to one maximum in the middle of the *A* dimension, has no maximum (is constant) along *B*, and has *n* maxima along the length *C*. For the dimensions *A, B, C* to be compatible with this mode, it can be shown that the following condition must be satisfied: $C=n\lambda_g/2$, where $\lambda_g=\lambda_0/(1-\lambda_0^2/4B^2)^{1/2}$ and λ_0 is the free-space wavelength of the microwaves. Thus in our case the wavelength of the microwaves resonant with the Cs transition is $\lambda_0=3.26$ cm; if we assume, for example, $B=2.5$ cm, then $\lambda_g=4.3$ cm and a choice of $n=48$ would make the length of the cavity 103.2 cm, appropriate for a fixed installation.

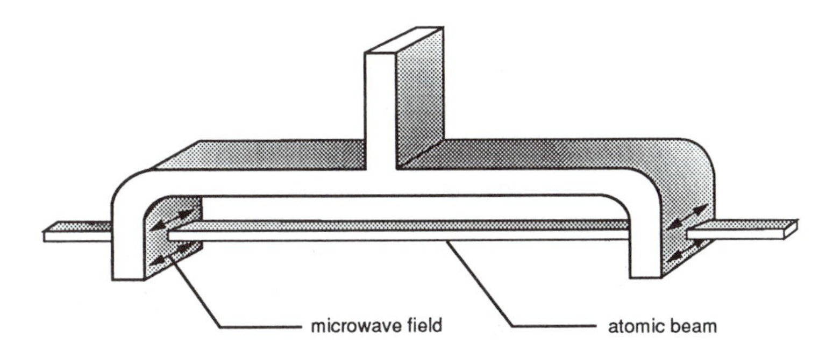

Figure 9.8 The Ramsey separated field atomic resonance cavity

To be effective in inducing transitions between the substates $F=4$, $m_F=0$ and $F=3$, $m_F=0$, it is not enough for the frequency of the microwaves to satisfy the conservation of energy condition $h\nu=\Delta E_{hfs}$, where ΔE_{hfs}, represents the difference in energy between the two hyperfine states; the microwave field must also have the correct directional properties, that is, polarization. We recall that to observe magnetic resonance between Zeeman substates in which m_F increases or decreases by one, the field inducing transitions must have an angular momentum component along the constant field axis to satisfy the conservation of angular momentum law as it applies to the combined system of atom and radiation field. To have such a component of angular momentum, the radiation field must have a component rotating about the constant field axis. Similarly here, since there is no change in m_F, being zero before and after the transition, and since only one quantum of radiation is involved, it must have zero component of angular momentum along the constant field axis. This will be the case if the radiation field oscillates parallel to the constant field; this determines the relative orientation of the microwave cavity and the coils producing the constant field.

To help gain a broader perspective on the use of separated fields we should mention a closely parallel case in radio astronomy of the use of two separated radio antennas, as shown in Figure 9.9, to increase the angular resolution in observing distant sources. By maintaining a common phase reference for the receivers at the two antennas, the system's ability to distinguish neighboring sources, that is, its *resolving power*, is made to approach that of a much larger antenna having a diameter equal to the distance between the two small antennas. To see this we must recall that even if an antenna were perfectly parabolic, so that rays coming in parallel to its axis would geometrically converge to a point focus, physically the reflected wave pattern does not converge exactly to a point; it approaches this ideal only to the extent that the *aperture*, that is, the diameter of the antenna, is large compared to the wavelength of the radio waves. The wave pattern near the geometric focus will be a series of maxima and minima resulting from the antenna cutting off the incoming wave at its outer rim, whence it spreads in a pattern dictated by interference from different parts of the aperture. The angular width of this diffraction pattern is set by the difference in the phase of an incoming wave across the aperture; the first minimum will occur when that difference in phase is on the order of 360°. The larger the aperture, the smaller will be the required increment in the direction of the incoming wave to produce that phase difference, and the greater the resolving power. We now see that by comparing the signals arriving at two widely separated antennas, a smaller difference in the angular position of distant sources is distinguishable because the longer base line magnifies the difference in the phase of the wave reaching the two antennas, as shown in Figure 9.9. The same principle is used in the much older "stellar interferometer" of A.A. Michelson, of the velocity of light fame. In this, two optically flat mirrors are mounted some distance apart to receive starlight and the light reflected from them combined through precise optics to a common detector whose output depends on the relative phase between the two interfering reflected beams (see Figure 9.10). By this interferometer Michelson was able to determine the (angular) diameter of stars that were smaller than could be resolved with telescopes available at the time.

Although in the years following its introduction in 1949 the principle of the separated field method has been applied in a variety ways in spectroscopy, it was originally developed to achieve greater accuracy in the measurement of atomic and molecular magnetic moments by the molecular-beam resonance method of I. Rabi. It is that application that provides the most visual explanation of the special properties of inducing transitions this way.

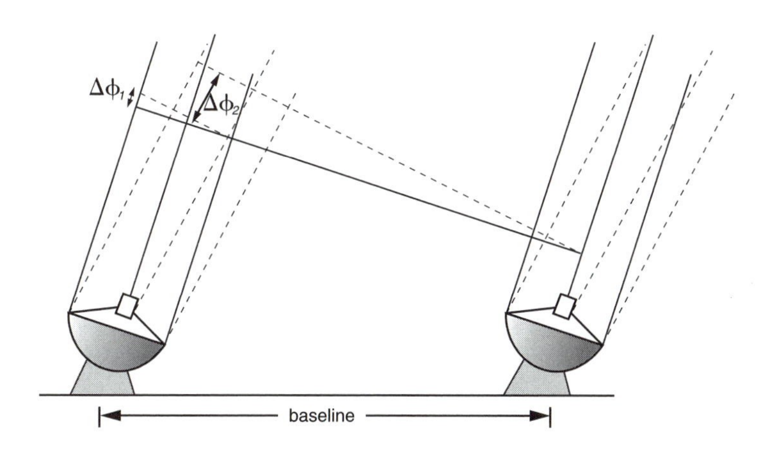

Figure 9.9 The use of separated antennae in radio astronomy to increase resolution

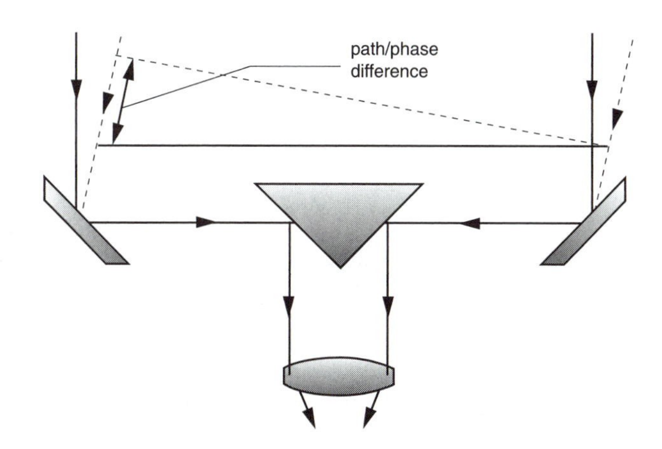

Figure 9.10 The Michelson stellar interferometer

We recall that in an earlier discussion of magnetic resonance we described the gyroscope-like precession of an atomic angular momentum (with an associated magnetic moment) about a static magnetic field, and how the application of a weak magnetic field oscillating at the frequency of precession will cause the axis of spin to tilt away from the static field so that the cone it sweeps out opens out to a larger apex angle. In the Ramsey arrangement, the atoms see an oscillating field in the first narrow transition region of sufficient strength to produce, for example, a 90-degree apex angle for atoms with the average thermal velocity. On leaving this first transition region, the atoms pass through a relatively long region free of any oscillating field, in which they continue to precess at the same frequency appropriate to the static uniform C-field. The atoms then enter the second narrow region, where they are again subject to an identical oscillatory magnetic field, which has a definite phase relationship with the first, usually the exact same phase. If the frequency of the oscillating fields is exactly the same as the average frequency appropriate to the static field, then the atoms will enter with the same phase as the field, and the direction of the spin axis will continue to tilt toward a cone angle of 180 degrees, corresponding to a complete reversal in the direction of the angular momentum. Note that the phase of the precessing moment relative to the oscillatory field will determine the direction and degree of tilt the latter produces; hence if the precession frequency in the C-field differs only slightly from the frequency of the oscillatory field, a large phase difference can develop in the intervening space, and the degree of tilt will be strongly reduced.

If the atoms all had precisely the same velocity, then there would exist a difference in frequency between the precession and the oscillatory field that will lead to a phase difference of exactly 360 degrees being developed between them in the space between the two transition regions; that is, the atoms would again enter the second region in phase with the field. In fact, the same would happen at frequencies leading to a phase difference of any multiple of 360 degrees. However, in reality, the atoms do not all have the same velocity, and these multiple "sidebands" occur at frequencies that depend on velocity, since a slow atom spends more time in the field and requires a smaller difference in frequency to develop the 360-degree phase difference than does a faster atom. The resonant frequency has the unique property of being independent of velocity; no phase difference can develop if the oscillatory field and the precession have the same frequency, no matter how long it takes an atom to reach the second transition region. Moreover, since there is a continuous distribution of velocity among the atoms, the sidebands form a continuum of reduced strength leaving a prominent central peak at exact resonance.

A quantitative analysis of the probability that an atom passing through the two separated field regions will emerge having made a transition to the other substate requires an exact quantum treatment of the problem, as was initially carried out by Ramsey. It would be inappropriate to attempt to reproduce that theory here; rather, we will try to gain some insight as to the shape of the resonance signal using a quantum result that is strictly valid only where the "perturbation" acting on an atom is weak. It is that the transition probability is proportional to the square of the Fourier amplitude of the field at the transition frequency. Although the perturbation of the atoms here is far from weak, nevertheless it serves to provide some general basis for understanding the system. In any event, since the exact results are known for this case, there is little danger of being misled by an invalid approximation.

The time dependence of the assumed separated field is shown with its Fourier spectrum in Figure 9.11. This should be compared with the Fourier spectrum of an oscillatory field extending the full length of the C-field, shown in Figure 9.7. What we find is that in fact the central maximum in the frequency spectrum is even narrower for the two separated fields than for the single extended one, a fact proved rigorously by Ramsey in a full quantum-mechanical treatment of the problem.

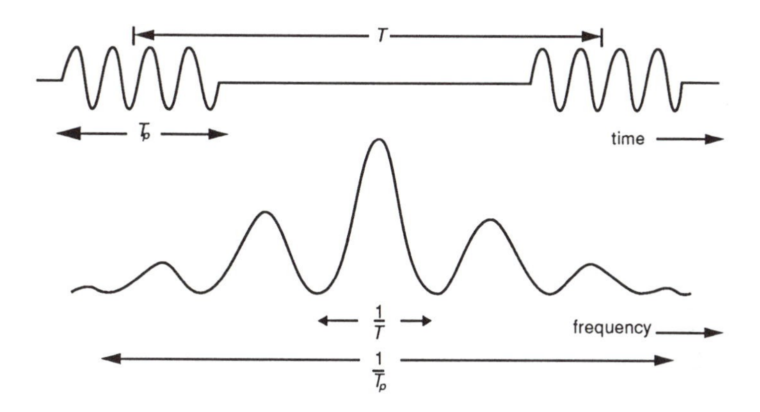

Figure 9.11 Field seen by an atom in the Ramsey cavity and its Fourier spectrum

Since the duration of a given atom's interaction with the separated fields and the time spent between them depends on the atom's velocity, the signal produced by a beam consisting of a large number of atoms having a thermal distribution of velocities is obtained by summing over the contributions from individual atoms. This has been analyzed rigorously by Ramsey, including the effect of introducing phase differences between the two separated field regions; the result for zero phase difference is shown in Figure 9.12.

The Ramsey arrangement alleviates another problem: that of ensuring a sufficiently uniform and constant magnetic C-field over an extended space. This would clearly involve a complex array of compensating coils and impose severe tolerances on the mechanical and electrical parameters, and particularly the shielding from external magnetic fields, etc. Fortunately, the phase difference that accumulates between the atomic moments and the oscillatory field, as the atoms travel between the two transition regions, depends on the spatial average of the C-field taken over the path of the atoms. It is reasonable to expect that this average fluctuates from atom to atom far less than the field strength itself along the path of any given atom.

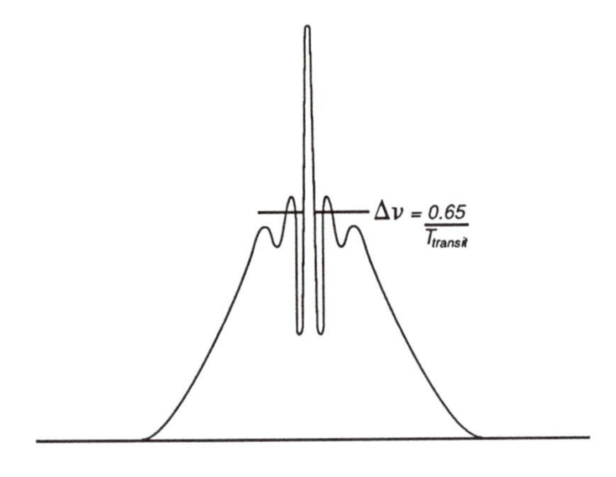

Figure 9.12 Theoretical signal shape for thermal atoms passing through Ramsey field (Ramsey, 1956)

9.5 Detection of Transitions

Next the atoms pass through another powerful deflecting magnet (the B-magnet), which serves to analyze the magnetic states of the atoms, allowing the occurrence of the desired microwave transition to change the number of atoms reaching the final element, the detector. Since the greatest challenge in the design of an atomic beam machine is to achieve a high signal-to-noise ratio, which means, because of shot noise, the highest possible beam intensity, it would seem advantageous to use focusing magnets for the A- and B-magnets. However, this presents a dilemma, since both magnets would focus atoms that are in the same state, so that ones that have made a transition in the C-region to the other state would diverge from the axis. If the detector is placed on the axis, then it would be exposed to the atoms that had *not* made a transition; to detect ones that *had* made a transition, it would have to accept atoms over an extended circular area. Neither option is particularly desirable, the first because it precludes the possibility of exploiting a narrower resonance with a weaker signal, since the shot noise would be due to the larger number of atoms that have not made a transition, and the second because the increased area may incur greater noise from background Cs vapor.

The availability of an efficient low-noise Cs detector played a critical part in the early development of the Cs beam resonance apparatus into an atomic standard. It is the *hot wire detector*, which is based on the phenomenon of surface ionization of the alkali atom, in which a Cs atom impinging on a pure tungsten surface (which must be maintained at high temperature to prevent surface layers of adsorbed gases) loses on electron to the metal and emerges as a positively charged ion. The phenomenon is permitted by the energy conservation law, since the binding of the outer electron in the Cs atom (3.87 electron volts) is less than the binding energy (the so-called *work function*) of the electron in the interior of tungsten metal, or a number of other metals, such as niobium and molybdenum. It is remarkable not only that the process occurs at all (the electron has to pass through a classically forbidden barrier to do it), but that it does so with a very high probability, so that very nearly all atoms reaching the surface of the metal become ions. The main difficulty in practice is that other ions, particularly those of potassium, are also emitted. Thus for the highest possible signal-to-noise ratio, not only are the purest materials used, but also a mass filter is incorporated into the detector design. This latter can take the simple form shown in Figure 9.13, in which the positive ions leaving the hot filament are accelerated by a negative voltage applied to a metal grid in front of it and emerge between a pair of focusing electrodes. They are then deflected by the field of a small, powerful magnet along circular arcs in a plane perpendicular to the field. This produces an angular separation of the ions according to their momentum and by proper adjustment of the accelerating potential deflects the mass 133 cesium ions through 60° to a focus on an ion detector, which in former times would have been a galvanometer.

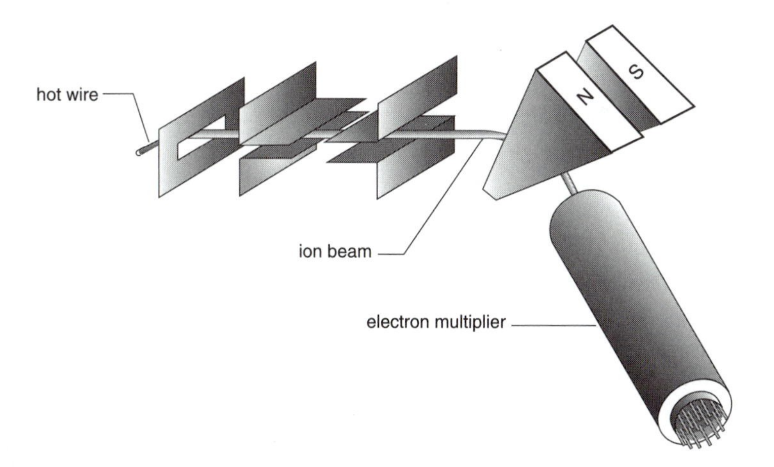

Figure 9.13 A simple ion mass filter for Cs beam detector

In view of the smallness of the rate at which ions arrive at the detector, optimum signal-to-noise dictates the use of an electron multiplier. This is a device, illustrated in Figure 9.14, in which an ion of sufficient energy falling on the cathode causes several *secondary* electrons to fly off and be focused onto the first of a succession of copper–beryllium *dynodes*. The secondary electrons from the first dynode are focused onto the next dynode, and so on; each time the number of secondary electrons flying off exceeds the number hitting the dynode. The result of this cascade process is that the electron current is multiplied by an enormous number before reaching the final stage. This makes possible the detection of single ions in the presence of the inevitable thermal and other types of noise in the output circuit. Another remarkable device working on the same principle is the photon multiplier, which is really an electron multiplier differing only in having a cathode that emits electrons when light, rather than energetic ions, falls on it. It puts out a pulse of many million electrons for every quantum of light (photon) that falls on its cathode. It can be used to count individual photons; however, since each pulse is necessarily of finite duration, there is a limit to how close in time two pulses can be before they overlap and are counted as one. For this reason its unique sensitivity is suitable only where the ion or photon flux is extremely low by ordinary standards.

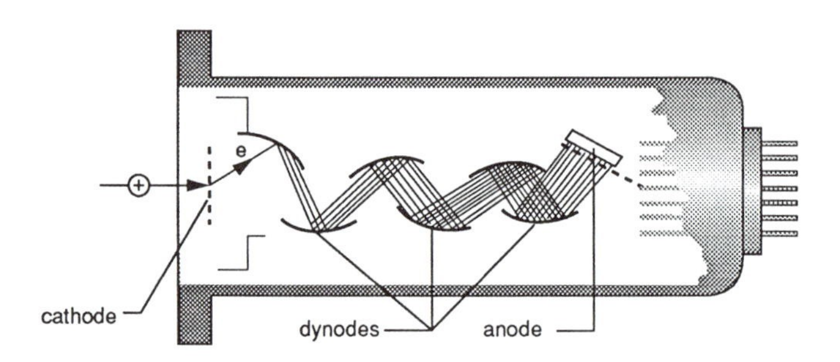

Figure 9.14 Electrode geometry of end-on electron multiplier

9.6 Frequency-Lock of Flywheel Oscillator to Cesium

Like the Rb standard, the Cs beam resonator is a passive device, which does not itself generate any microwave power but merely serves as a frequency reference. This reference can in principle be used in one of two ways: first, to make a manual comparison with a harmonic of the frequency of a source under test, or second, to automatically control the frequency of a *flywheel* oscillator by means of one or more servo loops. Both cases involve synthesizing from the given source a microwave frequency that can be applied to, or compared with, the resonance frequency of Cs. We will limit ourselves to the essentials of a Cs clock, in which a high-quality quartz oscillator, commonly operating at 5 MHz, has its frequency servo-controlled, so that a synthesized microwave frequency derived from it is locked to the center of the Cs resonance. The basic ingredients of such control circuitry have already been described as they apply to the Rb standard.

We recall that in order to obtain a measure of how far, and in what direction, an applied microwave frequency is away from the center of the resonance curve, we modulate the phase of the applied microwave very slowly, typically at some frequency, far from any harmonic or subharmonic of common frequencies, such as 37.5Hz. This, we may recall, produces an output from the detector at 37.5 Hz, which is zero at the center of the resonance curve (assuming it is symmetrical) and is of opposite phase on the two sides of resonance. This detector output then is passed through a phase-sensitive amplifier whose output voltage is a measure of the difference in phase between the incoming signal and a reference signal of the same frequency. By using as a reference the signal producing the phase modulation, we obtain an error signal that is a constant negative or positive voltage depending on whether the applied frequency is too low or too high. In the case of the Cs beam standard, the modulation of the phase of the oscillatory field poses a problem not encountered in the Rb standard: The transit time of the atoms between the two separated transition regions allows the phase of the field seen by the atoms to be different in the second region from what it was in the first. Now, such a difference in phase is known to cause a shift in the observed resonance; however, as the phase goes through a modulation cycle, the shift oscillates symmetrically about zero and merely changes the effective width and phase of the modulation over the resonance curve. Furthermore, there is a delay between the time a given atom passes the second transition region and the time it reaches the detector; these and other possible sources of phase shift dictate that a compensating phase shifter be included to adjust the reference phase in the phase-sensitive amplifier.

There are, not unexpectedly, numerous possible designs for phase locking a 5 MHz quartz oscillator to the Cs resonance signal. We give as an instructive example a brief outline of a typical early arrangement that utilized the output of a klystron microwave oscillator (rather than the direct output of a frequency multiplier) to induce the transitions in Cs. The object is to lock the output frequency of the klystron to the center of the Cs resonance and be able to relate its frequency to that of the quartz oscillator. In one such practical system, illustrated schematically in Figure 9.15, this can be done by first obtaining the 1,836th harmonic of the 5 MHz output of the quartz oscillator by a frequency multiplier chain that yields a coherent oscillation at 9,180 MHz. By mixing this frequency with that of the klystron output in a nonlinear diode designed for the purpose, a signal at the beat (difference) frequency is obtained. Now, if the klystron frequency is on the center of Cs resonance, for example at

9,192,631,770 + $\Delta\nu_m$ Hz, where we have allowed a difference $\Delta\nu_m$ from the defined value to account for the C-field and other calculable offsets, then that beat frequency should be 12,631,770 + Δf_m Hz. Now, this frequency can be synthesized coherently from the 5 MHz quartz oscillator by combinations of arithmetical operations on frequencies made possible by nonlinear diodes and crystal filters. Using this synthesized frequency as reference, the klystron frequency is controlled in order to phase-lock the beat frequency to it. These undoubtedly confusing frequency relationships can be made a little less so by the following equations:

$$\nu_K - 1836\nu_x = \nu_B \quad ; \quad (12.631770 + \Delta\nu_m)\nu_x / 5 = \nu_s , \qquad 9.1$$

where ν_K, ν_X are the frequencies (in MHz) of the klystron and the crystal oscillator, and ν_B, ν_S the beat and synthesizer frequencies. The control loop acting on the klystron forces the equality ν_B=ν_S to hold at all times, independently of the quartz frequency ν_X, making the klystron frequency track the quartz frequency according to the following equation:

$$\nu_K = (1836 \times 5 + 12.631770 + \Delta\nu_m)\nu_x / 5 . \qquad 9.2$$

A second control loop acting on the quartz oscillator, and using the Cs beam detector to provide the error signal forces ν_X to assume a value such that ν_K=(1836×5+12.631770+$\Delta\nu_m$), that is, (apart from the frequency offset $\Delta\nu_m$) the defined frequency of cesium: 9,192,631,770 Hz. In doing so, the quartz oscillator frequency is locked to the precise value of 5 MHz on the atomic Cs time scale. Outputs coherent with the 5 MHz signal at subharmonics such as 1 MHz, 100 kHz, down to 1 Hz are obtainable using what have become standard frequency-synthesizer techniques.

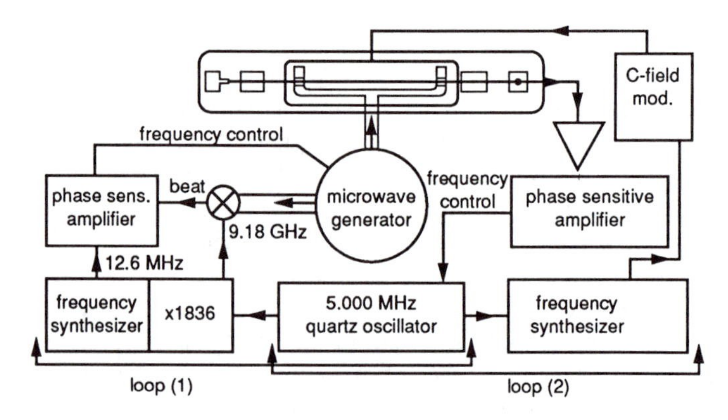

Figure 9.15 An example of servo loops used to phase-lock a quartz oscillator to the Cs resonance

9.7 Corrections to the Observed Cs Frequency

Prior to being adopted as the basis for the definition of the second, exhaustive studies had been conducted to identify and analyze possible sources of systematic errors in the observation of the Cs resonance. These, we recall, are persistent errors, arising from usually subtle factors, affecting the frequency at which the Cs resonance is observed in a deterministic way, rather than as an unpredictable random fluctuation. It is a primary quest of those responsible for the establishment of physical standards to seek out all conceivable sources of these errors and to attempt to correct for them. Having a number of systematic errors does not in itself detract from the acceptability of a standard, provided that they are well understood and calculable. However, the discovery of previously unsuspected sources would obviously be disastrous. In order to gain some appreciation of the complexity of establishing that what is measured is in fact what is called for in the definition of the standard, we list the following:

9.7.1 Magnetic C-Field

The transition frequency in weak magnetic fields, ν, is given by

$$\langle \nu \rangle_C = \nu_0 + 427 \langle B^2 \rangle_C, \quad 9.3$$

where $\langle \, \rangle_C$ represents the average taken over the C-field transition region. This equation, of course, applies only to Cs and specifically to our hyperfine transition. Typically, B is on the order of 50 milligauss, giving a correction of around 1 Hz.

9.7.2 Unequal Phases of Ramsey Fields

If the phases of the oscillating fields in the two separated transition regions of the Ramsey arrangement are not exactly the same, the detector output will not be at its maximum at resonance. To estimate the displacement in frequency, we note that 180° difference in phase reduces the detector output to its first minimum, that is, corresponding to a shift in frequency equal to the width of the resonance. If the latter is, say, 125 Hz, a 1° difference in phase would cause a shift in frequency of about 125/180 =0.7 Hz, which is not negligible in the context of a frequency standard. This asymmetry in phase and other asymmetries in the apparatus with respect to a reversal of the beam direction may have any number of causes; for example, inertial forces acting on the atoms or strains in the mechanical structure due to acceleration, such as might be experienced in a spacecraft. Where the environment dictates it, or where the utmost accuracy is sought, as in large standards laboratory installations, the effects of such asymmetries are corrected by providing for Cs beams to traverse the apparatus in opposite directions. Any spurious frequency shift due to asymmetry will reverse direction, and a corrected frequency can be obtained as the average of the two frequencies at which the signal is at its maximum.

9.7.3 Relativistic Doppler Shift

As we discussed in Chapter 7, the Doppler effect as manifested by electromagnetic waves is not accurately described by classical theory; the formula that is in accord with the principles of relativity cannot differ between situations where only the frame of reference is different. We saw that Einstein's theory yields the following:

$$\nu = \sqrt{\frac{1-\frac{V}{c}}{1+\frac{V}{c}}}\,\nu_0 . \qquad 9.4$$

In the present case, V/c is only on the order of 10^{-6}, and so in a power series expansion in V/c, terms beyond the second power are negligible; thus the Doppler correction to the frequency is given by

$$\nu - \nu_0 = -\frac{V}{c}\nu_0 + \frac{1}{2}\frac{V^2}{c^2}\nu_0 + \cdots . \qquad 9.6$$

The first term on the right, which involves V/c to the first power and is the dominant effect, agrees with classical theory and is called the linear Doppler effect. We have already seen how the Ramsey separated field technique circumvents this linear effect; however, at a level of accuracy on the order of 1 part in 10^{12}, the second term, involving $(V/c)^2$, becomes significant. The first thing we note about this second-order Doppler effect is that the shift does not change if the sign of V is changed; that is, it is the same whether the source and observer are approaching or receding from each other. Secondly, we note that as "seen" by the moving atoms, the microwave frequency is higher than would be observed if the atoms were not moving. Therefore, if a moving atom "sees" a microwave field that is resonant with its quantum transition, that same field would be below resonance for a stationary atom; that is, the observed frequency is lower than the "proper frequency" of the Cs transition.

9.7.4 Spectral Impurity of Microwave Field

An ideal standard would have a microwave field whose spectrum consists of a single, infinitely narrow line at a frequency that can be controlled to lock on to the maximum of the Cs resonance curve. In reality, the microwave field has a distribution of frequencies determined by the crystal oscillator and frequency multiplier chain from which it is derived. The most common contributors to this distribution are electronic noise and discrete sidebands spaced at intervals of 60 Hz, the commercial power frequency, due to modulation of the crystal oscillator frequency by the ubiquitous AC fields and possibly residual ripple on its DC power supply. This latter source of spectral impurity is aggravated by the high order of multiplication (×1836) to reach from the usual 5 MHz crystal frequency to the neighborhood of the microwave frequency, since the relative amplitude of the sidebands can be shown to increase with the order of multiplication. Serious error is incurred if the sideband amplitude distribution is not symmetrical about the central (unmodulated) frequency. Presumably, proper shielding and the use of battery power would largely eliminate this problem. Of course, just to be able to analyze the spectrum of microwaves at a frequency around 9 GHz with a resolution of a few hertz is no mean challenge; very special techniques are required involving a microwave source of known high spectral purity.

9.7.5 Neighboring Transitions

We recall that in addition to the desired (F=4, m_F=0)–(F=3, m_F=0) transition, there are atoms in the beam in neighboring $m_F=\pm 1$ states that contribute to the signal by making magnetic-field-dependent transitions in which the magnetic quantum number m_F changes by one unit. While the application of a uniform magnetic field in the transition region will separate these transitions from the desired one, there will nevertheless remain a finite probability of their contributing to the signal. Ideally, these transitions have an amplitude distribution that is symmetrical about the center of the desired one; however, in reality it can happen that the way the atomic trajectories fit within the magnet geometries leads to an asymmetric overlap between the desired transition and its neighbors. The consequence is a signal intensity distribution that is distorted, and whose maximum is displaced with respect to the true resonance frequency.

9.7.6 Residual Linear Doppler Effect

This arises from failure to meet the ideal conditions under which the linear Doppler effect is eliminated in the Ramsey cavity. In particular, if there is an asymmetrical flow of microwave power from the source to the two ends of the cavity through which the atoms pass, the resonance signal will be broadened asymmetrically and the maximum will be shifted. If we examine closely the microwave field at the shorted ends of the Ramsey cavity, we see that in the presence of power loss in the walls, the quasi-stationary field pattern can be analyzed into counter-traveling components of slightly different amplitudes. The (small) transverse component of velocity of the Cs atoms passing through the field will lead to a small *linear* Doppler shift; but because the two traveling components are of different amplitudes, the transition probabilities at the two Doppler-shifted frequencies will not be equal, leading to an asymmetry in the resonance signal profile and a shift in the maximum. The presence of such a shift and other possible power-related effects can be ascertained by varying the microwave power.

9.7.7 Final Word

The listing of all these possible sources of systematic error must not be allowed to leave the impression that this type of atomic clock is fraught with uncertainties. Quite the contrary, the sources listed and the many more subtle effects not listed merely show the exhaustive degree of scrutiny to which this standard has been subjected. As a primary standard, of course, the achievement of the highest possible accuracy and reproducibility requires that this be done.

Since the 1970s, when the classical beam machines described in this chapter had reached a high degree of development and general acceptance, thanks to the laser there have been fundamental developments that have radically changed the design of Cs standards, a subject we will take up in a later chapter.

Chapter 10
Atomic and Molecular Oscillators: Masers

10.1 The Ammonia Maser

The idea of a device using quantum transitions induced in molecules by a radiation field to achieve microwave amplification by stimulated emission of radiation, now familiarly known by the acronym *maser*, was first described by Gordon, Zeiger, and Townes, of Columbia University (Gordon et al., 1954) and independently proposed by Basov and Prokhorov, of the Lebedev Institute for Physics, in 1954. Townes, Basov, and Prokhorov shared the 1964 Nobel Prize in physics, "for fundamental work in the field of quantum electronics, which has led to the construction of oscillators and amplifiers based on the maser–laser principle."

The maser was conceived from the beginning as a high-resolution spectrometer for the microwave region of the spectrum, or as a microwave oscillator of great stability. It combines the techniques of molecular beams and microwave absorption spectroscopy, in which resonances in the absorption of microwave energy by matter are studied. The initial experimental implementation was on the so-called *inversion spectrum* of ammonia (NH_3), a molecule that because of the strength and abundance of its resonances in the microwave region of the spectrum played an important role from the beginning in the development of microwave spectroscopy. This field of study burgeoned in the early postwar period as a result of the rapid development of microwave techniques for radar applications during World War II. In fact, the first operational molecular frequency standard, as developed by Lyons and his associates at the U.S. National Bureau of Standards in 1948, was based essentially on the use of a strong absorption line in the ammonia spectrum as a reference to electronically stabilize the frequency of a quartz crystal oscillator. Following the initial successful demonstration of the maser principle, it was soon exploited in a greatly expanded range of applications, including solid-state microwave amplifiers and atomic beam maser oscillators, culminating in the hydrogen maser, about which a great deal more will be said in the next chapter.

10.2 Basic Elements of a Beam Maser

To illustrate the principle, we can do no better than follow the original description given by Gordon, Zeiger, and Townes of a molecular (or atomic) beam maser. Figure 10.1 shows schematically the essential parts of such a maser, with particular reference to ammonia.

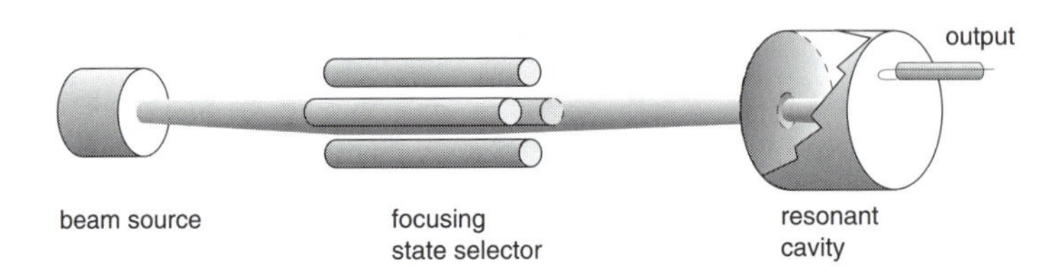

Figure 10.1 A schematic diagram of a beam maser oscillator

A beam of ammonia molecules is formed much as a Cs atom beam is formed, except that being a gas at room temperature, ammonia is supplied to the source at reduced pressure (around 100 Pa), and the molecules effuse through narrow channels into a vacuum sufficiently high to form a beam. This requires that the background concentration of gas be kept sufficiently low by high-speed vacuum pumps; otherwise, it would be meaningless to speak of a molecular beam. Since ammonia is a gas at room temperatures (boiling point –33°C), there is a greater burden on the vacuum pumps than for solid materials. In addition to the standard molecular diffusion pumps and turbo-pumps, based on the turbine principle, cryogenic pumping is practical in this case, since the freezing point is a readily achievable –78°C (it is near the temperature of dry ice, solid CO_2). Thus by providing for the quadrupole electrodes to be cooled with liquid nitrogen (–196°C), for example, ammonia molecules striking them would mostly be frozen on the surface, rather than accumulate as background gas.

The ammonia molecule, like all molecules, has a vast number of quantum energy states, corresponding to the complex rotational and vibrational motions of such a multi-atom system. Among these is a particular pair of states between which transitions fall in the microwave region and for which there is strong coupling to the radiation field, making it easier to observe and exploit in building a maser. However, from a source in equilibrium at ordinary temperatures, the populations of these states are very nearly equal, with the lower-energy state having a slightly greater population. Therefore, to observe a net transfer of energy to a resonant field requires a means of reducing the population of the lower-energy state in relation to the upper one. This is accomplished by having the beam pass through a quantum state selector analogous to the A-magnet in the Cs standard. In this case, however, the molecules are acted upon through an *electric* dipole moment, rather than a magnetic one as in Cs, and therefore the state selector consists of a strong electrostatic field, as described in more detail in Section 10.4. It is a focusing field with quadrupole symmetry, in which molecules in the upper energy state converge toward the axis and enter a resonant microwave cavity through a small opening on the axis, while the ones in the lower state diverge away from the axis and do not enter the cavity. Since the probability that the molecules will *spontaneously* make a microwave transition to the lower state is small, they will therefore be predominantly in the upper state inside the cavity.

The existence of a weak resonant microwave field in the cavity, introduced from an external source, will stimulate the molecules to emit radiation at the same frequency and in phase with the stimulating field, thereby increasing the amplitude and providing an amplified output. If the Q-factor of the microwave cavity is high, there will be a threshold number of state-selected molecules entering the cavity per unit time beyond which self-sustained oscillation will take place, and we have a maser oscillator.

10.3 Inversion Spectrum in NH_3

We will treat the ammonia maser in some detail, not only because of its obvious historical importance as the first molecular frequency standard, but also because of the broader application of some of the ideas involved in its development.

The quantum transition it uses at around 24 GHz occurs between energy states that are interesting in that they have no classical analogue; they can be pictured only as a superposition of two classical states, as if one had double vision! To make a little more sense of this statement, we have to study the quantum implications of the symmetry properties of the ammonia molecule. The geometrical arrangement of atoms in ammonia is known to be as shown in Figure 10.2; that is, the chemical bonds lie along the edges of a triangular pyramid; the three hydrogen atoms are in one plane at the vertices of an equilateral triangle, and the one nitrogen atom is on the symmetry axis perpendicular to that plane. If we imagine the chemical bonds to be elastic bands, we could pull the nitrogen atom along the symmetry axis through the center to a position diametrically opposite to where it was; that is, we can invert the molecule, and the resulting molecule would be indistinguishable from the original. The quantum description of the molecule must properly take this symmetry into account, so that no prediction based on that description will distinguish between these symmetrical states. Now, it happens that the minimum energy of the system (and therefore its ground state) occurs when the nitrogen atom is some distance (in either direction) from the symmetry plane passing through the H atoms. This means that if we imagine the nitrogen atom to be placed at different points along the symmetry axis, the energy of the system, as a function of the N-atom position, would have minima at equal distances from the H-plane and a local *maxi-*

mum in that plane. This system energy for an assumed (static) position of the N atom, which ultimately has its origin in the electrostatic forces between the fast moving electrons in the molecule, acts as a potential energy governing the relatively slow motion of the N atom. The consequence of the inversion symmetry is that the stationary quantum states must also reflect the same symmetry, and a configuration in which the N atom is located asymmetrically at just one of the potential minima cannot be a stationary state; that is, it will not remain in that state indefinitely. In fact, the two states that are stationary are represented by wave functions giving the N atom equal probability of being in either position! However, one stationary state is represented by an even wave function, and the other by an odd wave function with respect to inversion; that is, one is unchanged, and the other reverses sign when the molecule is inverted. This is analogous to having a pair of pendulums of exactly the same period of oscillation with a weak exchange of excitation between them. We have already seen that the only "stationary" states of motion are those where the two pendulums either swing together in step or exactly out of step; and if we start only one pendulum swinging while the other is initially at rest, soon they will exchange roles, with the first pendulum coming to rest and the second taking up the action, and so on. In the case of the N atom in ammonia, oscillation is possible about either of the two potential minima on opposite sides of the hydrogen plane; but there is quantum-mechanical *tunneling* by the atom through the potential hill separating them (which they would not be able to cross classically), providing a coupling between the two oscillations. If somehow we can place the molecule initially at the minimum of potential on one side of the plane, then in the course of time it will oscillate from one side to the other through the potential barrier.

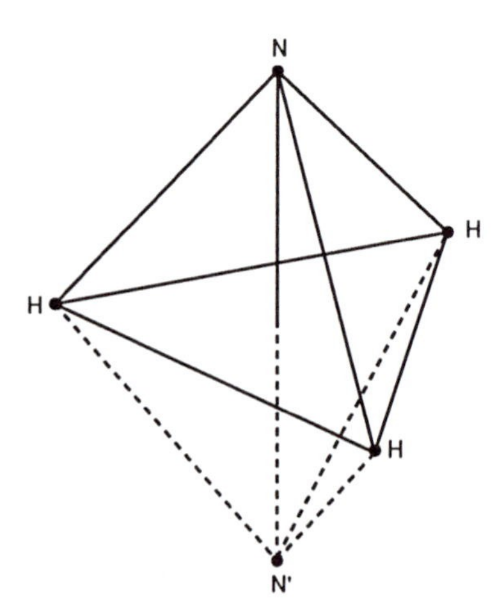

Figure 10.2 The ammonia molecule

The two stationary inversion states have different energies, and the transition frequency between them is, in fact, around 24 GHz, as we have already mentioned. If we recall that in quantum theory a frequency is associated with energy by the equation $\nu=E/h$ and that the state in which the N atom is on one side of the H-plane is a "mixture" of two wave functions of different frequencies, we would find that as time passes, the two wave functions go periodically in and out of phase at the *beat* frequency. Now, as illustrated in Figure 10.3, these two phase conditions yield total wave functions with a large amplitude first on one side of the H-plane, then the other; that is, the N atom oscillates back and forth across the H-plane at the beat frequency. But the transition frequency, being given by $\nu=(E_1-E_2)/h$, where E_1 and E_2 are the energies of the stationary states, can also be written as $\nu=\nu_1-\nu_2$; that is, the microwave transition frequency is equal to the beat frequency with which, we have just seen, the N atom oscillates across the H-plane.

Of course, this *hindered* vibration of the N atom along the symmetry axis is not the only *degree of freedom* we can visualize classically; in addition to the orbital angular momentum of the electrons and the nuclear spin of its constituent atoms, the molecule can rotate about its symmetry axis, which may itself rotate about the (constant) total angular momentum vector. As has become familiar, the conserved total angular momentum of the molecule is quantized, and it is conventionally designated by the quantum number J (if nuclear spins are excluded), so that the observable components along any axis fixed in space are an integer M_J in units of $h/2\pi$, where $M_J=-J, -(J-1),\cdots,+(J-1), +J$. Furthermore, the angular momentum of the molecule about its symmetry axis is designated by the quantum number K; since this is a component of the total angular momentum, clearly K cannot exceed J. Thus to specify the rotational part of the quantum state of a molecule, ignoring vibration and electronic excitation, we use the quantum numbers (J, K).

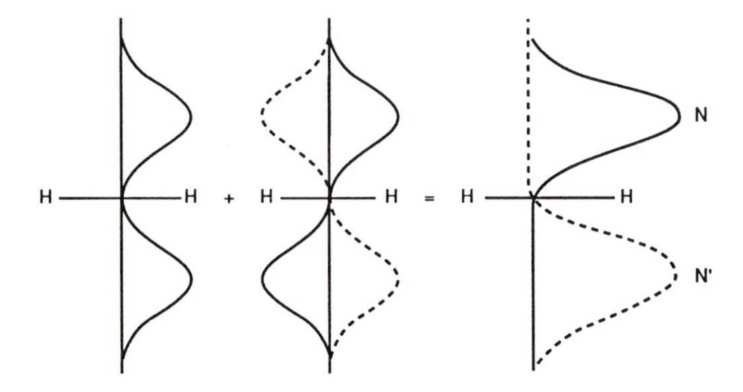

Figure 10.3 Inversion of ammonia as reversal of relative phase of wave functions

Since the maser action is based on stimulated radiation from the molecules in undergoing transitions from an upper to a lower energy state, it is clearly important that there exist states for which the probability of those stimulated transitions is sufficiently large. Unlike Cs, where the interaction with the radiation field was magnetic in nature, here the interaction is through the electric dipole moment of the molecule, which therefore involves the electric component of the radiation field. The chemical bonds between the nitrogen atom and the hydrogen atoms in the molecule are somewhat *polar*. This means that the center of the negative charges of the electrons is displaced relative to the positive charges of the nuclei; the measure of this polar property is the electric dipole moment. This will, in general, cause the energy of the molecular quantum states to be modified if the molecule is placed in an external electric field, the *Stark effect*. The detailed way in which the energy depends on the electric field intensity is interesting in that it confirms the peculiarly quantum nature of the description necessitated by the inversion symmetry. If the molecule had a "permanent" electric dipole moment in the usual classical sense, then the Stark effect shift in the energy of a molecular state would vary in a *linear* manner with the electric field intensity; in fact, it does not. Classically, the dipole moment is computed on the assumption of a nitrogen atom fixed on one side of the hydrogen atom plane; but then we must regard the dipole moment as rapidly reversing in direction as the nitrogen atom oscillates from one side to the other. The observed average of the dipole moment in any direction would then be zero. Of course, this bears on the question of the observability of the quantity we are calling the dipole moment itself. However, our real interest is only in how the energy of given quantum states of the molecule varies with the intensity of an electric field, and how strongly the molecule is coupled to a radiation field inducing transitions. It turns out that the state designated as J=3, K=3 exhibits the largest effective dipole moment, and therefore the strongest coupling to a resonant microwave field, in the inversion spectrum of ammonia.

10.4 The Electrostatic State Selector

Molecules issuing from a source in thermal equilibrium have not only a special distribution of velocity among them, but also a similar distribution among all the internal quantum energy states. This, according to the Boltzmann theory, means that there is a certain distribution of molecules among the energy states that through random collisions will eventually be reached no matter what their initial distribution might have been. Of course, any individual molecule will in the course of time be constantly changing its state, but the number of molecules in each of the quantum states will fluctuate about a value appropriate to it. Let E_n represent the energy of a molecular state, where the index n stands for a set of quantum numbers that identify the state, including J, M_J, K, the nuclear spin I of the H atoms, and a vibrational quantum number. The Boltzmann equilibrium distribution can now be given as follows:

$$p_n \sim \exp\left(-\frac{E_n}{kT}\right), \qquad 10.1$$

where p_n is the probability that a molecule is in a state with energy E_n, k is Boltzmann's constant, and T the absolute temperature. As is commonly the case, the states may fall into groups of *degenerate* states having the same energy; thus in the absence of an external field, for example, all the states that differ only in the value of M_J, and there are $(2J+1)$ of them, will have the same energy. If we now use the index q to differentiate only the energy, we can write the Boltzmann distribution as follows:

$$p_q \sim g_q \exp\left(-\frac{E_q}{kT}\right), \qquad 10.2$$

where g_q, the number of states having the same energy E_q, acts as a statistical weight.

It follows that the ratio of the populations N_1/N_2 of states having energy E_1 and E_2 is given by

$$\frac{N_1}{N_2} = \frac{g_1}{g_2}\exp\left[\frac{E_2 - E_1}{kT}\right]. \qquad 10.3$$

There are two physically important conclusions to be drawn from this: First, levels separated by energy in the microwave region are nearly equally populated at room temperature, since $(E_2 - E_1)/kT \approx 0.001$ in that case; and second, $N_2 < N_1$ if $E_2 > E_1$; that is, the lower-energy state has the greater population. The latter conclusion means that a gas in *thermal equilibrium* can never lead to maser (or laser) action. For maser amplification to occur there must be a *population inversion*; that is, there must be a nonequilibrium distribution of populations in which the upper energy level is more populated, rather than less than the lower energy level. Hence there is a need for a state selector, which in effect produces the inverted population by eliminating molecules in the lower-energy state.

A focusing electrostatic state selector is used based on the *Stark effect,* that is, the change in the molecular energy states due to an externally applied electric field. We will state without proof that the Stark effect for the rotation–vibration (inversion) states of ammonia computed to the second order of approximation is given by the following:

$$E = E_0 \pm \sqrt{\left(\frac{h\nu_0}{2}\right)^2 + \left(\mu \cdot E \frac{MK}{J(J+1)}\right)^2}, \qquad 10.4$$

where ν_0 is the inversion frequency, μ the electric dipole moment computed for a fixed N atom, **E** the electric field intensity, and (M, K, J) angular momentum quantum numbers specifying the molecular state. The plus sign applies to the upper and the minus to the lower energy of the pair of inversion states. We will not be making any quantitative use of this result beyond observing that the energy of the upper state increases with the electric field, while the energy of the lower one decreases. This is the basis of the electrostatic state selector, in which molecules in predominantly the upper state are selected to interact with the resonant field in the microwave cavity. It is based on the quantum-state-dependence of the electrostatic force that an ammonia molecule experiences in a static electric field having a steep gradient. Molecules in the upper states are deflected from regions of strong field towards regions of weak field, and conversely for the lower states.

A focusing electrostatic field having a quadrupole symmetry is ideally suited in this case; unlike the magnetic analogue in a magnetic resonance machine, only one state selector is involved, which simply focuses the desired upper state molecules into the microwave cavity. The field is produced by applying a high voltage (in the tens of kilovolt range) between two pairs of equally spaced parallel cylinders, as shown in Figure 10.4. To attain a pure quadrupole field distribution extending significantly away from the axis, the cylindrical electrodes should have a hyperbolic cross section. The field components are given by

$$E_x = -\frac{V_0}{r_0^2} x \, ; E_y = +\frac{V_0}{r_0^2} y \, , \qquad 10.4$$

where V_0 is the voltage applied between opposite electrodes and r_0 is the inner quadrupole radius. The resultant electric field depends only on the radius and increases linearly with it from zero on the axis to a maximum of $E_{max}=V_0/r_0$ at $r=r_0$.

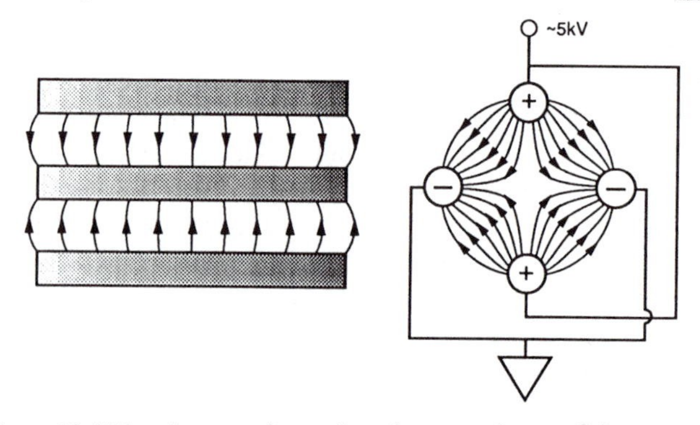

Figure 10.4 The electrostatic quadrupole state selector of the ammonia maser

In order to predict the possible trajectories of the molecules, we must compute the gradient in the Stark energy, which plays the role of potential energy in so far as the motion of the molecules is concerned. That energy function has already been cited; for moderate electric fields it can be approximated to give a simple quadratic dependence, which leads to a gradient, and hence force, that increases linearly with radial distance from the axis. In this approximation the molecules in the upper inversion states are drawn towards the axis as if by an elastic spring, causing molecules diverging from the axis at the source to follow trajectories that converge back on the axis at a point that depends on the applied voltage. By adjusting that voltage, these molecules are made to enter the microwave cavity. Molecules in the lower inversion states, on the other hand, will follow trajectories that diverge exponentially away from the axis. Of the molecules effusing from the source, those whose trajectories are bent to just graze the cylindrical electrodes define a critical direction of motion with respect to the axis, such that those with a greater initial angle will strike the electrodes and be lost to the beam, while those with a smaller angle will continue to be focused back towards the axis. This critical angle, the *acceptance angle*, corresponds to the radial part of the thermal kinetic energy of the molecules being equal to the Stark energy at the points of maximum electric field. Those that emerge from the source with greater radial kinetic energy than the maximum Stark energy will strike the electrode structure and join those making up the background gas in the surrounding space. The dynamics of the radial motion are analogous to those of a mass suspended by an elastic spring: If the mass is projected with a certain kinetic energy in a direction to stretch the spring, the mass will reach its maximum displacement when the (elastic) potential energy of the spring equals the initial kinetic energy of the mass.

The detailed analysis of the performance of the state selector is complicated by the fact that the molecules emerging from the source may be in any of a multitude of quantum states, each possibly having a different Stark energy, and their kinetic energy is spread over a wide range characteristic of their temperature. Of critical importance to the practicability of the technique is the ability to single out molecules in the one particular quantum state (J=3, K=3), from among the many in other states, to enter the resonant cavity in sufficient number per unit time. To estimate this rate we start with the fact that the ammonia gas in the source is in thermal equilibrium at some temperature, typically assumed to be 20°C.

In the case of the symmetric-top ammonia molecule, the determination of the degeneracy factors g_q, involving as it does the counting of the quantum states that are compatible with the symmetry of the molecule, would take us far beyond the compass of this book. However, we can attempt to convey the kind of reasoning that quantum theory invokes where symmetry and statistics are involved. First, the nuclei of the three H atoms (which are, of course, protons) have each an intrinsic spin of 1/2: This fact already imposes a restriction on the wave function that can represent the molecule, since according to the Pauli principle, which applies to protons as well as electrons, an exchange of the positions and spin states of any two H atoms should only reverse the sign of the function, and two such exchanges should leave the sign unchanged. If we label the atoms 1, 2, 3 and two exchanges are made in succession, for example $1 \rightleftarrows 2$ followed by $2 \rightleftarrows 3$, the final result is a rotation of the molecule through 120°, equivalent to increasing the angle coordinate about the axis of symmetry by 120°. In quantum theory the dependence of the wave function on this angle is determined by the value of the "conjugate" variable K, the angular momentum along the same axis. It is for K values that are multiples of 3 that the wave function returns to the same function with the same sign after the rotation. Moreover, it can be shown that wave functions having K a multiple of 3 can be constructed to have the correct symmetry under an exchange of a *single* pair of H atoms by combining functions with opposite sign for K, corresponding to inverted states. We recall that the states between which our transition occurs are superpositions of such states. When all the admissible functions having the proper symmetry are counted and their energy known, it becomes possible to find what proportion of the molecules in thermal equilibrium will be present in a given quantum state. Of particular interest to us is the finding that in ammonia, the states with K a multiple of 3 have twice the statistical weight that states with other values of K have. A combined measure of the fraction of molecules in the desired state, and the electric dipole moment coupling the two states, is the resonant absorption coefficient of microwave power $\gamma=8\times10^{-4}$ cm^{-1}, which is the highest among the microwave inversion lines in ammonia.

10.5 Stimulated Radiation in the Cavity

The state-selected molecular beam emerges from the quadrupole field and enters a microwave cavity tuned to resonance in a suitable mode of oscillation at the frequency of the 3–3 inversion line at around 23,870 MHZ. The scale of the resonant cavity dimensions is governed by the free-space wavelength of microwaves at this frequency, which is $\lambda=1.25$ cm. The geometry of the cavity is that of a cylinder, having either a circular or rectangular cross section, with ends closed, apart from a hole to admit the ammonia beam. An output waveguide is coupled to the cavity to draw out a part of the microwave power generated by the maser action. The essential criteria in the design of the cavity are first, that the molecules pass through regions where the electromagnetic field in the cavity is strongest, to ensure the greatest possible interaction; and second, to lengthen as much as possible the duration of the interaction by extending the length of the cavity, to reduce the width of the resonance. The first condition dictates the diameter of the cavity in relation to the diameter of the molecular beam, while the second requires that the cavity, regarded as a section of waveguide, is near *cut-off*; that is, the diameter of the waveguide is reduced to almost the free space wavelength, tending to extend the wave pattern along the length of the waveguide. Following the work of the Townes group, a figure of merit M may be defined as follows: $M=LQ_0/A$, where L, A are the length and cross-sectional area of the cavity and Q_0 its free (unloaded) quality factor, which, we recall, is defined as $Q_0=2\pi E/\Delta E$, where E is the energy stored in the cavity and ΔE is the energy lost in one cycle of the field, mainly as heat in the walls of the cavity. Among the possible resonant modes, with their characteristic field distributions, the TM_{010} mode, illustrated in Figure 10.5, has been shown by the same group to have a significantly higher figure of merit than the others.

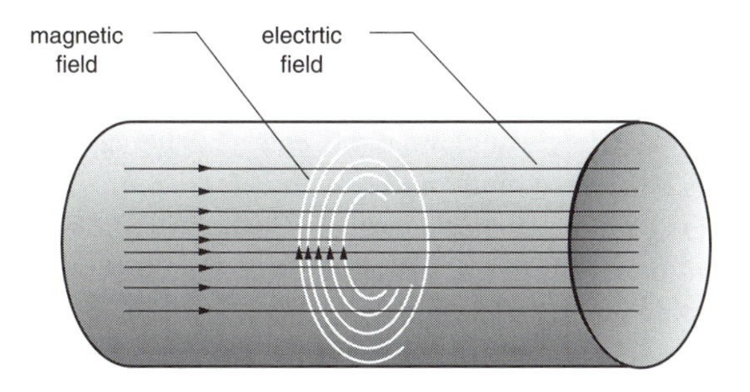

Figure 10.5 Cavity mode TM_{010} used in the ammonia maser

In order to achieve sustained oscillation, the power dissipated in the cavity must be compensated by the microwave power emitted by the ammonia molecules as they are induced to make quantum transitions from the upper energy state to the lower. We will not attempt to present the quantum theory of the radiative process that the molecules undergo, but will be content with a plausible approximation, which in fact leads to a result that is not far from the truth. We imagine that the *electric* dipole moment of a molecule is subject to a resonant electric field and thereby induced to radiate energy as it makes a quantum transition from the upper inversion state to the lower. It can be shown that the time dependence of the transition can be described as a rotation of the dipole at the rate of $2\pi\mu E/h$, where E is the electric field amplitude and μ is the dipole moment. If we call θ the angle of rotation as the molecule traverses the cavity, then to establish the correct correspondence with the quantum theory of the process, we must make the following assumptions:

$$p_1 + p_2 = 1;\ p_1 - p_2 = \cos\theta\,, \qquad 10.6$$

where p_1 and p_2 are probabilities of the molecule being in the upper and lower state, respectively. This is reasonable, since the probability of the molecule being in one or other of the states is obviously 100%. The second assumption, however, seems more arbitrary, but at least it leads to the expected results in the particular cases of θ=0°, 90°, and 180°, which correspond respectively to all molecules in the upper state, an equal number in the upper and lower states, and all molecules in the lower state. If we accept this interpretation of the angle θ, we easily find that as a molecule traverses the cavity, the probability of the molecule having radiated one quantum of energy and gone to the lower state increases from zero to (1/2)(1–cosθ), that is, $\sin^2\theta$.

10.6 Threshold for Sustained Oscillation

We are now able to derive an expression for the rate at which microwave power is radiated by the molecules in the cavity, assuming that, say, n molecules pass through the cavity per second. Since we are only concerned with the *threshold* condition for self-sustained oscillation, we note that initially, we may assume that E and hence θ are very small, and may therefore approximate $\sin\theta$ by θ (in radians). Now, the amount of energy radiated by a molecule when it undergoes a quantum transition is $h\nu$, and so it follows that the power radiated is simply given approximately by the following:

$$P_{\mathrm{rad}} = h\nu n\left(\pi\frac{\mu E}{h}\frac{L}{V}\right)^2, \qquad 10.7$$

Now we must express the power loss in the cavity in terms of the field strength in the cavity and its quality factor, Q. We will state without proof that the energy density in the field of the cavity is given by $E^2/2$, and therefore the total electromagnetic energy stored in the cavity is on the order of $V_0E^2/2$, where V_0 is the volume enclosed by the cavity and equals AL, where A is the cross-sectional area. Finally, from the definition of Q, we have for the power loss in the cavity

$$P_{\mathrm{loss}} = \pi\frac{AL}{Q}\nu E^2. \qquad 10.8$$

We can finally state the threshold condition on the strength of the molecular beam for sustained oscillation; we simply equate the power loss to the power radiated, to obtain after some reduction

$$n_{\mathrm{th}} = \left(\frac{h}{\pi}\right)\left(\frac{A}{QL}\right)\left(\frac{V}{\mu}\right)^2. \qquad 10.9$$

To obtain the level of oscillation as the beam intensity is raised above threshold, we simply retain in this approximate theory the probability p_2 as $\sin^2\theta$ rather than approximating it as θ^2. This leads to the following:

$$\frac{n}{n_{\text{th}}} = \frac{\theta^2}{\sin^2\theta}. \qquad 10.10$$

We note that as θ ranges between 0 and π (radians), n ranges from n_{th} to infinity. Thus according to this theory, θ and therefore the electric field in the cavity cannot exceed a certain limiting value obtained by setting $\theta=\pi$. This tendency of the microwave field amplitude not to continue increasing with the beam intensity is called *saturation.*

To gain an appreciation of the scale of the physical quantities involved, we draw on the early experimental results published by the Townes group on the relative merits of cavities oscillating in the TE_{011} and TM_{010} modes. For a cavity of Q=12,000 operating in the TE_{011} mode, the minimum state-selector voltage required to start oscillation was 11 kV for a source pressure of 800 Pa, whereas for a TM_{010} cavity with Q=10,000, the minimum voltage was only 6.9 kV at the same source pressure. Since for given source parameters, the number of molecules entering the cavity is roughly proportional to the square of the state-selector voltage, it follows that the threshold number of molecules per second, n_{th} for the TM_{010} mode cavity, is about one-third of the value for the TE_{011} mode cavity. The actual number of effective molecules per second in their experimental apparatus was $n=5\times10^{13}$ per second for a source pressure of 800 Pa and state-selector voltage of 15 kV.

10.7 Sources of Frequency Instability

10.7.1 Cavity Pulling

From the threshold condition for sustained oscillation it is evident that the higher the Q of the cavity, the lower is the threshold number of molecules that must pass through the cavity in unit time. Unfortunately, if the cavity is not tuned to precisely the molecular frequency, the actual microwave field in the cavity will not oscillate at the free molecular resonance frequency, but rather the frequency will be *pulled* by the cavity resonance to an extent that depends on the relative value of the Q of the cavity and the sharpness of the molecular resonance line.

To understand the underlying causes of this effect, we must recall two facts discussed in Chapter 2 concerning resonant systems and the conditions that determine the frequency of an oscillator. The first is that when a signal near the resonance frequency is injected into a resonant element such as a microwave cavity, the resulting output signal will have the same frequency but a shifted phase ranging from 0° to 180° as the frequency is swept through resonance, as shown schematically in Figure 2.2. The important feature to note is that the phase varies approximately linearly at the center frequency, between the maximum and minimum, a frequency range that can be taken as the line width of the resonance, $\Delta\nu$. It follows that an input signal whose frequency happens to be slightly displaced from the center of the resonance will be shifted in phase by an amount $\Delta\phi$ on the order of $180° \times (\nu - \nu_0)/\Delta\nu$.

Now, as we noted in the chapter on oscillators, the important condition that determines the frequency of oscillation is the phase change around the feedback loop. In the present case it is reasonable to assume that the oscillation of the maser will occur at such a frequency that the combined phase change in the cavity and the amplifying NH_3 molecules is zero. We can express this condition in practical terms using the Q-factor of the cavity Q_C and an equivalent Q-factor for the molecular resonance line Q_L, defined in a somewhat more general way than was done originally for a resonant circuit; namely in terms of the fractional frequency width $\Delta\nu/\nu$ of the resonance: $1/Q_L = \Delta\nu/\nu$. If we assume that the cavity is tuned to within the line width of the molecular resonance and that $Q_L/Q_C >> 1$, we arrive at a relationship for the frequency pulling that can be shown more rigorously to hold, namely

$$\nu - \nu_0 = A \frac{Q_C}{Q_L} (\nu_C - \nu_0), \qquad 10.11$$

where A is a constant that depends on the level of oscillation, $(\nu - \nu_0)$ is the deviation from the molecular frequency, and $(\nu_C - \nu_0)$ is the amount of mistuning of the cavity. This is a very general result, but it has particular relevance to the ammonia maser and, as we shall see, to the hydrogen maser, since the resonance line width, and hence the line Q, is not so large compared with the cavity Q that this frequency pulling is insignificant for a frequency standard. On the contrary, unless some scheme is used to correctly tune the cavity, the maser frequency will be dependent on all the factors that could affect the cavity tuning, for example temperature. For an ammonia standard that is supposed, to reach a stability beyond one part in 10^{12} this is intolerable.

One test of whether the cavity is properly tuned consists in varying the frequency width of the molecular resonance, that is, the line Q_L. When this is done, the maser oscillation frequency will vary also unless $(\nu_C - \nu_0) = 0$, in which case $(\nu - \nu_0) = 0$ no matter what value the line Q assumes. Thus a condition for the proper tuning of the cavity is that the maser frequency remain constant if the line Q is modulated.

In one ammonia beam maser described in 1961 by Barnes, Allan, and Wainwright, of the U.S. National Bureau of Standards (Barnes,1972), automatic tuning of the cavity was achieved by modulating the molecular resonance line width using the Zeeman effect. That is, a weak magnetic field is applied to the molecules in the cavity, causing a splitting of the ammonia line, as shown in Figure 10.6, effectively broadening it. By modulating the magnetic field at a low frequency, any mistuning of the cavity leads to a modulation of the output frequency of the maser at double the field modulation frequency; this doubling is due to the Zeeman broadening being identical for opposite directions of the magnetic field. For the purposes of this automatic tuning technique, the short-term stability of a high-quality quartz crystal oscillator is sufficiently high to serve as a reference in obtaining the modulation of the maser frequency. In the actual ammonia beam maser cited, the maser frequency is not compared directly with a synthesized frequency based on the crystal frequency, but rather through an intermediary klystron, phase-locked to the crystal-based frequency. A servo loop controls the tuning of the cavity by activating a motor-driven *tuning stub*, so as to annul the modulation of the maser frequency; this occurs when the cavity is tuned exactly to the molecular resonance. It might appear that in our attempt to construct a stable frequency standard, we have assumed that we already have one of equal stability to serve as a reference to detect the frequency pulling of a mistuned cavity. In fact, the frequency reference for the automatic tuning of the cavity must only have sufficient stability in the short term, that is, over periods on the order of 1/10 second. For such short-term stability, the quartz oscillator can be superior to other types of frequency standards. On the other hand, the maser is intended to have not only short-term stability, but more important for a time standard, long-term stability extending over periods of years.

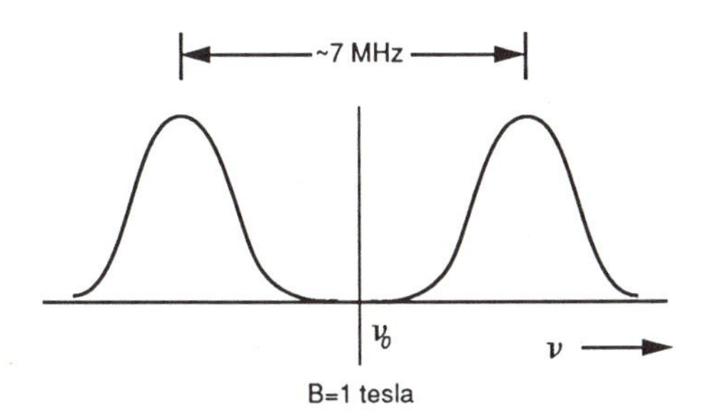

Figure 10.6 The Zeeman splitting of the ammonia line (Barnes, 1961)

10.7.2 Doppler Shifts

The ammonia beam maser is susceptible to a number of other systematic errors, which include Doppler shifts due to nonstationary field patterns in the cavity. Unbalanced traveling wave components in the cavity can arise from asymmetry in the position of the output coupling hole in the cavity, causing asymmetric power flow to the output. But a more insidious effect is the variation of the molecular emission of radiation along the path of the beam in the cavity, which gives rise to unbalanced traveling waves and a consequent shift in frequency.

Since the power emitted by the molecules is of the same order of magnitude as the output power, the frequency shift due to this effect will be comparable to that due to the asymmetric coupling to the output. To further complicate the matter, the distribution of the molecular emission along the beam in the cavity depends on the level of oscillation (saturation) in the cavity. For weak oscillation near threshold, the emission is more toward the end of the path in the cavity, in contrast to the case of high saturation, when the emission is mostly near the entrance to the cavity. Thus the direction of the unbalanced traveling wave depends on the flux of ammonia molecules in the beam and the corresponding power level in the cavity. The flux of molecules, in turn, depends on the pressure of ammonia gas at the source and the voltage on the quadrupole state-selector. In an attempt to minimize these frequency shifts, ammonia masers with two beams traversing the cavity in opposite directions have been studied.

10.7.3 Molecular Collisions

Another source of frequency shifts in the molecular resonance is the perturbing effect of collisions between ammonia molecules, either between those in the beam or with background molecules. As with any other atomic or molecular system, such collisions can cause a broadening of spectral lines as well as a shift in the center of those lines. Again, this would make the maser output frequency susceptible to fluctuations arising from possible instabilities in any of the parameters that affect the beam density or background pressure in the cavity. Of course, the same collision effects are present in the Cs beam standard; however, there the beam intensities can be very much lower since there is no requirement to go over a threshold value for oscillation.

10.7.4 Ambient Electric and Magnetic Fields

Again, as with all atomic and molecular spectra, the inversion spectrum of ammonia is affected by external electric and magnetic field; the Stark effect in ammonia consists not only of shifts in the energy of inversion states; it also induces changes in the coupling of the various nuclear spins to the molecular rotational motion. The transition of interest is between states in which the effect of the nuclear spins is the same for both the upper and lower inversion states, and the transition frequency is not altered on that account. Nevertheless, there remains the quadratic Stark shift in frequency, which is fractionally on the order of $(E\mu/h\nu_0)^2$, where E is the electric field strength, μ is the electric dipole moment, and ν_0 is the inversion frequency. Substitution of numerical values leads to shifts on the order of one part in 10^{12} for fields in the range of a few volts per meter. It is clear from this that some care must be exercised in preventing static electrical potentials from developing in the cavity. As for the Zeeman effect, the first-order effect produces a symmetric splitting that does not shift the center, but merely results in a broadening of the resonance, as already indicated in connection with the automatic tuning of the cavity. To second order of approximation there is a shift in line center having a quadratic dependence on the magnetic field intensity; however, it is about 10^6 times smaller than in the Cs atom, since the molecular magnetic moment arises from the nuclear moments rather than those of electrons, and is negligible.

Since it was the first molecular oscillator, a great deal of effort was invested in realizing its promise as the first of a new class of stable oscillators. However, as a primary standard it proved to have fatal drawbacks: its frequency depended on many operating parameters; and there was no fundamental prescription to define what values these parameters should be assigned. For example, the frequency depends on the source pressure and operating voltage of the state-selector; how is one to decide what values to use? If arbitrary choices are made, the definition of the standard would have to include in detail all the dimensions and operating conditions of the device. Furthermore, the standard would be subject to all the uncertainties and instabilities that all these parameters may have. Nevertheless, the ammonia beam maser ushered in the age of stable atomic and molecular oscillators and a new level of stability approaching one part in 10^{12}.

10.8 The Rubidium Maser

Efforts to realize an active, oscillating form of the rubidium standard in the 1960s are associated with the names of F. Hartmann, at the Ecole Normale Supérieure in Paris, and P. Davidovitts and R. Novick, at Columbia, who announced their success in 1966. Later, the same goal was pursued by J. Vanier, then at Laval University, and E.N. Bazarov, then of the Soviet Union.

The essential difference in the design of the rubidium maser, as compared to the passive gas cell resonator, lies in the need to allow the microwave radiation emitted by the atoms, as they make the reference hyperfine transition, to build up in a high-Q microwave cavity. Unlike the ammonia maser, this type of transition involves a magnetic interaction with the microwave field to stimulate emission, rather than the much stronger electric dipole interaction we have in the ammonia maser. This raises the threshold for oscillation and makes it necessary to achieve the greatest possible number of contributing atoms interacting coherently for the longest possible time in a high-Q resonant cavity.

Of the two isotopes of rubidium, Rb^{85} has the longer transition wavelength at around 10 cm, making it the easier of the two to achieve a high Q in its resonant cavity. The connection between cavity size and Q will be discussed at somewhat greater length in the next chapter.

The essential elements of an optically pumped gas cell maser are shown schematically in Figure 10.7. In the original experiments, which predate lasers, one of the greatest challenges was to achieve a sufficiently intense pumping light source, with the proper spectral distribution to achieve efficient and rapid pumping of the hyperfine state populations. The threshold value of atomic number density is so high that Rb–Rb collisions in which spin states are exchanged can cause rapid *relaxation* of the hyperfine populations, counteracting the pumping action of the light. More is said about spin-exchange collisions in the next chapter. As discussed in connection with the rubidium clock, another difficulty is that the reference frequency is shifted by the pumping light itself, an effect that is aggravated here because of the need to use high-intensity sources to reach the threshold.

Recent developments in optically pumped alkali vapor masers include using laser pumping sources and circumventing the light shifts using separated regions for pumping and cavity field interaction. Thus an *evacuated* double bulb design was proposed in 1994, in which an inert wall coating, a long-chain hydrocarbon *tetracontane,* rather than buffer gases, is used to prevent relaxation of hyperfine populations at the walls of the container. The atoms are free to travel throughout the combined volume of the bulbs, which are joined by a short passage tube. The atoms are optically pumped in one bulb and radiate inside a resonant cavity in the other.

Residual shifts in the oscillation frequency due to collisions with the walls, which are strongly dependent on detailed surface conditions, would still preclude this type of standard from being considered as a primary standard. Nevertheless, it was felt that a relatively high signal-to-noise ratio could be realized in the output of such standards, and that they would therefore have excellent short-term stability.

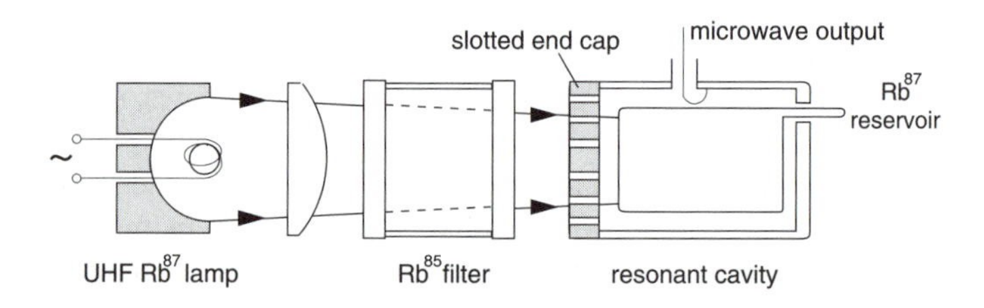

Figure 10.7 The essential elements of an optically pumped rubidium maser oscillator

Chapter 11
The Hydrogen Maser

11.1 Introduction

We come now to consider what proved to be the culmination of efforts to perfect maser oscillators: the hydrogen maser, one of the most stable of all present-day frequency standards. Few other microwave quantum devices approach its overall frequency stability. Conceptually, the H-maser was a natural outgrowth of the continuing experimental drive to improve the spectral resolution of atomic beam resonance machines by increasing the interaction time between the atoms and the resonant field. This was to be achieved by confining the interacting atoms within a space defined by inert walls; however, few would have predicted the degree of inertness exhibited by one fluorocarbon polymer named *Teflon* and the extraordinary length of perturbation-free interaction time it made possible.

We recall that in the atomic beam resonance technique, in which transitions are observed on atoms in free flight such as in the Cs beam standard, the frequency width of the resonance is determined by the length of time the atoms are free to interact with the resonant field. The natural line width, due to the limit on the observation time imposed by the process of spontaneous emission, is negligibly small for the microwave transitions involved here, because years may pass before such transitions occur! But there is obviously a practical limit to how far the time spent by the atoms interacting with the field can be lengthened simply by increasing the length of the apparatus. A way must be found to deflect the atoms from their straight-line path without disturbing their coherent response to the resonant field, so that the duration of that response can be lengthened in a confined space. The use of static electric or magnetic fields strong enough to confine *neutral* atoms is ruled out, if for no other reason than that the quantum state of the atoms would be severely disturbed. However, as we shall see in later chapters, laser radiation has been exploited to cool and trap atoms, with far-reaching advances in the design of atomic frequency standards. Prior to that, the "classical" approach was to constrain the motion of the atoms through elastic collisions with other inert atoms or molecules, either in a gaseous form or as a solid surface; clearly, in a beam apparatus, reflection from an inert surface is more compatible.

Accordingly, a series of experiments was carried out around 1958 in Ramsey's laboratory at Harvard University on a Cs beam apparatus in which reflecting surfaces of various materials were tried. It was realized that since the microwave transitions of interest involve changes in magnetic states, the surface atoms or molecules must not be capable of any magnetic type of interaction with the colliding atom; this rules out all metallic surfaces and any surface that has "free radicals" and unsaturated compounds, that is, chemical entities with unsatisfied bonds. As we pointed out in connection with the optically pumped Rb standard, where coatings for the absorption cells were sought for the same purpose, long-chain paraffins were studied as good possible candidates. Encouraging results were also found with a silicone compound (dimethyldichlorosilane) named Dri-Film. However, it was realized that Cs, a heavy alkali atom, has a high polarizability; that is, the one outermost electron, which in the ground state surrounds the inner shells in a spherically symmetric way, is easily distorted in a collision into a nonspherical state. In this state, the atom has angular momentum that has associated with it a magnetic field that will interact with the electron spin, thereby inducing transitions between the magnetic states of interest. The hydrogen atom, on the other hand, although it is alkali-like in having a single electron, has a very much smaller polarizability, since a great deal more energy is required to put the electron in the first available nonspherical quantum state. For hydrogen then, surface materials that lack "dangling" chemical bonds and interact only through a mutual polarization (van der Waals force) will cause very much smaller perturbation of the magnetic hyperfine states than in Cs. However, few would have predicted that hydrogen atoms can make on the average an incredible 100,000 collisions with a specially prepared Teflon surface before a radiating hydrogen atom loses coherence. This means, as we shall see, that atoms at ordinary temperatures contained in a bulb of convenient dimensions would remain with their internal quantum states undisturbed, on the average, a full second before leaving the bulb. In an atomic context this is a *long* time. The possibility of such a long storage time means not only a very sharp resonance, but also the practicability of observing sustained maser oscillation using the relatively weak *magnetic* dipole coupling to a resonant radiation field. This is in contrast to the ammonia maser, which is based on a far stronger *electric* dipole transition.

We recall that the short-term frequency stability of any quantum oscillator is determined as much by the signal-to-noise ratio as the width of the atomic or molecular resonance. The presence, for example, of the fundamental Johnson (thermal) noise introduces random fluctuations in the oscillation waveform that can be interpreted as amplitude and phase noise. It has been shown by Townes (Townes, 1962) that for the ideal case of an oscillator subject only to the fundamental thermal noise, the fractional standard deviation of the frequency is of the form

$$\frac{\sqrt{\langle\delta\nu^2\rangle_t}}{\nu_0}=\frac{1}{2Q_L}\sqrt{\frac{kT}{P\tau}}, \qquad 11.1$$

where P is the power radiated by the atoms well above threshold, τ is the averaging time for the $\langle\ \rangle$ average, Q_L is the Q-factor of the atomic resonance, and the product kT, as usual, is a measure of thermal energy. We see from this that whereas the ammonia maser has the advantage of more power than is to be expected of the hydrogen maser, the frequency fluctuations of the latter can nevertheless be smaller, since its Q_L-value is several thousand times larger. In practice, the fundamental limitation of thermal noise is rarely reached; many other sources of noise and instability are usually present, but on balance, the hydrogen maser is unmatched. As such (although as we shall see, not the most absolutely reproducible), the hydrogen maser will be described in sufficient detail to appreciate how it achieved that status.

11.2 The Hyperfine Structure of H Ground State

Its *active* medium is atomic hydrogen; not the diatomic molecular form H_2 in which it is ordinarily encountered. Hydrogen is the simplest of the elements, consisting of only one electron surrounding a nucleus that is simply a proton. There is another stable isotope, deuterium, or heavy hydrogen, with a nucleus consisting of a proton and neutron, but this does not concern us here. As a two-body problem the quantum energy states of hydrogen can be worked out to any desired degree of accuracy, taking into account not only the electrostatic Coulomb force between the electron and the nucleus, but also the magnetic interaction between them. This latter magnetic interaction arises from the fact that both the proton and electron have magnetic moments associated with their spin. The electrostatic interaction determines the gross features of the energy level structure of the atom, while its fine structure is due to the magnetic interactions between the magnetic fields produced by the orbital motion of the electron and its own magnetic moment; the weaker electron–nucleus magnetic interaction leads to a hyperfine structure. We have already treated the classification of the quantum states of the hydrogen atom as the starting point in discussing more complex atoms; we know the set of quantum numbers necessary to specify them. All but one of these quantum numbers, the principal quantum number, n, relate to angular momentum. In the absence of torques produced by external fields, the total angular momentum of the atom is conserved, and its component along any fixed axis obeys the space quantization rules with which we have by now become familiar. Since the electron has spin angular momentum of 1/2(units of $h/2\pi$) and the proton similarly has a spin of 1/2, then for an H atom in its (L=0) ground state, free of any external magnetic field, there are two possible values of total angular momentum: In the notation we have already introduced for the alkali atoms, these are F=1/2+1/2=1, and F=1/2–1/2=0. The corresponding observed components along any fixed axis are m_F=+1, 0, –1 and m_F=0, respectively.

Atoms in the F=1 state differ in energy from those in the F=0 state by what we know as the hyperfine splitting, the operating frequency of the hydrogen maser. The difference in energy comes, we recall, from the magnetic interaction between the proton and electron magnetic moments, leading to the F=1 state with the magnetic moments opposed being higher in energy than F=0, for which they are parallel. Note that because the proton and electron have opposite charge, states with parallel spin have the magnetic moments opposed, and vice versa. We have seen that classically the problem may be thought of as that of finding the potential energy of a small bar magnet embedded in a spherically symmetric magnetized medium, whose direction of magnetization is either with or against the direction of the bar magnet. Note that this is very different from the energy of two magnetic moments side by side. As mentioned in an earlier chapter, the quantum formula for the hyperfine splitting of the ground state of any hydrogen-like system was first given by Fermi around 1930; however, a more intuitive derivation based on a classical picture happens to yield the same result. The derivation hinges on the expression for the magnetic field inside a uniformly magnetized medium, which is given classically by the formula $\mathbf{B}=(2\mu_0/3)\mathbf{M}$, where μ_0 is a scale factor (called the permeability of free space, and defined as $4\pi-10^{-7}$), **B** is the intensity of the magnetic field (measured inside an imagined small spherical cavity), and **M** is the magnetization, defined as the magnetic (dipole) moment per unit volume. Recalling that the magnetic moment of the electron is one Bohr magneton, μ_B, we find

$$\mathbf{M} = g_e \mu_B |\psi(0)|^2 \mathbf{s}, \qquad 11.2$$

where **s**, the electron spin, is 1/2, and $|\psi|^2$ is the probability density of the electron. Here the one electron is pictured as a continuous (magnetized) charge cloud whose density is distributed around the nucleus according to the probability density for the ground state of the electron. We can now derive the intensity of the magnetic field acting on the proton: It is the field in a vanishingly small spherical cavity drawn around the proton, where the density in the immediate vicinity may be taken to be uniform and represented by $|\psi(0)|^2$; it has the value

$$\mathbf{B} = \left(\frac{2\mu_0}{3}\right) g_e \mu_B |\psi(0)|^2 \mathbf{s}. \qquad 11.3$$

Now, using the classical expression for the (potential) energy of a magnetic dipole in a magnetic field, $E=-\mu\mathbf{B}\cos\theta$, where θ is the angle between the direction of the dipole and the magnetic field, we find

$$E(F)=\left(\frac{2\mu_0}{3}\right)g_p g_e \mu_n \mu_B |\psi(0)|^2 \frac{\left(F^2-I^2-S^2\right)}{2}. \quad 11.4$$

Apart from the quantum numbers, this in fact is the correct Fermi formula for the electron–nucleus "contact" interaction, giving rise to the magnetic hyperfine splitting of the ground state. To correct the dependence on the angular momentum quantum numbers, we must follow the prescription for correcting results obtained using the vector model, namely, F^2 must be replaced by $F(F+1)$, etc.

As a first approximation, this correctly gives the frequency of transitions between these hyperfine states as 1,420 MHz, corresponding to a free-space wavelength of about 21cm. Thanks to the success of the hydrogen maser, the actual frequency of this hyperfine transition is arguably the most precisely measured quantity in all of physics: It has been measured to better than one part in a trillion! Such precision has naturally spurred ever-increasing refinement in the theoretical computation of the frequency, and in so doing has put the theory to an extraordinarily stringent test. The theory of quantum electrodynamics has thereby proven itself to be one of the greatest intellectual achievements of our time.

As the hyperfine interaction between the electron and nucleus considered here involves magnetic dipoles, we must take into account the unavoidable presence of external magnetic fields and the way they affect the frequency and line width of the hyperfine transition that is to be used as a standard. Figure 11.1 shows the dependence of the energy of the F=1, m_F=+1, 0, −1 and F=0, m_F=0 states of the ground state of hydrogen on the intensity of an applied magnetic field. We note as we did in the case of Cs that the graphs for the m_F=0 states are curves that start in a horizontal direction at B=0, indicating that their energy is stationary, that is, unchanging with respect to small changes in the magnetic field at that point. This, we recall, is the property on which the choice of the standard transition in Rb and Cs was based, because for B nearly zero, any small variation in its intensity over the region where the atoms interact with the resonant field will not broaden the resonance. This is in contrast with transitions involving the m_F=±1 state, where the energy changes in proportion to the field, and the lines involving them are broadened by any inhomogeneity in the magnetic field. Thus the maser operates on the "field-independent" F=1, m_F=0→F=0, m_F=0 transition near zero magnetic field. It is important to note that as with Cs, atoms in a given state will remain in that state as their energy varies with magnetic field, provided that the change in field intensity "seen" by an atom is sufficiently slow; this implies the absence of abrupt changes in the magnetic field either in space or time. We further note that in the limit of very strong magnetic fields, the graphs reduce to two pairs of parallel lines of opposite slope, which would be expected of the energy of the electron and nucleus independently pointing with or against the applied magnetic field. In this high field region, the interaction between the electron and nucleus is negligible compared with the interaction of each with the applied field. Again, following an argument we already made in connection with Cs, if atoms of hydrogen pass through a region where the magnetic field intensity varies from point to point, those whose energy increases with magnetic field will be deflected towards regions of weaker field, since the magnetic energy must be taken as potential energy in predicting the motion of the atoms.

Conversely, those whose energy falls with magnetic field will be deflected toward regions of stronger field. This is the basis of the magnetic state-selector used to separate atoms in the upper F=1 state from those in the lower F=0 state.

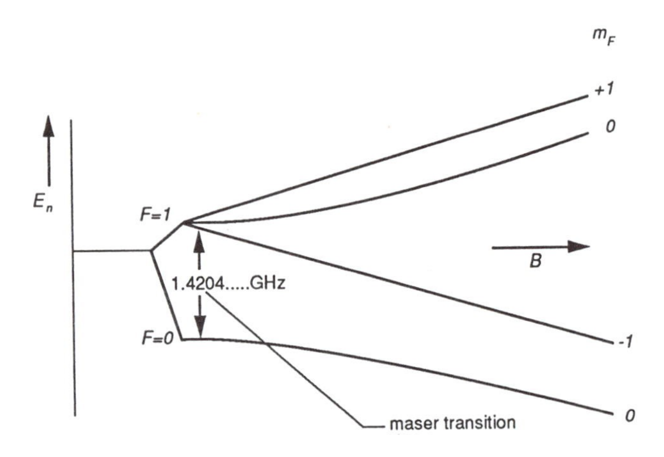

Figure 11.1 The energy dependence of the hyperfine states of H in the ground state

11.3 Principles of the Hydrogen Maser

As with the original ammonia maser, the essential features of the hydrogen maser, shown schematically in Figure 11.2, are a source of hydrogen atoms collimated into a beam; a quantum state selector, which removes most of the atoms in the lower of the two energy states between which maser action is to take place; a resonant cavity, in which the atoms, predominantly in the upper state, are induced to make transitions to the lower state by emitting radiation, thereby sustaining the field within it. This will occur when the power radiated by the atoms is sufficient to make up cavity losses as reflected in its loaded Q-factor.

The formulation of the conditions for oscillation and analysis of the properties of the maser, based on the treatment of the ammonia maser by Shimoda, Wang, and Townes, were published first in 1962 by Kleppner, Goldenberg, and

Ramsey, of Harvard University, followed in 1965 by an article coauthored with Vessot, Peters, and Vanier of Varian Associates (Kleppner et al.,1965), in which the theory is extended and experimental techniques are discussed. Contrary to intuition, there is not only a threshold rate at which atoms in the beam must enter the cavity in order for the maser to oscillate, but there is also an *upper* limit on the rate of atoms entering, beyond which oscillation will cease!

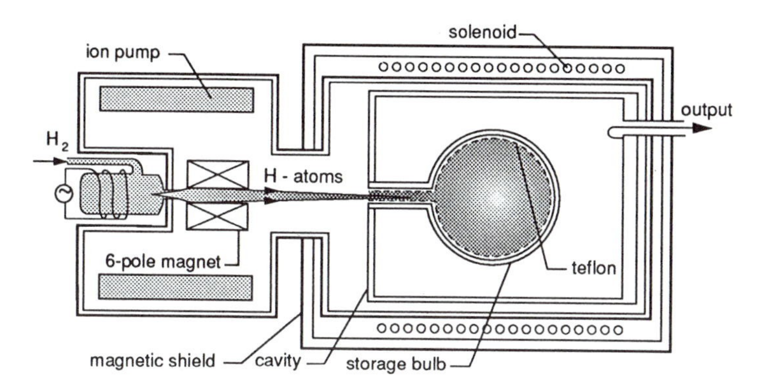

Figure 11.2 The basic elements of a hydrogen maser

11.3.1 Lifetime of Atoms in the Bulb

There are two fundamental processes, among others, that contribute to the shortening of the relaxation times. First is the escape of atoms from the cavity; this obviously limits their interaction time with the resonant field and contributes equally to T_1 and T_2. To find the probability (per unit time) of an atom escaping from the bulb through the opening by which it enters, we assume it makes many random collisions with the surface of the bulb and therefore has equal probability of occupying any (equal) element of volume within the bulb. Under the low pressure conditions in the maser bulb, the atoms are relatively free of collisions among themselves, and we may further assume that the atoms striking the opening by which the beam enters actually leave the bulb. If the cross-sectional area of the opening is represented by S, then in order to strike the opening in unit time the atom must be somewhere within a space defined by a cylinder of cross-sectional area S and length equal to the average distance it can travel in unit time perpendicular to S, that is, $V_{ave}/4$, where V_{ave} is the average 3-dimensional velocity. If we denote the volume of the bulb by V_b, it follows that the desired probability is given by $SV_{ave}/4V_b$, and therefore the probability of a given atom remaining in the bulb will decay exponentially as follows:

$$p(t) = \exp\left(-\frac{t}{T_b}\right) \quad ; \quad T_b = \frac{4V_b}{SV_{ave}}. \qquad 11.5$$

In terms of the absolute temperature T of the bulb, the average velocity is given by $V_{ave} = \sqrt{8kT/(\pi M)}$. For example at T=300K we have a mean thermal velocity of about 2.5×10^3 m/sec; hence for a bulb having a diameter of 0.25 m and a 4 mm diameter opening, the characteristic decay time is very nearly 1 second. Incidentally, in that time an atom would have made on the order of 20,000 collisions with the surface of the bulb. While the remarkable property of Teflon permits the approximations we have made, the actual system is more complicated, in that some adsorption and recombination into molecular hydrogen do occur on the bulb surface, and some atoms are scattered back at the opening.

11.3.2 Spin Exchange Collisions

The second inevitable contribution to the broadening of the resonance, which is essentially different from the first in that it depends on the number *density* of atoms, comes from collisions between atoms. These can both randomize the phase of the radiating atoms and cause nonradiative transitions, thereby reducing the relative number of radiating atoms. The type of collision between atoms that dominates under the ordinary operating conditions of the maser is one in which the electrons of the two colliding hydrogen atoms *exchange spin* directions. An exchange of spins manifests itself only when the electrons approach in opposite states, and the collision results in a mutual spin-flip; so our concern is with a process that we can write symbolically as follows:

$$A(\uparrow) + B(\downarrow) \rightarrow A(\downarrow) + B(\uparrow). \qquad 11.6$$

The duration of such collisions is short compared to the average time between collisions; so short, in fact, that the nuclei do not have time to be affected; that is, for the purposes of predicting the resulting states of the atoms, it is as if the hyperfine interaction between electron and nucleus were not there. The whole process is strictly a quantum-mechanical effect; the interaction of the electrons by virtue of their magnetic dipole moments is far too weak to explain the large cross section the atoms present each other in encounters that result in spin exchange. The interaction ultimately arises from the electrostatic Coulomb force between charges, which, although not spin-dependent itself, nevertheless manifests itself in a spin-dependent way because of a certain symmetry requirement on the two-electron wave functions, which we encountered earlier in connection with the Pauli exclusion principle. The likelihood of an atom undergoing a spin-exchange collision is expressed in terms of the effective cross section it presents as it travels through the cloud of other atoms, sweeping out a volume containing all those atoms with which it will make such a collision. If the number density of the hydrogen atoms in the bulb is represented by n, the cross section by σ_{ex}, and the mean relative velocity of the atoms is V_r then the volume swept out per second is $\sigma_{ex}V_r$, and the number of atoms in that volume will be $n\sigma_{ex}V_r$. This then is the average number of spin-exchange collisions a given atom will undergo per unit time, and the average time between collisions is the reciprocal of that, namely

$$T_x = \frac{1}{\sigma_{ex} n V_r} \qquad 11.7$$

showing explicitly the dependence of T_x on the number density of atoms in the bulb, and hence the rate at which atoms enter the bulb. The spin-exchange process contributes to both types of relaxations; however, not to the same extent. It can be shown that $T_1=T_x$, while $T_2=2T_x$.

11.3.3 Threshold for Sustained Oscillation

We begin by considering the power radiated by the atoms in the cavity. Following Ramsey et al., we use a result derived from a solution of Bloch's phenomenological equation for the response of the global magnetic moment of a group of atoms subject to a time-varying magnetic field. The resonant response of the H-atoms to a microwave magnetic field, inducing transitions between the hyperfine states, is through the magnetic dipole moment of the atoms, and to the extent that we may regard the transitions as occurring between just two quantum states, a vector description in terms of a spin 1/2 particle gives valid results. In the present case one finds that the net power P radiated by atoms entering the cavity in which the field amplitude is B_z with an excess of flux ΔI of atoms in the upper state over those in the lower state is as follows:

$$P = \frac{1}{2}\Delta I\, h\nu \frac{x^2}{\frac{1}{T_1 T_2} + x^2 + \left(\frac{T_2}{T_1}\right)\left[2\pi(\nu - \nu_0)\right]^2}, \qquad 11.8$$

where $x=2\pi\mu_A B_z/h$, ν_0 is the frequency of the atomic hyperfine frequency, and μ_0 is the magnetic moment of the atom. Note the presence of an x^2 term in the denominator of the expression for P, which has the effect that for amplitudes of B_z such that x^2 becomes dominant, the value of P reaches *saturation*, that is, no longer increases with the power level in the cavity. The graph of P has the shape of a typical resonance curve with a maximum at $\nu=\nu_0$ and a (full) frequency width at the half power level given by

$$\Delta\nu = \frac{1}{\pi}\sqrt{\frac{1}{T_2^2} + \left(\frac{T_1}{T_2}\right)x^2}\ . \tag{11.9}$$

The presence of the term involving x^2, and hence B_z^2, in the expression for the resonance width shows *power broadening* of the resonance, which can be thought of as due to the shortening of the time each radiating atom spends in a transition due to the strength of the inducing field.

Finally, we are in a position to understand the reason it is possible to have too large a flux of atoms entering the cavity to have oscillation. To that end we will simplify matters by including only the two fundamental contributions to the relaxation times and write

$$\frac{1}{T_1} = \frac{1}{T_b} + \frac{1}{2T_x} \quad ; \quad \frac{1}{T_2} = \frac{1}{T_b} + \frac{1}{T_x}. \tag{11.10}$$

This way of combining relaxation times is justified on the basis that for example, $1/T_1$ is the average probability per unit time of a certain event taking place. It follows that if there are two independent ways in which that event can come about, and we wish to have the probability of one *or* the other taking place, then we should add the individual probabilities. Now we see that since $1/T_x$ is proportional to n, it also depends on I_{tot}, the atomic flux entering the bulb. This is evident from the fact that under steady conditions the total number of atoms in the bulb, N, depends on the total flux according to the following:

$$\frac{dN}{dt} = I_{tot} - \frac{N}{T_b} = 0; \tag{11.11}$$

and hence the relaxation time T_x depends on I_{tot} as follows:

$$\frac{1}{T_x} = \frac{\sigma_{ex} V_r T_b}{V_b} I_{tot}. \tag{11.12}$$

The atoms entering the bulb from the state selector are with equal number in the states designated as (F=1, m_F=1) and (F=1, m_F=0), and ideally none in the other states. If we assume this ideal condition, then the difference in population between the two maser states is $I_{tot}/2$.

Now, the condition for a sustained level of oscillation can simply be stated as follows: The power radiated by the atoms must equal the total microwave power dissipated in the cavity and delivered as output. Using our definition of the Q-factor of the (loaded) cavity, this can be formulated as follows:

$$P_{\text{rad}} = 2\pi\nu\frac{E}{Q}, \tag{11.13}$$

where P_{rad} is the power radiated by the atoms at resonance and E is the electromagnetic energy stored in the cavity, energy that resides in the field and in classical theory can be expressed (in the SI, Système International, of units) in terms of the field amplitude B as follows:

$$E = \frac{1}{2\mu_0}\langle B^2\rangle_c V_c, \tag{11.14}$$

where the notation $\langle\ \rangle_c$ designates an average over the volume of the cavity, equal elements of volume being given equal weight. There remains the question of relating $\langle B^2\rangle_c$ to the appropriate average, over the volume of the bulb, of the field component responsible for stimulating transitions. This is done by defining a *filling factor*, given the symbol η, as follows:

$$\eta = \frac{\langle B_z\rangle^2{}_b V_b}{\langle B^2\rangle_c V_c}, \tag{11.15}$$

where V_b is the volume of the bulb and the notation $\langle\ \rangle_b$ denotes, as before, an average over the volume of the bulb. The ratio η is important in the analysis of the conditions for oscillation of the maser—this is hardly surprising, since the numerator is, in fact, proportional to the rate at which the atoms are stimulated to emit radiation, while the denominator is proportional to the energy stored in the cavity. The salient point in this definition of η is that the average in the numerator is taken to be $\langle B_z\rangle_b^2$, rather than $\langle B_z^2\rangle_b$, which would be true for fixed atoms. We will not attempt to pursue the question beyond saying that the explanation is to be found in the fact that the atoms rapidly cross the bulb back and forth many times during the emission process, randomly sampling the field throughout the space in the bulb. The atoms are confined to a space where the radiation field has almost constant phase but varying amplitude according to the field pattern of the particular mode. Under these conditions we further expect the Dicke effect to be observed in the atomic resonance line: a sharp line center free of (first-order) Doppler broadening on a broadened pedestal.

In terms of the filling factor, and the (loaded) Q-factor of the cavity, we can now relate $\langle B_z\rangle^2$, and therefore x^2, the saturation factor, to the output and loss power P; the result is a simple proportionality: $x^2 = \alpha P$, where α is a constant involving the physical parameters of the maser. We can now use this result in the formula for the power radiated by the atoms *at resonance* $(\nu - \nu_0) = 0$ and require that it be equal to the power P; that is, for the assumed constant level of oscillation we must have

$$P = \frac{1}{2}\Delta I\, h\nu - \frac{1}{\alpha T_b^2}\left(1 + \frac{3}{2}\frac{T_b}{T_x} + \frac{1}{2}\frac{T_b^2}{T_x^2}\right), \qquad 11.16$$

in which we have substituted for T_1 and T_2 in terms of T_x and T_b. Recalling that T_x is proportional to I_{tot}, we see already from the presence of the quadratic term $(T_b/T_x)^2$ that there may be two real solutions for I_{tot} to the equation obtained by setting P=0: one giving the threshold value for the appearance of sustained power and the other marking the cessation of it. This means that there may be, in addition to the threshold value of I_{tot} where oscillation begins, another where it ceases, as indeed proves to be the case experimentally. The range of values of atomic flux for which oscillation occurs depends on the coefficients of the quadratic equation, and it is useful to write these in dimensionless form. To that end an important design quantity q is defined as follows:

$$q = \frac{\sigma_x V_r}{4\pi\mu_0}\frac{h}{\mu_A^2}\frac{1}{\eta Q}\frac{I_{\text{tot}}}{\Delta I}. \qquad 11.17$$

This leads to an expression for the power P in a useful form:

$$\frac{2P/h\nu}{\Delta I_{\text{th}}} = \frac{\Delta I}{\Delta I_{\text{th}}} - \left[1 + 3q\frac{\Delta I}{\Delta I_{\text{th}}} + 2q^2\left(\frac{\Delta I}{\Delta I_{\text{th}}}\right)^2\right], \qquad 11.18$$

where ΔI_{th} is the threshold difference in the flux of the two maser states in the absence of spin exchange, given by

$$\Delta I_{\text{th}} = \frac{1}{2\pi\mu_0}\frac{h}{\mu_A^2}\frac{V_b}{Q\eta T_b^2} \qquad 11.19$$

This represents a theoretical limit; the actual threshold will always be higher because of the spectral broadening effect of spin exchange between atoms. We note the importance of a large perturbation-free storage time T_b in the bulb in realizing maser oscillation, in view of the smallness of μ_A. It can be shown that the roots are real if q satisfies in our case the following condition:

$$q \langle 3 - 2\sqrt{2}\,. \qquad 11.20$$

In Figure 11.3 is reproduced a set of graphs showing the dependence on q of the *range* of I_{tot} over which oscillation occurs. We note that as q approaches $3-2\sqrt{2}(\approx 0.171)$, the range tends to zero.

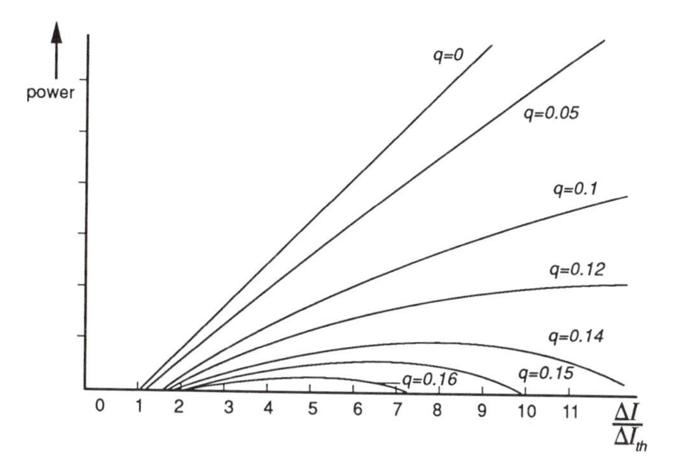

Figure 11.3 The oscillating range of the H-maser as a function of the parameter q (Kleppner, 1965)

11.4 Physical Design of the H-Maser

11.4.1 Atomic Hydrogen Source

We take up now a somewhat more detailed description of the physical apparatus. We begin with the source of the atomic hydrogen beam. Hydrogen gas naturally occurs in the diatomic molecular form H_2, and therefore the first task is to dissociate the two atoms that make up a molecule; this requires that the molecule be given an energy of about 4.4 electron volts to break the bond between the two atoms. This may be done in one of two ways: either by collisions with free electrons in an electrical discharge or collisions between molecules at high temperature.

The high temperature approach is relatively more predictable in design and performance; however, it is now mainly of historical interest in having been used in some classical experiments on the spectrum of the hydrogen atom. It is based on heating the gas to as high temperatures as available containment materials will allow; of the metals, tungsten has the highest melting point, at about 3,370°C. Even at this high a temperature the average kinetic energy of a molecule is only about 0.5 electron volt; however, the kinetic energy is distributed statistically, and some molecules may at times acquire kinetic energies far in excess of the average—and some far below the average. For a gas in thermal equilibrium at a given temperature, the distribution of kinetic energy (or equivalently, thermal velocity) is the Maxwell–Boltzmann distribution, which may be written as follows:

$$\frac{dN}{N}=\frac{2}{\sqrt{\pi}}\frac{1}{kT}\sqrt{\frac{E}{kT}}\exp\left(-\frac{E}{kT}\right)dE\,, \qquad 11.21$$

where dN/N is the fraction of the molecules having kinetic energy in the interval dE centered at the value E, T is the absolute temperature, and k=1.38×10^{-23} joules/degree is Boltzmann's constant. The distribution reaches a maximum at E=kT/2 and falls toward zero, but remains finite as E increases, reaching increasingly higher energies as the temperature is raised. Because of this fact there can be significant dissociation of molecular hydrogen in a heated tungsten oven, the actual degree of dissociation depending not only on the temperature, but also on the competing rates of recombination and influx of molecular hydrogen into the oven. In a practical implementation of this technique, as exemplified by the source used in some of the classical work on the hydrogen spectrum by Lamb and Retherford, a degree of dissociation of 64% was reported in a tungsten oven operated at 2,220°C and gas pressure of 101 Pa.

The thermal dissociation source has the advantages of ruggedness and long-term stability. Furthermore, as we have already indicated, the design principles involved are well understood, and the operating characteristics are theoretically predictable. Unfortunately, the atoms emerge from it at very much higher velocity than from the electrical discharge source we are about to describe. This has the serious effect of reducing the divergence angle of the beam accepted by the magnetic state-selector, thereby reducing the flux of atoms in the beam. In spite of this, there was at one time some effort devoted to developing high temperature ovens, because there was evidence that there may be a problem with the stability and aging of the electrical discharge sources then in use. However, these drawbacks have largely been eliminated, and the high-frequency electrical discharge source is universally used in hydrogen masers.

The earliest application of an electrical discharge for the production of atomic hydrogen was first described by R.W. Wood around 1920, and a discharge tube designed for that purpose was called a *Wood's tube*. It essentially consists of a long, narrow glass tube provided with internal metal electrodes at its ends and containing hydrogen at low pressure, typically around 100 Pa. The application of a high DC voltage between the two electrodes results in a glow discharge, much like the familiar neon signs. Apart from a small region near the negative electrode, the tube is filled with a glowing column, called the *positive column*, which consists of a neutral mixture of free electrons, positive ions, and neutral gas particles. The ions and electrons in this mixture are strongly "coupled" because of the long-range electrostatic force between opposite charges, causing them to act like a fluid; hence the name *plasma*. The ions in the plasma have a Maxwellian distribution of energy corresponding to a temperature only slightly greater than the neutral gas. The electrons, however, can have a very much higher temperature and are thus able to impart enough energy to the gas molecules through collisions to ionize them, and thereby replace ions that are neutralized on the walls of the tube. Collisions with plasma electrons will also raise any atoms or molecules present to higher quantum states, with subsequent giving up of this energy in the form of radiation, and hence the glow. Of particular interest here, of course, is the fact that the electrons will also have enough energy to cause dissociation of the H_2 molecules through collisions, mostly according to the following reaction:

$$H_2 + e \rightarrow H + H + e.$$

The resulting hydrogen atoms will, however, readily recombine to form molecular hydrogen, provided that the atoms come together in the presence of a third body, such as a solid surface. Two isolated atoms coming together in space cannot form a bound system unless some energy is removed from the system during the collision. The dynamical problem is analogous to rolling a ball across a horizontal surface with a deep depression in it; if the ball falls into the depression, it will come out again and continue moving away from it unless it loses some of its kinetic energy inside the depression. The probability of two hydrogen atoms coming together in the presence of a third particle is extremely small at the number densities contemplated here; however, many surfaces, including all metals, will have adsorbed layers of gas and other impurities that readily provide the "third body" and allow the atoms to stabilize in molecular form. Therefore, special care must be exercised in cleaning the inner surfaces of the glass tube to reduce the recombination rate.

The Wood's tube has been totally supplanted by the more efficient and compact electrodeless high-frequency discharge source. In this a glass bulb or tube no more than one or two centimeters in overall dimensions is placed in the high-frequency electric field of a resonant circuit tuned to a frequency in the range 100–200 MHz. The principal advantage is that a high electron flux can be sustained in the plasma without incurring great dissipation of heat caused by electrons hitting the walls of the bulb. In that UHF range, the resonant circuit consists only of a small coil of copper wire no more than around a centimeter in diameter. Great care must be exercised in the design and fabrication of the tube; it is cleaned, and evacuated according to standard vacuum practice, and connected to a source of hydrogen gas through a heated palladium–silver "leak." This consists of a thin-walled tube of the alloy closed at one end and sealed around its rim to the vacuum shell at the other. It separates the discharge tube from the hydrogen source and has the remarkable property when heated of acting as a filter admitting only pure hydrogen from the source. The pressure in the discharge, and hence the beam intensity, can conveniently be varied by controlling the temperature of the palladium–silver leak; typically the pressure is in the range 10 to 100 Pa. This capability of rapidly changing the beam intensity is, as we shall see, of great importance in the automatic tuning of the maser cavity. The detailed mounting of the tube in the coil, as well as the matched coupling of the coil to the source of the UHF power, are crucial factors in achieving a stable discharge. The UHF input power required is typically in the 10 to 20 watt range. There is a complicated interdependence between the level of excitation of the coil and the temperature of the bulb and the electrical characteristics of the plasma. This could easily lead to instability; however, in the absence of any useful analysis of the system, there have been many long-term experimental studies made, and apart from a disconcerting condition known as *the whites*, observed in some of the early work, this type of source has proven itself completely satisfactory. When the source is working normally, the discharge has a clear bright red color due to lines in the atomic hydrogen spectrum, whereas the presence of molecular hydrogen gives the glow a bluish-white hue. It is found that the Pyrex glass surfaces will become discolored due to chemical reactions after prolonged operation of the discharge; however, this does not seriously affect the operation of the source.

Unlike the other beam sources we have encountered, the beam-forming collimator in the hydrogen source, consisting either of a single narrow tube or a bundle of capillary tubes, is limited by the possibility of the atoms recombining into molecules through collisions on the capillary tube walls. The angular width of the emergent beam is typically very much wider than the acceptance angle of the state-selecting magnet that follows, making the utilization factor of the hydrogen no more than perhaps 0.01%. This results in the need for high-capacity vacuum pumps to maintain the background pressure in the main body of the maser at a tolerably low level, such as 10^{-6} torr. Fortunately, the background will be mostly molecular hydrogen because of the presence of surfaces where recombination readily occurs. Molecular hydrogen is so different from atomic hydrogen that it might as well be xenon; its presence will obviously not contribute to the populations of the atomic states in the resonant field, and apart from scattering the beam atoms or possibly causing small frequency shifts in the hydrogen hyperfine frequency through collisions, there are no objectionable effects resulting from its presence. Nevertheless, the system requires continuous pumping with a large pressure differential maintained between the source chamber and the rest of the system. For this purpose ion pumps are commonly used, since they are highly efficient in pumping hydrogen. However, their powerful magnets compound the problem of shielding the radiating atoms from external magnetic fields and add considerably to the weight of the system.

11.4.2 The Hexapole State-Selecting Magnet

The state-selector that is ideally suited to the hydrogen maser, and is commonly used, was originally proposed in 1951 by Friedburg and the Nobel laureate W. Paul for applications in magnetic resonance. It is the hexapole focusing magnet already mentioned in connection with the Cs standard.(See Figure 9.6.) The somewhat different focusing properties of the quadrupole magnet have also been exploited for the same purpose. The object is to separate atoms in the F=1 state from those in the F=0 state, and to allow only those in the upper F=1 energy state to enter the resonant cavity and be stimulated to radiate at the 1,420 MHz transition frequency.

We recall that the magnetic field distribution of the hexapole magnet has a 3-fold axis of symmetry; its components in polar coordinates r, θ can be written as follows:

$$B_r = kr^2 \cos 3\theta \quad ; \quad B_\theta = -kr^2 \sin 3\theta \, . \qquad 11.22$$

This shows that at a given radius, as the angle θ goes through a full circle, the field components will repeat themselves three times. The total field is given simply by $B=kr^2$, a function only of the radial distance of the field point from the axis. If we represent by B_m the field at the radius r_m that reaches the pole tips of the magnet, then the expression for the field at any point $r < r_m$ becomes $B=(B_m/r_m^2)r^2$. In practice, the magnetic field is made sufficiently strong that the energy of the atoms, which in low field start in the F=1, m_F=0 and F=0, m_F=0 states, is proportional to the field, as indicated by the linear graphs with opposing slopes in Figure 11.1. Thus in the high field region we can write the energy as $+\mu_0 B$ and $-\mu_0 B$ for these states, where μ_0 is an effective magnetic moment. Introducing the radial dependence of B, we find for the magnetic energy $\pm\mu(B_m/r_m^2)r^2$. Since this energy plays the role of potential energy in determining the trajectories of the atoms, it follows from the conservation of energy law that atoms in the upper state with radial kinetic energy $\frac{1}{2}mV_r^2=\mu_0 B_m$ will just graze the poles of the magnet before curving back towards the axis, as shown in Figure 11.4.

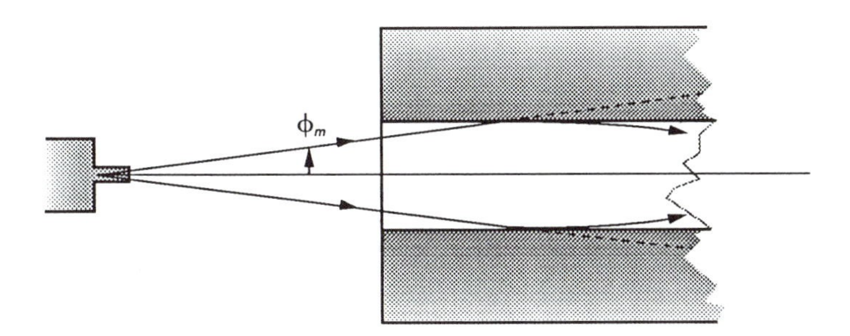

Figure 11.4 The acceptance angle of a hexapole magnet

Atoms with any greater initial radial energy will hit the poles and be lost from the beam. Those leaving the source at an angle of ϕ (radians) to the axis have a radial component of velocity of $V_r \approx V\phi$, and therefore the maximum (planar) angle ϕ_m accepted by the magnet is

$$\frac{1}{2}MV^2\phi_m^2 = \mu_0 B_m . \qquad 11.23$$

A more useful measure in determining the flux of atoms in the focused beam is the solid angle defined by a cone having a half-angle ϕ_m at its apex. We imagine a unit sphere drawn with the atom source as center; the solid angle is the area on the sphere enclosed by the circle in which the cone intersects the sphere. In our case $\phi_m <<1$, and the solid angle is simply $\Omega_m \approx \pi\phi_m^2$, that is,

$$\Omega_m = \frac{2\pi\mu_0 B_m}{MV^2} . \qquad 11.24$$

In arriving at this maximum acceptance solid angle we have assumed a magnet design that is universally used in practice, namely one in which the pole-pieces are straight cylinders, so that the magnetic field is constant along lines parallel to the axis. However, it can be shown that in principle by making the inner radius of the magnet r_m vary in a particular way with axial distance, the acceptance angle can be significantly increased. Unfortunately, fabricating the contoured pole-pieces of such a magnet is difficult, and the implementation of this refinement has not been seriously pursued. In practice, permanent magnets are used, and the limit on B_m is set by the saturation value in the magnetic material used for the pole pieces, typically in the range 0.5 T to 1.0 T. The total flux of atoms from a collimated source is on the order of 10^{16} atoms/sec, of which a small fraction, perhaps 5×10^{12} atoms/sec, in the upper state are focused towards the axis.

Now, the force on the atoms, being given by the rate of fall of the energy, will be $\pm 2\mu_0(B_m/r_m^2)r$, according as they are in the upper or the lower state. Thus atoms that in low field are in the upper F=1, m_F=0 state are drawn toward the axis with a force that is proportional to the distance from the axis, a force that produces simple harmonic motion about the axis. The atoms in the lower state, on the other hand, diverge exponentially away from the axis and are lost from the beam, becoming part of the molecular background. The question remains, however, as to whether the atoms in the upper state, issuing from the source in a narrow cone, are indeed focused back to a point on the axis. We will not attempt to give here a rigorous description of the focusing action of the hexapole magnet but will be content with some simplified general arguments. First we point to the simplifying fact that the acceptance angle of the magnet is in practice very small, since the magnetic (potential) energy of the atoms at the strongest point in the field is still a small fraction of the thermal kinetic energy of the atoms as they emerge from the source. Furthermore, we will assume that the atoms emerge from a point source rather than a significant area; this allows us to ignore skew off-axis trajectories and consider only the radial motion of the particles.

To show that atoms of a given velocity are focused to a point on the axis, we must show with reference to Figure 11.5 that the trajectories emerging from the magnet intersect the axis at the same point. Now, the trajectory of an atom in the hexapole field can be shown to be represented by the following:

$$r = r_0 \sin\left(\Omega \frac{z}{V} + \alpha\right) \quad ; \quad V_r = \Omega r_0 \cos\left(\Omega \frac{z}{V} + \alpha\right), \qquad 11.25$$

where r_0 and α are constants that must be chosen to conform to the given initial values of the particle position and velocity, and Ω is given by

$$\Omega = \sqrt{\frac{2\mu_0 B_m}{M r_m^2}}\,. \qquad 11.26$$

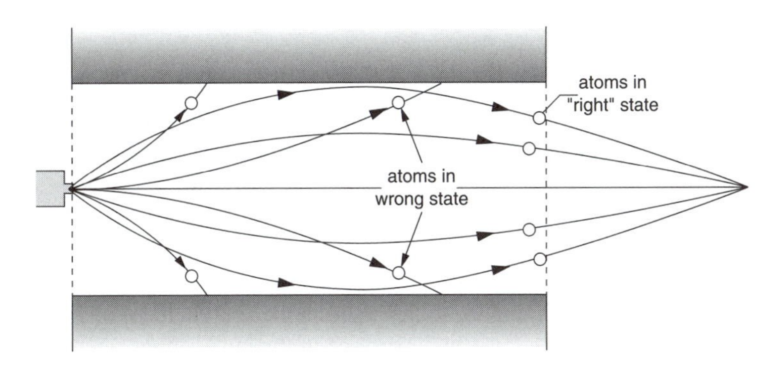

Figure 11.5 The converging and diverging paths of atoms in a hexapole focusing magnetic field

If all the particles can be assumed to originate from the same point on the axis, then we may take that point to be z=0, leading to the choice α=0, and the trajectories of *all* atoms having a given velocity V and $r_0 < r_m$ will return to the axis at the same point where $\Omega z/V=\pi$. This assumes, of course, that the particles remain in the hexapole field, whereas in fact, they must converge to a focus some distance away from the magnet, at the entrance to the cavity. In that event, $\Omega L/V < \pi$, where L is the length of the magnet. On emerging from the magnet they will of course travel in straight lines, and therefore, in order to intersect the axis at the same point, they must travel in directions at the exit plane of the magnet satisfying $r/(dr/dz)$=constant. This ratio in fact *is* constant for motion in a hexapole field, but only for a given velocity; there is a different focal point for each velocity. Of course, we know that they do not all have the same velocity; the velocity has a distribution characteristic of the source temperature. The dependence of the focal distance on velocity is analogous to the dependence in optics of the focal length of a simple glass lens on the color of the light, a deficiency called *chromatic aberration*. This leads to some broadening of the exit-beam image of the source produced by the magnet. It is tempting to draw on the analogy with optics to design an achromatic pair of lenses; unfortunately, for the atomic states of interest the hexapole magnet is necessarily analogous to a converging lens, and two such lenses cannot be made achromatic in the true sense.

Experimentally, the optics of the hydrogen beam, its profile as a function of distance along the axis, can be studied with a screen coated with white molybdenum oxide. The hydrogen atoms falling on the screen chemically reduce the oxide, resulting in a blue spot whose density gives some indication of the relative distribution of atoms in the beam. In this way the beam-profile at various crucial points can be analyzed to optimize the distances between the source, magnet, and cavity.

11.4.3 The Storage Bulb

The heart of the maser is the storage bulb, which confines the atoms to the central part of the interior of a microwave cavity tuned to the 1,420 MHz transition frequency between the hyperfine states. To minimize the loss of microwave power in the cavity through its conversion to heat in the material of the bulb, the latter is typically made of a low-loss dielectric material such as fused quartz. It is usually a spherical bulb, on the order of 15 cm in diameter, provided with a collimator similar to that used in the source, to admit the atomic beam. As already mentioned, the inner surface of the bulb is coated with an inert copolymer designated by the Dupont trade name FEP Teflon. The formation of a satisfactory coherent Teflon film on the quartz surface begins by a thorough cleaning of the quartz surface, once achieved with concentrated chromic and sulfuric acids, but now more commonly by the use of an organic solvent. After this it is wetted with a liquid suspension of the Teflon. The bulb is then dried by heating it while circulating clean, dry air through it. Then it is heated to 360–380°C to fuse the particles of Teflon into a coherent coating, while circulating clean air or oxygen through it to aid in oxidizing contaminants and removing the resulting gases. The bulb is kept at the fusing temperature for about 20 minutes. The collimating tube in the neck of the bulb may be a solid Teflon plug drilled with many holes or a Teflon coated Pyrex tube. From this description of the procedure for preparing the surface of the bulb it is apparent that we are dealing with a recipe that leads to a surface that is not absolutely defined in physical terms. We might speculate about using an electron microscope to determine the physical structure of the surface, but unless we can also specify a procedure that will always lead to that precise structure, it would fall short of establishing an absolute standard.

11.4.4 The Microwave Cavity

The microwave cavity in which the stored atoms interact with the resonant radiation field and under suitable conditions produce sustained maser oscillation must operate in a stationary mode such that the atoms move in an oscillatory field of constant phase; it would do little good to have a long interaction time if the conditions are such that the Doppler effect can shift and broaden the spectral line. The cavity commonly used is a right circular cylinder operated in the TE_{011} mode that resonates at the desired hyperfine frequency if both the length and diameter are about 27.6 cm. The axially symmetric pattern of the electromagnetic field in this mode is illustrated in Figure 11.6, which also outlines the space defined by the storage bulb.

As pointed out already in connection with the ammonia maser, while the Q-factor of the cavity must be high in order to get over the threshold for sustained oscillation, this introduces the undesirable possibility of cavity *pulling* of the oscillation frequency. The Q-factor of an isolated, unloaded cavity is determined by power loss in the form of heat on the inner surfaces of the cavity walls; a silver-plated cavity in the above mode has a theoretical Q-factor of 87,000. The presence of the quartz storage bulb has little effect on the Q-factor, but it will significantly detune the cavity, lowering its resonance frequency. Output microwave power is coupled to a 50 ohm coaxial cable by a hairpin loop mounted in one of the end plates, at a point of strong magnetic component in the cavity field. The plane of the loop must, of course, be set at right angles to the direction of the magnetic field for optimum induced current in the loop.

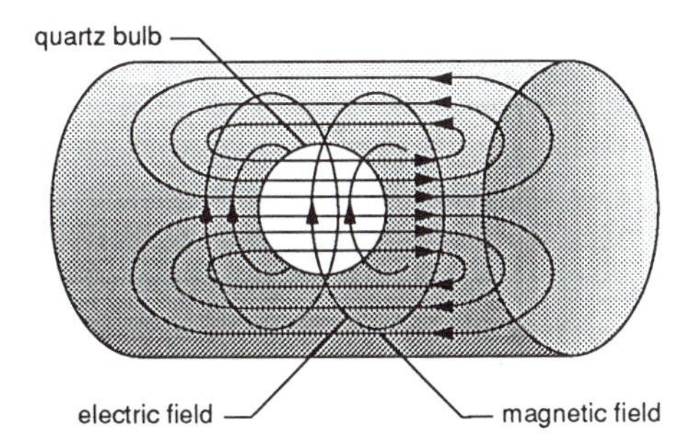

Figure 11.6 The magnetic field pattern in the TE_{011} mode of the H-maser cavity

The phenomenon of cavity pulling adversely affects the stability of the maser frequency by making it dependent on the tuning of the cavity, and hence on the changes of dimensions of the cavity due to temperature fluctuations. There are two approaches to minimizing this source of instability in the maser frequency. First is to use a material for the cavity whose dimensions are particularly insensitive to temperature changes; such a material is CER-VIT, the Owens-Illinois Company designation for a ceramic-vitreous material whose dimensions are remarkably constant with respect to temperature fluctuations. As with any other dielectric cavity material, a thin film of silver or gold on the inner surfaces provides sufficient electrical conduction, since at these high frequencies the fields penetrate only a microscopic distance into a metallic conductor. The second approach is to use a metallic cavity whose temperature is monitored at many points with ultimate sensitivity, and electronically stabilized with a fast-responding servo control loop. In general practice, however, fused quartz is used, whose coefficient of thermal expansion is 0.25×10^{-6} per °C, approximately 1/100 that of aluminum; nevertheless, temperature stabilization is still used in an electronically controlled oven.

11.4.5 Magnetic Shielding

As already pointed out, the (F=1, m_F=0)→(F=0, m_F=0) hyperfine transition frequency is "field-independent" only in the neighborhood of zero static magnetic field B_0. It is therefore necessary to reduce the ambient field at the storage bulb to the lowest possible value. However, there are two conditions on the static magnetic field that impose, in effect, a lower limit on the intensity at which the maser will operate. First, the static field must be in the direction parallel to the microwave magnetic field in the bulb, that is, for the TE_{011}-mode along the axis of the cavity. This condition on the polarization of the microwave field in relation to the magnetic field axis, which serves as the axis of quantization for the atomic states, arises from the fact that in the transition involved, m_F is unchanged. Therefore, the inevitable nonuniformities in the static field intensity must not involve a reversal of the direction of the field at any point; that is, the average field must be larger than the fluctuations. The second condition is a little more subtle: It is based on the fact that atoms moving through a field that varies from point to point can undergo Majorana transitions to other hyperfine states. The likelihood of such transitions diminishes if the magnetic field is sufficiently strong in comparison with the spatial variations in it. In practice, several layers of high magnetic permeability material, such as Moly Permalloy, are placed coaxially around the cavity to shield it from external fields. These materials will develop "hard spots" as a result of spot welding and shaping and must therefore be properly annealed; furthermore, they must be thoroughly demagnetized by taking them through cycles of magnetization by an alternating magnetic field whose amplitude is brought to zero. Inside the innermost layer a solenoid, wound on a nonmagnetic form, provides an adjustable uniform magnetic field over the storage bulb. The maser frequency depends on this magnetic field according to the Breit–Rabi formula, which in the present case takes the form:

$$\nu = \nu_0 \sqrt{1 + (19.72B)^2} \quad ; \quad (B \text{ tesla}). \qquad 11.27$$

For field intensities relevant to the maser, the quantity 19.72 $B \ll 1$, and we can expand the expression on the right to obtain

$$\nu = \nu_0 + 2.761 \times 10^{11} B^2 - 2.68 \times 10^{13} B^4 + \cdots \qquad 11.28$$

Clearly, for magnetic fields in the milligauss range ($\approx 10^{-7}$ tesla), terms higher than the third are negligible. In order to correct the observed frequency to the zero-field value, the field intensity is deduced from the frequency of transitions involving a change in the magnetic quantum number m_F, for example (F=1, m_F=+1)→(F=1, m_F=0). To induce these transitions requires typically an oscillatory magnetic field in the kilohertz range perpendicular to the cavity axis, a field readily generated by passing a current of that frequency through a wire loop. Since the static field inevitably has some residual variation from point to point, we should note that the $\Delta m_F = \pm 1$ magnetic resonance lines (observed by their effect on the maser oscillation) have center frequencies determined by the average of the magnetic field, rather than the average of the square of the magnetic field, as required for the hyperfine frequency correction. However, it is possible from the width of the magnetic resonance line to estimate the field inhomogeneity and ascertain that indeed we are permitted to assume $\langle B_0 \rangle_b^2 \approx \langle B_0^2 \rangle_b$.

11.5 Automatic Cavity Tuning

It is no exaggeration to say that the development of the hydrogen maser into a standard of exceptional long-term as well as short-term stability hinged on the ability to electronically control its cavity to remain precisely tuned to the atomic frequency, thereby avoiding *cavity pulling*. We have already encountered this in connection with the ammonia maser; in that case the molecular resonance is much broader than the hydrogen one and would lead to more serious errors. Nevertheless, in the drive to attain the highest possible levels of performance in the hydrogen maser, a great deal of attention has been paid to automatic cavity tuning.

The original method for the automatic tuning of the H-maser cavity followed the principle used in the ammonia maser; namely, to modulate the width of the atomic resonance and use a high-quality oscillator as a short-term frequency reference to detect any consequent change in the pulling of the maser output frequency. The formula for frequency pulling, we recall, is the following:

$$\nu - \nu_0 = \frac{Q_c}{Q_L}(\nu_c - \nu_0), \qquad 11.29$$

where as before, the fractional line width of the atomic resonance has been written in terms of a quality factor Q_L, and Q_c is the cavity quality factor. Accordingly, if the atomic line width, and hence Q_L, is modulated, the maser output frequency will remain constant and equal to the atomic resonance frequency if and only if $(\nu_c - \nu_0) = 0$; that is, the cavity is tuned to the atomic frequency.

In this method of automatic tuning, it is common to modulate Q_L by modulating the flux of atoms in the beam entering the cavity. As we have seen, this will vary the number density of atoms in the bulb and consequently the spin-exchange time T_x, which contributes to the transverse relaxation time T_2 and therefore the frequency width of the atomic resonance. There are three practical approaches to modulating the atomic beam: first is by an electronically activated mechanical shutter, second is by varying the supply of hydrogen to the source, and the last is by varying the excitation power of the UHF discharge. Faced with a number of alternatives, we judge their merits not only on the basis of their practicability but also on how selective they can be made in modulating just the desired quantity. Real systems are always far more complex than the simple models we are habituated to using. For example, changing the hydrogen pressure in the source will affect the electrical discharge conditions in a complex way, affecting not only the desired atomic flux in the beam, but also in all likelihood their velocity distribution. The same remark can be applied to modulating the UHF power input to the discharge. In fact, if we could ideally vary only the flux of atoms entering the cavity, all other properties of the beam remaining constant, we would thereby vary not only the atomic resonance width, but also to a smaller degree a shift in the center frequency, which also results from spin-exchange collisions. Fortunately, this is an added bonus, for it allows us not only to tune the cavity, but to do so in a way to compensate for the small spin-exchange shift in the atomic frequency. This source of frequency shift is not only another potential source of instability in the maser frequency, by making it depend on the flux of atoms entering the cavity, but would, if uncompensated, be another serious obstacle to the maser being accepted as an absolute standard.

The main difficulty in implementing this spin-exchange-based method of tuning the cavity is the requirement of an oscillator sufficiently stable to serve as a reference in detecting any change in the maser output frequency. Great care must be taken that the automatic tuning circuitry be designed in such a way that the frequency stability of the maser is not degraded by an inferior reference oscillator. Originally, it was found expedient to use a second hydrogen maser for the purpose; however, efforts were made to develop methods of automatic tuning that would circumvent the need for a second maser. One early approach was to develop special logic circuitry to determine the error signal in a digital automatic tuning system using a quartz oscillator as reference. In this way it was reported (H. Peters, 1968) that it is possible to achieve a maser frequency deviation due to line-width modulation an order of magnitude lower than the statistical deviation in the frequency of the reference oscillator. Nevertheless, for optimum performance, where this method of automatic tuning is used, an H-maser is used as reference.

To achieve the ultimate performance without the need of a second reference H-maser, it became apparent that some approach other than line-width modulation must be found. One such innovation, due to Harry Peters, is based on the fact that the coherence time (T_2) of the (induced) dipole moment of the radiating atoms is relatively long compared to the free oscillations of the cavity. It is therefore possible to modulate the resonance frequency of the cavity by electronic switching rapidly between two neighboring values, while the atomic moment preserves its frequency. This will cause the level of oscillation to be different between successive half periods of the modulation unless the cavity is properly tuned.

Another approach successfully designed by Audoin's group at the Laboratoire de l'Horloge Atomique, in France, is to inject a frequency-modulated stable signal near the resonance frequency (derived from a crystal oscillator phase-locked to the maser output) into the maser cavity. It can be shown that the maser output will provide an error signal if the cavity is not exactly tuned to the center of the atomic frequency. The tendency of the maser oscillator to lock on to the frequency of the injected signal must be avoided; it can be shown that with the proper Fourier spectrum of the injected signal (center frequency suppressed) this can be achieved.

11.6 The Wall Shift in Frequency

A far more serious obstacle, however, has already been alluded to: the inability to physically define and reproduce the wall coating of the storage bulb. This would be crucial for a primary standard, since the collisions of the radiating hydrogen atoms with that wall coating cause a residual shift in the atomic frequency, aptly called the *wall shift*. First let us recall that the atoms are confined in an enclosed space, assumed to be a spherical quartz bulb coated on its inner surface with Teflon. It is provided with a small opening through which the beam enters, and atoms effuse out, after making many collisions with the Teflon surface. In order to estimate the wall shift in the atomic resonance frequency, it is necessary to know the relative *time* a given atom spends interacting with the wall as a fraction of its free time between wall collisions. If then we can estimate the shift in frequency that occurs *during* a collision, we will be able to derive the average phase shift each collision causes, and ultimately the mean frequency shift. Now, the distance of free travel between collisions should clearly be proportional to the radius of the bulb; in fact, one can show that it is $4R/3$, and the mean free time between wall collisions is therefore $\tau_c = 4R/3V_{ave}$.

Surfaces are generally complex physical and chemical structures with adsorbed layers of whatever gaseous material they have been exposed to, not to mention other unintended "impurities." In addition, the solid substrate itself is not necessarily uniform throughout. Therefore, it is not only difficult to develop a satisfactory detailed analysis of the wall shift, it would have to be based on such an idealized model as to be of very limited value. It is possible, however, to make some general observations that are useful in understanding some of the factors on which that shift depends.

When a hydrogen atom approaches the Teflon surface, one of several types of collisions can ensue; but whatever the outcome, it is fairly certain that it will dwell on the surface for a finite time, trapped in a *potential well*, shown schematically in Figure 11.7. It is formed by a force of attraction as the atom approaches the surface, which however turns repulsive as the atom penetrates into the surface. The length of time it stays trapped strongly depends on its thermal energy, $3kT/2$, compared to the depth of the well. It must not be assumed that during this time the atom remains fixed at the point of impact; it will eventually either form a chemical bond at some surface site or escape back into the volume. If it forms a chemical bond, it may do so irreversibly with a surface impurity and remain there, or it may encounter a free radical previously formed by another hydrogen atom and exchange spins with it. The first possibility, which simply leads to a loss of radiating atoms, shortens the relaxation times T_1 and T_2 an equal amount, which is independent of the number density of atoms in the bulb. On the other hand, the encounter with a radical formed by a previous atom will increase in probability in proportion to the number density of atoms, making the effect on the relaxation times have a corresponding dependence.

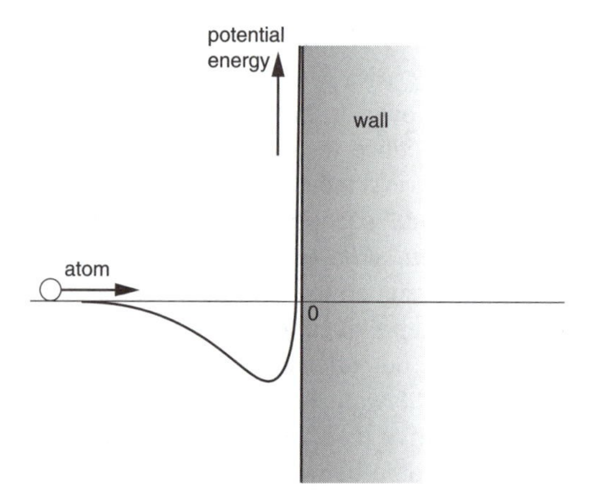

Figure 11.7 Schematic representation of the potential well at a solid surface

If, on the other hand, it does not get involved in a chemical bond, but rather only suffers the strong electric forces at the surface derived from the potential well alluded to above, the electron distribution around the nucleus is distorted, thereby changing the electron density at the nucleus, and hence the hyperfine frequency interval, that is, our maser frequency. It turns out that the frequency is reduced during that part of the atom trajectory when the atom is approaching the surface, where an attractive *van der Waals* force acts on it, due to mutual electric polarization of the atom and surface. On the other hand, the frequency increases when the distortion is due to the repulsive force that prevents it from penetrating farther into the surface. The duration of these forces is extremely short, being on the order of the range of intermolecular forces divided by the average thermal velocity of the atom, that is, on the order of 10^{-13} sec. In reality, the dwell time at the surface can be considerably longer than this, depending on the depth of the potential well and the degree to which the atom loses kinetic energy during the collision, since it has to climb out again by a set of favorable gains in thermal energy from the surface. Nevertheless, it is only a fraction of the period of oscillation of our atomic frequency; hence the disturbance of the atomic state is in the nature of a sharp impulse. It can cause real nonradiative transitions between the hyperfine states, but more seriously, it causes a small shift in the phase of a radiating atom. This phase shift not only leads to dephasing of the radiating atoms, but also, more significantly, to a shift in their frequency. The dephasing comes about from the statistical nature of the collisions; not all atoms experience the same phase shift, nor make precisely the same number of collisions in a given time. In fact, it can be argued that if a large number of atoms make an average of n collisions in a given time, there will be a spread in the actual number of collisions made by the different atoms amounting to $2\sqrt{n}$. Such a spread means that their phases are also spread by an amount $2\sqrt{n}\Delta\phi$, which, upon reaching the order of π radians, corresponds to complete loss of phase coherence. The average time it takes to reach this condition gives us the contribution to transverse relaxation, T_2, at the wall.

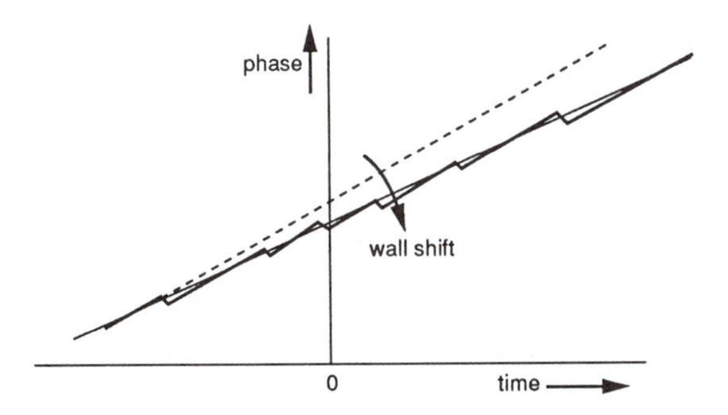

Figure 11.8 Frequency shift due to accumulation of phase shifts at wall collisions

Finally, the most serious effect of the small phase shifts the atoms experience at each collision is that they accumulate in time, leading to a shift in the center frequency of the resonance. We can readily see this if we accept the simplifying assumption that a given atom makes a collision at regular intervals equal to the average time between collisions; in fact, of course, the time between collisions varies statistically about the mean. On the basis of that assumption, the graph of the phase with respect to time is a "staircase," which can be regarded as the sum of a straight "ramp" and a periodic angular graph, symmetric about the ramp. If we assume that each collision causes a jump in phase of $\Delta\phi$ and the mean time between collisions is τ_c, then the slope of the linear ramp is $\Delta\phi/\tau_c$, and the average phase of a radiating atom over many collisions will therefore be given by

$$\langle\phi\rangle = 2\pi\nu_0 t + \frac{\Delta\phi}{\tau_c} t\,. \qquad 11.31$$

The second term on the right increases linearly with time and represents a shift in frequency amounting to $\Delta\nu=\Delta\phi/2\pi\tau_c$, which may be rewritten in terms of the bulb geometry as follows:

$$D\Delta\nu_0 = 3V_{ave}\frac{\Delta\phi}{4\pi}, \qquad 11.32$$

where $D=2R$. It has been experimentally confirmed that $D\Delta\nu_0$ is temperature dependent and ranges from 0.38 Hz·cm at 31.5°C to 0.35 Hz·cm at 40°C. Furthermore, studies have indicated that in certain coatings the wall shift passes through zero at a particular temperature.

11.7 The H-Maser Signal Handling

As already indicated, the power output of the hydrogen maser is very low, typically on the order of 10^{-12} watt; therefore, to obtain a useful signal, the maser output is phase-locked to a synthesized frequency derived from a high-quality 5 MHz quartz crystal oscillator. The challenge is to do this without significantly affecting the free oscillation of the maser, a situation analogous, of course, to the problem of the escapement in mechanical clocks. Whatever is connected to the output port of the maser cavity must cause a minimum of detuning and additional noise. This requires that the maser be presented with a constant low-noise load, that is, a circuit element drawing little power and with minimal changes in its input capacitance or inductance. Great care is therefore taken to ensure temperature stability of the solid-state circuitry used in the processing of the maser output. A selected low-noise preamplifier is typically used to amplify the maser output and to isolate the maser from subsequent circuits. The figure of merit of such amplifiers, the *noise figure*, is roughly defined as the ratio (on a decibel scale) of the actual noise power at the output of the preamplifier to what it would be if only the fundamental Johnson noise were present. Solid-state preamplifiers are available that operate at the desired frequency of 1,420 MHz with a noise figure around 3 db; that is, the noise power is $10^{3/10} \approx 2$, or about twice the ideal. In systems where a second maser is used as reference in the tuning of the cavity, extraordinary steps must be taken to isolate the one maser from the other; otherwise, frequency locking will take place, in which the two masers pull each other to a single common frequency. To avoid this, microwave devices called *circulators*, based on a special property of ferrites, may be used as isolators.

The detailed design of the receiver and synthesizer necessary to phase-lock a 5 MHz quartz oscillator to the maser output in order to obtain useful standard signals can, of course, vary widely. However, we can illustrate the general principles by describing briefly the pioneering system developed by H.E. Peters for NASA satellite tracking stations. The system is shown schematically in Figure 11.9. The output of the precision 5 MHz quartz oscillator is multiplied to 1,400 MHz and mixed with the preamplified output of the maser to give a heterodyne frequency at an intermediate frequency (IF) of 20.405 MHz. This is passed through a tuned IF amplifier, followed by two further heterodyne stages with IF frequencies at around 405 kHz and 5 kHz. The amplified signal at 5 kHz is connected to a phase detector (comparator), whose reference phase is derived from the 5 MHz quartz oscillator through a digital frequency synthesizer. The frequency of the latter, which can be advanced in steps of 0.0001 Hz, determines the precise frequency to which the quartz oscillator will be stabilized, since its actual setting depends on the corrections for the magnetic field and wall shift. The output of the phase comparator, which constitutes the error signal for the control loop, then passes through a filter to ensure stability before closing the loop by connecting it to the frequency-control input of the quartz oscillator. From the phase-locked 5 MHz quartz oscillator, frequency dividers provide standard outputs at 1 MHz and 100 kHz as well as one-second pulses to drive a clock.

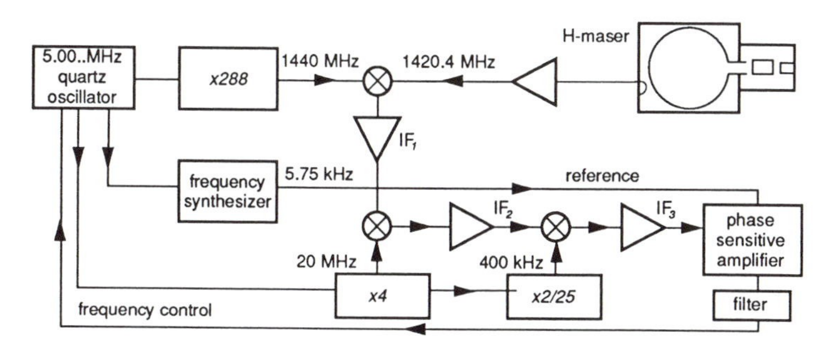

Figure 11.9 An example of an H-maser receiver–synthesizer (Peters, 1969)

Finally, we consider the frequency stability actually achieved in masers built at national standards laboratories around the world. In an earlier chapter we saw that a conventional "time domain" definition of frequency stability (really instability) is the so-called Allan variance of the frequency deviation. A typical plot of this quantity as a function of the measurement period for an actual hydrogen maser is reproduced in Figure 11.10.

The required stability in the standard used as a reference in obtaining the deviations in frequency is so great that only another hydrogen maser can qualify. As already emphasized, great care must be taken to isolate the masers from each other. Otherwise, they will lock on to the same frequency, and the "deviation" would always be zero! Looking at the graph in Figure 11.10, we cannot but be impressed by the extraordinary stability this device exhibits; a clock stable to one part in 10^{15} will gain or lose only 1 second in about 32 million years!

When all the anticipated systematic corrections have been made to the observed maser frequency, such as the magnetic field correction, the wall shift, and the second-order Doppler effect, we get the following for the hydrogen hyperfine frequency in terms of the international atomic time scale as defined by the Bureau International de l'Heure:

$$\nu_H = 1{,}420{,}405{,}751.778 \pm 0.003 \text{ Hz}.$$

This is arguably the most precisely measured quantity in all of physics.

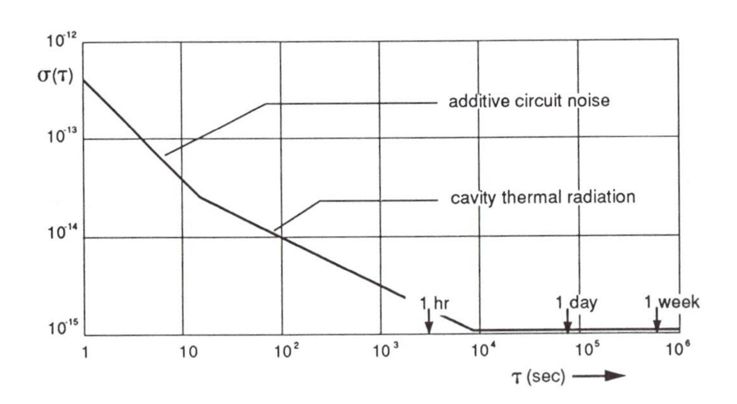

Figure 11.10 A plot of the Allan variance in frequency of a typical H-maser (Peters, 1992)

While the hydrogen maser set new standards of performance in the metrology of frequency and time, standards that will not be easily surpassed, nevertheless, for certain applications it suffers from two deficiencies: First, it lacks portability, and second, the wall shift limits its absolute accuracy. The size and mass of the maser are a necessary consequence of the relatively large wavelength (21cm), which sets the scale of its dimensions. This is further aggravated by the need for elaborate shielding from ambient magnetic fields and the need for a large vacuum shell and massive vacuum pumps to maintain the required degree of high vacuum in Earth-based systems. The size would not, of course, be objectionable for laboratory installations, but it renders it unsuitable for mobile systems. In the one application where portability is not an important requirement, namely as a primary standard, the wall shift unfortunately limits its absolute accuracy.

11.8 Hydrogen as a Passive Resonator

If we pursue the idea of eliminating these objections by simply making some changes in the detailed design of a system based on the same beam maser concept, we soon find that the hydrogen maser is difficult to surpass; the choice of hydrogen with its design consequences is a particularly fortunate, if not a unique, one. We have already seen the reasons we expect hydrogen atoms not to have their internal energy states as perturbed as heavier alkali atoms in collisions with a solid surface. And even if a suitable transition is found whose wavelength is significantly shorter, the smaller resonant cavity (aside from exotic designs based on superconductors, etc.) will have a smaller quality factor Q, making it more difficult to reach the threshold for oscillation. This can be understood in a general way in terms of two relevant factors: First is the penetration depth of the radiation into the metal surface of the cavity wall, and the second is the area of the surface of the cavity in relation to the volume it encloses. The first is important because it is the loss of field energy through resistive heating of the metal surface that determines the Q-factor. At the shorter wavelengths the penetration is less and the energy dissipated is less; unfortunately, this is more than offset by the increased ratio of surface area to volume enclosed. The importance of this geometric ratio depends on the oscillation mode in the cavity, that is, the field distribution in the cavity; but it is reasonable to assume that it is a rough measure of the ratio of the energy of the field at the surface to that stored in the volume enclosed by it. Thus on balance, the Q-factor of resonant cavities oscillating in the same mode and made of the same material is expected to be smaller for shorter wavelengths.

This might suggest that the idea of a maser oscillator be abandoned in favor of a hydrogen-based *passive resonator*, for then a smaller cavity can be made to resonate at the hydrogen frequency by introducing dielectric material into the cavity, thereby "loading" it to lower its resonant frequencies, which would otherwise be too high. This inevitably lowers the Q-factor and makes the threshold for operation as an oscillator too high. The performance of such passive hydrogen resonators has been compared both theoretically and experimentally with the standard maser and such resonators have been commercially produced for applications where portability is important.

Aside from greater portability, however, they offer no particular advantage over the maser. Moreover, from the point of view of accuracy, the smaller cavity will, of course, dictate a proportionately smaller bulb in which to contain the atoms; this will cause the wall shift to be further aggravated because the collision rate with the walls will be greater. We seem to be faced with an unpleasant choice: either a large size and small wall shift, or a small size and large wall shift, with some doubt as to whether the latter can be achieved at all. As pointed out in the last chapter, the real problem with the wall shift is that it is difficult to stipulate an absolute physical definition of a complex surface structure; simply defining a procedure for preparing the surface is not really enough. Even if such a physical definition could be given, it would be impractical to implement.

Chapter 12
The Confinement of Ions

12.1 Introduction

The development of atomic standards based on quantum resonance in neutral atoms confined by diffusion through a buffer gas, or collisions with inert walls, culminated in the hydrogen maser, a standard of astonishingly high stability. However, as explained in the last chapter, for certain applications the hydrogen maser suffers from two deficiencies: First, its lack of portability due to its size and the need for elaborate magnetic shielding, and second, the wall shift that limits its absolute accuracy, and disqualifies it as a primary standard.

We recall that the use of strong electric or magnetic fields for the purpose of deflecting and confining otherwise free *neutral* particles was ruled out on the basis that they could perturb the internal quantum states, on whose energy separation the frequency standard depends. This would certainly be unacceptably so if static or low-frequency fields were applied to ordinary atoms, which are overall electrically neutral. However, for ions, which have a charge imbalance produced by the parent atom having gained or lost one or more electrons, the situation is quite different. In this case, particularly if the ion kinetic energy is low, relatively weak electric and magnetic fields can significantly deflect their motion. But the point, which sets this apart from the use of reflecting surfaces on neutral atoms, is that the fields can be precisely defined and measured, and the perturbation they cause, unlike the wall shift, can be calculated to any desired degree of accuracy.

We may recall that the hydrogen maser evolved as a frequency standard from the effort to push to the limit the resolution and sensitivity of beam machines for the purposes of magnetic resonance spectroscopy. In the same way, the use of fields for particle confinement has its origin in the drive to increase the resolution and accuracy of radio-frequency spectroscopy on charged particles. By using the restraining force that a suitably designed field exerts on the particle, it can ideally be suspended essentially in almost total vacuum for an indefinite time. The ultimate success of this approach hinged on an essential technological development: the ion vacuum pump, which we have already encountered. This is simply another example of a general truth that scientific advances can occur only as technology makes them possible, and conversely. The ion pump made possible the attainment of pressures below 10^{-10} Pa, a pressure region called ultrahigh vacuum. In this pressure range, an ion would travel on the average 1,000,000 km before colliding with another particle, were it not for collisions with the walls of the vacuum chamber!

12.2 State Selection in Ions

This extraordinary degree of isolation can be exploited, however, only if a suitable method of quantum-state selection can be devised that results in the stored ions being preferentially in certain states and not others. Furthermore, a method must be available for monitoring the evolution of those states in response to a resonant field causing transitions among them. The magnetic state selectors used in the Cs standard and the hydrogen maser, which are based on the deflecting force of a strong magnetic field gradient acting on the magnetic dipole moment of a neutral atom, are not practical for ions. The reason is that an electrically charged particle experiences a force due to its motion in a magnetic field, the *Lorentz force*, which is so much stronger than the force arising from any magnetic moment it may have that it completely masks the latter. To illustrate this numerically, let us assume that an ion has a velocity of 3×10^3 meters per second in a field of 1 tesla (10^4 gauss). The Lorentz force that would act at right angles to its direction of motion is as follows:

$$\mathbf{F} = q\, \mathbf{V}\times\mathbf{B}, \qquad 12.1$$

where q is the charge, which for an ion is that of one electron: 1.6×10^{-19} coulomb. This yields a force of 4.8×10^{-16} newtons. This should be compared with the force, for example, on the magnetic moment of an electron (one Bohr magneton) in a magnetic field gradient of, say, 1 tesla per millimeter, which comes to about 9×10^{-21} newtons. We see in this case that the Lorentz force, which does not discriminate between internal magnetic states of the ion, is 50,000 times larger than the force that does. Of course, in making this comparison we have assumed ions with a certain velocity, which implies a certain kinetic energy; clearly, if the energy is vastly lower than we have assumed, the situation would be very different. In fact, at very low absolute temperatures, below, say, 1° Kelvin, one may indeed contemplate a magnetic state selector based on having an axially symmetric magnetic field with a steep gradient along the axis, more or less parallel to the direction of the ion velocity rather than perpendicular to it. In this way ions in magnetic states whose energy increases with field intensity would be slowed down, and eventually even stopped, if their initial energy was low enough, while those whose energy is negative would continue with an accelerated motion. To our knowledge this type of state selector has been exploited only for free electrons.

The application, then, of the field confinement of ions to high-resolution spectroscopy, which, as we have seen, was the scientific precursor to the development of atomic frequency standards, required a new method of state selection. By the middle of the 1950s two possibilities were demonstrated: first, optical pumping of the sort used in the Rb standard, and second, spin-dependent collisions with state-selected neutral atoms in a beam passing through the ion-storage field.

We have already described optical pumping at some length; it is the method used to preferentially populate the upper hyperfine state in the Rb standard. We recall that it relies on the fact that the absorption and emission of photons by atoms are governed by quantum selection rules, which are ultimately based on the laws of conservation of angular momentum and energy for the system comprised of the atom plus photon.

The method based on the spin-dependence of collisions between the stored ions and state-selected atoms assumes that we have alkali-like ions and atoms, having a single electron outside a spherical electronic core. If we neglect at first any nuclear magnetic moment the colliding particles may have, then the situation is similar to that two of hydrogen atoms (without nuclear moments) colliding. We have already seen that the most likely result of such a collision at low energy (in addition to an obvious deflection) is the exchange of spin directions between the two outer electrons. This may be looked on as a spin-dependence of the elastic scattering of the particles, in which each preserves its initial kinetic energy, merely being deflected from its original path in an angular pattern that depends on their spin state. When we previously discussed spin exchange, it was from the point of view of its role as a relaxation mechanism, which broadens the atomic resonance in the hydrogen maser. Here we are interested in it from the point of view of generating, among the magnetic states of an ion, population distributions that are far from the nearly uniform distribution characteristic of thermal equilibrium. To fix our ideas let us suppose the ion is that of the helium atom, namely He^+, and that the atoms of an alkali element, such as Cs, pass through the ion storage region in the form of a beam, undergoing spin exchange collisions with the ions. Now, the removal of one electron from the helium atoms leaves only one electron outside the nucleus, and therefore apart from having a net positive charge, it differs from a hydrogen atom only in having a different nuclear structure. Now let us suppose that by a magnetic state selector or by optical pumping, the Cs atoms in the beam have their electron spins polarized prior to entering the ion storage space; that is, they are caused to preferentially populate magnetic substates designated by m=1/2 rather than m=-1/2. This polarized state we interpret physically as having a net global angular momentum pointing in the direction of a magnetic field as axis. Associated with this angular momentum there is a magnetic moment, which because the electron carries a negative charge points in the opposite direction to the field. If for simplicity we ignore the nuclear spin of the Cs atoms and represent the electron spin states of the particles symbolically by arrows, as we did in the last chapter, we see that the following two exchange processes are equally likely:

$$\mathrm{Cs}(\uparrow) + \mathrm{He}^+(\downarrow) \rightarrow \mathrm{Cs}(\downarrow) + \mathrm{He}^+(\uparrow),$$
$$\mathrm{Cs}(\uparrow) + \mathrm{He}^+(\uparrow) \rightarrow \mathrm{Cs}(\uparrow) + \mathrm{He}^+(\uparrow),$$

since the helium ions are assumed to be initially unpolarized, having equal probability of being "up" as "down." We see that the net result is a transfer of the polarization to the helium ion, since they end up pointing the same way.

To observe magnetic resonance, we need, in addition to producing a polarization, a means of monitoring that polarization in order to know when a resonant field is inducing transitions that affect it. The number of ions contributing to the global magnetic polarization is far too small to enable direct detection through the magnetic field it produces, as in nuclear magnetic resonance on liquid or solid samples. One of the important advantages of the optical pumping technique is that the same process that produces the desired differences in populations also provides the means of monitoring their evolution in time.

In the case of methods based on exchange collisions with polarized atoms in a beam, the obvious suggestion of monitoring the polarization of the atomic beam after it has traversed the ion storage space unfortunately is not practical. However, other collision processes between the ions and the polarized neutral atoms are possible. The author has exploited, for example, the spin-dependence of the neutralization of the helium ion by the capture of an electron in collisions with polarized Cs atoms, according to the following reaction:

$$\mathrm{Cs} + \mathrm{He}^+ \rightarrow \mathrm{Cs}^+ + \mathrm{He}^* + \Delta E,$$

where He^* represents a neutral helium atom in an upper energy state. The quantity ΔE, called the *energy defect* and included to balance the energy, happens to be small but significantly different for the two relative directions of the electron spins, corresponding to possible 2-electron spin states of the resulting He^*: For one it is +0.1 eV and for the other –0.25 eV. Now, the probability of the reaction taking place is sensitive to the value of ΔE, and therefore to the spin state of the colliding partners. This results in an increase in the rate of loss of ions from the storage field to an extent dependent on the polarization of the ions. If that polarization is changed by means of a resonant magnetic field, it creates an observable change in the number of ions remaining in the storage field.

The monitoring of the ion polarization can also be reduced to the measurement of another observable quantity, namely the energy distribution of the ions. This is exemplified by the process that has been successfully used to study the magnetic resonance of free electrons in vacuum using field confinement. It depends on the spin dependence of *inelastic* collisions between the stored electrons and polarized sodium atoms, by which the electrons lose kinetic energy at a rate dependent on their polarization.

All these methods allow alkali-like ions to be polarized and that polarization monitored while the ions are moving about, confined in the storage field; otherwise, it would introduce a good deal of complexity to separate spatially the regions where the ions undergo polarization, resonance, and detection. To see this, and to appreciate the power and limitations of using ion traps in the field of frequency standards, we will now address the question of ion motion in various field configurations.

12.3 The Penning Trap

First let us recall a theorem in the theory of electrostatics, namely, *Earnshaw's theorem*, which states; A charged body placed in an electrostatic field cannot be maintained in stable equilibrium under the influence of electric forces alone. This can be shown to be a consequence of the fundamental equations governing the electrostatic field: In charge-free space the electrostatic potential energy of a test charge cannot be a minimum at an isolated point in space. This means that there is no point from which a displacement of a test charge in any direction would cause an increase in potential energy; at most, we can have the potential energy increasing in some direction but decreasing in another. Since electrostatic potential energy is analogous to the potential energy of an object acted on by gravity, our statement on the absence of a minimum is analogous to saying that it is impossible to have a bowl-shaped valley, but at most a saddle-shaped one, such as we would find between two peaks. In the neighborhood of a saddle point in the potential distribution, the charge would be restrained in one direction but repelled in the other. This would seem to be fairly discouraging to someone trying to trap an ion; but in fact, we are not limited to static electric fields; we simply have to look to either nonstatic fields or combinations of electric and magnetic fields.

12.3.1 The Field Configuration

We will begin with the field configuration that has come to be called the *Penning trap*, because it is reminiscent of the Penning vacuum gauge in its arrangement of electrodes and the use of a magnetic field. The gauge, first described by Penning in 1936, is illustrated schematically in Figure 12.1. It extends the range of operation to much higher vacuum than previous types by the use of a strong magnetic field, in conjunction with high voltage electrodes to maintain an electrical discharge in what is called the "blackout" vacuum region. In ordinary discharges, such as we have in a neon sign tube, blackout occurs when the pressure of neon is reduced below a certain point. The action of the magnetic field is to cause the electrons in the plasma to move in tight spirals, thereby increasing their path length, and hence the probability of ionizing collisions with background gas molecules, before striking and losing energy to the electrodes and other surfaces.

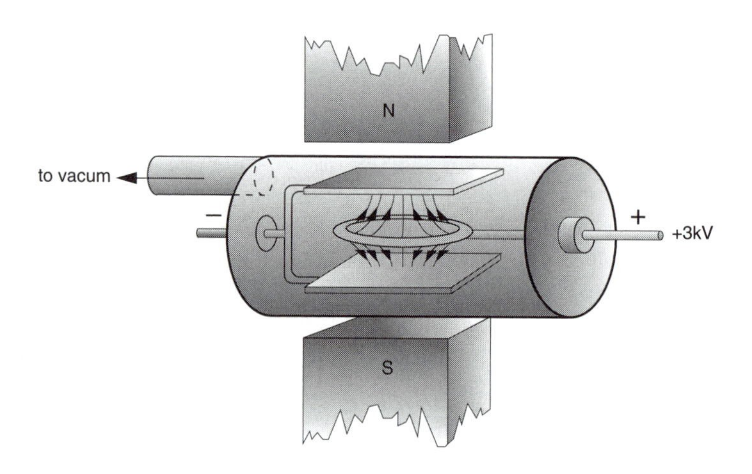

Figure 12.1 The field configuration of a Penning vacuum gauge

The pure quadrupole electric field geometry illustrated in Figure 12.2, commonly used for ion confinement, was in fact first described by J.R. Pierce, of the Bell Telephone Laboratories (Pierce, 1954). For the confinement of positively charged particles, an electrostatic field is produced by applying a negative voltage to an hourglass-shaped cylinder, with coaxial bowl-shaped end caps carrying a common positive voltage. By placing the electrode system between the pole pieces of a magnet, a strong axial magnetic field is superimposed on this electric field. If the charges to be confined are negative, then of course the polarity on the electrodes would have to be reversed; the direction of the magnetic field along the axis is immaterial. In terms of circular cylindrical coordinates r, z with the z-axis along the axis of symmetry of the system, the electric potential of the field between the electrodes has the form

$$V = \frac{V_0}{2r_0^2}\left(2z^2 - r^2\right). \qquad 12.2$$

The surfaces of equal potential in this field are figures of revolution about the axis with hyperbolic cross section; the field can be generated by having hyperbolic conducting surfaces coinciding with a set of equipotential surfaces. Note that we have chosen a potential field symmetric about the origin, in the sense that the potential there is zero, and the potentials on the electrodes are $\pm V_0/2$; the total voltage applied between the cylinder and the end caps (the two sheets of one hyperbola) is V_0. Along the z-axis (r=0) the potential varies as z^2, increasing as we go in either direction, and reaching the maximum of $V_0/2$ at the electrodes where $z=\pm r_0/\sqrt{2}$.

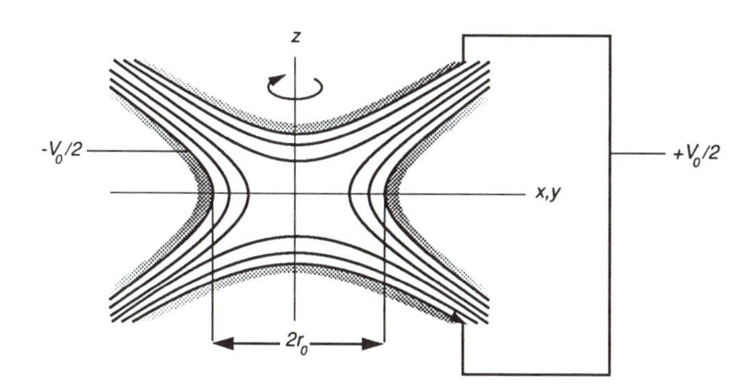

Figure 12.2 The quadrupole electric field geometry

12.3.2 Ion Motion

If the magnetic field is truly uniform and everywhere in the axial direction, then the ion motion parallel to the axis is not affected by the magnetic field, since the Lorentz force comes into play only if the particle has a velocity component perpendicular to the magnetic field. The equation of motion for the z-coordinate of a particle having charge e and mass M at any point (r, z) is therefore as follows:

$$\frac{d^2z}{dt^2} = -\omega_z^2 z \quad ; \quad \omega_z^2 = 2\left(\frac{e}{M}\right)\frac{V_0}{r_0^2}, \qquad 12.3$$

which is the equation for simple harmonic motion. The particle therefore oscillates with a finite amplitude about the origin with a frequency $\nu_z=\omega_z/2\pi$. To avoid hitting the end caps, the amplitude of the particle must be less than z_0, since otherwise the radial motion of the ion would cause it to strike an end cap.

It follows that the maximum energy an ion can have by virtue of its axial motion is $qV_0/2$, so that $V_0/2$ is the depth of the "potential well" for the axial motion. The radial component of the motion is considerably more complicated and is more easily described in terms of Cartesian coordinates x, y. If the magnetic field were absent we, would have for the equations of motion of the x- and y-coordinates the following:

$$\frac{d^2x}{dt^2} = +\left(\frac{e}{M}\right)\frac{V_0}{r_0^2}x \quad ; \quad (B_z = 0), \qquad 12.4$$

with an identical equation for the y-coordinate. Because of the plus sign on the right-hand side, the solution of this equation is the exponential function, which rapidly goes to infinity. This shows explicitly that in the absence of a magnetic field, the x- and y-coordinates would continue to increase until the particle hits an electrode. If the sign of V_0 were reversed in order to make the motion along the x- and y-axes oscillatory, then the motion along the z-axis would diverge exponentially. The same applies, of course, if the sign of the charge on the particles is reversed; from this we get the important conclusion that the Penning arrangement does not trap both positive and negative particles simultaneously.

The effect of a magnetic field is to introduce a Lorentz force that causes ions having a radial component of velocity to swing around in more or less cycloidal orbits. A cycloid is the geometric figure traced out in space by a point on the rim of a rolling wheel. In our axially symmetric field geometry, the wheel must be assumed to lie in a plane perpendicular to the axis and roll around on a circle in that plane centered on the axis. Thus we see that there are two periodic motions involved: The wheel turns about its center, making a certain number of revolutions per second, while its center revolves with uniform speed around the axis of the system. If the chosen field strengths are such that the magnetic Lorentz force is dominant over the electrostatic one, then the former motion of the wheel about its center can be shown to have nearly the frequency ν_c given by

$$2\pi\nu_c = \frac{eB}{M}, \qquad 12.5$$

which is simply the frequency with which a charged particle executes a circular orbit in a uniform magnetic field in the absence of any electric field. Since this frequency is a central quantity in the design of the cyclotron particle accelerator, it is referred to as the *cyclotron frequency*. The cyclotron is made possible by the fact that this frequency is independent of the velocity of the particle (provided that it is much smaller than the velocity of light) or the size of its circular orbit. This allows an oscillatory electric field at that frequency to remain in synchronism with the particle motion as it gains energy from the field and its orbit expands.

The other, slower, motion of the center of the wheel around the axis of the system can be shown to occur at the velocity at which the Lorentz force in the magnetic field is balanced by the electrostatic force. Since the electric field is proportional to the radial distance, a balancing Lorentz force requires the (linear) velocity to increase the same way; this implies a constant angular velocity around the axis. The frequency of this motion, sometimes referred to as the *magnetron* motion, is as follows:

$$2\pi\nu_m = \frac{V_0}{Br_0^2}. \qquad 12.6$$

The magnetron is a high-power microwave tube used in radar transmitters and microwave ovens. Its field configuration differs from the Penning trap in having a radial electric field between a tubular electron-emitting cathode and a coaxial copper ring forming the anode. A strong axial magnetic field causes electrons emitted by the cathode to curve around and across the openings of a series of microwave cavities machined out of the anode, thereby inducing oscillations in them at microwave frequencies. Note that this frequency ν_m does not depend on the properties of the particles, but only on the geometry and field intensities of the trap.

This separation into a fast cyclotron motion on which is superposed a slower magnetron motion is only an approximation valid when $\nu_c >> \nu_m$. Moreover, there are other possible orbits, namely, circular ones around the axis as center; curiously, these also have two possible frequencies. The general result for the two frequencies is as follows:

$$\nu^{\pm} = \frac{\nu_c}{2} \pm \sqrt{\left(\frac{\nu_c}{2}\right)^2 - \nu_r^2}, \qquad 12.7$$

where ν_r would be the radial frequency if the electric field were acting alone. From this result we see that only if $\frac{1}{2}\nu_c > \nu_r$ are the frequencies $\nu^{\pm}$ real in a mathematical sense and the motion oscillatory with a finite amplitude. This establishes the equivalent "binding" potential of a magnetic field, that is, the equivalent electrical potential well depth created by a magnetic field. If the condition $\nu_c >> \nu_r$ is met, the two frequencies $\nu^{\pm}$ may be approximated as follows:

$$\nu^{+} \approx \nu_c - \nu_m \quad ; \quad \nu^{-} \approx \nu_m, \qquad 12.8$$

where

$$\nu_m = \frac{\nu_r^2}{\nu_c} \quad ; \quad \nu_r^2 = \frac{eV_0}{4\pi^2 M r_0^2}. \tag{12.9}$$

We should note that ν^+ and ν^- are in a sense frequencies of the "normal modes" of vibration of the ion, and further that ν^+ is a mixture of cyclotron and magnetron motions.

Since the fields acting on the confined particles are static, in the absence of collisions the system has the property of being *conservative*, which means, among other things, that the kinetic energy of a particle at any point is determined by the electrostatic potential at that point, no matter how the particle got there. This has the important consequence that if a particle enters the field with some energy through a hole in one of the electrodes, it will not remain in the trap, but pass through to strike another electrode, or even to return to strike the same electrode with the same energy. In order to be trapped in the field, an ion must either sustain a sufficient loss of energy in the trap or be created in the trap with sufficiently small initial energy. The situation is analogous to rolling marbles on a smooth surface at a depression in that surface; they will not settle in the depression unless they lose some kinetic energy, for example by friction, as they cross the depression. In the case of atomic or molecular ions, capturing them is readily accomplished by one of two approaches: First, injecting the ions into the trap, where some make inelastic collisions with particles in the trap, losing energy to the colliding partner; or second, having the parent atoms or molecules fill the trap at a very low pressure, and passing an electron beam through one of the end caps along the axis to ionize them through collisions. The latter method is more convenient, but it has two objectionable aspects: First, the electron beam adds its own electric field, causing a deviation from the proper field distribution; and second, the presence of the parent gas, as we shall see, limits the lifetime of the ions in the trap through collisions. The disturbing effect of an electron beam can be addressed by separating in time the operation of filling the trap from that of observing the ion spectrum. This can be done simply by using a pulsed rather than a continuous beam, making the pulse duration much shorter than the ion lifetime in the trap. The other objection, having to do with the reduction of the ion lifetime in the trap, can be overcome by simply providing a fast ion vacuum pump to quickly reduce the pressure to an ultra-high vacuum level after each electron beam pulse. During each pulse, sufficient gas may be liberated by the electrons impinging on the metal electrodes to create ions by subsequent electrons. There are three essential parameters that characterize an ion trap: First, the maximum number of ions that it can contain; second, the maximum energy they can have; and third, the length of time they remain trapped. Let us consider these in that order.

12.3.3 Maximum Ion Number

The limit on the number of ions that can be contained is set by the mutual electrostatic repulsion between the charged particles. The difficulty in analyzing this type of system is that the field disturbance the repulsion produces depends on the distribution of the charges, which in turn depends on the same field disturbance. One way to break the circle is to assume some general particle distribution with variable parameters, compute the electrostatic field it produces, and then fix the parameters to be consistent with this field. It would go well beyond our present interest to attempt such a calculation; instead, we will make the crude assumption that the ions are uniformly distributed over a spherical volume concentric with the trap, of radius z_0 (the distance to the end caps). If N is the total number of ions, each carrying a charge e, then from the theory of electrostatics (Gauss's law) this produces an electric field within the charge distribution, whose potential has the same quadratic dependence on the coordinates r and z as the trapping field—with one important difference; namely, the term in z^2 is negative, opposite to the applied field. Clearly, the limit is reached when the potential of the ions cancels the applied potential. For a typical trap we find N_{max} on the order of 10^6 as the theoretical upper limit to the number of ions that can be confined. In practice, the number actually trapped would rarely reach 1/10 of this number, which on the atomic scale of things is extremely small; nevertheless, this has not prevented important applications of the technique to fundamental studies on free electrons and the development of ion frequency standards.

12.3.4 Maximum Ion Energy

We next consider the limit on the energy that the ions may have and still be confined within the space defined by the electrodes. We should note first that in any practical system, the ions will always be involved in some collisions with other background gas particles, no matter how infrequently. Therefore, there will be random exchange in kinetic energy between motion parallel to the z-axis and motion perpendicular to it. This means we must assume that the ion energy cannot exceed the lesser of the "binding" energy along the two directions if it is not to collide with an electrode. For motion along the z-axis, the appropriate binding energy is just the (positive) potential energy of an ion at the end caps, which is simply $eV_0/2$. In the plane perpendicular to the z-axis, we require the confining action of the magnetic field expressed in terms of its equivalent electrical potential barrier against which the particle moves. Earlier, we saw that the condition for stable oscillatory motion in the radial direction, at a "real" frequency, was that $(\nu_c/2)^2-\nu_r^2$ be positive. Now we can interpret this same condition as a statement concerning potential energy, since in fact, ν_r^2 is proportional to V_0; a sufficiently strong magnetic field, relative to the electric field, has the effect of making the potential become more positive rather than negative with increasing distance from the axis, leading to stable confinement. For example, we find for protons (mass M=1.67×10^{-27} kg) in a trap with r_0=1 cm, B=0.1 tesla, and V_0=1 volt, the depth of the equivalent potential is on the order of 10 volts.

12.3.5 Average Trapping Time

Finally, we consider the expected average time that in practice, an ion can be contained in the trap. The implication is that in a practical system there is a mechanism for the ions to diffuse ultimately out of the trap; in fact, collisions with background particles do just that. To see this we have to analyze what happens to a particle's trajectory as a result of a collision, which may affect both the magnitude and direction of its motion. If the background gas consists of the parent atoms or molecules, then the most probable result of a collision is charge exchange in which the particles emerge with the roles of parent and ion exchanged, according to the reaction

$$\mathrm{X} + \mathrm{X}^+ \rightarrow \mathrm{X}^+ + \mathrm{X}.$$

Such a process can take place with little change in the momentum of each of the colliding particles. Since the parent atom or molecule had the relatively low kinetic energy characteristic of a gas near room temperature, this type of collision produces ions that start a new trajectory with a thermal velocity. This does not mean that their total energy is thermal; they also have potential energy in the electric field. If we assume that an ion has a finite kinetic energy E_k, then the cyclotron radius varies inversely as the intensity of the magnetic field approximately as follows:

$$r_c^2 = \frac{2ME_k}{e^2B^2}. \qquad 12.10$$

Depending on where the ion happens to be in its cyclotron orbit when a collision occurs, it may be put in an orbit displaced to a larger radial distance or, with equal probability, to a smaller one. Hence the mean position of the ion in the trap jumps at every collision with equal probability in opposite direction along the r-axis. This situation corresponds to the classical "random walk" problem in statistics (remember the drunkard?). If we simplify the problem by assuming that each collision results in the same displacement equal to $\langle r_c \rangle$, then after n collisions, the mean square displacement $\langle D^2 \rangle$ is $\langle D^2 \rangle = n \langle r_c \rangle^2$. If we set $\langle D^2 \rangle$ equal to r_0^2, the corresponding n will be the average number of collisions n_{esc} required for the ion to reach the electrode. The average time to make that many collisions is approximately given by the following:

$$\tau_{esc} \approx \frac{E_c(r_c = r_0)}{E_k}\tau_{coll}, \qquad 12.11$$

where $E_c(r_c=r_0)$ is the cyclotron energy corresponding to a cyclotron radius equal to the electrode radius, and τ_{coll} is the average time between collisions. To illustrate the scale of the physical quantities involved, let us again assume that we have protons in a trap with r_0=1 cm, B=0.1 tesla, and E_k=0.025 electron volts. Substituting these values, we find for the lifetime $\tau_{esc}=2000\tau_{coll}$. Now, at an ultra-high vacuum pressure of, for example, 10^{-8} Pa, the average time between collisions τ_{coll} is probably on the order of a thousand seconds, from which we predict a lifetime of over a fortnight!

In all of the above we have not considered any mechanism for extracting energy from the ions, that is, cooling them; with sufficient cooling, of course, the ions will remain trapped indefinitely. With the availability of powerful tunable lasers, the cooling of trapped particles has become a reality, about which much more will be said in a later chapter.

From the point of view of its exploitation as a new approach to achieving a portable frequency standard, the Penning configuration is probably unsuitable for an absolute standard, since the frequency of all atomic transitions will depend to some degree on the strength of the magnetic field on which its operation depends. Displacements of atomic energy states that may not ordinarily be regarded as significant may well be intolerable at accuracies of parts in 10^{14}. However, if absolute accuracy is not sought, but rather a portable standard with extreme stability, then it is only necessary to select a quantum transition whose frequency is stationary (maximum or minimum) with respect to small variations in the magnetic field around some suitable value. In the cases of the Cs standard and hydrogen maser this was achieved by operating near zero magnetic field; other atomic systems have such stationary points at higher magnetic field intensities. We recall that the need to operate at such a point arises from the broadening of the atomic resonance that would otherwise occur due to inevitable variability in the ambient magnetic field.

12.4 The Paul High-Frequency Trap

Fortunately, the other important type of ion trap, the Paul trap, named for Nobel laureate Wolfgang Paul, does not require a magnetic field; rather it uses a high-frequency alternating electric field. Its precursor, first described by W. Paul and M. Raether in 1955, was an ion beam mass filter using a high-frequency electric field between cylindrical quadrupole electrodes, as shown in Figure 12.3. Ions in a narrow mass range are focused by the field, while all others diverge exponentially from the beam and are lost.

Figure 12.3 The Paul high-frequency ion mass filter

It was early realized that the focusing action of the mass filter field in two dimensions could be used to trap charged particles in three dimensions. The successful realization of such a device and a comprehensive description of its properties appeared in a government report in 1958 written by W. Paul, O. Osberghaus, and E. Fischer (Paul et al., 1958), followed by an account by E. Fischer in the Zeitschrift für Physik, 1959. Figure 12.4 illustrates the device.

12.4.1 The Field Distribution

The electrode geometry is identical to the one used in the Penning configuration: an hourglass-shaped cylinder between two end caps. However, instead of a constant voltage being applied to the electrodes, with a polarity appropriate to the sign of the charge of the particles to be confined, the polarity alternates at high frequency, so that the sign of the charge is immaterial. The effect is that along any given coordinate axis, a charged particle experiences a force alternately towards the center and away from it. If the strength of the electric field were the same at all points, then the motion of a particle would be simply an oscillatory one driven by the field, superposed on any original uniform motion. Thus it is clear that a uniform high-frequency field would not do, since the original uniform motion continues undeterred.

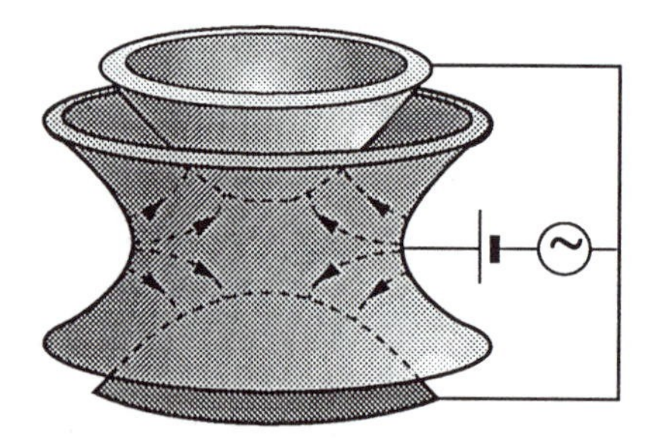

Figure 12.4 The Paul high-frequency 3-dimensional ion trap

The essential property of the quadrupole field is that it is *not uniform*, having a minimum at the center; the fact that the field strength varies simply in proportion to the distance from the center has the advantage that it leads to ion motion that lends itself to exact mathematical analysis. A charged particle placed in such an alternating quadrupole field will, along any given coordinate direction, experience a force that alternates but is not entirely symmetric between directions away and toward the center; it happens that the net result is an average force toward the center. This can be anticipated on the basis of something called the *strong focusing principle*, which was originally enunciated in terms of a set of static ion "lenses," alternating between focusing and defocusing, positioned sequentially in space, so that a particle passing through them experienced a time sequence of focusing and defocusing forces. This evolved into the single "lens" alternating in time. It can be made plausible that such an alternating focusing–defocusing sequence will yield a net focusing action by noting that on the average the defocusing starts nearer the center where the field is weaker, while at the end of the defocusing half cycle, when the focusing begins, the ion is farther from the axis where the field is stronger.

The analysis of the ion motion in the Paul trap begins with the definition of the field, which now includes a periodic time-dependence of the voltage applied between the quadrupole electrodes; the components of the electric field are as follows:

$$E_r = \left(U_0 - V_0 \cos\Omega t\right)\frac{r}{r_0^2}; \quad E_z = -2\left(U_0 - V_0 \cos\Omega t\right)\frac{z}{r_0^2}, \qquad 12.12$$

where U_0 is a constant voltage, and V_0 and Ω are respectively the amplitude and (angular) frequency of the high frequency voltage.

12.4.2 Ion Motion in the Paul Field

The motion of the ions is governed by Newton's **F=ma** equation of motion, which in this case takes on the following form for the r-coordinate:

$$\frac{d^2 r}{d\theta^2} + \left(a_r - 2q_r \cos 2\theta\right) r = 0, \qquad 12.13$$

where

$$a_r = \frac{4eU_0}{M\Omega^2 r_0^2}; \quad q_r = \frac{2eV_0}{M\Omega^2 r_0^2}; \quad \theta = \frac{\Omega t}{2}. \qquad 12.14$$

The a and q coefficients for the z-equation are given by $a_z=-2a_r$ and $q_z=2q_r$. Differential equations of this form are referred to as *Mathieu* equations after the French mathematician E. Mathieu, who in 1868 published his study of the vibrations of an elliptical membrane, in which this form of equation arises. In fact, this equation arises in many important practical applications, for example parametric amplifiers, among other things. We note that it has the form of the equation for a simple harmonic oscillator, in which, however, the frequency-determining parameter is itself an oscillatory function of time. The "pumping" of a child's swing is an example of such a parameter being modulated at twice the frequency of the swing and leading to *parametric excitation*, as we saw in Chapter 2. It is not our purpose, of course, to delve into the mathematical properties of this equation and its solutions, but rather to state the results pertinent to the design of the trap.

The most important property of the Mathieu equation for our purposes is that its solutions are stable or unstable depending on the values of the parameters a and q. Here the question of the stability of a solution refers to whether or not there is an upper bound on how far a particle may move away from the center. All the solutions are oscillatory (but not necessarily periodic) about the center. However, the unstable ones have an amplitude that increases in time without limit. If the values of a_r and q_r are plotted with respect to a set of Cartesian axes, then the plane is divided into areas where the equation has stable solutions and areas where they are unstable, as shown in Figure 12.5. Whether a solution for given values of a_r, q_r is stable or unstable depends on whether the point with these coordinates lies in a stable or unstable region. In nearly all applications, only the first stability region has been used, since it corresponds to the least amplitude V_0 of the high-frequency field.

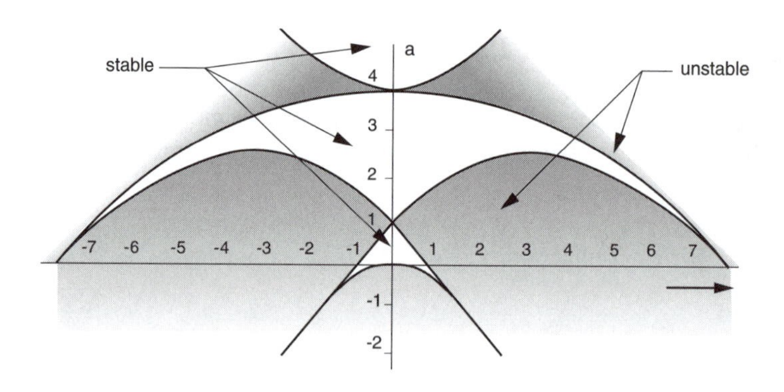

Figure 12.5 The a–q stability diagram of the Mathieu equation

For the particle to be confined in all three dimensions it is necessary that not only a_r, q_r lie in a stable region, but a_z, q_z also. Since the latter differ only by a factor of −2, it is convenient, following Paul, to make a composite plot of the stability boundaries in which those for a_z, q_z are drawn to half scale and inverted along the a-axis, as in Figure 12.6. Inverting along the q-axis produces no change because of symmetry. If then the point a_r, q_r lies in the region of overlap between the r- and z-stability regions, then the motion will automatically be stable in three dimensions. As a numerical illustration, suppose we choose a point well within the stability region a_r=0.01 and q_r=0.2, and we wish to operate a trap with radius r_0 =1 cm at a frequency of 500 kHz, then to trap (say) mercury ions (mass number 199), the voltages we would require are U_0=5 volts and V_0=200 volts.

From the theory of the Mathieu differential equation, for a, q in a stable region, the general solution shows that the spectrum of the ion motion consists of a discrete set of frequencies given by

$$\omega_n = \left(\pm n + \frac{\beta}{2} \right) \Omega , \qquad 12.15$$

where n is an integer and the relative amplitudes and the constant β are functions of the operating point a, q. (We have for brevity dropped the r and z subscripts and will do so in all cases where it would not cause ambiguity.)

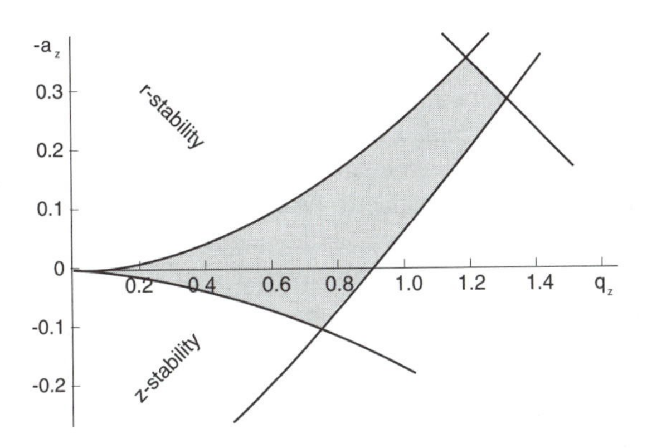

Figure 12.6 A composite plot of the first stability region for both r- and z-stability

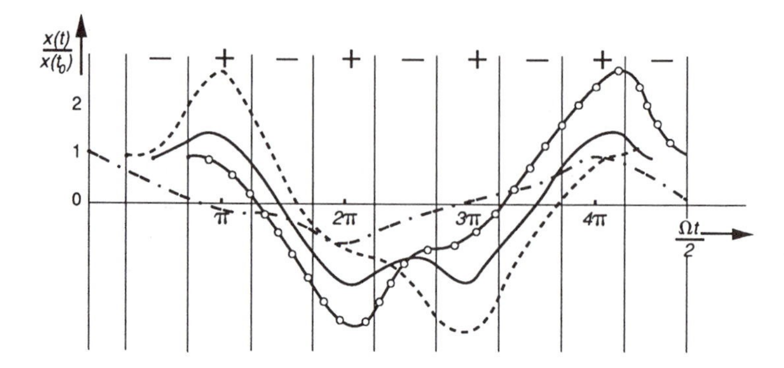

Figure 12.7 Particle oscillation in the Paul high frequency quadrupole electric field (Paul, 1958)

To illustrate the particle trajectories typical of those described by this solution, we reproduce from the work of W. Paul et al. plots for the special case of a trap operating at the point a=0, q=0.631, β=0.5, and ions created with zero initial velocity (see Figure 12.7). The plus and minus signs indicate the focusing and defocusing half cycles of the field. It is seen, for example, that the ion created when the phase of the high-frequency field is $\pi/4$ is first defocused and then focused back at a point farther from the center where the field is stronger and is thereby caused to swing back more strongly. This results in a finite oscillation of complicated form reflecting the presence of many frequency components in the Fourier spectrum. However, one frequency component is evident: the lowest frequency at $\beta\Omega/2$, which for β=0.5 corresponds to one-fourth the frequency of the field. We should note also the strong dependence of the amplitude on the phase of the field at the time the ion is created.

Unlike the Penning trap, a magnetic field is not required for stable confinement of ions in the Paul trap. However, when used for frequency standards based on a magnetic hyperfine transition, a weak magnetic field is essential for the proper functioning of the standard. From the point of view of the motion of the trapped ions, the effect of a weak axial magnetic field can be predicted from the *Larmor theorem*, which we encountered in the discussion of magnetic resonance. It states that the effect of an external field of strength B on the motions of identical ions moving about in finite orbits is entirely equivalent to what would be observed if the motion was referred to a frame of reference rotating uniformly with an angular velocity of $eB/2M$ about the field axis. The component of the motion along the field axis, taken as usual to be the z-axis, is not affected by the magnetic field; however, the radial component behaves as if there were a centrifugal "potential," with a quadratic dependence on r simply added to the constant potential term U_0 applied to the Paul trap. In consequence, the parameter a_r, which is a measure of the constant potential, is shifted as follows:

$$\tilde{a}_r = a_r + \left(\frac{2\omega_L}{\Omega}\right)^2, \tag{12.16}$$

where $\omega_L = eB/2M$ is the Larmor frequency. The spectrum of the ion motion is also made more complicated; as implied by the Larmor theorem, the equivalent rotation (or *precession*) frequency is added to the spectrum of radial frequencies.

It can be shown in general that if the parameters a and q are much less than one, the amplitudes of the higher frequencies in the motional spectrum rapidly become negligibly small as n increases beyond $n=1$. Therefore, a reasonable approximation is obtained by retaining only oscillations at frequencies corresponding to $n=0$ and $n=\pm1$; in this case, the theory shows that we have the following approximate solution:

$$r(t) = A\left(1 + \frac{q_r}{2}\cos\Omega t\right)\cos\frac{\beta_r\Omega}{2}t; \quad \beta^2 = \left(a + \frac{q^2}{2}\right), \tag{12.17}$$

which shows that indeed the motion is bounded, with an upper limit of $A(1+q/2)$.

This result can be interpreted as representing a slow *secular* oscillation (since under our assumption $\beta \ll 1$) at a frequency $\beta\Omega/2$, with a *jitter* at the frequency of the field, having an amplitude that increases with the low-frequency displacement.

This last result could have been derived much more simply if the assumed condition that a and q are very small is used from the beginning, instead of starting with the general solution for all values of those parameters. To do this we must recognize that from its definition, q equals the amplitude of oscillation (as a fraction of r_0) that a charge would have in a *uniform* high-frequency field of intensity V_0/r_0, which is the maximum it reaches in the trap, at $r=r_0$. Thus a small value of q means that the field causes only a small high-frequency jitter; that is, the motion can be analyzed as a superposition of two motions, a fast oscillation at the field frequency and a possible slow motion of the center of that oscillation. Under conditions where this separation of the motions is justified, it has been shown that a general solution is possible, and not just the specific case of a quadrupole field. If a charged particle is acted on by a high-frequency electric field $E_0(x,y,z)\cos\Omega t$, whose amplitude is a slowly varying function of space, such that it varies little over the particle jitter, then its motion in the field can be written in the form

$$r(t) = R(t) - \frac{eE_0(R)}{M\Omega^2}\cos\Omega t, \tag{12.18}$$

in which the high-frequency jitter is about a point $R(t)$ that moves according to the following equation of motion:

$$M\frac{d^2R}{dt^2} = -e\frac{dU_0}{dR} - \frac{e^2}{4M\Omega^2}\frac{d\left(E^2(R)\right)}{dR}. \qquad 12.19$$

To obtain the particular solution for the quadrupole field we would substitute $E_0 = V_0 R/r_0^2$. The result agrees with the earlier theory; the oscillation frequency of $R(t)$ is the lowest frequency $\beta\Omega/2$ in the spectrum, where $\beta^2 = a + q^2/2$. We should note that although the jitter is at the field frequency, the spectrum of the motion does not contain that frequency Ω itself, since its amplitude is modulated by the low frequency oscillation; in our approximation only $\beta\Omega/2$, $(\Omega-\beta/2)$, and $(\Omega+\beta\Omega/2)$ are present.

An instructive result, one that gives some insight into the behavior of the particle motion in a high-frequency field, is obtained by computing the average kinetic energy over many cycles of the field. Under conditions where the approximation we used above is valid, and assuming that only a high-frequency field acts on the particle, so that $a=0$, we find that the total kinetic energy averaged over many cycles of the high-frequency field remains constant; it merely goes over into the high-frequency jitter form as the center of that oscillation slows down, and it continues to alternate between the two forms of motion as the particle executes its slow oscillation between regions of strong and weak field intensity.

An analogous exchange of kinetic energy between different components of particle motion occurs in an axial magnetic field that has a weak axial gradient in its intensity. In this case, a charged particle with an axial component of velocity will have that component reduced or increased as the particle cyclotron motion around the axis gains or loses kinetic energy, depending on whether the gradient is positive or negative. This is the basis of the "magnetic bottle," an ion confinement device consisting of an axial magnetic field, uniform over a certain length but becoming more intense at the two ends. For ions moving along certain angles with respect to the axis there exist conditions when they would not only be slowed down in approaching the more intense magnetic fields at the ends, but will in fact be reflected back and forth between the two ends.

12.4.3 Initial Entrapment of Ions

Unlike the Penning configuration, the question of what initial position and velocity an ion may have and still be confined in the available space is more complicated in the Paul trap. The two most important complicating circumstances are first, the particle trajectory depends on the phase of the high-frequency field when the ion is created, and second, the particle trajectories are not even simply periodic. However, limits can be found to the amplitude of the motion as a function of initial ion position and velocity at any given phase of the high-frequency field. For a given initial phase it can be shown that the particle trajectory will have a given upper limit, provided that the initial position and velocity are related by a certain quadratic expression; if we plot this initial velocity versus position, the resulting graph is an ellipse whose parameters depend on the phase when the ion was created, as shown in Figure 12.8. For each phase, those initial values of velocity and position that fall *within* the ellipse lead to an amplitude smaller than would reach the electrodes.

Changing the operating point in the a–q stability diagram results in significant changes in the ellipses, as might be expected. From these graphs, no matter what the initial velocity or phase may be, the ion must be created within the space defined by the electrodes. It is useful in practice to know what fraction of ions created uniformly throughout the trap, say by electron collisions or ionizing radiation acting on the parent atom or molecule, will in fact be trapped. Since the energy of ions resulting from these processes is not expected to be much above the thermal energy of the parent particle, an energy negligibly small compared with the hundreds of volts present in the trap, we may, without appreciable error, assume the initial velocity to be nearly zero. In terms of the velocity–position ellipses, all points representing ions actually created will lie in a narrow band along the position axis. It becomes a simple matter to determine for each ellipse the fraction of the number of points falling inside it.

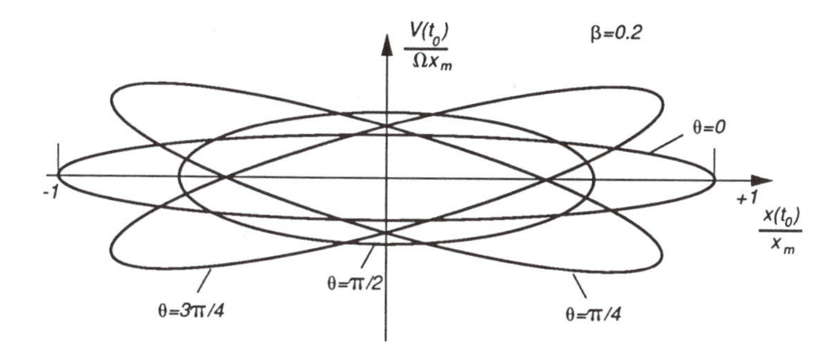

Figure 12.8 The initial positions and velocities leading to confinement in a Paul trap (Paul, 1958)

12.4.4 The Maximum Number of Ions

The total number of ions that can be accumulated in the trap is limited by the same mutual repulsion between like charges we have already discussed with respect to the Penning trap. Here again on the basis of the simplifying assumption that the charged cloud of ions has a constant density and is bounded by a sphere, the electrostatic potential field produced by the ions has the same quadratic dependence on the coordinates as does the quadrupole field, and therefore its effect is fully taken into account by increments (of the same sign) in the values of the parameters a_z and a_r.

Since these parameters have opposite signs for the *r*- and *z*-coordinates, it follows that one will be decreased while the other increased. The effect is that the boundaries of the stability regions in the composite *a–q* diagram are narrowed, as shown in Figure 12.9. The amount of shift can be obtained using the result for the potential we found for the Penning trap. The usefulness of this approximation is limited to establishing at what ion number the electrostatic repulsion will compete appreciably with the confining field, rather than as a quantitative model for accurately predicting ion behavior at high number density. In practice, the maximum number of ions confined is far below the number predicted by the simple theoretical model we have assumed. The experimental difficulties of working with small numbers of confined ions, which plagued the technique in the beginning, have largely been removed since the advent of lasers and the vastly improved signal-to-noise ratios they have made possible in spectroscopic measurements on ions.

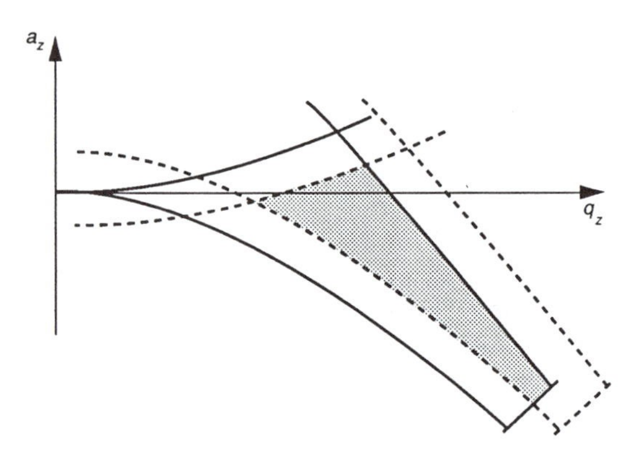

Figure 12.9 The effect of space charge on the boundaries of the stability region

12.4.5 The Effect of Collisions

Ideally, an ion following a stable orbit within the trap will continue to do so indefinitely; however, in a real system an ion will eventually collide with other background gas particles in the trap. Generally, the number density of ions is so low that the probability of collisions with other ions is negligible; they are most likely to occur with the parent atom or molecule, or other background gas particles in the vacuum system. As we noted earlier in this chapter, collisions with the parent atom or molecule can lead to "resonant" charge exchange with little change in momentum. In such a collision, a high-velocity ion and an atom or molecule with thermal energy exchange roles, leaving an ion with nearly the original low velocity of the neutral particle. It must not be thought, however, that this necessarily leaves the ion in a lower-energy orbit; on the contrary, depending on the phase of the high-frequency field at the time of the collision, it may go into a *higher*-amplitude orbit. If it is valid to assume that the probability of a collision is independent of the velocity of the ion, then it will occur with equal probability at all phases of the field. In that case, it can be shown that the phase of the low-frequency motion at the time of the collision determines whether there is a gain or loss in energy. Clearly, the collisions continuously redistribute the energy among the ions. Although charge exchange collisions are the most likely in the presence of the parent particle, collisions with other gas particles may be important, depending on the degree of ultrahigh vacuum in the system. Among the possible collision processes that may occur, those conserving the total (translational) kinetic energy, that is, elastic collisions, are the most likely. In this case the question of whether a collision results in a gain or loss of energy on the part of the ion depends on the relative mass of the collision partners. If an ion is *lighter* than its collision partner, it will be scattered with an increase in kinetic energy derived ultimately from the high-frequency field. This can be readily argued on the basis of the approximate separation of the fast and slow oscillations, in the limit of elastic scattering from particles that are so heavy as to be immovable. In that case, collisions will cause successive reversals in direction to occur at random phases of the field; this can be shown always to lead to a gain in the kinetic energy. The average increase in energy per collision is expected to be in proportion to the initial energy, and as a result of successive collisions to continue to increase at a rate dependent on the frequency of collisions. This is a well-known heating effect of collisions in any system of charged particles subjected to any high-frequency field, whether uniform or not. Since electrons are much lighter than atoms, this applies to them particularly, and the effect is called simply *rf heating*. The situation is quite different if the collision partners have equal mass; then it can be shown that on the average there is no gain or loss. Finally, if the collision partner has a smaller mass, there is a net loss of kinetic energy; this is familiar in limiting cases where a heavy object is subject to numerous weak collisions with very much lighter particles, as in air drag on flying aircraft or the damping of oscillation of a pendulum due to air resistance. In some designs of the mercury ion frequency standard, as we shall see, a light buffer gas, such as helium, is used to cool the ions. However, where other methods of cooling are used and the highest degree of vacuum is sought, statistical averages on energy gain and loss are not particularly useful, since the number of events is not always large enough as a statistical sample. In some cases only a few collisions may put an ion on a trajectory that intersects an electrode. In some of the early work it was estimated that typically it took on the average tens of collisions before an ion gained enough energy to reach an electrode and be lost.

12.4.6 The Detection of Ions

For the development and application of ion traps, a method is required of observing the ions and measuring their energy; and several techniques have been used for the purpose. The original studies of Paul et al. were mainly motivated by their application to mass spectrometry. This obviously requires a method that discriminates between ions of different mass (or more precisely, charge/mass ratio) and combines mass resolution with accuracy in the measurement of ion number. The method they developed is analogous to that used in nuclear magnetic resonance: the resonant absorption of energy by the ions from an oscillating electric field. The method is implemented by subjecting the ions to an electric field applied between the two end caps at the low oscillation frequency $\beta\Omega/2$. In the first stability region in the a–q diagram, where one commonly operates, the value of β lies between 0 and 1; on the two boundaries of that region β is constant with $\beta=0$ on one and $\beta=1$ on the other. Elsewhere in that region its exact value, and therefore the detection frequency, is a complicated function of a and q, which approaches the simple formula derived above only in the limit of small a, q. Therefore, in practice it is useful to have a graphical representation in which points in the a–q diagram having the same value of β are joined to form iso-β lines; such a diagram is shown in Figure 12.10. Now, from the definitions of the parameters a and q, we can easily verify that the ratio $a/q=2U_0/V_0$ for all ions, independent of their charge or mass.

Therefore, for a given U_0/V_0 ratio, a plot of the parameter a versus q for different masses must lie on a straight line through the origin with a slope of $2U_0/V_0$, intersecting the iso-β contours at points that depend on the charge/mass ratio of the ion. Therefore, in principle, if the detection field frequency is swept, ions of different mass will come into resonance with it, producing a mass spectrum. Furthermore, if we wish to study the number of a particular species of ion in the trap, we can, for example, vary U_0 slightly while keeping V_0 constant. Then ions of the given mass will have their β value varied as the operating point shifts parallel to the a-axis, keeping q constant. For the typical values $a_z=-0.1$, $q_z=0.5$, $\beta_z=0.47$, and $\Omega/2\pi=1$ MHz, the lowest frequency in the spectrum $\beta_z\Omega/4\pi$ would be 235 kHz.

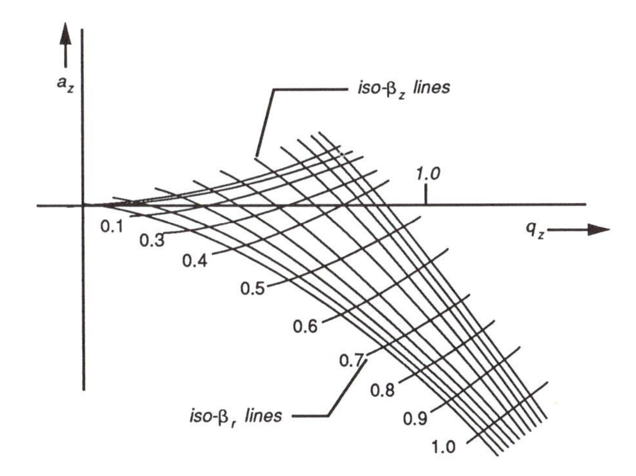

Figure 12.10 The iso-β lines in the stable region of the *a–q* diagram

The ion detection frequency is so much lower than Ω, the quadrupole field frequency, that it is possible to generate the detection field by connecting the detection frequency signal across a tuned parallel inductance-capacitance (L-C) circuit. The tuned L-C circuit presents nearly a short circuit for both DC and to the high-frequency voltage, thereby keeping the end caps at the same DC and high frequency-voltage as required. This is because the capacitance that tunes the inductor to resonance at the detection frequency will have a very much smaller reactance at the quadrupole frequency. A typical circuit arrangement is shown schematically in Figure 12.11. The complicated oscillatory motion of the confined ions is effectively coupled to the outside circuit through the charges that they induce on the end caps, which act like the plates of a capacitor. Thus it is predominantly the component of the motion along the z-axis that induces current in the outside circuit, a current that will have the frequency spectrum of motion along that axis. For the simple model of a single ion moving between a pair of plane parallel capacitor plates, the current would be $\mathbf{i}=(e/d)\mathbf{V}$, where d is the separation of the plates and $\mathbf{V}$ is the instantaneous velocity. In reality, of course, the end caps are curved, so that even oscillation in the radial direction can bring the ions closer and farther from an end cap, inducing in the outside circuit a much weaker current, at twice the radial frequency. Since the motions of individual ions are uncorrelated, the net current they produce would fluctuate as random *shot* noise. However, if by exciting an oscillation in the tuned L–C detection circuit from a suitable outside generator they are subjected to a common oscillatory electric field, a coherent global oscillation will be superimposed on the random oscillations of the ions. This will induce a much stronger current to flow in the detection circuit, since the ions are then moving in concert, at least in their response to the external excitation field.

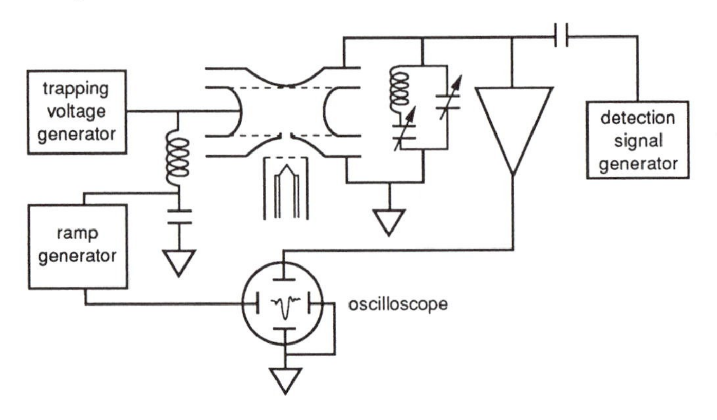

Figure 12.11 A schematic diagram of ion resonance detection circuit

To display the resonant response of a particular species of ion, it is necessary to sweep the frequency of the field, or equivalently the frequency of the ions, through the resonance condition. This is most conveniently done by imposing a sawtooth modulation on U_0 so that the ion frequency passes periodically through resonance with the detection field. By synchronizing the sawtooth waveform with the time-base of an oscilloscope, a stationary display of the resonance signal is obtained. Because of the loss of ions in the detection process and their finite lifetime in the trap due to collisions, it is necessary to "fill" the trap at the beginning of each cycle by a pulsed electron beam or by some other means.

The signal appearing on the detection circuit will depend on several factors, principally the amplitude of the detection field and the sweep rate of the ion frequency. In order to use the resonance signal quantitatively to determine the number of trapped ions of a given species, we must have a theory of line shape that can be used to properly take into account these various factors. An approximate theory can be modeled on a damped harmonic oscillator driven by a periodic force; however, there is one fundamental complication: Here the ion number is not constant in the detection process. As the ion frequency comes into resonance with the detection field, the amplitude of the ion oscillation begins to build up, causing more and more of the ions to reach the electrodes and be neutralized. Apart from the variable ion number, the system is analogous to the resonance absorption method in the nuclear magnetic resonance technique. Let us follow the motion of the center of the ion cloud as the ion frequency approaches resonance with the detection field, goes through the resonance condition, and continues away from resonance. During the approach to resonance, the center of the ion distribution, which was initially at rest, begins to oscillate at the detection frequency with an amplitude that increases rapidly as resonance is approached; at resonance the amplitude begins a linear increase, till finally, as the frequency sweep continues past resonance, the signal will exhibit beats between the current induced by the ions and the external excitation current of that circuit.

In order to follow the complicated interaction between the ions and the detection field, and in particular to analyze the way that interaction is manifested in the detection circuit, it is fortunately possible to incorporate the effect of the oscillating ions as equivalent circuit elements, which enable standard circuit theory to be used. Under conditions where the model of a driven harmonic oscillator is useful, it can be shown that the effect of the presence of the oscillating ions on the external circuit can be fully taken into account by assuming a (high) resistance parallel to the tuned L–C circuit to account for the power absorbed by the ions at resonance, and in series with that resistance, a capacitance and inductance to account for the resonant behavior of the ion current. Typical numerical values of the equivalent resistance for 10^6 ions in a 1 cm trap are on the order of $R_{equiv} \approx 1 M\Omega$. Since this equivalent resistance is assumed connected between the two end caps, its high value simply indicates that the ion current is small, as is the power absorbed by the ions. Therefore, to observe this small power absorption by the ions, the L–C circuit supplying the excitation to the end caps must present a comparably high resistance. From circuit theory we know that a high-Q parallel L–C circuit becomes resistive (voltage and current in phase) at resonance, with a value of $R_{res}=Q\omega L$, where ω is the (angular) frequency.

Therefore, what is required is a high-Q coil of large inductance with the smallest possible parallel capacitance, since for a given resonance frequency ω we must have LC=constant. The resonant excitation of ion oscillation by the detection field draws power from this tuned circuit, effectively lowering its Q-value, and thus its resistance. If the excitation signal originates from a constant current (high resistance) source, then the signal level on the tuned circuit falls, giving the typical signal waveform illustrated in Figure 12.12. Note the "ringing" that occurs on the trailing edge of the signal due to the beat phenomenon.

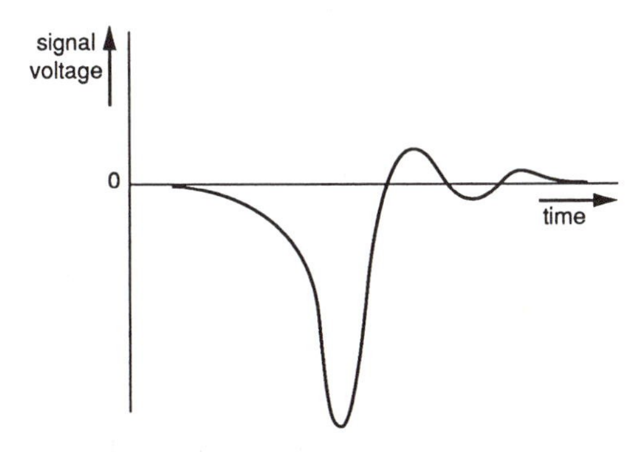

Figure 12.12 The ion detection signal waveform

A more fundamental condition on the required properties of the detection circuit comes from considerations of electrical noise. The ultimate aim is to reduce all sources of noise below the fundamental shot noise due to the oscillating charge being made up of indivisible ions; it depends on the average ion current as follows:

$$\left\langle i_n^2 \right\rangle = 2e\langle i \rangle \Delta\nu , \qquad 12.20$$

where $\langle i \rangle$ is the mean signal current and $\Delta\nu$ is the frequency bandwidth in which the mean square noise current, $\langle i_n^2 \rangle$ is determined. In addition to environmental sources of fluctuation, this means ultimately that the fundamental Johnson (thermal) noise must be less than the shot noise. The Johnson noise is given by the following:

$$\left\langle i_n^2 \right\rangle = \frac{4kT}{R} \Delta\nu , \qquad 12.21$$

where T is the absolute temperature and R is the resistance through which the noise current flows. It follows that ideally the value of R should be determined by the condition that $R\langle i \rangle \rangle 2kT/e$. But $R\langle i \rangle$ is simply the voltage drop across the resistance R in the presence of a current $\langle i \rangle$, and at room temperature, $2kT/e \approx 0.05$ volts. This is compatible with the signal expected according to our approximate model from the current induced by 10^6 ions in a 1 megohm resistance. To achieve the desired high resistance, a coil tuned to, say, 250 kHz would have to have a Q-value of 200 and inductance of a few millihenries; that is, we require a very high-Q RF coil with large inductance and low interwinding capacitance. This means winding a special coil with thin multistrand wire with low-loss insulation on a ferrite ring and coupling it to the source of excitation through a very small capacitance.

The detection frequency signal on the tuned L–C circuit is coupled through a high-input impedance, low-noise preamplifier whose output is matched to a 50 ohm cable to another amplifier, then a diode detector (demodulator), and finally to an oscilloscope.

If sufficient care is taken in the construction of the trap and detection circuit to ensure adequate mechanical and electrical stability and isolation from environmental sources of electrical noise, the ideal limit of observing individual ions can be approached even for heavy ions. The fundamental thermal noise at room temperature on the detection circuit, in a 10 Hz bandwidth, is on the order of 3×10^{-13} amp, whereas the current induced by a single ion oscillating at 250 kHz is on the order of 2×10^{-13} amp. These numbers of course represent theoretical limits not easily achievable in practice; it makes it all the more remarkable that evidence of the discreteness of the charge of trapped ions was found early in the development of the technique for heavy ions by Rettinghaus in Paul's laboratory. The evidence is established by analyzing the statistical distribution of the height of resonance signals that indicate the ion population in the trap. At noise levels sufficiently low that an increment of one ion is discernible, the ion signals should fall into discrete levels distributed according to the *Poisson* distribution about the average.

The resonance absorption method of detecting the ions is not the only one that has been used, although its relative simplicity and mass discrimination commend it. Two other methods deserve mention: The first extracts the ions from the trap, accelerates them, and counts them with an electron multiplier; and the second uses the resonance scattering of laser light, which is collected with wide-angle optics and detected with a photon counter. These methods are more complex but have an assured ability of detecting individual ions.

Chapter 13
The NASA Mercury Ion Experiment

13.1 Introduction

The exploitation of ion confinement for the development of a new kind of portable atomic frequency standard was first proposed by Major in 1969 as an approach that promised extraordinary accuracy in a light, compact device suitable for aerospace applications (Major, 1969). It was predicted that a microwave resonance of unprecedented spectral resolution was possible at around 40.5 GHz on mercury ions of isotopic mass 199, when observed under the perturbation-free environment of field confinement in vacuum, where free observation times of tens of seconds were routinely obtainable. Moreover, the relatively large mass of the mercury ion has the further advantage of a small (second-order) Doppler width for a given distribution of kinetic energy, a subject we will consider in greater detail later. The resonance is at the high end of the microwave region of the spectrum, where for a given line width the resonance Q is high, and yet falls in a range that can conveniently be reached by common frequency-synthesis techniques. Finally, the short wavelength of the resonant microwave field (7.4 mm) permits the physical dimensions of the microwave components to be correspondingly small. For all these reasons it was argued that the choice of mercury is particularly suited to fully exploit the new technique of field confinement in the development of a spacecraft clock.

The observation of radio-frequency and microwave transitions in field-confined ions had previously been successfully demonstrated on the singly charged ions of helium (He^+) and molecular hydrogen (H_2^+). These experiments were motivated strictly by the intrinsic scientific interest in these atomic and molecular systems; nevertheless, they are mentioned here as the first successful experiments in which the Paul trap is used in the field of ion spectra rather than the original application of mass spectrometry. They are noteworthy in having solved the problems not only of producing the difference in population between quantum states necessary to observe their resonance spectra, but also of adapting the electrode design of the trap to the needs of microwave circuitry. To achieve the necessary population difference between the states involved in the resonant transitions, use was made of spin-dependent scattering processes with polarized atoms or photons, which traversed the trap in the form of a beam.

In late 1972 at a NASA laboratory, by successfully applying the technique of optical hyperfine pumping, a microwave resonance was observed by Major and Werth in mass 199 mercury ions that was, as predicted, so sharp as to be beyond measurement by commercially available equipment at the time (Major et al., 1973). The estimated fractional line width had to be less than the best available standard; less than one part in 10^{10}. This work was discontinued by NASA in 1973, and through relocation of the original investigators, activity in this field moved to Orsay, France, and Mainz, Germany. It was not until a decade later that significant progress was made towards a prototype frequency standard based on the mercury ion. Since then it has been pursued by a number of laboratories around the world and is under commercial development.

Some of the relevant properties of the isotope 199 of mercury are shown in Table 13.1.

Table 13.1 Properties of Hg^{199}

Isotopic abundance	16.9%
Ion ground state	$(^2S_{1/2})$
Nuclear spin	$1/2(h/2\pi)$
Nuclear magnetic moment	$+0.504117\mu_n$
Ion hyperfine frequency	40.507 GHz (ground state)
Ion resonance (UV) wavelength	194.2 nanometers

The element mercury has many stable isotopes ranging from mass 196 to 204; however, only the two with odd mass number are known to have a nonzero nuclear magnetic moment associated with a nuclear spin: mass 199 and 201. It has historically been a popular element in spectroscopic studies, partly because of intrinsic interest in its rich spectrum, and partly no doubt because of industrial applications as a source of ultraviolet radiation and illumination. Many of the fundamental experiments on the technique of optical pumping of magnetic state populations were first conducted on the isotopes of mercury. It was also used to demonstrate the selective population of upper magnetic states by excitation with a collimated beam of electrons near threshold energy.

13.2 Ground State Hyperfine Structure of Hg^{199}

The mercury atom and its ion are obviously complicated systems, with respectively 80 and 79 electrons outside the nucleus. The electronic structure of the ion, which results from the removal of one electron from the atom, consists of 6 principal groups of shells characterized by different principal quantum numbers n=1, 2, 3, 4, 5, 6, which have to do with the sizes of the orbits; all except n=6 contain filled shells. We recall that there is one general result that radically simplifies the picture: Electrons in a filled shell form a spherically symmetric distribution with zero angular momentum. It follows that in the ion only the single n=6 electron, in the zero orbital angular momentum state 6s, has a spin unpaired with another of opposite direction. Therefore, the ground state of the ion is labeled spectroscopically as $^2S_{½}$, which is the same as, for example, the alkali atom Cs. Like that atom, the single "outer" electron has a finite probability of being at the nucleus, and those isotopes whose nuclei have a magnetic moment will have a magnetic hyperfine splitting of the ground state energy, reflecting the two ways the electron spin (S=½) and nuclear spin (I=½) are coupled according to quantum theory to produce a conserved total angular momentum (F), namely $F=I+½$ or $F=I-½$. As already pointed out in connection with the other microwave standards, these two possible states differ in energy because of the two possible orientations of the magnetic moment of the nucleus relative to the direction of the magnetic moment of the electron cloud in which it is immersed. This is the now familiar magnetic hyperfine splitting of quantum energy levels, between which the standard frequency transitions occur. The fact that the nucleus of the mass 199 isotope has a small nuclear spin I=1/2 and yet a relatively strong magnetic interaction with the outer electron is an added important advantage weighing in favor of choosing the ion of Hg^{199} as the "working substance." The reason is that it gives a simple quantum-level structure, one that is similar to the simplest of all, the hydrogen atom; hence there are fewer competing states in trying to place the ions in the one required to observe the resonant transition. Figure 13.1 shows the relevant hyperfine quantum levels for the ground state of the $(Hg^{199})^+$ ion. As is by now familiar, the transition used as reference is the so-called field independent hyperfine transition between the states with m_F=0.

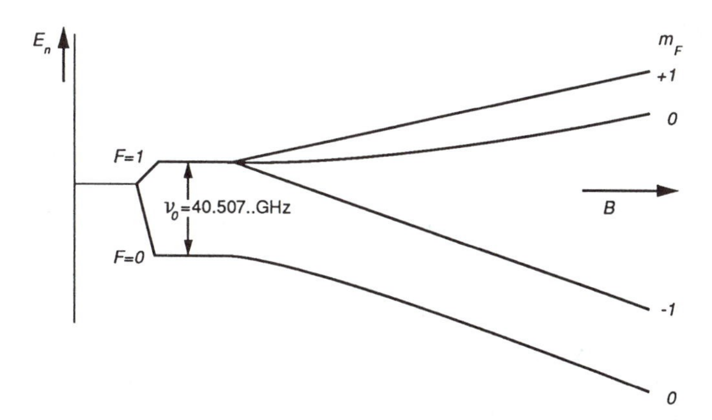

Figure 13.1 The hyperfine structure of the ground state of the mass 199 Hg^+ ion

13.3 Hyperfine Optical Pumping

13.3.1 Use of Hg^{202} Lamp

In order to be able to observe transitions between these states with m_F=0, a significant difference in the populations of these two states must be created. This, it will be recalled, is due to the fact that a resonant magnetic field will stimulate upward and downward transitions in each ion with equal probability; a net change in populations can be observed only if there is a difference in the numbers of ions making these transitions. Of the techniques we might consider for creating a difference of population, the use of deflecting magnets is ruled out because of the overwhelming force on the charge, and spin-dependent collisions with spin-polarized atoms will not directly produce the desired effect.

The one fortuitous circumstance that provided an effective compact solution, and which was as critical to the success of the mercury ion standard as the effectiveness of Teflon to the hydrogen maser, involves the ultraviolet *resonance* emission line in the spectrum of mass 202 mercury ions. In Figure 13.2a are shown the pertinent low-lying quantum levels for the mass 199 and 202 mercury ions. Figure 13.2b shows a plot of the hyperfine components of the ultraviolet resonance transition between the ground state and the first excited state, such as might be obtained with a high-resolution ultraviolet spectrograph. Naturally, the relative heights of the peaks depend on the particular lamp design, but their precise relative positions on the wavelength scale are very nearly constant. We note that in the case of mass 199 ions, the ground state and first excited state are "doublets" because of the hyperfine splitting. The separation of the hyperfine components in the first excited state is very much smaller than that for the ground state because the distribution of the p-electron responsible for the splitting in the excited state, unlike the ground state s-electron, has little "contact" with the nucleus. The ground state, on the other hand, has a relatively large frequency splitting of around 40.5 GHz, a value deduced at the time of the original microwave experiment only indirectly from spectroscopic data in the ultraviolet. It is now known to over twelve significant figures. Isotopes with an even mass number, such as 202, have zero nuclear spin and magnetic moment, and therefore their levels are not split. Now, because of the difference in nuclear structure and mass between the two isotopes, there is what spectroscopists call an *isotopic shift* in the spectrum of one isotope relative to another. From the classic work of Mrozowski on high resolution spectroscopic studies of the shifts in the spectra of the different isotopes of mercury, it was indirectly deduced that the UV emission line at λ=194.2 nanometers for mass 202 ions accidentally falls much nearer in wavelength to one component in the mass 199 hyperfine doublet than the other. Furthermore, the Doppler broadening of the UV spectral lines of ions in the trap, though large, amounting to perhaps 8 GHz for ions of energy 2.5 electron volts, is still very much less than the hyperfine splitting. This, coupled with the greater overlap of the coincident lines brought on by the same

Doppler broadening, provides in a mass 202 lamp a relatively simple, compact source for *hyperfine pumping* mercury 199 ions between the desired hyperfine states.

While the use of a mercury 202 lamp remains the most suitable pumping source where a compact, portable standard is required, there are enormous signal-to-noise advantages to using laser techniques to generate the desired wavelength. A typical approach would be to triple the frequency of a high-power stabilized laser operating at three times the wavelength (582.6 nm) in the yellow region of the spectrum. More will be said about the application of lasers to more recent developments in ion standards in a later chapter.

The principle of the pumping process is illustrated in Figure 13.3. Mercury 199 ions confined in a Paul trap under ultrahigh vacuum are irradiated with a beam of UV light from a specially designed mercury vapor lamp containing the enriched 202 isotope of mercury. The mass 199 ions in the trap can absorb radiation to any significant degree only if the photon polarization and energy match an allowed transition between its quantum states. Since the light from the mercury 202 lamp is unpolarized, that is, contains equally all states of polarization, there will be no restriction on what transitions the ions will make on that account. However, the interaction between the photon and ion is a resonant phenomenon—it is not just a matter of the photon having enough energy, it is a matter of having the *right* energy. Therefore, only those mass 199 ions that are in the F=1 state can absorb the UV resonance light from mercury 202 ions in the lamp and make transitions to the upper electronic state; the others, in the F=0 state, are undisturbed. Ions in the upper electronic state will, after an extremely short time, reradiate the excitation energy by making downward transitions to both F=1 and F=0 hyperfine states. Since once an ion reaches the F=0 state it no longer responds to the pumping light, eventually all the ions in the trap will end up in that state, creating the desired difference in populations between the two hyperfine states.

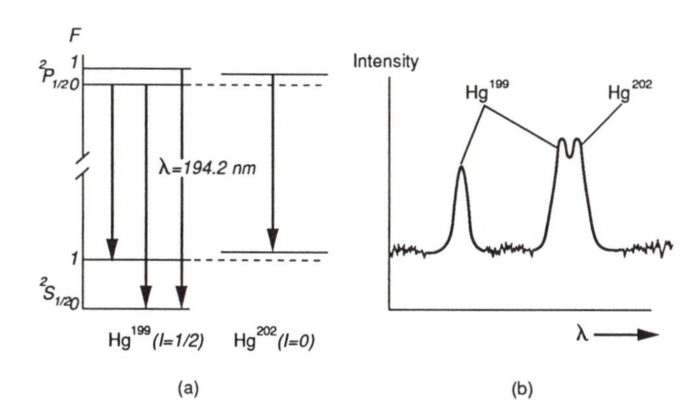

Figure 13.2 (a) The resonance transitions in mass 202 and 199 mercury ions (b) the observed hyperfine components of 194.2 nm resonance line

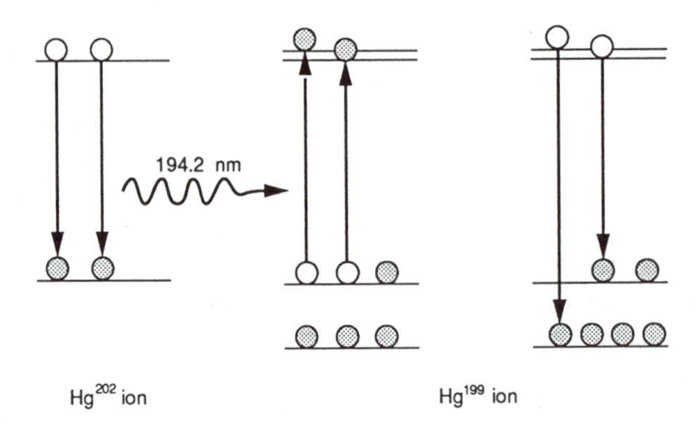

Figure 13.3 The quantum transitions involved in the optical pumping of Hg^{199} ions hyperfine populations using a Hg^{202} lamp

13.3.2 Design of the Lamp

To anticipate the requirements the lamp must meet, we must consider several aspects: First, the intensity required to pump the ion population at a sufficient rate, and the related question of the spectral linewidth of the 194.2 nm light output; second, the tolerable proportion of mercury 199 light and other undesirable wavelengths in the output; and finally, the long- and short-term stability in the light output. We will treat the first question here and deal with the others later in the chapter.

Since we are assuming that the pumping light spectral profile overlaps and is therefore resonant with that of the absorption line in the mercury 199 ion, the process is the same resonance fluorescence we encountered in the Rb standard. In order to estimate the required light intensity from the lamp, we need to relate the intensity of the pumping beam to the probability per unit time that an ion will absorb a photon from it. This requires a knowledge of the strength of interaction between the electrons in the ion and the resonant ultraviolet light. As we saw in dealing with the identical question with respect to the optical pumping of Rb, this we can obtain from the process inverse to absorption, namely emission, since the ion-dependent part of the interaction is the same for both radiation processes. In fact, we recall that stimulated emission and absorption are treated on an equal footing in quantum theory; and spontaneous emission can be thought of as emission stimulated by vacuum "zero point" oscillation in the light field. We further recall that the probability per unit time A_{nm} for spontaneous emission from a quantum state labeled n to one labeled m, such as is involved in resonance fluorescence, is easily obtained from the natural spectral line width $\Delta\nu_n$; in fact, $A_{nm}=2\pi\Delta\nu_n$. The natural line width is a measurable spectroscopic quantity obtainable through high-resolution studies under conditions of small Doppler and other types of spectral line broadening. If $B_{mn}\rho_\nu$ represents the probability per unit time of the inverse process, namely, the absorption from a resonant radiation field of spectral density ρ_ν (that is, energy density per unit frequency range), then we know that the ratio A_{nm}/B_{mn} does not depend on the ion but is a universal constant, given by the Einstein relationship

$$\frac{A_{nm}}{B_{nm}} = \frac{8\pi h\nu^3}{c^3}. \qquad 13.1$$

If the ions are irradiated by a parallel beam of resonant radiation of intensity I_ν, so that $j_\nu = I_\nu/h\nu$ is the spectral flux density of photons, that is, the number of photons crossing unit area per second per hertz, then the probability per unit time of an ion absorbing a photon can be estimated from the following useful result:

$$\frac{1}{\tau_p} = \frac{\lambda^2}{4}\Delta\nu_n j_\nu. \qquad 13.2$$

The form of this result suggests that $\lambda^2/4$ is the cross section that the ion presents to the beam in its resonant response, a cross section that curiously reflects the wavelength of the radiation rather than the physical size of the ion. This fact has an important implication for the use of the scattering of a beam of particles as a method of probing the state of other particles: Resonance scattering can have a far greater cross section, in our case where $\lambda^2=3.7\times10^{-10}$ cm^2, compared with (say) spin exchange, whose cross section is on the order of $\pi(2r_{atom})^2 \approx 10^{-15}$ cm^2.

To gain some appreciation of the numerical scale of the quantities involved, particularly the required light output from the pumping lamp, assume that $\tau_p=0.1$ sec is the average time between photon absorptions by a given ion, and $\Delta\nu_n=3\times10^8$ Hz; we find a required photon flux density of $j_\nu=3\times10^6$ photons/m^2·sec·Hz. For a typical beam geometry and spectral width of the lamp output, this leads to a required beam power in the desired spectral line of around 3×10^{-7} watt. A more useful way to specify the light intensity is to give the photocurrent that would be obtained from an ideal photocell; in the present example this figure is 5×10^{-9} amp. Of course, this represents only a small fraction of the total output of the lamp—the light output of any real lamp will contain many other unwanted wavelengths. Depending on the "speed" of the optics used to form the beam and the efficiency of the lamp, the input power will usually be on the order of several watts. The ability to project the output of the lamp into the trapping region is limited fundamentally because the lamp output comes from an extended surface, and any attempt to reduce the beam diameter with some form of optics inevitably leads to an increase in the angular spread of the rays making up the beam: This is a consequence of the *brightness theorem*, which we have already encountered. Since the acceptable solid angle spanned by the rays entering the trap and the cross section of the beam they define are limited by the electrode geometry, a consequence of the theorem is that the amount of light that can be made to pass through the trap depends only on the brightness of the surface of the lamp. This is a far more significant statement than it may appear: It means that there is a fundamental limit set by the lamp brightness to the benefit one can achieve by "improving" the optics. Furthermore, the smaller the solid angle and beam cross section defined by the electrodes, the lower is that limit. In the past it was an all-consuming pursuit of researchers to overcome the limitations of conventional light sources; now the challenge is to design a suitable laser pumping source, which, when successful, can totally eliminate such difficulties.

13.4 Detection of Microwave Resonance

The optical hyperfine pumping technique, as we have observed before, not only produces the necessary difference in populations of the hyperfine states, but also allows that difference to be monitored, thereby permitting the resonant transitions induced by a microwave field to be detected as a function of the frequency of that field. Unlike the Rb standard, however, in which hyperfine pumping is also used, the detection of resonance is not made through the change in the intensity of the transmitted pumping beam, but rather through the fluorescence, that is, reradiated light. This alternative mode of detection is dictated by signal-to-noise considerations. The relatively small number of ions in the trap means that a given photon in the beam has a very small probability of being absorbed by an ion; consequently, the total number of photons passing through the trap is little affected whether the ions are in an absorbing hyperfine state or not. Even under ideal conditions where the noise is predominantly the fundamental shot noise, the small change in the transmitted light due to resonant hyperfine transitions would be swamped by that noise. In a real system the situation is even worse, since there may be residual instabilities in the lamp output and other sources of noise.

The fluorescent UV light on the other hand, which is radiated in all directions (but not necessarily isotropically) can be detected with enhanced signal-to-noise ratio because photons which are not absorbed and reemitted as fluorescence, are discriminated against by suitable optics and do not contribute as much to shot noise, or to any other noise originating from the lamp. Wide-angle optics are needed to concentrate the radiated fluorescence onto the finite area of the detector. If the number of ions in the trap contributing to the fluorescence is N, then N/τ_p is the average number of fluorescent photons radiated per second into a 4π solid angle, and hence if the optics accept a solid angle Ω, the number of photons reaching the detector would be $N\Omega/4\pi\tau_p$. An experimentally important quantity is the ratio of the number of fluorescent photons radiated per second to the total number of photons in the beam passing through the trap per second. Using the expression we have found for $1/\tau_p$, this ratio is as follows:

$$\frac{I_D}{I_0} = N\frac{\lambda^2}{4A}\frac{\Delta\nu_n}{\Delta\nu_L}\frac{\Omega}{4\pi}, \qquad 13.3$$

where $\Delta\nu_L$ is the spectral width of the light from the lamp and A is the cross-sectional area of the beam. If, for example, we set $N=10^6$, $A=10^{-4}$ m^2, $(\Delta\nu_n/\Delta\nu_L)\approx 0.1$, $\Omega/4\pi=0.01$, we find that the ratio is very small, amounting to less than 10^{-7}. This shows the severity of the problem of adequately reducing the amount of spurious light scattered from electrode surfaces or the vacuum shell and preventing it from reaching the detector. In the ideal case, where the dominant type of noise present is shot noise, the signal-to-noise ratio for a fluorescent photon count of n photons is $\sqrt{n}$. If this is accumulated at a constant rate over a time interval T, the signal-to-noise improves with counting time as $\sqrt{T}$. In reality, there will always be other sources of noise, particularly from background radiation when a conventional lamp is used. The other important source of noise is the *dark current* in the photomultiplier tube used to count the photons. This noise problem was expected to be minimized by the choice of the mercury ion with its resonance line well in the ultraviolet region of the spectrum. This circumstance allows the use of photomultipliers (so-called *solar-blind* tubes) whose cathodes are sensitive only to UV and shorter wavelength photons, and not to the light from the hot cathode or other ambient light.

13.5 Microwave Resonance Line Shape

Like the Cs beam standard and the Rb gas cell standard, the mercury ion standard is a passive compact resonator, responding sharply to a particular frequency of excitation by an externally provided microwave field. Its compactness makes it akin to the Rb resonator, but a resonator without the perturbing effects of collisions with a background gas of uncertain composition, and one in which the frequency shifts induced by the pumping light can far more easily be eliminated. Since the ions can be contained in a space comparable to the wavelength of the resonant microwave field, the conditions exist for the Dicke effect to be observed, a Doppler-modifying effect of great importance to the Rb sand H-maser standards. We recall that the Doppler effect, which generally is a source of broadening in optical spectra, has a radically different effect on the observed spectrum when the particles are not free to move distances much greater than the wavelength. We recall that under these conditions, a moving particle interacting with a radiation field sees a frequency-modulated field with a small index of modulation, and consequently one whose Fourier spectrum has a large amplitude at the undisplaced frequency and a number of sidebands. The resonance spectrum then will have a central sharp line, free of first-order Doppler broadening; however, its shape and the displacement of the center frequency due to the second-order (relativistic) Doppler effect depend on the velocity distribution of the ions. It is reasonable to assume that the ion distribution center in the trap remains fixed as the individual ions collide with other particles and eventually reach the electrodes and are neutralized; hence there should be no first-order Doppler shift. To the extent that this is true, the microwave field need not be stationary; the phase need not be constant over the ion distribution. This simplifies the technical problem of providing the resonant microwave field; all that is necessary is to irradiate a suitable hole in an electrode with the desired microwaves from a horn.

In addition to the central undisplaced frequency, the oscillatory motion of the ions will result in the resonant field seen by the ions being frequency modulated because of the Doppler effect. However, since their motional spectrum is discrete, the frequency modulation of the field is at discrete frequencies, and therefore its Fourier spectrum will consist of well separated discrete frequencies, called the *Doppler sidebands.* It is interesting that in introducing his analysis of the way the Doppler effect is manifested in the spectra of gases in thermal equilibrium, Dicke begins with the example of a particle constrained to move back and forth at constant speed between two fixed points and considers the way the spectrum depends on the distance between the points in relation to the wavelength; then he proceeds to consider the random motion of atoms in thermal equilibrium. In our present case the confined ions in fact are more like his introductory example; their oscillations contain theoretically an infinite series of frequencies, but under actual operating conditions only three dominant frequencies. In general, β is not a rational number (expressible as a ratio of two whole numbers), and therefore strictly speaking, the motion is not periodic. Nevertheless, the motional spectrum is discrete, with frequencies $\beta\Omega/2$, $\Omega-\beta\Omega/2$, and $\Omega+\beta\Omega/2$. Now, if we assume that the microwave field can be resolved into plane waves traveling in random directions, then the spectrum of one such plane wave as seen by an oscillating ion will, on account of the first-order Doppler effect, consist of the undisplaced frequency of the wave ν_0 and Doppler sideband frequencies at $\nu_0 \pm l\omega_r$, $\nu_0 \pm m(\Omega-\omega_r)$, $\nu_0 \pm n(\Omega+\omega_r)$, etc., where $\omega_r=\beta_r\Omega/2$, and l, m, n are integers. Similar frequency components arising from the axial motion are also present. The relative amplitudes of these frequency sidebands depend on the distribution of amplitudes (or energy) of the ions and the spatial distribution of the field. Naturally, it is desirable that the central, undisplaced frequency have the dominant amplitude with only weak sidebands; this will occur if the amplitudes of the ion oscillation are not much larger than the wavelength. Apart from this requirement, the presence of the sidebands may be ignored; the main concern is the central line and its frequency width. An important contribution to this width is the second-order Doppler effect, which varies as $(V/c)^2$ and, as we have seen in an earlier chapter, requires the special theory of relativity to give the correct expression for it. It is not difficult to estimate the size of the second-order Doppler broadening of the central line if we assume that it is legitimate to apply the relativistic Doppler formula, which is derived for uniform motion, to an oscillating ion. In that case, for example, the instantaneous frequency of the field seen by an ion oscillating at an angular frequency ω with amplitude A is given by:

$$\nu = \nu_0 \left[1 + \frac{V}{c} \cos\alpha + \frac{1}{2}\left(\frac{V^2}{c^2}\right) + \cdots \right], \qquad 13.4$$

where the velocity $V = A\omega\cos(\omega t)$, and α is the angle between the direction of the wave and the line of oscillation of the ion. The second-order Doppler shift can be written as a fractional shift amounting to $\langle E_k \rangle_{ave}/Mc^2$, that is, the mean kinetic energy divided by Einstein's rest mass energy Mc^2. In the case of ions in a Paul trap operating with $a \ll 1$, $q \ll 1$, the average total kinetic energy is, as we have seen, constant throughout the ion motion, merely alternating between the high-frequency jitter and the low-frequency oscillation. For a mass 199 mercury ion having an average kinetic energy of 2.5 electron volts, the fractional frequency shift is about 1.4×10^{-11}. This is evidently a significant effect, which must be reduced by using lower-energy ions; otherwise, a detailed knowledge of the energy distribution of the ions must be found, on which a theory of the line shape can be based. The first is, of course, to be preferred. It has become possible through the availability of laser pumping sources not only to greatly improve the signal-to-noise ratio, making it possible to work with the smaller number of ions that a shallow trap can contain, but actually to cool the ions, as we shall see in the next chapter.

The alternative approach of determining the energy distribution of the ions is also possible in principle. It can be done by making use of those same Doppler sidebands we were assured could be ignored as not affecting the central frequency; their relative amplitudes, in fact, contain all the information needed to determine the amplitude distribution of the ions, if the latter move in a resonant field of known pattern. For example, the trap electrodes can be designed to support one of the stationary modes of oscillation of the microwave field, that is, to serve as a cavity resonator. This approach is made practicable by the fact we have already mentioned, that the electrode geometry can be simply a right circular cylinder with planar end caps. This supports well-known resonant modes, among which a convenient mode with the right polarization for inducing the desired transition in the trapped ions is the TE_{011} mode. Given the spatial dependence of the field in such a mode, it is possible to compute the amplitudes of the sidebands for each assumed amplitude and frequency of oscillation of an ion and ultimately to use the observed amplitudes to find the amplitude distribution. Armed with this the second-order Doppler correction to the resonance line shape can be carried out.

13.6 The Magnetic Field Correction

Of the two field configurations we discussed in the last chapter for confining ions, namely the static Penning trap and the high frequency Paul trap, the latter is better suited in cases where the resonance is in magnetic transitions between quantum states sensitive to a magnetic field, as is typical of the microwave standards. The reason is that unlike the Penning configuration, it does not require an intense magnetic field for stable confinement in three dimensions, a field that would displace the resonance frequency. On the other hand, as with all standards based on the magnetic hyperfine transition, there is a need to prevent any significant overlap between the desired transitions between the $m_F=0$ sublevels and those involving the magnetic field-dependent $m_F=\pm1$ sublevels; this dictates that the ions be immersed in a *weak* uniform magnetic field. Such a field has, as we saw in the last chapter, a predictable effect on the ion motion. For a massive ion such as that of mercury, the Larmor frequency $\omega_L=eB/2m$ in a field such as 10^{-4} tesla (1 gauss) is very small, amounting to only 25 rad/sec, or about 4 Hz. This means that $\omega_L/\Omega \ll 1$ and the effect on the trapping parameter a_r is negligible; in any case, an axial magnetic field tends to enhance radial confinement and has no effect on the axial motion. Moreover, its presence is beneficial in collimating an axial electron beam, if such is used to produce the ions from the parent atoms.

However, the magnetic field will displace the reference frequency. But because the hyperfine frequency for the Hg^{199} ion is so much larger than, for example, the hydrogen value, the magnetic field correction will be far smaller. This follows from the Breit–Rabi formula, which in this case leads to the following:

$$\nu = \nu_0 + 9.7\times10^9 B^2\,, \qquad 13.5$$

where B is in teslas. Since the applied field strength need only just exceed the residual inhomogeneity of the magnetic field over the space occupied by the ions, the correction is below the order of 10^{-2} Hz. As usual, the value of the magnetic field intensity B can be determined by observing the field-dependent transitions involving $\Delta m_F=\pm1$ and $\Delta F=1$, whose frequencies are given with sufficient accuracy by

$$\nu_{m=\pm1} = \nu_0 \pm 1.4\times10^{10} B + 9.8\times10^9 B^2\,. \qquad 13.6$$

For a field of 10^{-7} tesla these frequencies lie above and below the zero field hyperfine frequency by around 1.4 kHz.

13.7 The Physical Apparatus

Having dwelt at some length on the principles underlying the operation of the mercury ion standard, we will now consider some of the salient experimental problems.

13.7.1 The Quadrupole Electrodes

For the Paul quadrupole field to be established, the electrodes need not be hyperbolic, as has been assumed in the theory mainly to enable an analytical solution to the ion motion. In fact, for almost any electrode geometry having the proper symmetry, the field distribution approaches the pure quadrupole form as we approach the center. We may therefore choose a right circular cylinder with planar end caps with arbitrary geometric proportions. However, the detailed dimensions, particularly the diameter of the cylinder in relation to the distance between the end caps, will greatly affect the detailed field distribution. It is to be expected that there is an optimum choice of proportions that will lead to a field most closely approximating the Paul field over the largest space. Naturally, the parameters r_0 and z_0 that appear in the specification of the Paul field no longer directly apply to the physical dimensions of the cylindrical cavity. In the particular case of the cylinder and plane end caps forming a closed cavity, the problem of finding the field distribution can be solved analytically in cylindrical coordinates. If the cylinder is assumed to have a potential $V=0$, and for the end caps $V=V_0$, the field distribution can be expressed in the form of a series expansion using certain cylindrical functions as base functions. We will not pursue this any further beyond mentioning two significant results concerning the cylindrical electrode geometry: First, the particle motions in the r- and z-directions are no longer independent, as they are in the true quadrupole field; this makes for a far more complex motion. Second, for a cylindrical trap of radius r_1 and half-length z_1 having the ratio r_1/z_1 the same as the Paul trap ($\sqrt{2}$), the approximate quadrupole field in the vicinity of the center is as if the applied voltages had been reduced by a factor of 0.55 and applied to hyperbolic electrodes with $r_0=z_1$ and $z_0=z_1$.

Figure 13.4 shows schematically the general arrangement of the parts that make up the mercury ion resonance apparatus. In the original NASA design the electrodes were stainless steel, precisely machined to the proper hyperbolic geometry with $r_0=1.13$ cm and $r_0/z_0=\sqrt{2}$. The cylinder is drilled with diametrically opposite holes to admit the pumping light and the resonant microwaves. Similarly, one end cap is slotted to allow the fluorescent light to pass to the detection system, while the other end cap has a hole to admit the ionizing electron beam. UV grade fused quartz or sapphire windows are incorporated in the vacuum shell for the transmission of the pumping light and fluorescence. The vacuum is maintained in the ultrahigh vacuum region below 10^{-7} Pa with an ion pump. The 83% enriched mercury 199 is most conveniently obtained in the form of the oxide HgO, and the mercury is reclaimed by heating the HgO *slowly* to break down the compound into oxygen gas and metallic mercury (remember Lavoisier?). The copious volume of oxygen must be pumped away and the pure mercury condensed in a side arm, whose temperature can be controlled to control the vapor pressure. Since the equilibrium vapor pressure is relatively high at room temperature (10^{-1} Pa), the problem is not to generate a sufficient density of atoms in the trapping region but rather to keep it under control at the desired low pressure.

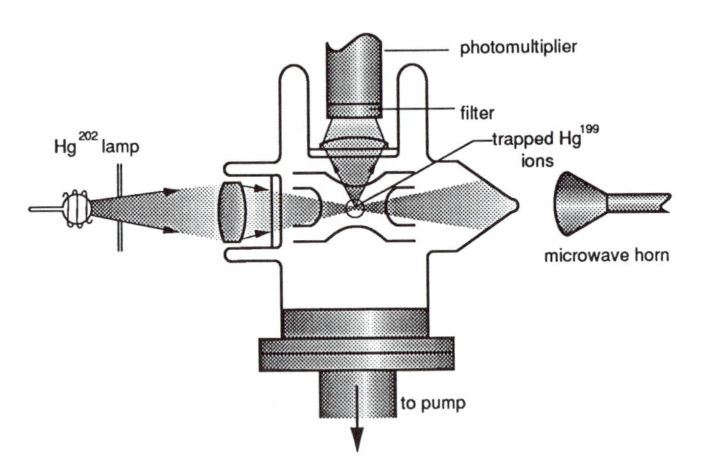

Figure 13.4 The original NASA Hg^+ ion microwave resonance system

The electron gun essentially consists of a thermionic electron-emitting cathode and a simple cylinder forming an electrostatic lens, which can also serve as a control electrode to vary or stop the electron beam from reaching the trap. For experimental purposes the most convenient form of electron emitter is a "dispenser" heated cathode, which allows it to be reactivated after exposure to the atmosphere. It typically consists of oxides of various alkaline earth elements, usually barium and calcium, which impregnate a porous tungsten plug in the end of a molybdenum cylinder surrounding a heater wire insulated with a high-temperature material such as alumina. Such a cathode must be activated under high vacuum by raising the temperature to around 1150°C (bright red) to cause pure barium to diffuse to the surface. During this process an electric field is established between the cathode and the control electrode to draw electrons from the cathode, producing an electron current that gradually rises to a saturation value, beyond which it no longer increases. Saturation currents of several hundred milliamperes are obtainable from a 3 mm diameter cathode. At the normal operating temperatures around 1000°C, the light emitted, although mostly at a wavelength for which the photomultiplier is "insensitive," nevertheless can add to the background photon count unless precautions are taken to avoid it. Fortunately, it is easy to separate the electrons from the photons emitted by the cathode simply by using electrostatic fields to deflect or focus the electrons onto a small aperture, leaving the photons to continue in their path towards an absorbing surface. The electron gun is usually mounted coaxially with the trap behind one of the end caps, which has a small aperture at its center for admitting the electrons into the trapping region.

13.7.2 The Mercury 202 Lamp

Perhaps the most critical hardware problems center on the Hg^{202} lamp, optimizing its output at the desired wavelength, and the wide-angle optics necessary to detect the fluorescence from the ions while suppressing spurious background radiation. The lamp is generally a fused quartz (UV grade Supersil) sphere typically 1.5 cm in diameter with a cylindrical hollow stem for ease of mounting and control of the mercury vapor pressure through the temperature of the condensed mercury in it. The 74% enriched Hg^{202} isotope, like the Hg^{199}, is obtained in the form of the oxide. After the quartz lamp blank is thoroughly baked out at 500°C under high vacuum, the mercury is introduced by gently heating the HgO to cause it to break down into oxygen gas and mercury. The gas is pumped away while the mercury is condensed in a cold trap, from which it is distilled into the lamp. The successful operation of the standard depends critically on the lamp providing an intense, stable output at the 194.2 nm resonance wavelength, with a manageable amount of the strong line at 253.7 nm emitted by the more numerous neutral mercury atoms. Ordinary mercury vapor lamps in fact emit this latter wavelength with great efficiency; and do so in spite of the fact that it is classified spectroscopically as an "intercombination line," which, according to one approximation, is forbidden. General Electric and other companies have long exploited this fact in the design of the highly efficient fluorescent lamps; it is the UV emitted by mercury that causes the phosphor lining of the lamp walls to emit that ghastly glow. The presence of almost any background gas, whether purposely introduced or such as might evolve from the walls of the lamp, tends to favor the emission of this 253.7 nm line over the desired ionic line. This is unfortunate, since the presence of a "carrier" gas, typically a noble gas used in metal vapor lamps, tends to stabilize the lamp and make it far easier to ignite. As a source of the ionic line at 194.2 nm, the lamp is most brilliant when operated in an *arc* mode in high vacuum. Like the Rb lamp, the mercury lamp is excited by an electrodeless discharge in a UHF coil, but with much greater power, typically 25 watts. The sole drawback to the vacuum lamp is the formation after a few hours of operation of a grayish deposit on the inner surface, causing a deterioration in the output intensity at 194.2 nm. Even with a vacuum lamp, the emission at 253.7 nm can still degrade the fluorescence signal from the trapped ions, since it contributes nothing to the signal but simply adds noise through spurious scattering from other parts of the apparatus. The ionic line is favored by high vapor density, presumably because of the "imprisonment" of the 253.7 nm radiation by repeated absorption and reemission by neutral mercury atoms; the ion density remains well below the point where multiple absorptions can take place. To further reduce the relative intensity of the 253.7 nm line, an external neutral mercury vapor filter, consisting of a long cylindrical absorption cell operating around 200°C, can be used.

13.7.3 Detection of UV Fluorescence

The suppression of stray scattered UV pumping light requires that very sharp edged apertures be used in selecting the pumping beam, with the window leading into the vacuum chamber as far removed as possible from the ion region and made of high-quality UV grade quartz. After passing through the ion region, the beam must be totally absorbed, a problem classically solved with a *Wood's horn*, a curved and gradually constricted tube lined with an absorbing material. Naturally, any such material used in the vacuum chamber must be compatible with maintaining ultrahigh vacuum; a material particularly useful not only as an absorber of light but also mercury vapor (through the formation of an amalgam) is *gold black* (sic). This is very finely divided gold deposited by vaporizing gold in an inert atmosphere to form a "cloud," which results in a thin layer of fine gold particles being deposited on all surfaces. It is an efficient light absorber and hence looks black simply because on a microscopic scale the surface is like a dense forest, and light falling on it has little chance of escaping.

The fluorescence light detector is typically a solar-blind photomultiplier with a Cs–Te photosensitive, semitransparent end-on cathode. Such a photomultiplier may have a cathode quantum efficiency approaching 10%; that is, one in ten photons falling on the cathode results in one electron being ejected and accelerated to the first of perhaps fourteen dynodes. The material of the dynode (often beryllium–copper) is such that the impact of the electron causes several secondary electrons to be ejected, which in turn are accelerated to the next dynode; thus multiplication in the electron number occurs at each stage by this process of secondary electron emission. Of course, not every time is precisely the same number of secondary electrons emitted; hence there will be some statistical fluctuation in the current gain. However, this is not an additive noise that could swamp a small signal; rather it is a fluctuation in the multiplication factor, and the smaller the number of photoelectrons emitted by the cathode, the smaller will be the output fluctuation. A far more important source of noise in photomultipliers, and one that does set a limit on their ability to detect small changes in the number of photons falling on their cathode, is the *dark current* and its attendant shot noise. Even when a photomultiplier is completely shielded from all external sources of radiation, if it operates at finite temperatures, there will inevitably be some random thermal "evaporation" of electrons from the cathode surface. The extent to which this occurs depends on the amount of energy required to free an electron from the surface, relative to the mean energy of thermal excitation, which of course is proportional to the absolute temperature. One of the important advantages of the mercury ion is that the fluorescence photons, being in the UV region of the spectrum, have a relatively high energy, and therefore the cathode material can be chosen to have a large "work function," that is, energy to free an electron from the surface with very low dark current at ordinary temperatures. Furthermore, as already noted, such a cathode material will be insensitive to smaller energy photons in the visible region of the spectrum. Experimentally, this is of critical importance, since it makes the detector quite insensitive to ambient room light, but more especially to the glow from the heated cathode of the electron gun used to produce ions in the trap. Further discrimination against 253.7 nm and other visible light is achieved by placing in front of the photomultiplier an interference filter with a pass-band centered on 194.2 nm and blocked against the visible region of the spectrum.

13.8 Hg^+ Ion Frequency Standard System

13.8.1 Choice of "Flywheel" Oscillator

The system for generating the microwave field for inducing transitions begins with a *flywheel* oscillator having high short-term stability and low noise. This may take the form of a hydrogen maser in laboratory installations or a high-quality quartz crystal oscillator in portable units. While the detailed design of the system is obviously not unique, nevertheless there are some remarks that apply in general; for example, the desired microwave frequency is generated using coherent frequency synthesis techniques starting (say) with a 5 MHz quartz crystal oscillator. The extraordinary sharpness of the mercury ion resonance places a far greater demand on the spectral purity of the synthesized frequency-inducing transitions than, for example, the Cs standard. In view of the degradation in spectral purity that occurs in a frequency multiplication chain, a critical design problem is either to minimize this or, alternatively, optimize the operation of microwave oscillators, such as a klystron or Gunn diode, to obtain a spectrally pure low-noise output, which can be phase-locked to the synthesized crystal-based frequency. The klystron, we recall, is an electron tube based on the use of a velocity-modulated electron beam; the Gunn diode, on the other hand, is a solid-state device based on a phenomenon called the *Gunn effect,* exhibited, for example, in a uniformly doped n-type GaAs crystal. In such a crystal, the application of a steady voltage beyond a certain threshold value results in a fast pulsating current. This odd behavior finds explanation in the fact that when an applied electric field exceeds a certain threshold value the electrons responsible for conduction in this particular crystal can make quantum transitions to an upper band of quantum states characterized by an effective electron mass that is *larger* than the effective mass for the lower conduction band. The consequence of this is vaguely reminiscent of electron bunching in the klystron: As the threshold field across the diode is passed, transitions occur at the negative electrode, putting some electrons in a more slow-moving state while the others continue with a higher drift velocity. This creates a narrow region of rapid change in electron concentration called a "domain," which advances with fewer electrons in front and a "pileup" of electrons in the back. When the domain reaches the positive electrode, there is a pulse of current in the outside circuit, the electric field is again as it was before the creation of the domain, and the cycle is repeated. Depending on the transit time of the domain across the diode, the duration of the cycle can equal a period of oscillation in the microwave region of the spectrum.

The synthesis of the microwave frequency at 40.507348... GHz can be achieved in many different ways; for example, the 5 MHz from the voltage-controlled crystal oscillator (VCXO) first passes through a multiplier chain to reach the 1,620th harmonic at 8,100 MHz. This can then be combined in a crystal mixer with a signal at around 1.4696... MHz from a synthesizer driven by the same crystal oscillator, to obtain a signal at the sum of the two frequencies at 8.1014696... GHz. Finally another multiplication stage yields the fifth harmonic of that frequency, to give us the desired frequency.

13.8.2 Resonance Signal Handling

In order to observe the precise frequency at which the peak of the resonance occurs in the mercury ions requires first, that the ions be observed freely interacting with only the microwave field, and second, that an optimum scheme be designed for averaging the fluorescence signal and its correlation with the microwave frequency. The first requirement is met simply by interrupting the pumping light during the "interrogation" of the ions; this may be done either by modulating the power of excitation of the lamp or, preferably, by an electro-optic shutter, which would allow the lamp to remain operating in a stable condition. To maintain the ion population in the trap, the electron beam is also pulsed by a suitable voltage on the control electrode; this is preferred to operating with a continuous beam, albeit a weaker one, which would perturb the ions through spin exchange collisions and add spurious light from particles in the trap excited by electron collisions. At the beginning of the detection cycle the electron beam is turned on for a short time to "fill" the trap with ions, while the pumping light beam remains on for an additional period to set up the hyperfine population difference. Then the light is turned off while the microwave field of fixed frequency is turned on for a period to induce transitions between hyperfine states "in the dark"; and finally, the pumping light is turned back on to observe the degree to which transitions have occurred by counting the fluorescence photons detected by the photomultiplier. The electron gun is again pulsed on, and the cycle is repeated.

The difficulty of having a weak fluorescent signal, which is aggravated by the presence of stray background radiation and the need for using a "time bridge" to compare photon counts taken at different times to determine the extent to which the microwave field is inducing transitions, makes the statistical handling of the photon counting data particularly important. The ion number decays in time, and the initial number produced by successive electron pulses will fluctuate, giving rise to fluctuations in the intensity of fluorescence. There will also be fluctuations in the intensity of the pumping light, with possibly a longer-term drift. Clearly, each detection cycle must be looked on as an independent repetition of the same measurement, and the photon count data must be taken and reduced in a way that yields the resonant frequency of the ions, independent of the number of ions or the intensity of the pumping light. One way is to program the synthesizer to step the microwave frequency between two values, slightly above and below a center frequency, which is itself ultimately servo-controlled to lock on resonance with the ion hyperfine frequency. Just how far above and below the center frequency the two frequencies should be set is clearly determined by the criterion that the photon count be most sensitive to changes in frequency at those points, that is, the frequencies at which the resonance curve has the greatest slope. For an ideal (Lorentzian) resonance curve, such as would result just from lifetime (or "natural") broadening of the transitions, the greatest slope occurs at $\nu=\nu_0 \pm \nu_m$, where $\nu_m=\Delta\nu/2\sqrt{3}$ and $\Delta\nu$ is the width at half amplitude of the resonance line. This width being on the order of 1 Hz shows the extraordinary spectral purity implied in the microwaves that must be used, that is, freedom, for example, from residual frequency-modulation noise. In this scheme a fluorescence photon count is obtained (after transitions are induced in the dark) at the two frequencies for each detection cycle. Under ideal conditions, the two photon counts are identical only when the two frequencies are symmetric about the center of the resonance curve. By subtracting one from the other for each detection cycle and averaging over many cycles, we would have a signal that would pass from a negative value through zero to a positive value as the microwave frequencies are swept across the resonance. This is precisely the kind of error signal needed for a servo loop to phase-lock the crystal to the ion resonance, since it is zero when the frequency is just right, and it distinguishes between being too low in frequency and too high. We know, however, that as the ions slowly escape from the trap there is a slow decrease in the number contributing to the fluorescence within the period of each cycle. For times short compared to the mean ion lifetime, the rate of decrease is nearly constant, and a first-order correction is obtained by taking the second differences between the photon counts. This requires that a third count be obtained within each cycle, after the field frequency has been switched back to its initial value. Symbolically, this amounts to computing the signal as follows:

$$S=\left[N_1(\nu_1)-N_2(\nu_2)\right]-\left[N_2(\nu_2)-N_3(\nu_1)\right]. \qquad 13.7$$

Since these counts are accumulated over equal intervals of time, it follows that any linear decay in the ion number or drift in the pumping light, or even the improbable aging of the photomultiplier cathode, will result in $N_3(\nu_1)$ having twice the correction of $N_1(\nu_1)$ because of the additional delay in obtaining it. From the expression for the second difference we see that such linear variations in the photon counts will automatically drop out.

To allow for the possibility of fluctuations in the initial number of ions between different cycles, or variations in the lamp output on a time scale approximating the cycle repetition time, it would be desirable to "normalize" the counting data by dividing by the total number in each cycle.

Since the data comes in digital form, its digital processing is readily carried out: the arithmetical operations to obtain $\langle S\rangle$ and the algorithm for its averaging over the desired number of cycles. That average is converted to an analog signal, and before it is applied to the controlling element in the crystal oscillator, it must, for a stable servo-control loop, be passed through an analog integrator. Such an integrator produces a voltage ramp in the output for a constant voltage input; the slope of the ramp is ascending for a constant positive input, descending for a negative one, and flat for zero input. Needless to say, extreme stability is required of the integrator; any drift in the output voltage for a given input will cause a frequency offset from the true ion resonance.

When the servo-control loop is closed, any error signal from the photon-count computer will cause a voltage ramp to appear at the crystal oscillator control element, tending to change the oscillator frequency in the direction of reducing the error, until finally the error reaches zero and the integrator output is constant, fixing the oscillator frequency. At this point the microwave frequency that is coherently derived from the crystal oscillator is locked on to the center of the ion resonance.

If we compute the figure of merit, as previously defined for the Cs standard, namely F=(Signal/Noise)/$\Delta\nu$, assuming that the noise is solely due to the photon shot noise, we find the following:

$$F = \frac{(N_r - N_0)}{\sqrt{(N_r - N_0)}\Delta\nu}, \qquad 13.8$$

where N_r, N_0 are respectively the photon counts at resonance and far off resonance (background counts). This is an important figure, since in fact the mean fractional deviation in the frequency is proportional to $1/F$; thus, in terms of the total counting time τ we can write

$$\sigma(\tau) \approx \frac{\Delta\nu}{\nu_0}\frac{\sqrt{R_r + R_0}}{R_r - R_0}\frac{1}{\sqrt{\tau}}, \qquad 13.9$$

where the R's are the photon counting rates.

With the potential of realizing resonance line widths on the order of 0.1 Hz and a signal/noise ratio of 10^3 for say a 100 second averaging time, the mercury ion standard promises a fractional deviation in frequency of only 2.4 parts in 10^{15}, equaling the stability of the hydrogen maser, with the potential for matching the absolute accuracy of the Cs standard. All this in a device that would fit in a shoebox, the attribute that first motivated the experiments at NASA. In recent years it has come under commercial development, and prototype models have been manufactured and subjected to long-term comparison with other time standards at the U.S. Naval Observatory. The commercial models use helium at low pressure as a background gas in the quadrupole trap to produce a viscous drag on the ion motion, thereby tending to slow it down and reduce the second-order Doppler broadening of the resonance. As noted in the original NASA report, since the helium atom is only 1/50 the mass of the mercury ion, the motion of the latter is affected only in small increments when collisions occur with the helium atoms. Not every collision results in a loss of kinetic energy, even if the helium gas is much colder, but over a complete period of oscillation of the ion there will be an average transfer of a small fraction (proportional to the mass ratio) of the ion kinetic energy to the helium gas, where it is degraded into heat. Unfortunately, the use of helium compromises the whole idea of completely isolating the atomic system responsible for setting the frequency standard. Although the density of helium needed to cause an appreciable cooling of the ions is small (typically no more than 10^{-3} Pa), nevertheless, at the level of accuracy being contemplated here (parts in 10^{15}) we should expect it to cause a significant (positive) shift in the hyperfine frequency. But perhaps more importantly, since the collision rate depends on the temperature of the helium gas, it makes the ion frequency susceptible to changes in the environment. Although it has not so far been attempted, it may not be impracticable to admit the helium in the form of a beam that can be programmed to be shut off and the remaining gas rapidly pumped out of the trapping region prior to the "interrogation" of the ions during the observation cycle.

In spite of the helium compromise, a decision based on the desire to develop a unit with the best possible performance consistent with capturing a broad market, the commercial prototypes have exhibited impressive performance. This is accomplished in part by the care taken in stabilizing the ambient conditions of the units, made necessary by the use of the gas to cool the ions. To circumvent this requires some other method of narrowing the energy distribution of the ions; a method has become possible through the availability of highly stable tunable lasers.

The development of the laser, which we shall take up in the next chapter, brought the fulfillment of the promise of isolated ions as the basis for the ultimate frequency standard closer to reality. Laser technology brought immediate, powerful solutions to both experimental difficulties encountered in the mercury ion standard: First, the degradation of the signal/noise ratio due to stray scattering of the optical pumping beam, and second, the broad energy distribution of the ions, leading to a (relatively) large second-order Doppler broadening of the resonance. The two outstanding attributes of laser light, namely directionality and spectral purity, are precisely those needed for achieving the optimum signal/noise ratio for the resonance fluorescence signal from the ions. The brightness of a laser source is incomparably greater than a conventional lamp, even a mercury lamp operated in a brilliant "arc" mode, as is done in the mercury ion standard. This means that radiant energy can be projected in a laser beam with such small divergence that it has become proverbial.

In designing the optics for the pumping beam it greatly simplifies matters that the intensity profile over the cross section of a laser beam falls off very rapidly with distance from the central axis. Furthermore, any beam-defining apertures that are used can be very far removed from the ions, without the severe loss of beam intensity that inevitably occurs with a conventional source.

The remarkable power and elegance with which laser technology removed the severe difficulties that had to be overcome initially to observe the spectra of isolated ions illustrates that perhaps Shakespeare's "There is a tide in the affairs of men..." applies equally to the conduct of research!

Chapter 14
Optical Frequency Oscillators: Lasers

14.1 Introduction

The first international conference at which papers were presented on the subject of "optical masers," as they were then called, was at Ann Arbor in June 1959. The principal topic of the conference was not lasers, but the optical pumping method of observing magnetic resonance in free atoms, a technique that had recently been introduced by Kastler in Paris. The session devoted to lasers was a "miscellaneous session" in which papers on theoretical aspects of laser oscillation in gas discharges and ruby crystals were presented by Gould, Javan, and Schawlow, among others.

The workings of a laser do not involve any physical theory or even practical technique that was not already known for some time. The quantum theory of the absorption and emission of light by atoms and molecules was well established, there was abundant spectroscopic data such as wavelengths and line intensities, and the theory of light wave optics and the techniques of optical interferometry were well advanced. The study of electrical discharges through rarefied gases and crystal optics had been pursued since the 19th century. This undoubtedly explains the veritable explosion that occurred in the number of reports of laser action once the first appeared.

The special properties of lasers as quantum oscillators in the infrared and optical regions of the spectrum will now be treated in the broader context of quantum oscillators in general. Their impact on the design and performance of the microwave standards (except the H-maser), which has literally transformed these standards, and their development as frequency standards in the infrared and optical regions of the spectrum will be treated in succeeding chapters.

14.2 The Resonance Line Width of Optical Cavities

We have seen in the case of atomic and molecular beam masers that the spectral line width of the resonant response of the atoms or molecules is far sharper than the resonant modes of the cavity in which the particles interact with the radiation field. Since these cavity modes are well separated, this means that oscillation occurs in a unique mode of oscillation of the field in the cavity: the one tuned to resonance with the particle frequency.

In contrast, at optical frequencies it is the "cavity," which may in fact consist of only two small parallel mirrors some distance apart, whose resonance modes have narrow spectral widths compared to those of the "active" atoms or molecules. The arrangement of two highly reflecting, precisely parallel surfaces to form an optical resonator has as precursor a high-resolution spectroscopic device called a *Fabry–Pérot* interferometer, whose introduction in 1899 far predates the era of modern optics. Since it was originally applied to conventional light sources with limited coherence, the spacing of the mirrors was typically on the order of millimeters, rather than centimeters or tens of centimeters as is typical of gas laser cavities. Strictly speaking, such an open Fabry–Pérot arrangement of mirrors does not have discrete resonance frequencies, unlike a completely closed cavity with reflecting walls. However, detailed computations on such a cavity have shown that there exist more or less discrete sets of modes with the optical field localized along the axis between the two mirrors. For these modes the loss of optical energy from the cavity due to diffraction, the inevitable spreading out of a wave, is small. We can readily show this is plausible if we accept the result from wave theory that a plane wave reflected by a circular plane mirror of diameter D (large compared to the wavelength) will be diffracted at an angle of about λ/D (radians), as shown in Figure 14.1. Such a wave traveling to another similar mirror a distance L away will partly fall outside the rim of that mirror and suffer a fractional loss of $4\lambda L/D^2$, provided that we make the crude assumption that the energy of the beam is spread uniformly over the expanded area. Clearly, even if the mirrors were perfectly reflecting, there would still be a decay in the cavity field energy due to diffraction, with a consequent broadening of the resonance spectrum. The fractional loss of $4\lambda L/D^2$ occurs at each mirror and repeats at intervals equal to the transit time of the light wave between the two mirrors, namely L/c, where as usual, c is the velocity of light. It follows that the average fractional loss of energy per unit time is $4\lambda c/D^2$. Now we can apply this result to the two counter-traveling waves of equal amplitude whose sum is one of the stationary axial modes, belonging to the quasi-discrete set supported by the cavity. These longitudinal modes are analogous to vibrations on a string at frequencies such that the phase of the wave after traveling $2L$ is a whole number of cycles: Symbolically, this means that $(2L/c)\nu_n=n$, where n is a whole number, and ν_n is the frequency of oscillation of the optical field in the nth mode. Then the Q-factor for that mode is as follows:

$$Q_n = \frac{2\pi\nu_n}{\left(\frac{1}{E}\frac{dE}{dt}\right)} = \frac{\pi D^2}{2\lambda_n^2}, \qquad 14.1$$

where $\lambda_n = c/\nu_n$ is the wavelength of the nth mode. For example, if we take for the mirror diameter D=2.5 cm and λ=632.8 nm (the common wavelength of the He–Ne laser output), we find for that pure axial mode $Q \approx 10^{10}$. This is significantly larger than is typical of microwave cavities and is even higher than most optical atomic transitions. Actually, the situation is far more complicated, in that the intensity profile of the beam is far from uniform over the cross section of the beam; rather, it has a radial distribution that can be analyzed in terms of certain radial modes. These are designated as TEM modes (transverse electromagnetic) with indices specifying the order and hence the number of zeros in the intensity distribution. For example, if we take the mirror axis as the z-axis of a coordinate system, then in the TEM_{21} mode, the field intensity has two zeros along, say, the x-direction and one zero along the y-direction. Beam profiles for some of the lower-order radial modes are shown in Figure 14.2. The least complicated mode (TEM_{00}) has only one maximum, which occurs on the axis and is described by the form $\exp(-r^2/r_0^2)$, called a Gaussian function.

The output beam of a laser oscillating in this mode is called a Gaussian beam, and the theory of the action of optical elements such as lenses and mirrors on such beams is called Gaussian optics. It differs from ordinary ray optics, which takes no account of the wave nature of light; it is characterized by the absence of sharply defined focal points and beam profiles. The usual lens and mirror formulae of ray optics are not valid.

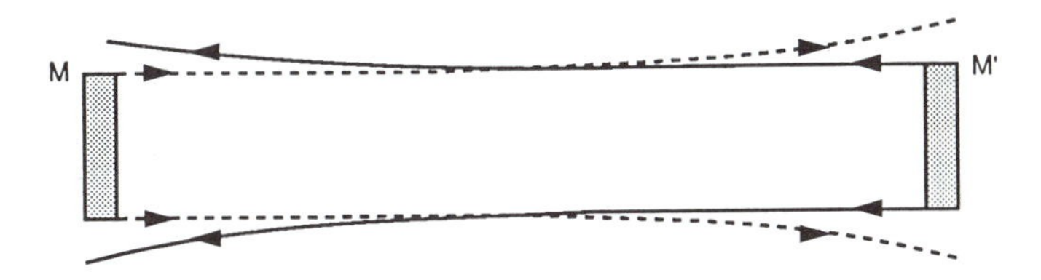

Figure 14.1 Diffraction loss at the mirrors of an optical cavity

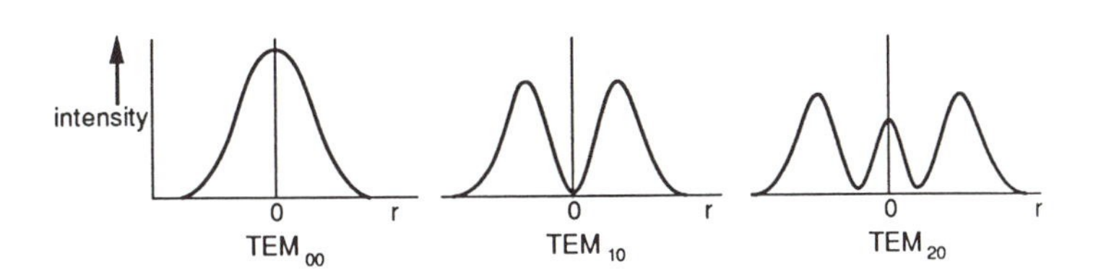

Figure 14.2 The intensity distribution for some low order-radial modes in an optical cavity

Optical wave theory shows that the diffraction loss is radically smaller if instead of plane mirrors, concave mirrors are used in a *confocal* arrangement, in which the focal points of the two mirrors coincide at the midpoint between them, as shown in Figure 14.3. The results of detailed computations on the diffraction loss for various oscillation modes in an optical two-mirror cavity are reproduced in Figure 14.4, in which the fractional loss is plotted against the parameter $D^2/4\lambda L$ (Boyd, 1961). Note that according to the approximate theory outlined above, the graph for the plane mirrors should be linear with a slope of -1. We see from these results that even for the unfavorable case of plane mirrors, fractional losses as low as 10^{-4} are attainable (assuming the mirrors are perfectly aligned) for $D^2/4\lambda L \approx 100$, a practical figure. This points to the limit on the Q of these cavity modes not really being set by diffraction, but rather by the imperfect reflectance of the mirror, which in practice rarely exceeds 99.9%, corresponding to a fractional loss of 10^{-3}. In any event, to be useful, a laser oscillator must provide an output beam, and therefore at least one of the mirrors must be partially transparent, with a consequent power loss from the cavity.

Since an optical cavity commonly has dimensions large compared to the wavelength, the various resonance modes, with their characteristic stationary field distributions, have frequencies that differ fractionally very little from each other. Thus for the purely axial modes we find the following result for the difference between consecutive modes:

$$\frac{\nu_{n+1} - \nu_n}{\nu_n} = \frac{\lambda_n}{2L}, \tag{14.2}$$

which, by our assumption that $\lambda_n << L$, proves our assertion that the fractional difference between mode frequencies is small. This, of course, need not be true for microwave cavities, since in that case the cavity dimensions can be of the same order of magnitude as the wavelength, and the lower mode frequencies are relatively far apart. If the optical cavity consisted of a truly closed cavity with reflecting walls, analogous to a microwave cavity, then if its dimensions were large compared to the wavelength, there would be a very large number of modes lying within any given frequency band, all having an appreciable Q-value. This can be seen from the one-dimensional example given above, where the fractional frequency difference between consecutive axial modes was shown to decrease as the wavelength is made smaller compared to the cavity dimensions. This means that there will be, in a given frequency range, an increasing number of modes as the wavelength becomes smaller. In fact, that number in the general 3-dimensional case appears in the theory of blackbody radiation and is given by the following:

$$\Delta N = \frac{8\pi V}{\lambda^3}\frac{\Delta\nu}{\nu}, \tag{14.3}$$

where ΔN is the number of modes having a frequency in the interval $\Delta\nu$ centered on the frequency ν, and V is the volume of the cavity, which classical theory has shown can have any shape, provided that its dimensions are very much larger than the wavelength. To illustrate just how large ΔN can be, let us assume λ=500 nm, V=100cm^3, and $\Delta\nu/\nu=10^{-7}$; substituting into the formula yields $\Delta N \approx 2\times10^9$! It is indeed fortunate that quasi-discrete Gaussian modes do exist in a wide-open Fabry–Pérot resonator, with only a few radial modes having a high Q-value, to restrict the number of modes into which the stimulated emission from the atoms occurs. Not only does sustained laser action in these modes become possible, but it yields the extraordinary directionality of the laser output beam, and with proper selection of axial modes, great spectral purity.

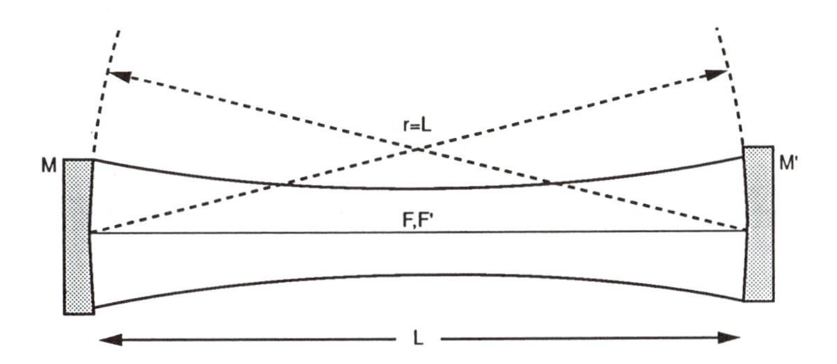

Figure 14.3 A confocal optical cavity

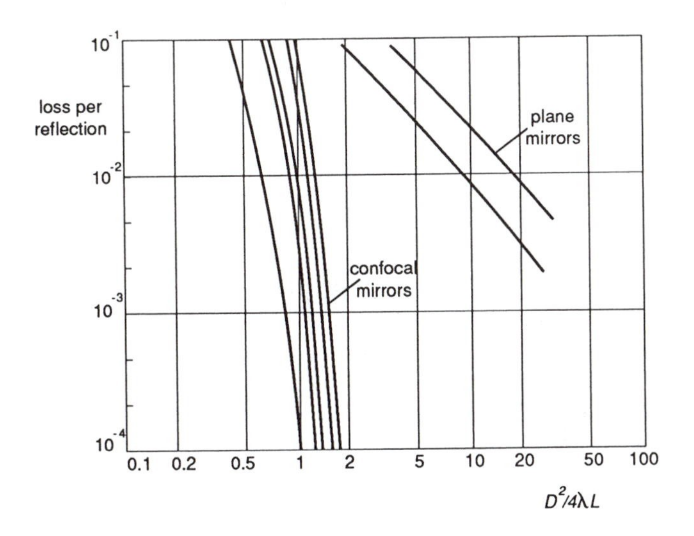

Figure 14.4 A plot of the optical loss as a function of $D^2/4\lambda L$ (Boyd, 1961)

14.3 Conditions for Sustained Oscillation

As with beam masers, sustained optical frequency oscillation of the field in a resonator is possible if atoms or molecules are present that through stimulated emission yield a net gain in the field energy sufficient to make up for all losses, including that represented by the output beam. However, while the general principles are identical for microwave and optical frequency oscillators, there are many important practical differences that give them quite different physical aspects. In addition to the obvious differences attendant upon the very different wavelengths, the kind of coupling of the atoms or molecules with the optical field is also different: The atomic beam masers involve magnetic dipole transitions, while lasers involve the much more strongly induced electric dipole transitions. Furthermore, while in the magnetic dipole transitions the field acts on a permanent atomic magnetic moment, the existence of a permanent electric dipole moment is excluded on symmetry grounds in free atoms. This follows from the fact that insofar as all the forces acting *within* the atom are concerned, it is indistinguishable from a transformed atom in which the positions of all the particles are diametrically opposite to where they were. For a permanent electric dipole moment to exist, the nucleus would have to be displaced relative to the center of the electron distribution, in contradiction to the symmetry under inversion we have just stated. The symmetry is broken when an external electric field is introduced; in that case, oppositely directed forces are exerted on the positive nucleus and negative electrons, resulting in a dipole moment. Transitions result from the action of the electric component of the optical field, which induces an oscillating electric dipole moment in the atom. The amplitude of that dipole moment depends on the dynamical response of the particular atom; we recall the classical model of an atom having elastically bound electrons used by Lorentz in his theory of optical dispersion to predict that response.

To achieve a net gain of power from an atomic or molecular system, it must be prepared with a preponderance of population in the upper of two quantum energy levels, between which transitions are to be stimulated. This, it will be recalled, is simply because the probability per atom per unit time for stimulated emission is exactly the same as for absorption. The achievement of this preferential population of the upper energy state in the case of optical transitions is complicated by two circumstances: First, unlike the microwave case, an optical quantum (photon) has considerably greater energy than the mean thermal energy of particles in equilibrium at ordinary temperatures. This, it will be recalled from Boltzmann's theory, implies that in thermal equilibrium the number of atoms of a gas in the lower of the two states will be far greater than the number in the upper state. Thus according to the theory, if the number in the upper quantum state is N_1 and the number in the lower state N_2, then we have in equilibrium at absolute temperature T,

$$\frac{N_1}{N_2} = e^{-\frac{h\nu}{kT}} . \qquad 14.4$$

The temperature T is, of course, positive; hence in equilibrium we must always have $N_1 \langle N_2$. For example, if T=300°K and $\nu=6\times10^{14}$Hz, we find that, on an average, only one atom in 10^{41} is in the upper state! Clearly, then, for laser action we require very nonequilibrium conditions; in fact, we require what is called *population inversion*, or a "negative (absolute) temperature."

The second essential difference caused by the greater energy of the optical photon is that the probability (per atom) per unit time for *spontaneous* emission is far from being negligibly small, as it was in the microwave case. We can see this from the Einstein expression for the ratio of his A- and B-coefficients

$$\frac{A_{nm}}{B_{nm}} = \frac{8\pi h\nu^3}{c^3} , \qquad 14.5$$

if we recall that the probability per unit time for stimulated emission is $B_{12}\rho_\nu$, where ρ_ν is the spectral energy density of the optical field causing the transitions, given by $\rho_\nu = I_\nu / c$, for a parallel light beam of spectral intensity I_ν. At ordinary light levels such as might exist in a conventional lamp, where I_ν is perhaps on the order of 10^{-8} watt/m^2·Hz, we find that the probability of spontaneous emission is about 2000 times greater than that of stimulated emission. This shows why stimulated emission plays an insignificant role in the operation of a conventional lamp; in fact, all ordinary sources of light, from the common tungsten lamp to the sun, are examples of spontaneous emission. However, in a lasing medium the energy of the optical field is concentrated in a narrower spectral width, which means a far larger spectral energy density and the emergence of stimulated emission as an important process. It is important to recall that spontaneous emission, in contrast to stimulated emission, is indifferent to whether an optical field is present and is induced with *random phase* by "zero-point" quantum fluctuations in the optical field. Stimulated emission/absorption, on the other hand, results from induced electric dipole oscillation in the individual atomic systems and is correlated in phase to the common stimulating optical field. In quantum theory, stimulated emission and absorption are different consequences of the same process; which one is manifested is simply a matter of whether the initial state is the one with lower energy or higher energy. From what has been said about the Boltzmann distribution of populations in a system in thermal equilibrium, it follows that a light beam passing through a medium in thermal equilibrium will always suffer absorption at those frequencies in its spectrum that are resonant with transitions in the system. In such a case, the intensity of a monochromatic light beam becomes weaker as it passes through a medium with a resonant transition. On the other hand, if by some means, such as an electrical discharge or intense optical pumping, a population inversion is sustained, then that monochromatic light beam *increases* in amplitude; it is amplified, as shown schematically in Figure 14.5.

To express these notions more quantitatively, suppose a monochromatic parallel beam of intensity I_ν watts/m^2 passes through an atomic/molecular medium with n_1 particles/m^3 in the upper quantum state and n_2 particles/m^3 in the lower; and let them have a frequency response (resonance line shape) $g(\nu)$, so that according to the definition of B_{12}, the probability of the optical field stimulating emission (in unit volume) at the frequency ν is $n_1B_{12}(I_\nu/c)g(\nu)$ per unit time, with an identical expression for absorption, except that n_1 is replaced by n_2. In practice, because of the various spectral broadening mechanisms such as the Doppler effect in gases, the function $g(\nu)$ will have a bell-shaped graph, generally broader than the spectral width of the light, so that the assumption of a monochromatic beam having the single frequency ν is not an unrealistic one. It follows that since each transition involves the exchange of one quantum of energy $h\nu$, the *net* rate of energy exchanged coherently with the field (per unit volume) is $(n_1-n_2)B_{12}(I_\nu/c)h\nu g(\nu)$. If we choose the direction of the beam to be the z-axis of a coordinate system and balance the energy flow in the beam with the amount emitted (or absorbed) by the atoms, we are led to the following:

$$\frac{dI_\nu}{dz} = (n_1 - n_2)h\nu B_{12}g(\nu)\frac{I_\nu}{c}. \qquad 14.6$$

This has a solution of the form

$$I_\nu(z) = I_\nu(0)e^{\gamma z}, \qquad 14.7$$

where

$$\gamma = (n_1 - n_2)B_{12}\frac{h\nu}{c}g(\nu) \qquad 14.8$$

is the (exponential) gain constant. As expected, this shows that for the light beam to be amplified, we must have $(n_1-n_2) > 0$, that is, population inversion with more atoms in the upper state than the lower. In a system in thermal equilibrium, as already emphasized, we have necessarily $(n_1-n_2) < 0$, and the intensity falls exponentially, in agreement with the classical experimental law, sometimes called Lambert's law. The two cases are illustrated in Figure 14.5. We can usefully rewrite the gain constant γ in terms of Einstein's A-coefficient, since the latter is related to the mean lifetime of the upper state against spontaneous emission, a lifetime that can be deduced from the empirical "natural" width $\Delta\nu_n$ of the emission line. The result is as follows:

$$\gamma = \frac{1}{4}(n_1 - n_2)\lambda^2\Delta\nu_n g(\nu). \qquad 14.9$$

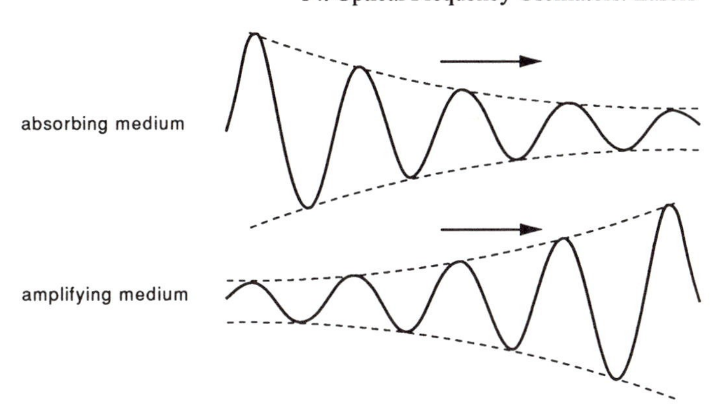

Figure 14.5 A schematic illustration of a light wave passing through absorbing and amplifying media

The line shape factor $g(\nu)$, which gives the spectral response of the atoms to the optical field, may result from a number of different processes. For some applications it is important to draw a distinction between two different types of broadening mechanisms: homogeneous and inhomogeneous. As we saw in Chapter 7, the distinction applies to a group of atoms: If a broadening mechanism affects all atoms *identically*, such as the natural lifetime of the radiating state or collisions with other particles that interrupt the radiation process, then it is homogeneous. Lifetime broadening, for example, we know leads to a Lorentzian lineshape:

$$g(\nu)=\frac{1}{\pi}\frac{\frac{\Delta\nu}{2}}{(\nu-\nu_0)^2+\left(\frac{\Delta\nu}{2}\right)^2}, \tag{14.10}$$

where $\Delta\nu$ is the width of the $g(\nu)$ versus ν curve at half its maximum, which occurs at $\nu=\nu_0$ and has a value there of $2/\pi\Delta\nu$.

On the other hand, it can happen that each individual atom in the group has its own slightly different frequency because, for example, the atoms have different velocities, and therefore different Doppler shifts in their frequency, or perhaps because each atom sees a slightly different environment; in this case we say the broadening is inhomogeneous. For the case of Doppler broadening in a gas in thermal equilibrium, we found

$$g(\nu)\approx \exp\left(-4ln2\frac{(\nu-\nu_0)^2}{\Delta\nu^2}\right). \tag{14.11}$$

In order to set up the conditions for sustained oscillation at optical frequency, we combine the essential elements of a feedback oscillator by placing the amplifying atomic medium inside an optical resonator. And as with any other feedback oscillator, the threshold condition for oscillation to break out is that the feedback be regenerative and the loop gain G=1. To obtain an expression of these two conditions explicitly in terms of a specific model, assume that we have a Fabry–Pérot resonator filled with a population inverted gas acting as a distributed amplifier. Let R_1 and R_2 represent the ratios of reflected to incident light intensity at the two mirrors, and let α be an absorption constant to account for all distributed loss of intensity due to interaction with the gas, so that we can write the condition on the loop gain as follows:

$$R_1 R_2 \exp[(\gamma - \alpha)2L] = 1, \tag{14.12}$$

from which we deduce the threshold value of γ to be

$$\gamma = \alpha + \frac{1}{2L} ln\left(\frac{1}{R_1 R_2}\right). \tag{14.13}$$

The condition on the phase is a little more complicated, since the light travels through an amplifying medium that is dispersive; that is, the velocity of a light wave through it depends on the frequency of the wave. The interaction of the light with an atomic medium near a resonance can strongly affect the velocity of the light wave in a frequency-dependent way. If we define $c/n(\nu)$ as the velocity of light in the medium, then for a light wave starting from any point in the cavity, to have the same phase after making a complete round trip between the two mirrors requires the following phase condition:

$$\frac{2L}{c/n(\nu)} \nu = m, \tag{14.14}$$

where m is a whole number. In the absence of the atoms, $n(\nu)$=1 and $\nu=\nu_m$, the cavity resonant frequency in the mth order axial mode. We will not attempt to derive the expression for $n(\nu)$ but merely state the important result that it involves the frequency dependence of absorption (or in our case the stimulated emission), and its substitution in the phase condition leads to the following approximate result for the actual frequency of oscillation:

$$\nu = \nu_m \left\{1 - \frac{(\nu_0 - \nu_m)}{\Delta\nu} \frac{\gamma}{k}\right\}, \tag{14.15}$$

where $k=2\pi/\lambda$ is the magnitude of the wave vector. This shows that oscillation does not take place exactly at the resonant mode frequency ν_m of the cavity: an example of *frequency pulling* such as we already encountered in the hydrogen maser. It can be shown that this result can be rewritten to show the same dependence on the relative line widths of the atomic and cavity resonances; here, however, it is the cavity that has the sharper resonance, rather than the atoms.

14.4 The Sustained Output Power

The threshold conditions alone do not, of course, tell us anything about how and to what level the optical power builds up in the laser cavity. As with all feedback oscillators, once the threshold is passed, oscillations will start from ever-present incoherent zero-point excitations of the field or, as in this case, the spontaneously emitted light. To predict the further buildup of the optical field, we must take into account the dependence of the population difference itself on the amplified field it generates. This involves taking the theory of the interaction of the atoms with the optical field to a higher order of approximation, beyond the approximation so far implied. This was done by W.E. Lamb, who by developing the quantum theory of interaction between atom and field to the third order in the field amplitude gave explicit expressions for the coefficients α_n and β_n in the "equation of motion" for the field amplitude E_n of the nth mode:

$$\frac{dE_n}{dt} = \alpha_n E_n - \beta_n E_n^3, \qquad 14.16$$

which applies to a loop gain $G > 1$ (Lamb, 1964). This leads to a steady state when $dE_n/dt=0$, which occurs for $E_n^2=\alpha_n/\beta_n$. For the stationary field modes assumed in the theory, Lamb found that the linear gain factor α_n as a function of tuning has a Gaussian shape arising from Doppler broadening, while the nonlinear "saturation" factor β_n is much less Doppler broadened, exhibiting mainly lifetime broadening. This causes the overall frequency dependence of the laser output to exhibit what is now called the *Lamb dip*, a local minimum at the peak of the Doppler line shape, as shown in Figure 14.6. This sharp spectral feature, whose width reflects the homogeneous line width of the atoms, rather than the much broader Doppler width, has proved very useful in stabilizing the frequency of oscillators in the infrared and optical regions of the spectrum.

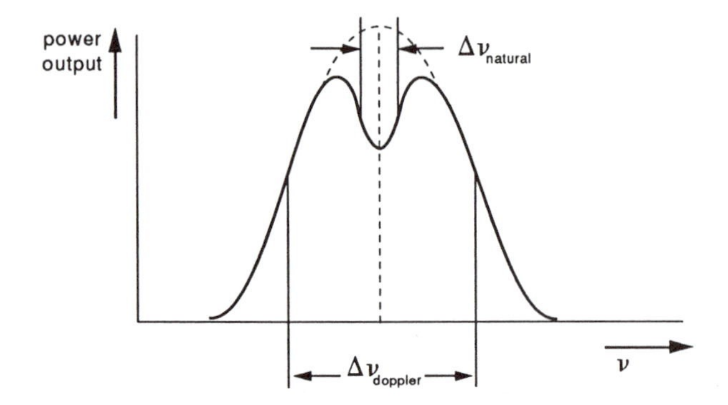

Figure 14.6 The Lamb dip in the output of a gas laser

14.5 Laser Optical Elements

Components such as high-reflectance mirrors, windows, and other optical elements for use with laser light are characterized by precision on the order of fractions of a wavelength in order to preserve and exploit its coherence properties. An "optically flat" surface will typically be specified as $\lambda/20$, meaning that at all points on the surface the mean departure from a geometric plane is less than one-twentieth of an optical wavelength. Furthermore, laser-grade optical surfaces have a higher degree of polish as specified by the "scratch and dig" figures, which indicate the "visibility" and number of scratches and pits in the polished surface.

In classical optics, mirrors were almost universally made by depositing a film of silver onto the desired surface either from a chemical solution (Rochelle process) or more commonly now by deposition of silver vapor in vacuum. Of all metals, silver has the highest reflectance in the visible region (aluminum is better at the violet end) of the spectrum, reaching about 93% at a wavelength of 600 nm. Currently, however, much higher values of reflectance are readily available using a radically different approach. This is made possible by the fact that for the most part, laser light is nearly monochromatic, and therefore the reflectance needs to be (and in some cases is preferred to be) high only for a very narrow wavelength range. The new mirrors are called *multilayer dielectric* mirrors, formed by vapor deposition onto an optically flat substrate (usually quartz or sapphire) of many thin layers of highly transparent dielectric materials, with values of refractive index alternating between high and low values. The principle underlying this type of mirror is that of superposition of light waves and the phenomenon of interference; this implies the need for a high degree of polish and flatness in the substrate. Suppose we have a set of plane parallel films of alternating refractive indices n_1 and n_2, and let a monochromatic beam of light fall perpendicularly on them. To find the reflectance of such an arrangement we recall Fresnel's formulas for the reflection and refraction of light waves at boundary surfaces between different media. Originally derived on the basis of the "ether vibrations" theory of light, which predates Maxwell's electromagnetic theory, these formulas, with some reinterpretation, remain valid. For the particular case of a light wave in a medium with refractive index n_1 falling perpendicularly on the boundary surface with a medium of refractive index n_2, the (amplitude) reflectance for such a light wave is given by

$$r = \frac{n_1 - n_2}{n_1 + n_2}. \quad 14.17$$

We note that if $n_1 < n_2$, the wave suffers a change of phase of 180° (a change of sign), whereas if $n_1 > n_2$, there is zero change in phase. In a multilayer dielectric mirror, the thicknesses of the films are chosen such that it takes half a period of oscillation for the wave to traverse the thickness of any film in both directions. That is, since the velocity of light is c/n_1 and c/n_2 in the two media, we require the following:

$$\frac{2d_1 n_1}{c} = \frac{2d_2 n_2}{c} = \frac{\tau}{2}, \quad 14.18$$

where d_1 and d_2 are the layer thicknesses, and $\tau = 1/\nu$ is the period of the light wave.

Now, referring to Figure 14.7 we see that light waves reflected from any boundary surface are in phase with those reflected from any other surface; that is, there is constructive interference. To derive the reflectance of a large number of dielectric layers would quickly become hopelessly complicated if we were to try to follow the progression of a wave as it underwent multiple reflections in each layer; instead, one makes use of the boundary conditions that the total electric and magnetic components of the light wave on either side of any boundary must obey. These conditions are a statement of the fact that the components of the electric and magnetic field along the dielectric surface do not change abruptly as the wave crosses the boundary. By applying these conditions of continuity and allowing for the transit delays between boundaries, it is possible to relate the field components at one boundary to those at succeeding boundaries. The result of such an analysis we give without proof; for ideal nonabsorbing dielectrics it is as follows:

$$r = \frac{\left(\frac{n_2}{n_1}\right)^{2N} - 1}{\left(\frac{n_2}{n_1}\right)^{2N} + 1}. \quad 14.19$$

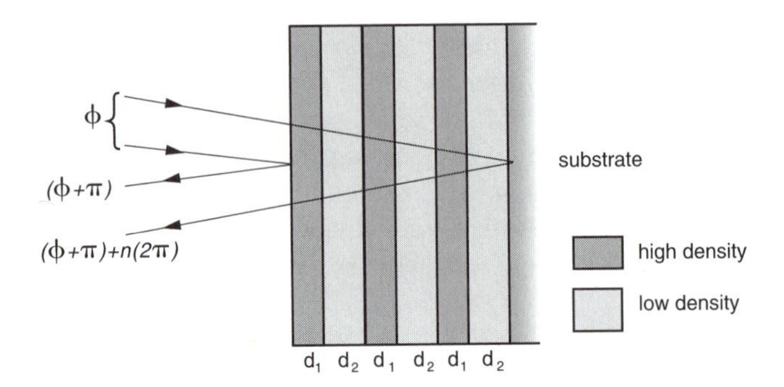

Figure 14.7 A section of a multilayer dielectric high-reflectance mirror

For example, if magnesium fluoride (n=1.35) and zinc sulphide (n=2.36) are used in a 14-layer (N=7) mirror, the intensity reflectance r^2=99.8%. This high value, it must be emphasized, obtains only for light of a single wavelength, one for which the optical thickness of each layer is one-fourth the wavelength. Of course, in reality, the achievable reflectance is ultimately limited not only by absorption in the media, but also by scattering from irregularities in the boundary surfaces and within the media. Currently, there are commercially available mirrors fabricated using the most advanced polishing and coating technology that are claimed to have a reflectance as high as 99.99%. Of course, a few particles of dust could easily put that last decimal place in question!

If a smaller number of layers is used, for example N=5, we obtain r^2=98.4%; hence, since the absorption and scattering are assumed negligible, we have a partially transparent mirror with a transmittance of $t^2=1-r^2$; that is, t^2 = 1.6%. Such a mirror would be useful as the output mirror in an optical cavity.

While on the subject of the optics of thin dielectric films, perhaps we should mention two other very important applications: antireflection coating of lenses and narrow pass-band interference filters. As indicated above, light crossing a boundary surface between media of differing refractive indices will be partially reflected, and in the case of optical instruments having many surfaces, as in high-quality camera lenses, the intensity loss may be significant. In its simplest form an antireflection coating of an optical surface consists of a single layer of a medium having a refractive index intermediate between that of the lens material and air, usually magnesium fluoride (n=1.35). Unlike the case of a large number of layers, where there is a sharp wavelength dependence, here it is very broad, and the thickness may be chosen to be one-fourth of the wavelength of light in the middle of the visible range and the coating will still be effective over much of that range. If three layers of dielectric are used, the added flexibility makes it possible to design for zero reflectance at two different wavelengths, and an average less than 0.25% over the visible range.

The other important application that is critical to the success of both the rubidium and the mercury ion standards is the *interference filter*. In contrast to the multilayer dielectric mirror, it is high transmittance rather than reflectance that the filter is designed to have, and this in as narrow a wavelength range as possible. The simplest bandpass filter is really a Fabry–Pérot cavity with the space between the parallel mirrors filled by a dielectric layer with an optical thickness nL equal to half the wavelength at the center of the desired band. The two mirrors are commonly in the form of the multilayer dielectric type described above, which, combined with the half-wavelength layer between them, form one integral unit.

Another optical element that has become important in laser optics is the *Brewster window*, an optically flat transparent plate with parallel faces set at the Brewster, or *polarizing*, angle to the direction of light falling on it. We recall that a light wave, being transverse, can be polarized so that, for example, the electric field oscillates in one plane all along the wave. While polarization effects are generally associated with crystals such as calcite, it has been known since Malus that light can be polarized simply by reflection, a fact easily confirmed now by looking at sunlight reflected from water through polarizing sunglasses. In 1812 Brewster discovered experimentally the law that bears his name, giving quantitatively the angle at which light reflected from a dielectric is completely polarized. If an unpolarized beam falls on a boundary surface between two media, even though its electric field is randomly oriented about the beam direction, we can express the field at any instant as the resultant of components parallel and perpendicular to the plane of incidence. This is the plane that by the first law of optics passes through the incident, reflected and refracted beams, as well as the perpendicular to the surface. Brewster's law states that the amplitude of the component in the reflected wave whose electric field is parallel to the plane of incidence will be zero when the angle of incidence θ_B satisfies the condition

$$\tan\theta_B = \frac{n_2}{n_1}. \qquad 14.20$$

For an air–glass interface $n_2/n_1 \approx 1.5$; thus $\theta_B \approx 56.3°$. If we recall Snell's law, $n_1 \sin\theta_1 = n_2 \sin\theta_2$, we find that for an angle of incidence equal to the Brewster angle, $\sin\theta_2 = \cos\theta_1$, and the refracted beam is at right angles to the reflected one, as shown in Figure 14.8. It follows from Brewster's law that if an unpolarized beam falling on a glass surface at that angle to the perpendicular will have the small fraction that is reflected completely polarized, since it will contain only the component with the electric field perpendicular to the plane of incidence. But of greater interest for laser applications is the fact that there will ideally be no reflected beam if the incident beam itself is polarized parallel to the plane of incidence, and it falls on the boundary surface at the Brewster angle. The use of Brewster windows to avoid reflection losses is particularly important in low gain gas lasers such the helium–neon laser. In this case, if the plasma tube is to be separate from the optical cavity, its ends would be sealed with precisely oriented Brewster windows. When such a polarization-sensitive element is incorporated into a laser cavity, the result is that oscillation will take place with the field polarized in the direction having the least loss.

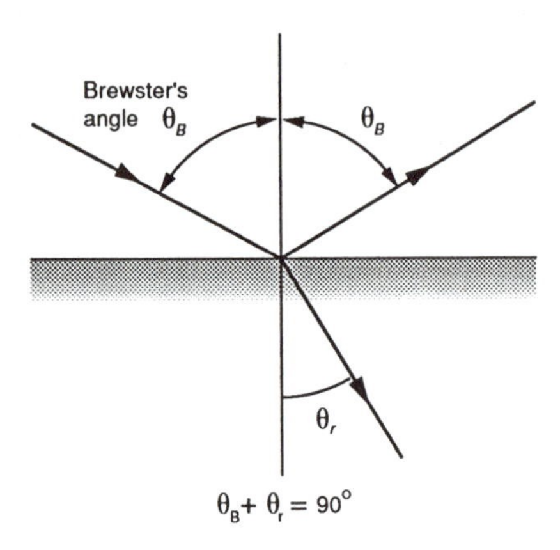

Figure 14.8 The angle relationships at the Brewster polarizing angle

14.6 The Ruby Laser

We will now consider specifically lasers that typify each of several important classes. Historically, the ruby laser occupies a special place as the first successful physical embodiment of emerging ideas on how to extend the concept of a maser oscillator to optical frequencies, ideas associated mainly with the names of Townes, Schawlow, Basov.

In an article entitled "Stimulated optical radiation in ruby masers," published August 1960 in the British journal *Nature*, Maiman described the first experimental evidence of stimulated emission at optical frequency (Maiman, 1960). According to an article in *Physics Today* (October 1988) a brief article on the subject submitted by Maiman two months earlier to *Physical Review Letters* was turned down by its then editor, Goudsmit. Maiman's historic laser is drawn schematically in Figure 14.9. The "active atoms" are chromium ions in a matrix of crystalline alumina (Al_2O_3). In pure form, crystalline alumina is colorless. However, when "doped" with chromium by the addition of a small amount of chromic oxide to the powdered alumina fed into the crystal-growing furnace, it becomes artificial ruby, whose red color varies in intensity according to the concentration of chromium. It may be of interest to note in passing that if instead of chromic oxide, an oxide of titanium, for example, is used, one gets artificial blue sapphire. In Maiman's experiments a low concentration of about 0.035% by weight was used, which results in a pink ruby; higher concentrations of chromium are needed to produce the deep red color of the precious stone. He was very familiar with this material, since he had been active in the field of solid-state masers, where *paramagnetic* ions, which have a permanent, as opposed to an induced magnetic dipole moment, such as Cr^{+++} in ruby, are used. These masers are characterized by the use of high magnetic fields, at low temperatures, and the application of a pumping microwave field to achieve a population inversion and the conditions for broad-band maser amplification at lower microwave frequencies.

To understand the special properties of ruby that enabled its successful application to reduce to practice the idea of an *optical* frequency "maser," we must recall a few points about the atomic structure of elements such as chromium. They are called *transition* elements because of their place in the periodic table of the elements. They are in a sense interlopers, interrupting the "normal" progression of chemical properties among the "representative" elements in each period of the table. They correspond to the filling up of an *internal* electron shell, the 3d shell; their ions, in solution either in liquids or crystalline solids, are characterized by absorption in the visible part of the spectrum, making them appear colored. There is, however, strong broadening of the spectrum due to the crystal fields and thermal vibrations in the host crystal, suggesting the need to work at very low temperatures.

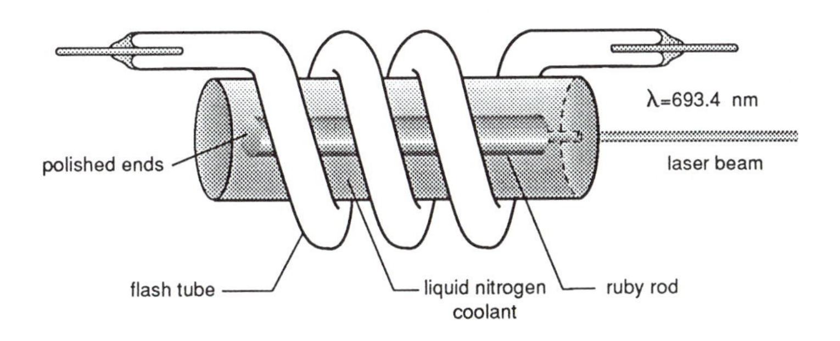

Figure 14.9 Schematic diagram of the first laser: Maiman's ruby laser

Chromium is among the transition elements in the 4th period of the table, a group of elements of particular importance in modern alloys. The chromium ion Cr^{+++} has three electrons outside closed shells; as each electron has a spin of ½, the maximum spin of the ion is 3/2. This maximum value, according to a spectroscopic rule known as Hund's rule, would be the spin of the lowest energy state of a free ion. But of course, the Cr^{+++} ions are far from being free; they are subject to strong electrostatic crystal fields having the geometrical symmetry of the alumina crystal. While the total spin remains a "good" quantum number and may therefore be used to label the quantum states, the orbital angular momentum that is bound up with the spatial part of the electron distributions is radically affected by the crystal field, and the proper representation of the states is quite different. Luckily, it is not necessary for our purposes to understand the difficult theoretical basis for the nomenclature used to specify the states; it is sufficient to know the structure of the quantum energy levels and the transition probabilities between them.

The relevant energy levels are shown in Figure 14.10. The levels between which quantum transitions give rise to laser action, which we will henceforth refer to briefly as the "laser levels," are the excited state labeled ^2E and the ground state, labeled $^4A_2$. For our purposes this spectroscopic notation serves only to label the states. The ^2E state is actually composed of two closely spaced levels labeled $2\bar{A}$ and $\bar{E}$, from which transitions to the ground state lead to a pair of spectral lines referred to as R_1 at λ=694.3 nm and R_2 at λ=692.9 nm. The radiative lifetime of the upper laser level $\bar{E}$ is a relatively long 3 milliseconds. The $^4A_2$ ground state actually consists of two very close levels only 12 GHz apart. The most important feature of this level structure from the point of view of attaining the prerequisite inversion of populations between the laser levels is the presence of the broad energy bands labeled $^4F_1$ and $^4F_2$, to which transitions from the ground state result in strong absorption of light in the violet and green regions of the spectrum, causing the red color of ruby. Furthermore, ions in these energy bands undergo rapid nonradiative transitions to the upper laser level ^2E within an average time of only about 5×10^{-8} seconds. This provides an effective means of optically pumping ions from the ground state to the upper laser level: An intense flash lamp, whose output is spread over a wide spectral range, can nevertheless have a significant fraction of its energy absorbed by the ions and lead to the desired population inversion. Conventional light sources are limited as to spectral intensity, that is, the power contained in a given narrow spectral range. The ability to use such a source effectively here is due to the circumstance that the upper laser level can be reached through the intermediary of broad absorption bands. On this account, the ruby laser is classed as a *three level laser* (the two bands are here regarded as one "level").

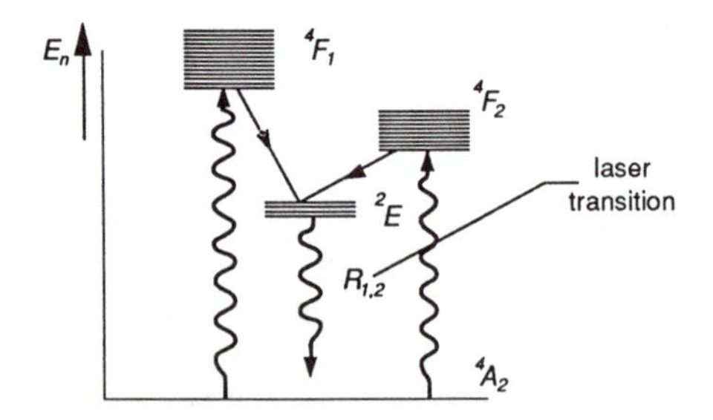

Figure 14.10 Quantum levels pertinent to ruby (Cr^{+++}) laser

In order to estimate the attainable degree of population inversion as a function of the intensity of the pumping light source, let us assume a model system with just three quantum energy states with energies E_0, E_1, and E_2, as shown in Figure 14.11. Assume further that a pumping beam of intensity I_0 and wavelength $\lambda = ch/(E_1 - E_0)$ is resonantly absorbed by ions in the ground state, causing transitions to the upper E_2 state with a probability per second (per ion) that is proportional to I_0 and will be represented by αI_0. (We are here assuming conditions where phase coherence among the ions in response to the pumping light can be neglected, and the use of a constant transition probability per unit time is valid.) Some of the ions may make a transition directly back to the ground state; others may cascade down by way of level E_1. Then as the pumping continues, the populations of the three levels will tend towards a steady state in which the number of ions entering a given level equals the number leaving it. In terms of level populations and transition probabilities, the dynamics of the system are described by three *rate equations*, typified by the one for the excited state population N_2, which below oscillation threshold is as follows:

$$\frac{dN_2}{dt} = \alpha I_0 N_0 - (\omega_{20} + \omega_{21}) N_2 \,. \qquad 14.21$$

The steady-state populations are obtained by setting $dN_2/dt=0$, etc., leading to algebraic equations from which we find that in order to produce a population inversion, the following condition must be satisfied:

$$\alpha I_0 \rangle \frac{\omega_{10}}{\omega_{21}} (\omega_{21} + \omega_{20}), \qquad 14.22$$

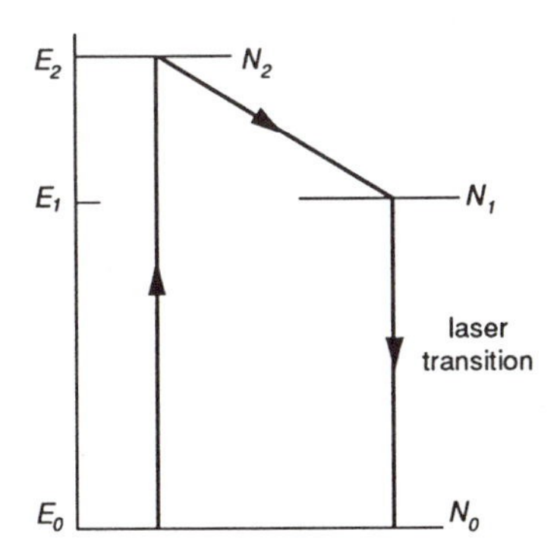

Figure 14.11 The quantum levels of a 3-level laser system

which confirms what is otherwise intuitively clear, namely that the transition rate into the upper laser level should be large compared to the rate of its decay to the ground state. This is certainly satisfied in the case of ruby, for which $\omega_{21}/\omega_{20}\approx10^5$, making it a particularly favorable laser material. This illustrates in a striking manner how a 3-level system makes optical pumping practicable; a 2-level system, on the other hand, does not, since the pumping light would have to compete directly with the spontaneous emission from the upper state. Another major advantage of ruby is that the density of ions in ruby (1.6×10^{19} ions/cm^3) is far greater than, for example, in an ionized gas under ordinary conditions. Hence a much larger optical gain γ is possible, and therefore greater losses can be tolerated, and a shorter optical length in the amplifying medium required to reach the threshold for oscillation.

The physical design of a ruby laser consists typically of a cylindrical rod a few centimeters in length, with a circular cross section a fraction of a centimeter in diameter. The ends are polished flat to better than 1/20 of a wavelength, parallel to a few seconds of arc, and coated with silver to form the mirrors of an optical cavity. One end is coated so as to have a transmittance of around 2% to allow the emergence of the output beam. The pumping source commonly used is a high-power helical or straight photoflash lamp, usually containing xenon gas at high pressure. In the case of a helical lamp, the ruby rod would be mounted along the axis of the helix, while for a straight lamp it would be parallel to the lamp along the focal lines of an elliptical reflecting cylinder. The flash is produced by the discharge through the lamp of a capacitor bank, charged to a potential of several kilovolts, a discharge triggered by a high-voltage pulse applied to a lamp electrode. The resulting intense flash has an energy of several hundred joules and lasts about one millisecond; hence the peak power is on the order of hundreds of kilowatts. The peak laser output power, on the other hand, is on the order of kilowatts; consequently, the energy conversion efficiency is less than 1%, and most of the lamp energy is ultimately dissipated as heat. Since the spectral lines of the Cr^{+++} ions are broadened by thermal vibrations in the host crystal, any heating of the ruby rod is intolerable. It is on this account that the ruby laser is limited to pulsed rather than continuous operation, and in fact, some of the early experiments were done on ruby rods cooled to the temperature of liquid nitrogen (−196°C).

The ruby laser breaks into oscillation when threshold is reached some time during the pumping flash. However, once the laser action begins, stimulated emission depopulates the upper laser state much faster than the optical pumping can sustain, leading to a cessation of oscillation and a buildup of population inversion once again until the oscillation threshold is reached and the cycle repeats. The resulting output is a series of spikes each lasting about one microsecond.

Another optically pumped solid-state laser, one that has greater efficiency than the ruby laser, is based on the rare earth ion Nd^{+++} (neodymium ion) as the active element. It is most commonly in the form of an impurity in the crystal *yttrium aluminum garnet* (YAG) or in certain glasses. Unlike the ruby laser, it is a *four-level* laser, in which the lower lasing quantum level is above the ground state in energy and consequently remains almost unpopulated during the pumping process, making it easier to achieve population inversion between the lasing states. The output wavelength of the Nd:YAG and Nd:glass lasers is in the near infrared at 1.06 microns. These lasers are usually operated in a repetitive pulse mode reaching kilowatts power at the peak; the Nd:glass lasers particularly are characterized by their resistance to high power densities.

14.7 The Helium–Neon Laser

A much more familiar laser to the general public is the helium–neon gas laser, first successfully made to oscillate at the Bell Telephone Laboratories in 1960 by Javan, Bennett, and Herriott (Javan, 1961). It is the first CW (continuous wave) laser, achieved by an electrical discharge in a mixture of helium and neon gases as the amplifying medium, with an output wavelength in the near infrared at λ=1.1523 μm; the familiar thin red beam came somewhat later. It is not that this particular gas mixture has some unique property that sets it apart from all other gases; indeed, immediately after the announcement of its success, innumerable other reports of laser action in other substances came in quick succession.

The active element in the He–Ne laser is neon, which, like helium, is a noble gas. It has six *equivalent* electrons in its outer shell, which occupy all the states in that shell permitted by the Pauli principle; hence its ground state has zero spin and zero orbital angular momentum and is designated spectroscopically as a $^1S_0$ state. The lower excited states must therefore involve an electron going into the next shell, leaving a vacancy in the previous one, which must therefore have a net angular momentum equal in magnitude to that of one electron. The resulting possible total angular momentum states and their fine-structure splittings are complicated, and so we will content ourselves with accepting the spectroscopic notation simply as a way of labeling the states without further inquiry. Some of the relevant lower excited states are shown in Figure 14.12. The laser states, between which transitions give rise to the familiar red beam at λ=632.8 nm, are in the groups of fine-structure levels labeled 3s and 2p; in addition, lasing action occurs in the infrared at λ=3.39 μm between states in the 3s and 3p groups, and at λ=1.15 μm between states in the 2s and 2p groups. Although excitation of the Ne atoms into the upper laser state is possible through collisions with electrons in the plasma sustained by the electrical discharge, these collisions will unfortunately also excite the lower state, making it difficult if not impossible to create the requisite population inversion. It is true that the radiative lifetime, for example, of the upper 3s state for the λ=632.8 nm transition is considerably longer ($\approx 10^{-7}$ seconds) than that of the lower 2s state ($\approx 10^{-8}$ seconds), a circumstance that favors the development of population inversion. Nevertheless, direct electron impact on the Ne atoms would do little to reinforce that trend. This is the reason for the presence of helium atoms in relatively large numbers: They act as carriers of excitation energy that preferentially raise the Ne atoms to the upper laser states by a process called *collisions of the second kind.* Although originally applied to collisions between excited atoms and free electrons in which the excitation energy of the atom is given to the electron in the form of increased kinetic energy, it is now used more broadly to include any collisions in which the excitation energy of one particle is transferred to another particle. In our present case, the helium atoms are first excited by collisions with electrons in the plasma to higher electronic energy states, from which many cascade down to accumulate in excited *metastable* states. These are states from which an atom has little probability, because of quantum selection rules governing changes in angular momentum, of making spontaneous transitions to lower energy states by emitting a photon.

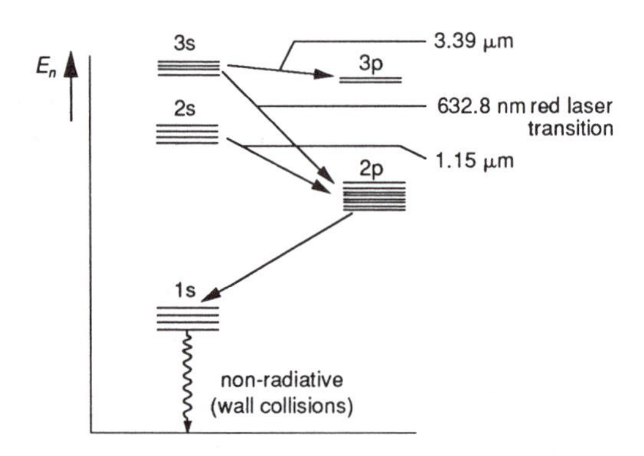

Figure 14.12 The lower-lying quantum states in neon, showing the laser transition

The quantum states of the helium atom, which has two electrons completing the first shell around the nucleus, can be divided into two classes: those in which the spins of the electrons are coupled in parallel to give $S=½+½=1$ and those in which the spins are opposed to give $S=½-½=0$. According to the quantum selection rules for radiative transitions, it is extremely improbable for an atom in one class to go over to a state in the other class by emitting or absorbing one photon. The rarity of spontaneous transitions from one class to the other is such that atoms in the two classes have been distinguished by name: one is orthohelium (S=1) and the other parhelium (S=0). In the ground state the atom is parhelium, with the spectroscopic designation 2^1S_0, and therefore the first excited state, which is the lowest orthohelium state ($^3S_1$), is metastable, since it is highly improbable that it will make a spontaneous transition to a lower state by radiating one photon. It even has a greater chance of making an unusual transition in which two photons are emitted! This two-photon process is highly unusual in that when it occurs, the emitted photons have a continuous spectrum rather than the single-line discrete frequency familiar with one-photon processes. The first excited parhelium state 2^1S_0 is also metastable, since the only state below it is also an S-state, and single-photon radiative transitions to it are strictly forbidden. Again, the most probable radiative process is the rare two-photon spontaneous emission. Of course, these statements about radiative transitions apply to isolated atoms; if an atom collides with another body, thereby dynamically altering its state during the impact, then it can give up its excitation energy to the other body without radiating. Thus in a low-pressure discharge through helium gas, it is expected that at least at some distance away from the walls of the container there will be a significant number of atoms in the 2^3S_1 state at 19.8 electron volts above the ground state, and atoms in the 2^1S_0 state at 20.5 electron volts above the ground state.

There remains one important condition for the collisional transfer of excitation between the metastable helium atoms and the neon atoms: For such a collision process to occur with a large cross section, there must be near resonance between the colliding atoms. That is, the excitation energy of the helium atom must be nearly equal to the excitation energy of the desired state in neon. The combination of these two particular gases was undoubtedly chosen precisely in order to fulfill this requirement. There is indeed a near match between the 2^3S_1 level in helium and the upper laser level 2s in neon, as well as between the 2^1S_0 level in helium and the 3s level in neon, as shown in Figure 14.13. The collision process can be represented as follows:

$$He^*(2^1S_0) + Ne \rightarrow He + Ne^*(3s) - 0.05eV.$$

The discrepancy in the energy is made up from the kinetic energy of the center-of-mass motion of the colliding atoms.

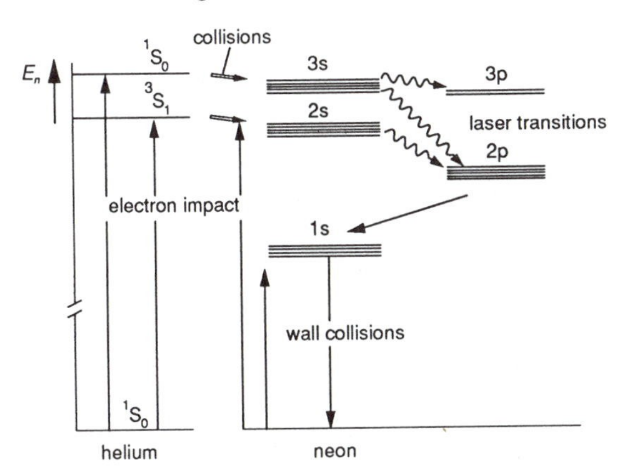

Figure 14.13 The matching of metastable levels in helium with the upper lasing state in neon

The pumping of the Ne atoms into the laser states proceeds then as follows: Electrons in the plasma collide with helium atoms, exciting some of them to higher quantum energy states, from which they cascade down into two metastable states 2^3S_1 and 2^1S_0. The concentration of these metastable atoms falls to zero at the walls of the discharge tube, where de-excitation can occur, but there will be a significant number, depending on the discharge current, along the axis of the tube. Metastable helium atoms collide with ground state neon atoms, transferring to some of them their excitation energy, thereby leaving them in one of the upper laser states; which one is determined by the near resonance condition we mentioned earlier. It is this condition that leads to the preferential excitation of the upper levels, as required for population inversion. Of course, there will also be electron collisions with neon atoms, which can excite them to the lower laser state, tending to reduce the population inversion. However, as already stated, the radiative lifetime of this state is much shorter than the upper state; this eliminates atoms from the lower state faster than from the upper one. Moreover, the inversion is favored by having a sufficient number density of helium atoms that the probability of excitation of a neon atom is far greater through collisions with helium than with electrons. A typical laser tube might contain helium at a pressure of 200 Pa and neon at only 10 Pa. A meaningful measure of the relative numbers of electrons and helium atoms is their flux density, that is, the average number of particles crossing an imagined unit surface area per second. For the assumed helium pressure at room temperature, the helium flux density is $nV/4 \approx 1.5\times10^{21}$ $cm^{-2}s^{-1}$; for comparison, if the electron flux density were this high, it would correspond to 240 $amp.cm^{-2}$, over an order of magnitude larger than in a typical laser. Of course, the collision rates also depend on the cross-sectional area a neon atom presents to these colliding partners; but it is not expected that the cross section for excitation by electron collisions is much larger than a resonant energy transfer from helium.

It is not only the helium atoms that have metastable states of concern to us; the neon atoms themselves are metastable in the state we have labeled 1s, and they can accumulate there as a result of rapid transitions from the lower laser 2p state. Unlike ground state neon atoms, which are subject to excitation through collisions with metastable helium, those in the metastable 1s state can be removed only through collisions with the walls of the tube. Their presence is an obstruction to the pumping cycle and could cause a serious reduction in population inversion and hence optical frequency gain. For this reason the inside diameter of the central section of the laser tube is generally small, on the order of one millimeter, to shorten the diffusion time to the walls, where they are de-excited.

As already indicated, laser oscillation can occur at three different wavelengths, two of which are in the near infrared. Oscillation at the infrared wavelength 3.39 μm, which shares the same 3s upper laser state as the red 632.8 nm transition, is in fact favored over the visible transition, partly because of the shorter lifetime of its lower 3p state, allowing a much greater population inversion to develop, and partly because the natural line width at λ=632.8 nm is not sufficiently larger than at λ=3.39 μm to offset the λ^2 factor in the formula for the optical gain γ. It has been reported that the 3.39 μm transition may have a (small-signal) optical gain as high as 50 dB/m, far exceeding the other transitions. Because of this high gain, a laser with an optical cavity of equal Q-factor at both wavelengths would reach the threshold for oscillation at 3.39 μm before 632.8 nm and would tend to suppress the latter from ever reaching threshold. The reason is that once laser oscillation at 3.39 μm builds up, the population of the 3s upper state it shares with the 632.8 nm transition can no longer rise; any tendency to do so merely changes the level of oscillation at 3.39 μm. If oscillation at 632.8 nm is desired, the optical cavity must have a Q-factor sufficiently low, at 3.39 μm compared to 632.8 nm, that threshold is reached first by the latter. This can be done in practice by incorporating windows or mirrors into the optical cavity that are preferentially absorbing at the infrared wavelength. In Figure 14.14 is shown schematically the arrangement of a typical He–Ne gas laser. It consists of a plasma tube, which because of the relatively low power density required to achieve oscillation can be made of common borosilicate glass. The geometry is generally a coaxial one consisting of a capillary plasma tube along the axis of a larger glass cylinder, which provides mechanical support to the capillary tube, the cathode, and possibly a reservoir of helium. The latter may be provided since Pyrex at elevated temperatures is slightly permeable to helium gas, and its pressure in a sealed tube will fall in the course of time. If DC excitation is used, the electrodes through which the external voltage source is connected to the plasma must be in direct contact with the plasma inside the tube, and connection is made to them through vacuum feedthroughs. Otherwise, if high-frequency excitation is used, the electrodes can be external to the plasma tube; however, in this case, local heating and strong fields, where the high-frequency current enters the tube, may aggravate the loss of helium. Most portable lasers make the optical cavity an integral part of the plasma tube assembly; the mirrors are sealed permanently at the ends of the tube. If the mirrors are to be adjustable, it is possible either to mount them on bellows or to seal the tube with flat, optical-quality windows and mount the tube between two external mirrors.

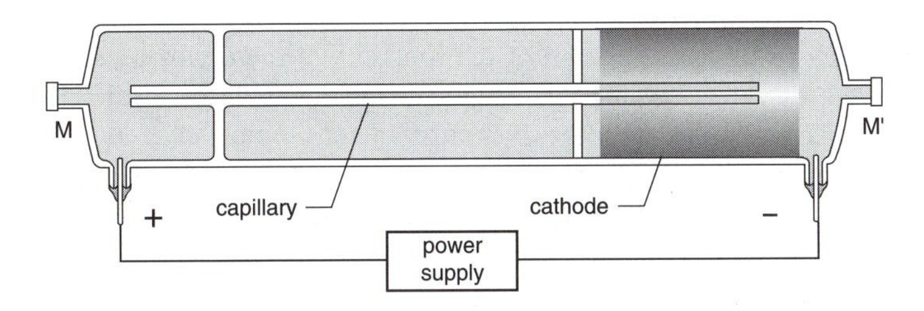

Figure 14.14 A typical He–Ne laser design

Although the helium–neon laser has superior spectral purity and directionality, it has two important limitations: First, the output wavelength can be varied only over a minute range; that is, it is practically untunable; and second, it is fundamentally a low-power laser, seldom exceeding 100 milliwatts, and more typically rated in the 1–10 milliwatt range.

14.8 The Argon Ion Laser

Among the gas lasers, the noble gas ion lasers using Ar^+ or Kr^+ ions as the active medium can be designed to have the greatest CW output power at several wavelengths in the visible part of the spectrum, making them among the most useful. Argon is a noble gas whose atom has a full complement of six electrons in its 3p outer shell. It requires an energy of about 12 electron volts just to raise it to its first excited state (hence its chemical inertness) and 15.7 electron volts to completely remove one electron and thereby form the A^+ ion. The ground state of the ion has a single vacancy in the 3p shell, giving it an angular momentum equivalent to one p electron (orbital momentum one unit) and a fine structure splitting into two close-lying levels. The relevant excited states of the ion, between which the laser transitions occur, require a further 19 electron volts to be reached from the ion ground state. They arise from electron configurations in which one of the 3p electrons goes into the n=4 s-shell. With so many electrons involved in the determination of the energy states of the ion, it is not surprising that the spectrum is complex, and it provides a wealth of transitions on which to obtain laser oscillation. The strongest laser outputs are at wavelengths 514.5 nm and 488.0 nm in the green–blue region of the spectrum, which arise from transitions between fine structure sublevels in the $3p^4 4p$ ($^4D^0$) state and the $3p^4 4s$ (^2P) state. The radiative lifetime of the upper laser level is a relatively short 9 nanoseconds (9×10^{-9}sec), while that of the lower is even shorter (as it must be for laser action), at 1.8 nanoseconds. This circumstance makes it necessary that the pumping of ions into the upper levels be correspondingly fast; but in the process, of course, the output laser power is much higher. The laser may be classed as a 4-level laser, since the initial pumping through electron collisions takes the ion to levels above the upper laser level, to which some cascade back, building up its population. Also, the lower laser level has a very short lifetime and is far above the ground level of the ion, so that there will be a far greater density of ions in the ground state than in the lower laser level. Hence the pumping process begins with the excitation of ions from the ground state and not the lower laser level. On the other hand, since the excitation produced by electron impact is not particularly selective as to the final state as long as energy is conserved, there will be some excitation to the lower laser level, opposing the desired buildup of population inversion.

The ions are produced, and then excited by electron impact in a high current (some tens of amperes) arc discharge through pure argon gas in a plasma tube, as shown in Figure 14.15. Since the mean energy of the electrons is on the order of a few volts, it follows that many collisions with a given ion are required to reach the laser levels. The electron flux required in these lasers is vastly greater than for the helium–neon laser; consequently, every effort is made to reduce the heating of the plasma tube walls. This is done by applying a strong magnetic field coaxial with the tube, to confine the plasma and slow down the diffusion of ions to the walls. This field is generated by passing a current through a coaxial solenoid. In spite of this, there may still be many kilowatts of power dissipated as heat, requiring water cooling and a plasma tube made of a high-temperature refractory material such as beryllia (beryllium oxide), with possibly metal fins for efficient heat exchange. For ease of igniting and maintaining a high current DC arc, the plasma tube has a hot electron emitting cathode. To maintain the highest possible electron energy, the gas in the arc must be free of molecular contaminant gases evolved from interior surfaces of the tube, since such contaminants will tend to have lower ionization and excitation energies, and the electrons give up their energy before reaching values high enough to excite the desired Ne levels. In a sealed tube this is accomplished by providing a getter, such as barium, to combine with such contaminants and remove them from the arc. Finally, it is found that in a DC arc there is a need to provide a return path for the argon gas outside the arc from one end of the tube to the other.

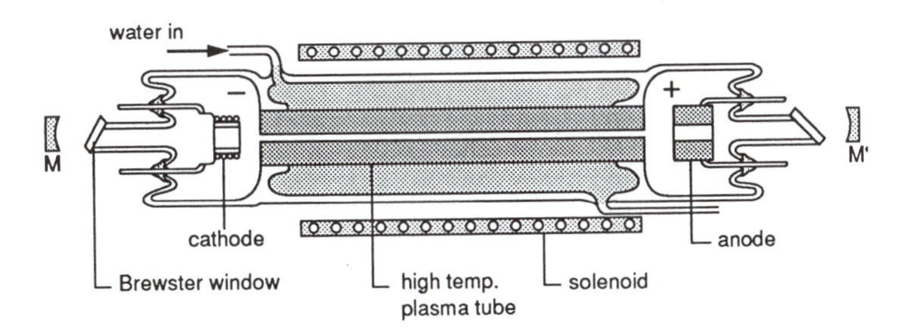

Figure 14.15 A typical Ar^{+} ion laser plasma tube

In general, if a simple optical cavity is used, an argon ion laser will oscillate simultaneously on more than one transition; these are commonly at the wavelengths 514.5 nm and 454.5 nm. In order to have oscillation on a single transition, an intracavity prism is incorporated into the optical cavity design; by a small tilt of the prism, the cavity can be tuned with sufficient finesse to select laser transition wavelengths. However, because of the relatively short radiative lifetime of the laser levels, the Doppler broadening due to the wide thermal velocity distribution of the ions in a high-temperature plasma, and the Zeeman effect due to the magnetic field used to constrict the plasma, the gain curve will encompass many longitudinal modes. The bandwidth of the output is in the range of 4–12 GHz; however, it can be drastically reduced by selectively allowing only one longitudinal mode to oscillate. This is done by introducing a sufficiently narrow transmission filter inside the optical cavity, a filter that has very high transmittance only over a frequency band narrow compared with the frequency interval between longitudinal modes. Such a filter is a Fabry–Pérot type of arrangement consisting of an optically flat plate of glass or fused quartz, with highly reflective coatings, referred to in this connection as an *etalon* (from the French word meaning *standard*, a reference to its original use in connection with standards of length). If the thickness of the etalon is t and the refractive index of its material is n, then it will have sharp transmission peaks spaced at intervals of $c/2nt$ in frequency; thus for t=1 cm, the spacing is around 10 GHz, the precise value depending on the refractive index and the angle of tilt (if any) between the etalon and the light wave. This spacing is much larger than that of the longitudinal modes of the cavity, which is $c/2n_0L$ and is typically in practice around 0.15 GHz. By adjusting the system so that one of the maxima of transmission of the etalon falls at the center of the gain curve, the number of modes that can oscillate is radically reduced. Furthermore, once the laser breaks into oscillation in one of the longitudinal modes, the rapid buildup of the optical field accelerates the stimulated emission and reduces the population inversion, hence the gain, thereby suppressing oscillation at all but the adjacent modes. These modes are close enough that it is possible for a fully operating laser to exhibit *mode competition*, in which buildup of oscillation at one mode frequency comes at the expense of gain for another mode. This can manifest itself in *mode hopping* if, for example, the etalon position fluctuated slightly. But all these limits on output frequency stability are still below the megahertz range on an optical frequency around 6×10^{14} Hz, a remarkable figure in an instrument that can provide a CW output of several watts of laser light.

14.9 Liquid Dye Lasers

The next important class of lasers we shall consider is that of organic liquid dye lasers. Apart from the obvious differences due to the fact that the active medium is a liquid rather than a gas, these lasers are distinguished by their wide range of continuous tunability; in fact, by using different dyes, the entire visible range of the spectrum, from violet to red, can be covered. Since laser action in an organic dye solution was first reported in 1966 (Sorokin,1966) the technology and applications of dye lasers have vastly expanded; they occupy a unique place by virtue of their broadband gain and tunability.

The active medium in liquid dye lasers consists of complex organic dye molecules, with names such as rhodamine 6G, coumarin, and stilbene. They are commonly in a crystalline powder form and are dissolved in some solvent such as water, ethanol, or ethylene glycol.

These dye solutions are characterized by a strong absorption band in the ultraviolet or visible region of the spectrum. When irradiated with light matching these bands in wavelength, the dye molecules strongly absorb it and re-emit light in a broadband shifted to lower frequency, in an intense display of what is called *fluorescence*. Previously when we used that term, it was in connection with atomic *resonance* fluorescence in alkali atoms, where the re-emitted light had the same wavelength as the absorbed light. Here the fluorescence is shifted to lower-energy photons, the balance having been taken up and dissipated by the molecules. Figure 14.16 shows the absorption and fluorescence spectra of rhodamine 6G dissolved in ethanol. The broadband nature of the absorption and fluorescence reflects the complexity of the polyatomic dye molecule, with each of its electronic states a complex of rotational and vibrational levels, as depicted schematically for a typical dye molecule in Figure 14.17. The interaction of the dye molecules with the solvent is so strong that collisions cause the closely spaced rotation–vibrational sublevels to be *homogeneously* broadened to such an extent as to form a continuum, and consequently, the absorption and fluorescence spectra are broad bands. The fact that the broadening is the same for all molecules, and therefore homogeneous, is significant, since under ideal conditions it allows *all* molecules to contribute to oscillation at a single frequency. Of course, for a given population inversion the optical gain is lowered by the broadening, but threshold is reached simultaneously by all molecules at a single frequency, unlike inhomogeneous broadening, where different groups of molecules reach threshold at different frequencies.

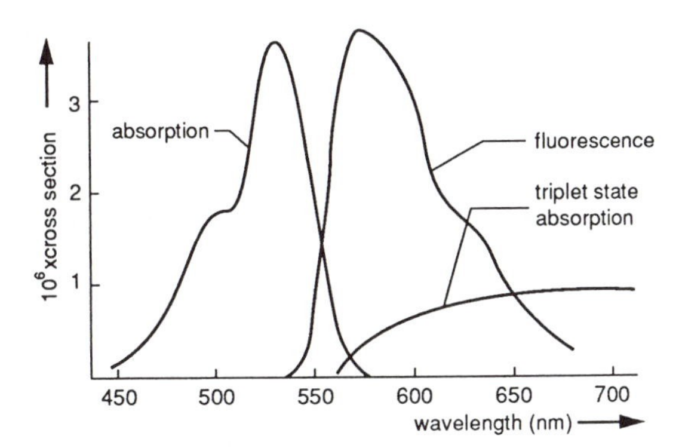

Figure 14.16 The absorption and fluorescence spectrum of the dye rhodamine 6G

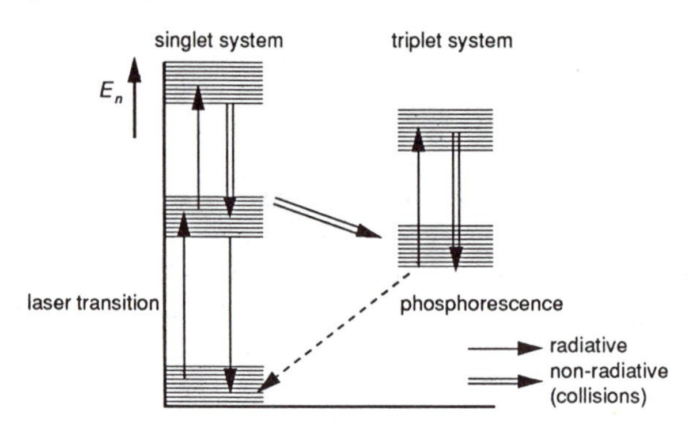

Figure 14.17 Schematic illustration of the quantum level structure in organic dyes

There are two systems of electronic states, designated as *singlet* (S_n), and *triplet* (T_n), which are marked by a difference in their molecular spin state, the singlet having S=0 and the triplet S=1. In consequence, radiative "intersystem" transitions between them are quantum-mechanically "forbidden." However, transitions between certain levels within each system are not forbidden and will generally occur very rapidly. The ground state of the molecules is a singlet electronic state S_0, made up, as already mentioned, of a continuum of rotational and vibrational sublevels, among which the molecules are distributed in a manner characteristic of thermal equilibrium at the liquid temperature. The absorption of light by ground state molecules is through transitions to the complex of rotation–vibrational sublevels of the first excited singlet state S_1. This results in excited state molecules whose energy distribution does *not* correspond to thermal equilibrium; therefore, through thermal agitation in the liquid environment, there are rapid downward nonradiative transitions within that complex to reestablish thermal equilibrium. This equilibrium distribution is reached in a matter of picoseconds (10^{-12} sec), with the largest population going into the lowest vibrational sublevel, which is the upper level of the fluorescence, and hence also of the laser transitions. This permits a buildup in the population of this laser level, since its lifetime against fluorescence, that is, transitions to some sublevel in the ground state (S_0) complex, the lower laser level, is on the order of nanoseconds (10^{-9} sec), that is, very much longer. The final phase of the pumping cycle is reached when the molecules very quickly return, again by nonradiative processes, to a thermal distribution among the sublevels of the ground state, with most molecules going into the lowest vibrational sublevel, leaving few in the lower laser level. This pumping scheme can be modeled as involving a 4-level system, if we regard the sublevels of S_1 above the upper laser level as one "level," and similarly for those in S_0 below the lower laser level. For given transition rates between these levels, it is possible to derive the steady-state population inversion and hence optical gain before oscillation. However, we will not pursue this beyond noting that the fluorescence quantum efficiency (the ratio of fluorescent to absorbed photons) for some dye solutions can approach 100%, a figure strongly dependent on the molecule and solvent.

It is because the observed fluorescence results from spontaneous transitions from the lowest sublevel in S_1 to upper sublevels of the ground state, an energy change smaller than that involved in absorption, that the fluorescence is shifted in wavelength relative to the absorption band. Fortunately, there is a spectral range (see Figure 14.17) in which the absorption has fallen to zero while the fluorescence remains strong; this, as we shall see, makes a dye laser possible.

Unfortunately, in addition to the desired fluorescence, other nonradiative processes can also de-excite molecules in the S_1 state, thereby degrading the fluorescence quantum efficiency. These may lead to a return to the S_0 ground state, or what is worse, they may lead to a crossing to T_1 in the other system, where they remain for a much longer average time, since there is no "allowed" radiative transition to a lower state. As the pumping proceeds, the molecules will consequently accumulate there, presenting a serious problem, since absorbing transitions in the triplet system $T_1 \rightarrow T_n$ partially overlap the wavelength of the fluorescent photons, and therefore also of the photons involved in laser action. This unfortunately becomes a determining loss factor that prevents continuous laser action by a given group of molecules. Ultimately, they will return through "spin-forbidden" transitions to the ground S_0 state, possibly accompanied by weak light emission, called *phosphorescence*. There are two approaches to getting around the difficulty of the triplet state: The first is to operate in a pulsed mode, and the second is to achieve CW operation by continual renewal of the dye molecules by making the dye solution flow through the pumping beam.

The former option of pulsed operation can be achieved either by pumping with a pulsed laser, such as the nitrogen laser, or a xenon flash lamp. The preferred source is a (molecular) nitrogen laser, which is remarkable in having an output whose wavelength is in the near ultraviolet at 337.1 nm and is in the form of very short (10 nanosecond) high-power pulses, on the order of 100 kilowatts peak. The quantum-level structure of the N_2 molecule is such that the population of the lower laser level can build up rapidly, destroying the population inversion needed for continued oscillation, and the laser is self-terminating. To make it oscillate, the excitation (pumping) must be extremely fast; this is achieved in practice by passing almost instantaneously an intense electron current at high voltage laterally through a column of gas at a pressure of around 10^4 Pa. This requires a special high-voltage (tens of kilovolts), low-inductance circuit for discharging a capacitor through the gas. The maximum pulse repetition frequency is ultimately limited by the rate at which the heat generated by the discharges can be dissipated, typically 60 pps.

The nitrogen laser is eminently suitable as a pumping source, since not only does its output wavelength at 337.1 nm overlap the absorption band of many dyes, but what is equally important, the pulse widths are far shorter than the mean time for T_1 buildup through $S_1 \rightarrow T_1$ intersystem transitions. Furthermore, the high peak power makes it possible to reach threshold population inversion in those dyes that because of a relatively low fluorescence quantum efficiency cannot be operated with a flash lamp.

For CW operation, the dye solution must flow rapidly through the region of interaction with the laser pumping beam, most popularly an Ar^+ ion laser. Owing to the high power density of a focused laser beam, the windows through which the beam enters and leaves the dye stream are a common source of trouble in CW operation. A method of circumventing this problem is to use an unenclosed fast-flowing jet in the form of a thin ribbon, produced using a specially designed nozzle as part of a circulating system for the dye solution. In order to establish a sufficiently stable jet, it has been found that the viscosity of the solvent is important; ethylene glycol is often used. The ribbon of dye solution is intercepted at the Brewster angle (55° for ethylene glycol) by a CW pumping laser beam, which therefore irradiates continually different molecules. Since the pumping laser beam can be focused down to a fraction of a millimeter, with flow rates of 1–10 m/sec, transit times across the beam can easily be made shorter than the T_1 population buildup time. Circulating the dye solution also permits it to be cooled and filtered.

The optical design of a CW dye laser must accommodate the pumping laser as well as a provision for narrowing the frequency bandwidth of the optical gain around the desired wavelength. There are obviously numerous possibilities, generally classified according to whether the pumping beam is parallel to the laser beam in the cavity or perpendicular to it. Two critical optical requirements must be met: The pumping beam must be sharply focused onto the active dye medium, where a cavity mode must have a small beam waist. Figure 14.18 shows an example of a commercial design using a three-mirror folded cavity with a coarse tuning wedge and a fine tuning etalon. In this design the focusing of the pumping laser onto the tilted dye jet is corrected for astigmatism by the tilted input mirror, and the waist of the cavity beam is made small at the dye jet by the short distance between the concave mirrors. (In a confocal cavity, the beam waist at the center of the cavity can be shown to depend on $\sqrt{L}$, where L is the distance between mirrors.)

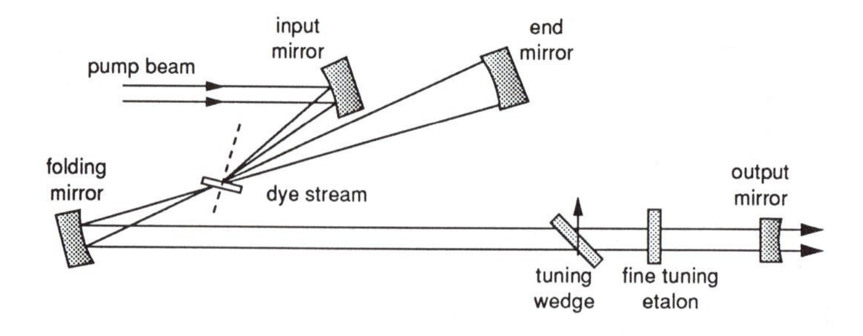

Figure 14.18 A typical commercial dye laser design using a 3-mirror folded optical cavity

The features that set dye lasers apart from other lasers and make them so important are their high quantum efficiency and gain over broad frequency bands. This means not only that they are tunable to yield an output at any desired wavelength, but also that they are capable of fast modulation. Thus by locking the phases of the cavity modes in a technique called *mode locking*, a pulsed output may be obtained with pulse durations in the picosecond (10^{-12} second) range, and by building on that technique pulse widths in the femtosecond (10^{-15} second) range! Each dye covers only a limited part of the optical spectral range, and different dyes with their appropriate high reflectance mirrors must be used to cover different parts of the spectrum. However, by doing so, it is possible to generate useful power over the full visible range, from 400 nm to 800 nm. Figure 14.19 reproduces the output power curves for different dyes covering different portions of the spectrum. It is seen that rhodamine 6G and oxazine will yield an output of over 0.5 watts for a pumping power of 4 watts from an Ar^+ ion laser.

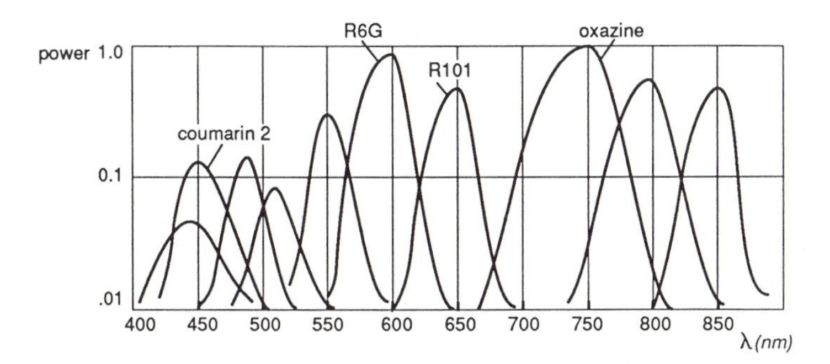

Figure 14.19 Typical output power as a function of wavelength for different dyes

14.10 Semiconductor Lasers

14.10.1 The p–n Junction

The last of the laser classes it is appropriate to include here is that of semiconductor junction lasers. As tunable coherent sources in the red and infrared regions of the spectrum, semiconductor lasers are important not only in micro-optics and as transmitters in optical fiber communication systems, but also in the present context of frequency standards in that region of the spectrum.

We take up first the fundamental process of light emission by a junction of p-type and n-type semiconductors when charge carriers are *injected* into it by the application of an external voltage across it. Figure 14.20a illustrates schematically the energy band diagram of an isolated p–n junction showing the equilibrium condition where E_F is the same on both sides of the junction. To achieve this, the energy bands are displaced relative to each other by the formation in the junction (or *depletion*) region of a "dipole layer," consisting of positive donor ions and negative acceptor ions, caused by the diffusion across the junction of electrons and holes respectively. If now an external voltage source is connected to the junction with the p-region connected to the positive side of the source and the n-region to the negative, that is, *forward biased*, the equilibrium is disturbed, and the Fermi levels are no longer aligned across the junction; the result is that the step in potential energy across the junction is lowered, as shown in Figure 14.20b. The majority carriers, conduction electrons in the n-region, and holes in the p-region are now able to penetrate the junction region in far greater numbers than for an unbiased junction. The result is that in that same region we now have electrons occupying higher-energy states predominantly near the bottom of the conduction band, and vacant lower-energy states near the top of the valence band. The possibility therefore exists of electrons making downward transitions between these states, with the emission of radiation. However, for this radiative process to occur, the laws of conservation of energy and linear momentum must be obeyed. Now, the energy is conserved simply by the emitted photon energy $h\nu$ being equal to the change in energy of the crystal as a result of the transition: $h\nu=(E_1-E_2)$, while the conservation of linear momentum requires that the linear momentum **p** carried by the photon, as given by the de Broglie formula $\mathbf{p}=(h/2\pi)\mathbf{k}$, should equal to the (vector) change in the momentum of the electron making the transition. If we substitute actual numerical values, for example, for GaAs, whose band gap is about 1.44 electron volts, we quickly find that an emitted photon has a wave number $k \approx 7.25\times10^6$ m^{-1}. The wave number of the electron, on the other hand, is on the order of π/a, where a, the crystal lattice spacing, is on the order of 0.1 nm; hence for the electron, $k\approx3\times10^{10}$ m^{-1}, that is, about 4000 times greater. This means that the electron wave number must not change more than one part in 4000 in a transition. Since the electrons are concentrated near the bottom of the conduction band and the holes near the top of the valence band, a significant rate of radiative transitions occurs only when the minimum of the conduction band comes at the same k value as the maximum in the valence band, that is, in crystals like GaAs having what we have called a direct energy gap. Crystals that fail to satisfy this condition, such as silicon and germanium, have radiative efficiencies orders of magnitude lower than GaAs, because in order to conserve linear momentum, the emission of a photon must be accompanied by the improbable simultaneous emission in the crystal of a quantum of acoustic energy, called a *phonon*.

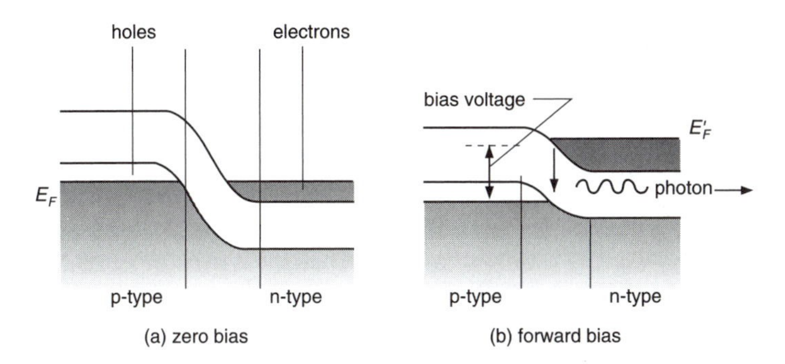

Figure 14.20 The energy bands in (a) an unbiased p–n junction and (b) a forward biased junction

14.10.2 The Gallium-Arsenide (GaAs) Laser

If no provision is made to meet the feedback conditions for laser action, passing a current through a p–n junction in a direct semiconductor simply leads to the emission of incoherent radiation, which in the case of GaAs is around λ=870 nm in the near infrared. As such, it would be called a *light emitting diode* (LED), a member of a class of devices that includes diodes using other combinations of elements such as GaP and ternary (3-element) combinations such as $GaAs_{1-x}P_x$ to have emission at other wavelengths. They are of great importance as light sources in fiber-optic communication systems and are also generally familiar in LED displays.

The physical design of a basic GaAs diode laser is illustrated schematically in Figure 14.21. Because of the relatively large number density of electron–hole pairs, the linear optical gain is high; the optical "cavity" is therefore very short, amounting to perhaps a few microns, with cross-sectional dimensions of a few hundred microns. Moreover, the reflecting surfaces forming the cavity need not have very high reflectance; in fact, it is common simply to use the reflection from cleaved (or polished) faces of the crystal perpendicular to the junction plane, as shown in Figure 14.21. The refractive index of a GaAs crystal at the wavelength of the emitted light is $n\approx3.6$, and thus using Fresnel's formula for the (intensity) reflectance at perpendicular incidence, we find $R\approx32\%$, a small figure indeed, but sufficient for this type of laser. The threshold current density to achieve oscillation is very high in the kind of p–n junction we have been describing, called a *homojunction*; it is on the order of 20,000 $A{\cdot}cm^{-2}$ at low temperatures, and more than twice that at room temperature, so that the heat generated limits operation to a pulsed mode. The principal reasons for this high threshold current have to do with the factors that determine the "population inversion," which in this context means the density of electrons that can be maintained in the junction region against nonradiative electron–hole recombination. The average number density of conduction electrons in the junction region, below threshold, is determined by the current density injected into the junction, the thickness of that region, and the average lifetime of an electron against nonradiative recombination. In fact, on the basis of a simple model, we can obtain the following relationship:

$$\rho_e = \frac{J\tau}{ed}, \qquad 14.23$$

where τ is the average electron lifetime, d is the thickness of the active region, and e as usual is the electronic charge. This shows that a large ρ_e is favored by having τ as long as possible and d as small as possible. The former requirement is met by making the crystal as free of impurities and imperfections as possible, while the latter suggests that it would be advantageous if the electrons could be confined to a smaller space in the active region.

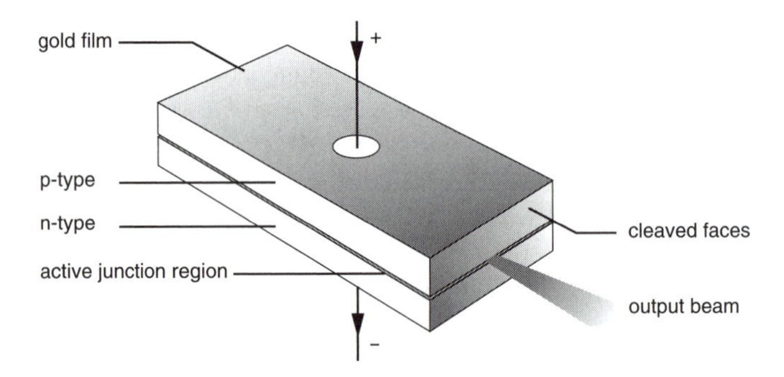

Figure 14.21 A gallium arsenide diode laser

14.10.3 The Heterojunction Lasers

It was in 1969 that reports were first published of the successful development of a new class of semiconductor lasers having greatly reduced threshold currents at room temperature (Panish, 1969). These were called heterostructure lasers but are now more commonly referred to as *heterojunction* lasers. Instead of a simple junction between p- and n-types of the same semiconductor, this new class has one or more layers on either side of the GaAS, of sharply different energy gaps between the valence and conduction bands, as shown in Figure 14.22. The effect of the sudden jump in band gap is to present the injected electrons with a potential barrier that tends to concentrate them in the active region, even as the temperature is raised, and also to act as a wave guide that reduces the loss of stimulated radiation. For example, a double heterojunction laser is composed of four layers: n–GaAs, n–$Ga_xAl_{1-x}As$, a thin p–GaAs layer, and a p–$Ga_xAl_{1-x}As$ layer. These layers are formed epitaxially; that is, the 3-element solid solution $Ga_xAl_{1-x}As$ forms a crystalline layer that continues, without a significant break, the crystal lattice of the GaAs. For this to succeed, it is obviously necessary that the lattice of the substrate crystal and the would-be epitaxial layer be matched as closely as possible to avoid lattice dislocations, which could serve as nonradiative recombination centers. It was expected that this match should exist in the case of $Ga_xAl_{1-x}As$ and GaAs, since it was known that the lattice parameters of GaAs and AlAs are nearly the same. To achieve a sharp interface between the layers, they are grown using liquid-phase epitaxy rather than by diffusion from the vapor. The heterojunction design dramatically reduces the threshold current density to the neighborhood of 1000 amperes·cm^{-2}, and permits CW operation at room temperature.

The output of these lasers as the threshold current is approached shows a rapidly increasing intensity called *superradiance* with a broad spectrum, until as threshold is passed, the output spectrum breaks up into a number of intensity peaks equally spaced in frequency. These peaks correspond to the various longitudinal modes the optical cavity can support within the gain curve of the laser diode. Further increase of the current results in the intensity becoming increasingly peaked around one of these modes. For true single-mode operation, a number of different approaches have been studied; in one interesting design, distributed feedback is used to select a particular wavelength. In this a regular array of corrugated ridges is formed in the interface between layers to act in an analogous way to an interference filter, allowing a narrow wavelength band to resonate with low loss. The output frequency of these layers is sensitive to temperature, a fact that detracts from their usefulness where extreme frequency stability is required; however, it does allow them to be tuned over a narrow range, and they may therefore be locked to a standard.

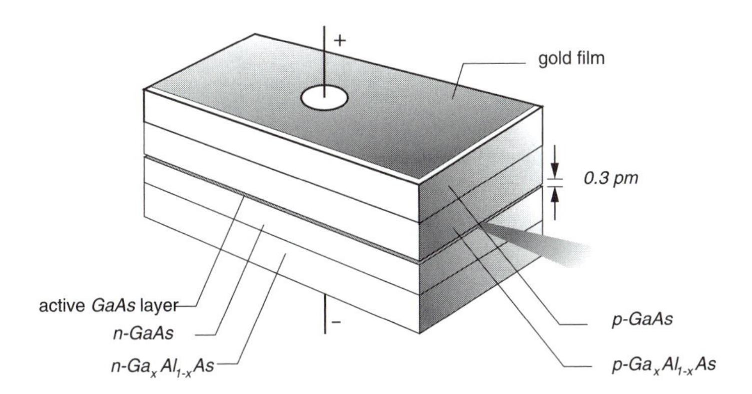

Figure 14.22 A Ga–Al–As heterojunction laser

Chapter 15
Laser Cooling of Atoms and Ions

15.1 Introduction

As generators of intense, spectrally pure radiation at optical frequencies, lasers serve not only as stabilized sources for frequency standards in that region of the spectrum, but equally importantly, they have changed the entire prospects of the *microwave* standards. They provide a means of slowing down the thermal motion of atomic particles. This is critically important when the resonance frequency of a transition in an atom or ion is used as a reference, since it is essential that the Doppler frequency shifts due to the particle motion be eliminated as far as possible. This can be accomplished by what is now called *laser cooling*, a truly remarkable technique, which we take up in this chapter. It is a technique that has made it possible to reach particle velocities corresponding to temperatures only a small fraction of a degree above absolute zero, where all thermal motion ceases.

An understanding of how laser cooling of atomic particles is possible begins with the realization that the interaction of a light beam with such a particle cannot only affect the internal motions within the particle, but also its center-of-mass motion. This ultimately derives from the fact that a light wave not only conveys energy but also carries with it momentum, both linear and angular. Long before the famous $E=mc^2$ formula of Einstein's theory of relativity was established, giving the equivalence of mass and energy, Maxwell's classical electromagnetic theory of light predicted that light falling, for example, upon a reflecting surface exerts a pressure on it, proportional to the intensity of the light. Even before that, the question of whether a light beam carried momentum was a subject of experimentation and philosophical debate for two centuries. The central issue was whether a light beam was a wave in the *ether* or a stream of minute particles, as held by the so-called *corpuscular theory*. Adherents of the latter, which incidentally included Isaac Newton, believed that the correctness of their theory would be confirmed if it could be shown that a light beam possesses momentum, as would a stream of any material particles. To resolve the question, many investigators directed powerful light beams at delicately suspended objects and looked for any minute movement in those objects that could be attributed to the impact of particles. In one such experiment carried out in mid-18th century, a very thin sheet of copper was delicately suspended and a beam of sunlight was directed at it by means of a mirror; a deflection of the copper sheet was observed, and although the presence of heating of the air near the surface of the copper was recognized as contributing to that deflection, there were some who believed that nevertheless the observations confirmed the mechanical impact of the light beam on the surface. Subsequent experiments by other investigators failed to show any movement in the irradiated object distinguishable from the effects of heat; this shows, it was concluded, that light cannot consist of particles and must therefore be vibrations in some medium. Among those taking the apparent absence of a deflection as an argument favoring the wave theory was Thomas Young, whose light interference experiments perhaps clouded his judgment. For it had previously been pointed out by Euler, the great Swiss mathematician, that light pressure might just as reasonably be expected from the wave theory as from the corpuscular theory. It is a remarkable historical fact that Euler not only correctly postulated the existence of light pressure but also, following a suggestion by Kepler, developed a theory to account for the tail of a comet based on the pressure of light from the sun acting on small dust particles in its head.

15.2 Light Pressure

An argument can readily be made for the existence of light pressure on the basis of Maxwell's electromagnetic wave theory of light. Thus a light beam falling perpendicularly, for example, on a reflecting metal surface must be pictured as subjecting the conducting surface to mutually perpendicular electric and magnetic fields oscillating at optical frequency in a plane parallel to the surface. The electric current produced in the metallic surface by the electric field will be acted on by the magnetic field, producing a Lorentz force perpendicular to both fields, that is, in the direction of the incident wave. The physical origin of this force is identical to the force on a current-carrying conductor in a magnetic field, as in an electric motor. The magnitude of the pressure this causes on the surface is derived in Maxwell's original theory in terms of the mechanical stress transmitted through the *ether* the postulated universal medium. However, with the abandonment of the concept of a universal medium, as required by Einstein's theory, and in order to preserve the conservation law of linear momentum, Maxwell's field equations are now interpreted as leading to the electromagnetic field *itself* carrying momentum. From this it follows that to deflect a light beam requires a force to be exerted, with an equal reaction on the object providing that force; hence the radiation pressure on light-scattering objects. The amount of momentum (per unit volume) represented by M, which we must attribute to the field in a light beam of intensity I (watts·m^{-2}), can be shown according to classical theory to be given by

$$M = \frac{I}{c^2}, \qquad 15.1$$

where as usual, c is the velocity of light. We can interpret the intensity I as the flow at the velocity c of energy residing in the field and distributed with density ρ_E (joules·m^{-3}), with a similar interpretation for the amount of momentum M. The classical result suggests that we must attribute to the field a mass density ρ given by the following:

$$\rho = \frac{\rho_E}{c^2}. \qquad 15.2$$

This is recognized as a special case of Einstein's $E=mc^2$, which applies to any form of energy. That it should appear in classical electromagnetic theory is not surprising, since Einstein's theory was constructed on the basis that the way Maxwell's equations transform from one coordinate reference frame to another should be true of all physical laws, including those of mechanics. It was this that necessitated the revolutionary changes in the concepts of space and time that are characteristic of Einstein's theory.

If a light beam of intensity I falls on a perfectly absorbing surface, so that the directed energy of the beam is converted to random thermal motion with zero net momentum, there is a continual change in momentum of the beam, which must be taken up by the absorber; hence by Newton's laws of motion there will be a pressure P exerted on the surface equal to the rate of change of momentum. Since the amount of momentum change that occurs per unit time over unit area is Mc, this being the amount of momentum carried by photons in a cylinder of unit cross section and length equal to the distance light travels in one second, the pressure is simply given by the following:

$$P = Mc = \frac{I}{c}, \qquad 15.3$$

where I (watt·m^{-2}) is the beam intensity and c is the velocity of light.

The same result immediately follows from the de Broglie formula, which applies to all particles, including photons, namely $p=h/\lambda$, where p is the linear momentum of the particle. If we suppose the light beam consists of a stream of photons with a flux density (number of photons crossing unit cross-sectional area per unit time) represented by j, then clearly we have $I=jh\nu$. Furthermore, the rate at which momentum is lost from the beam over unit area of the surface of an ideal absorber, which by Newton's law is the pressure, is given by $P=jh/\lambda$, and the relationship $P=I/c$ follows.

Numerically, the size of radiation pressure is extremely small on the ordinary scale of things; for example, the pressure due to direct sunlight, whose intensity is on the order of 1000 watts·m^{-2}, is only about 3×10^{-6} newton·m^{-2}, that is, about $3\text{x}10^{-11}$ times atmospheric pressure. It is little wonder that so much experimental difficulty was encountered in finding unambiguous evidence of its existence. A sensitive device, now a scientific curiosity called "Crookes' radiometer," was devised to demonstrate radiation pressure by William Crookes in 1875, the year that Maxwell's treatise was published. It consists of light metallic vanes blackened on one face and shiny on the other, delicately balanced and free to rotate inside an evacuated tube. It does indeed spin when exposed to a sunbeam. However, it soon became evident that the observed spinning of the vanes had more to do with thermal effects due to the residual gas surrounding them than to any radiation pressure. Later experiments, however, did confirm the essential validity of the theory.

15.3 Scattering of Light from Small Particles

With the advent of the laser it became possible to exploit a different approach to studying light pressure, with far-reaching ramifications. The ability to achieve high intensities of monochromatic light by focusing a laser beam down to diameters on the order of microns meant that the motion of individual small particles could be observed. Moreover, this can be done in highly transparent nonabsorbing materials, simply relying on the reflection and refraction of light to bring about a change in the particle's momentum; this alleviates the problems associated with heating effects. In one such experiment, reported by Ashkin in 1970, small transparent spheres of latex having diameters on the order of one micron (10^{-6} m) were freely suspended in pure water. An Ar^+ ion laser beam passed through a glass cell containing the suspension of latex spheres and was manipulated to converge to a radius of about 6 microns on an individual sphere, as shown in Figure 15.1. The strong scattering from the spheres enabled their motion to be observed visually by means of a microscope. It was found that not only will a sphere be driven in the direction of the beam by radiation pressure, but also, if the sphere is initially off the beam axis, it will be drawn toward it. This Ashkin explained using an argument based on the theory of light pressure and the approximate concept of light rays. This is valid to the extent that the wavelength of the light is very much smaller than the radius of the sphere; otherwise, a proper wave solution must be sought. The solution to the problem of the scattering of electromagnetic waves from small particles is of considerable importance in such fields as atmospheric optics and optical astronomy; the case of plane waves scattered by a homogeneous sphere permits a general solution, first given by Mie around 1908. For the purposes of an approximate, qualitative explanation of the behavior of the particles in the laser beam, there should be little error in assuming ray optics. First we note that the refractive index of latex at the wavelength of the laser light (n=1.58) is greater than that for water (n=1.33), and therefore the light falling on a sphere will be converged on the emergent side, as by a thick convex lens, as shown Figure 15.2. This means that part of the linear momentum of the beam is deflected away from its axis and the remaining component along the axis reduced; therefore, to conserve momentum the latex sphere must experience a force along the axis in the direction of the initial beam. Moreover, if the sphere is off-axis and the intensity of the beam varies over its surface, there will be an asymmetry in the amount of momentum deflected toward the axis versus the amount away from the axis; the result is a net force on the sphere in the direction of increasing intensity, that is, toward the axis. The result would have been in the opposite direction had the sphere been of a lower refractive index than the medium, for example an air bubble.

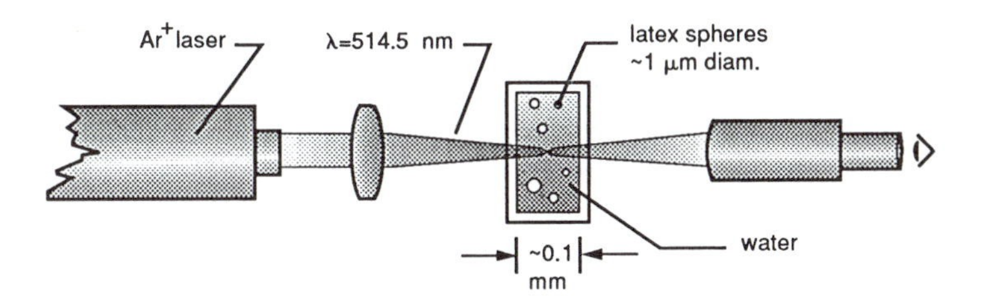

Figure 15.1 Ashkin's experiment to study forces on latex spheres due to a laser beam

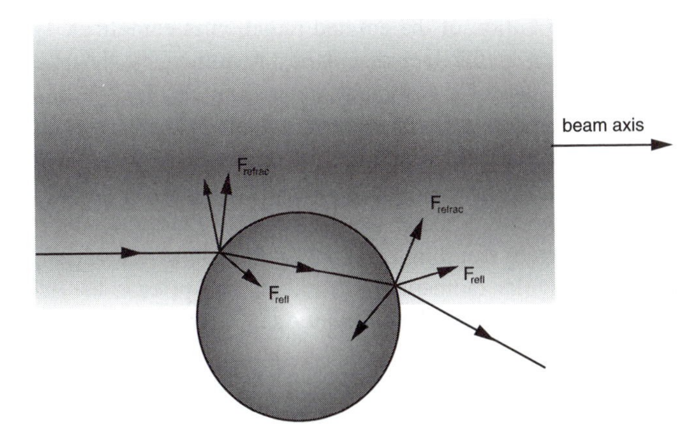

Figure 15.2 The balance of radiation forces acting on a dielectric sphere according to Ashkin

15.4 Scattering of Light by Atoms

In extending the idea of linear momentum exchange to that between a laser beam and free atoms or ions, we are faced with the need for an altogether different and more complicated quantum description of the interactions involved. One of the important processes that takes place has already been met in the optical pumping of free atoms and ions, namely resonance fluorescence. In this, the atomic particle will interact strongly only if the radiation has a frequency resonant with one of a set of discrete frequencies in its spectrum. Thus resonant absorption of a photon from a beam is accompanied by a quantum transition to a higher energy state, from which the particle ultimately falls back to the ground state, radiating its excitation energy in all directions in a definite angular pattern. It is as though photons were taken out of the beam and scattered in all directions away from the atom or ion. The net transfer of linear momentum resulting from this process is obtained by applying the law of conservation of linear momentum between the initial state of the beam–atom system prior to the interaction, and a final state, long after the interaction, consisting of the remaining beam continuing in the original direction and any photons radiated from the atom. The relative probability of the photon being radiated in the different directions, that is, in the language of antenna engineers, its radiation pattern, must conform to the symmetry laws governing the system. In particular, a free atom spontaneously emitting a photon does *not* do so preferentially along any *one* particular direction; this assumes that in the process of spontaneous emission the atomic particle has no "memory" of the excitation by the laser beam. This certainly would not be the case for stimulated emission. It does not mean that the radiation pattern must be isotropic, with equal intensity in all directions; however, it does mean that for any given direction the intensity will be equal to that along the opposite direction, as shown in Figure 15.3. The detailed angular distribution of the emitted radiation is complicated, as we saw in the rubidium case, by the fact that it is determined by the change in the *angular* momentum state of the atom or ion accompanying the transition. We recall that a photon is endowed with an intrinsic spin of one unit of $h/2\pi$, and the photon–atom system exchanges angular momentum as well as linear momentum. If after the absorption and re-emission of a photon the atom is left in a different angular momentum state, then in order to conserve angular momentum, the radiation pattern resulting from the photon–atom interaction must carry the balance of the angular momentum the photon had prior to the interaction. It is this type of exchange on which Kastler's optical pumping technique for magnetic resonance in free atoms is based. In any event, as already stated, there can be no preferred direction of spontaneous emission, and therefore the expected total linear momentum carried off by the *scattered* photon is zero. It follows that in order to conserve linear momentum, the linear momentum of the absorbed photon must be taken up by the atom or ion. Thus as photons are absorbed out of a beam by the atomic particle, its momentum changes incrementally in the direction of the beam as if it was impelled by a succession of impulses, producing an average force.

In order to establish the relative scale of these effects, a knowledge of which is essential to distinguish between what is observable in practice from what may be only of theoretical interest, let us assume that a laser beam of intensity I watts·m^{-2} is directed exactly in the direction opposite to the velocity of an otherwise free particle, so that the flux density of momentum carried by the photons is, as we have seen, I/c. If a given particle presents an absorption cross section to the beam of σ, then it will suffer, on the average, a change in momentum of $(I/c)\sigma$ units per second, which by Newton's law is the measure of the force acting on it. Now, if we assume that the conditions are such that this force continues unchanged while the particle is slowed down (this assumes that the spectrum of the light is broad compared to the maximum Doppler shift of the particle resonance frequency), then the distance traveled by the particle before it is brought to rest can be obtained simply by using the conservation of energy: If the stopping distance is D, then the work done on the particle by the radiation pressure is $F_{rad}D$, and therefore $F_{rad}D=\frac{1}{2}MV_z^2$; that is, finally we have the following:

$$D=\frac{1}{2}MV_z^2\frac{c}{I\sigma}=\frac{1}{2}kT\frac{c}{I\sigma}, \qquad 15.4$$

for particles in thermal equilibrium at temperature T. If we substitute realistic numerical values, for example $I=10^2$ watts·m^{-2} and $\sigma\approx4\times10^{-14}$m^2, we find for T=300K that $D_{stop}\approx50$ cm, a very practical figure.

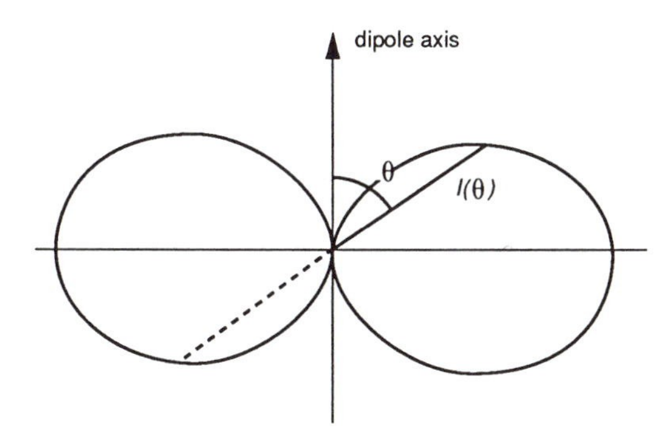

Figure 15.3 The polar diagram giving the radiation pattern from an atom undergoing an electric dipole transition

Of course, the beam can be focused down to a smaller cross section to increase the intensity at the ion, thereby reducing the braking distance. However, we should recall that it takes a finite time (the *radiative lifetime*) for the emission of the photon from the excited state, and therefore obviously the rate at which the absorption–emission cycle can be repeated is limited by this lifetime. As the intensity is raised to the point where the probability of absorption becomes comparable to the that of spontaneous emission, the particle will be increasingly in the excited state, and stimulated emission will become significant. The radiation pattern from particles undergoing stimulated emission is radically different from that of spontaneous emission, ultimately because unlike the latter, stimulated photons are phase coherent with the photons in the beam. The photon resulting from stimulated emission in fact propagates in the same direction as the photon stimulating the emission; consequently, the atom momentum is unchanged in an absorption-stimulated emission cycle. The result is that the simple description we have given for momentum exchange with spontaneous emission does not apply even approximately to the case where stimulated emission enters the picture.

15.5 Optical Field Gradient Force

There is another mechanism by which an atomic particle interacting with an optical field may experience a force affecting its center-of-mass motion; it is manifested only when the intensity of the optical field is strongly *inhomogeneous*. The physical origin of this force, sometimes called the *gradient force*, and some of its features can be understood in general terms using classical field concepts rather than the proper quantum description. We recall that the positive (nuclear) and negative (electronic) charges in an atomic particle are displaced relative to each other by the electric component of the optical field against their mutual attraction, thereby inducing an electric dipole moment. By virtue of its position in the electric field, this induced dipole possesses potential energy proportional to E^2, where E is the electric field component in the light wave. The dynamical response of the charges to the oscillating electric field is a quantum problem; but suffice it to say that it exhibits resonant behavior at certain frequencies. It is found that the dipole energy is positive or negative depending on whether the optical frequency is above or below a resonant frequency of the particle. This behavior can be understood in terms of the reversal in the phase of the charge displacement (and hence direction of the induced dipole) relative to the field, which occurs for field frequencies on opposite sides of the resonant frequency. Although the field oscillates at the optical frequency, the energy of the polarized atomic particle, being proportional to E^2, does not average to zero over time, as we saw in connection with the light shifts in the energy separation of the hyperfine states in Rb. As a consequence, the particle has an average potential energy that varies with the intensity of the field where it happens to be located. As with the AC Stark effect leading to the light shifts, this potential energy is positive or negative depending on the optical frequency in relation to the resonance, in strict analogy with an object having potential energy by virtue of its position in Earth's gravitational field above or below, say, sea level. If the intensity of the optical field varies in space, as in a laser beam with a Gaussian intensity profile, there will be a strong gradient in the particle energy, with the result that, to continue the analogy, it tends to fall like a rock down a gradient.

We will not pursue the gradient force any further and will limit ourselves in what follows to radiation pressure; later in the chapter we will learn of an exciting new mechanism involving *polarization gradients*.

15.6 Doppler Cooling

If an atomic particle had an initial velocity component in a direction opposite to a laser beam tuned to exact resonance, that component would be steadily reduced and ultimately pass through zero and increase in the reverse direction. That clearly would not lead to a stationary state in which the particle motion has been slowed, corresponding to a lowering of temperature; what we would have is a propulsion mechanism, rather than a cooling one. To achieve a slowing down of the particle such as might be caused by a frictional force, it is necessary that the particle experience a force opposing its motion, no matter in what direction that motion happens to be. An obvious solution, one might think, is to have *two* laser beams, one resisting motion in one direction and the other in the opposite direction. Unfortunately, the effects of the two beams would cancel each other out, unless the radiation pressure exerted by a given beam depends on its direction relative to the motion of the atomic particle. For if we assume for simplicity that the particle experiences a force only when it is moving in the direction opposite to that of a given beam and no force otherwise, then it is clear that having two opposed laser beams will cause a retarding force to act on the particle by either one beam or the other, depending on which direction it is moving in. This crucial condition on the laser–particle interaction can be met if the *absorption* probability depends on whether a particle moves *in* the direction of the beam or *opposite* to it. But in fact, we know that this is in general the case: the Doppler effect shifts the frequency of the field seen by a moving particle, and therefore only if the laser frequency is tuned precisely to the center of the resonance in the particle frame of reference will the probability of absorption be the same for either direction of motion.

Suppose then that an atomic particle is placed in two opposing collinear laser beams, and for simplicity we restrict our attention to the component of the particle motion along the common beam axis. The frequency profile of the atomic absorption will be assumed to have a certain natural line width and be symmetrical about the maximum at its center. Let us consider the interaction of the particle with the laser beam directed opposite to its velocity component, causing a Doppler displacement of the laser frequency seen by the atom towards a higher frequency. It is possible then that by tuning the laser so that its frequency would fall on the *low* side of the resonance profile if the particle were at rest, the Doppler shift would result in the absorption becoming *more* likely, because of the upward shift in frequency toward the maximum, than it would be if the particle were at rest. And similarly, the Doppler shift would cause the absorption to be *less* likely if the atom were moving in the same direction as the laser beam. The result is that such an atom experiences a much greater force opposing a component of its motion directed opposite to the beam direction than one accelerating a component in the same direction as the beam. The total effect in the presence of both laser beams is that there is a net force opposing the motion in either direction, as desired. Ideally, therefore, the continued absorption of photons from the laser beams, with the frequency offset we have described, will cause an atom to be slowed down much like the effect of a frictional force.

In order to make this last statement a little less confusing and a little more quantitative, let us assume that the atomic particle is free (aside from the laser beams of course) and therefore has a Lorentzian resonance line shape characteristic of natural broadening due to the finite lifetime of its excited state. If we represent the full width of the resonance line by $\Delta\nu$ and assume that the two laser beams are tuned to the same frequency $\nu_L = \nu_0 - \delta$, that is, *below* the center of the line, then a plot of the relative strength of the *net* force exerted on the particle as a function of its velocity will be as shown in Figure 15.4. We note that for particle velocities V such that the Doppler shift $(V/c)\nu_L$ is less than δ, the dependence of the force on velocity is nearly linear and can be approximated for those values of V by the equation

$$F_{rad} = -\alpha V , \tag{15.5}$$

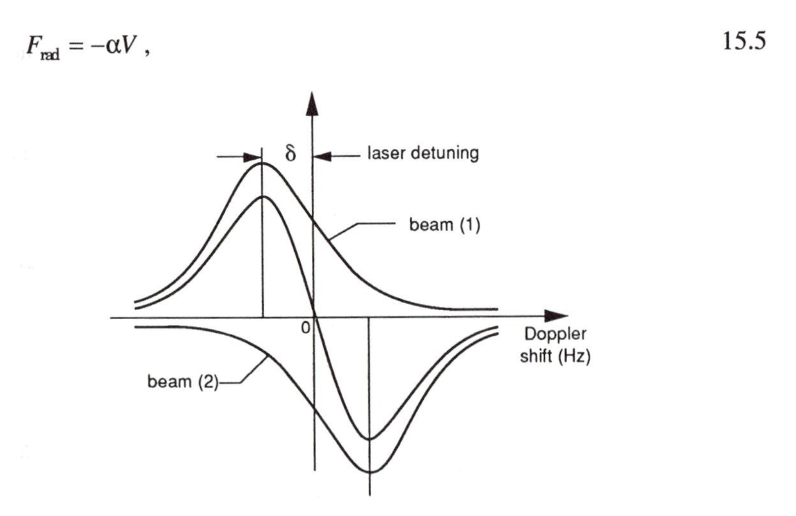

Figure 15.4 The net force on an atomic particle in two opposed laser beams tuned below resonance

where α is a positive constant analogous to the damping factor used to model friction in a mechanical system. The optimum choice of δ to obtain the greatest damping factor is $\delta \approx \Delta\nu/2$. Thus an atomic particle whose initial velocity produces a Doppler shift in the range between $+\Delta\nu/2$ and $-\Delta\nu/2$ (called the capture range) will experience a viscous drag tending to slow down its motion along the beams' axis. For beam intensities below the extreme where the rate of absorption becomes comparable to the rate of spontaneous emission, the value of the coefficient α is proportional to the intensity of the laser beams. However, for intensities beyond that point, the proportionality is no longer valid, and α cannot be increased indefinitely. Now we can estimate the rate of loss of kinetic energy by the atomic particle due to radiation pressure by assuming that the force F is the resultant of the two forces exerted by the opposing beams. If as before, we assume that the beams are tuned to $\nu_0-\Delta\nu/2$, and recall that the (linear) Doppler frequency shift can be written $(V/c)\nu=kV/2\pi$, then for a scattering cross section having the natural Lorentzian line shape, we can write for the resultant force the following:

$$\langle \mathrm{F} \rangle = \frac{I}{c}\left(\sigma^{+}-\sigma^{-}\right); \quad \sigma^{\pm}=\sigma_0\left(\frac{\Delta\nu}{2}\right)^2 \frac{1}{\left(\frac{\Delta\nu}{2}\pm\frac{kV}{2\pi}\right)^2+\left(\frac{\Delta}{2}\right)^2}, \tag{15.6}$$

where σ_0 is the scattering cross section at resonance, and k here represents the wave vector $(2\pi/\lambda)$. In the range of small velocities such that $kV \ll \Delta\nu$, we can approximate the expressions for $\sigma^{\pm}$ to obtain simply

$$\sigma^{-}-\sigma^{+}=\sigma_0\frac{kV}{\pi\Delta\nu}. \tag{15.7}$$

Hence the net force can be written

$$\langle F \rangle = -2I\sigma_0\left(\frac{\nu}{c^2\Delta\nu}\right)V. \tag{15.8}$$

Now, the average rate at which this force does work on the particle in resisting its motion, and hence the rate of *loss* of kinetic energy by the particle, is simply $\langle FV \rangle$. It follows that in time t the average loss of kinetic energy is given by

$$E_k = 2I\sigma_0\left(\frac{\nu}{\Delta\nu}\right)\frac{V^2}{c^2}t. \tag{15.9}$$

15.7 Theoretical Limit

It might be thought that the slowing-down process would continue until the particle came to complete rest; but this is not the case. As the particle is slowed down, the Doppler shift tends toward zero, and the differential force between particle motion in one direction and the other is diminished. But the fundamental limit is ultimately set by the discrete quantum nature of the momentum exchange attending the absorption and emission of photons: The time-dependence of these events is defined only statistically by certain probabilities; and the direction of spontaneous emission is also statistically defined. At each of these random events the atomic particle recoils with a finite jump in momentum in a random direction. The result is that the atomic particle has a residual random motion similar to the erratic zigzag motion, called *Brownian motion,* of small particles caused by random collisions with surrounding molecules. First observed under a microscope by the English botanist Brown, who was working with a suspension of plant pollen in a liquid, the name is now applied more generally to include, for example, aerosol particles. Its existence is historically important in the kinetic theory of gases, since it demonstrated directly that thermal energy is, on a molecular scale, the kinetic energy of their random motion. Since the suspended particle may suffer collisions from any direction, the average of its displacement from a fixed point is zero; however, since it executes a "random walk" in the manner we discussed in a previous chapter, it diffuses out with a *mean square* displacement that increases linearly with the number of collisions, and therefore with time. In our present one-dimensional case, where the particle is subjected to impulses randomly distributed in time and direction, each imparting (h/λ) units of momentum through photon scattering from two opposing laser beams, the mean square momentum $\langle p^2 \rangle$ *increases* with time as follows:

$$\langle p^2 \rangle = 2\left(\frac{h}{\lambda}\right)^2 \frac{I}{h\nu}\sigma_0 t\,, \qquad 15.10$$

where I (watts·m^{-2}) is the intensity of *each* beam, σ_0 the (resonance) atomic cross section, and t the time.

Finally, if we make the reasonable assumption that these processes will lead to an equilibrium where the gain and loss of energy exactly balance each other, then according to our model this occurs when the kinetic energy is the following:

$$\frac{1}{2}MV_{\min}^2 = \frac{1}{4}h\Delta\nu\,. \qquad 15.11$$

Expressed in terms of an equilibrium temperature, this mean kinetic energy along one dimension must be equated to ½kT, where k is the Boltzmann constant; thus we obtain the lowest temperature T_{min} attainable by this Doppler technique:

$$T_{min} = \frac{h\Delta\nu}{2k}. \quad 15.12$$

If we substitute for $\Delta\nu$ the atomic resonance line width for cesium, that is, approximately 5×10^6 Hz, we find that the lowest attainable temperature is on the order of 120 μK, that is, 120 millionths of a degree above absolute zero! At this temperature the average linear momentum of a Cs atom reaches within an order of magnitude of the momentum of a *single* photon, the fundamental limit to cooling by this method. At such a super-cold temperature the second-order fractional Doppler shift (E_k/Mc^2) in the cesium resonance frequency would be entirely negligible at only one part in 2.5×10^{19}! In practice, of course, this degree of accuracy is meaningful only if a multitude of other sources of systematic frequency shifts, which no doubt become significant at this level, could also be as elegantly reduced or taken into account.

15.8 Optical "Molasses"

So far, the discussion has been limited to one dimension; clearly, for free particles lacking any mechanism for exchanging energy between velocity components in three dimensions, multiple sets of laser beams are required. A configuration consisting of three mutually perpendicular pairs of opposing lasers has been successfully used to cool a cloud of free neutral atoms. In one set of experiments at the Bell Telephone Laboratories, sodium atoms were observed not only to be cooled by the action of the lasers but also to manifest another effect of the viscous drag their motion is subjected to, namely a reduction in their diffusion rate. That is, a given group of atoms occupying a certain space take very much longer to diffuse out into a larger volume and are in this sense "confined" in what was dubbed "optical molasses."

Initial attempts to study these effects quantitatively, to measure the temperature of the atoms in the "molasses," yielded results that seemed to be compatible with the theory as outlined. It was assumed therefore that the well-understood processes invoked in the theory were adequate to explain fully the phenomena involved. However, it soon developed that with the application of a more refined technique to measure the velocity distribution of the cooled atoms by W. Phillips and his group at the U.S. National Institute of Standards and Technology, there was revealed a conflict between theory and observation. The temperature was, in fact, considerably lower than the lower limit predicted by the theory. Moreover, contrary to the theory, the lowest temperature was reached when the lasers were tuned not one half a line width, but several line widths from resonance. There was understandable skepticism at first about these contrary results; not only because of the well-established nature of the processes assumed in the theory, but also because experimental results generally fall short of theoretical limits and certainly should not exceed them! Nevertheless, further experiments on sodium and other atoms at the Ecole Normale Supérieure, in Paris, and other laboratories in the U.S. confirmed that the low temperature limit of Doppler cooling had been broken and an explanation had to be sought elsewhere.

15.9 Polarization Gradient Cooling "The Sisyphus Effect"

To find an explanation for the unexpectedly low temperatures, we must take a much more detailed view of the interaction of the sodium atoms with the laser beams. First, the alkali atom is not a two-level atom with a simple ground state and an excited state. Like the other alkali atoms, rubidium and cesium, which we discussed in earlier chapters, sodium's ground state has the spectroscopic designation $^2S_{1/2}$ with a spin angular momentum $J=1/2$, which couples to a nuclear spin $I=3/2$, giving rise to a hyperfine frequency of 1.771 GHz, which is, incidentally, less than one-third the value for Rb^{87}. In a weak magnetic field it has Zeeman substates with magnetic quantum numbers $m_J=-2, -1, 0, +1, +2$ for the $F=2$ hyperfine state and $m_J=-1, 0, +1$ for the $F=1$ state. Further, we recall the vast improvement in the observation of magnetic resonance in free atoms made possible by an optical method called by its originator, Kastler, *optical pumping*, which involves the absorption of circularly polarized resonance light followed by spontaneous re-emission. This technique exploits the selection rules governing optical transitions imposed by the law of conservation of angular momentum applied to the photon–atom system. By continued repetition of the optical pumping cycle, atoms are transferred predominantly into that substate in which the angular momentum is oriented in the direction of that of the absorbed photon. In general, then, we can expect that the detailed distribution of atoms among the different magnetic substates will depend on the type of polarization of the absorbed resonance light, since that determines the component of its angular momentum along the beam axis.

Following Cohen-Tannoudji and Dalibard (Cohen-Tannoudji, 1989) we look to one of the subtle effects in optical pumping long studied theoretically and confirmed experimentally, namely the displacement of quantum energy levels by the action of the pumping light (Cohen-Tannoudji, 1962), as providing a new cooling mechanism. We have already met this before in connection with the optically pumped rubidium standard: *light shifts*. The precise energies of the quantum states of an atom interacting with the pumping light beam are displaced slightly with respect to a free atom, and the displacement is in proportion to the intensity of the beam. Moreover, the energy displacements appear to an extent dependent on the probability of the transitions, which occur according to the selection rules applied to the type of light polarization present.

According to Cohen-Tannoudji and Dalibard, we will find that under suitable conditions, rapid spatial variation in the polarization of the light can lead to a lower temperature limit than that of Doppler cooling. To do this consider the idealized example of a one-dimensional "molasses" consisting of two opposing frequency-offset laser beams having mutually perpendicular linear polarizations and equal intensity. Because each laser beam is spatially coherent, that is, there is a well defined phase relationship between the optical field at different points along the beam, the fields of the two laser beams will, in the span of half a wavelength, have a phase difference that varies from zero to 360°. Thus if we let E_x and E_y represent the two counterpropagating light waves, we have the following:

$$E_x = E_0 \sin(\omega t - kz);\; E_y = E_0 \sin(\omega t + kz), \qquad 15.13$$

and therefore at a fixed point along the axis z_0, the two components will have a difference of phase $\Delta\phi = 2kz_0$. This means that at points within each half wavelength along the common beam axis, the phase difference between the two equal and perpendicular field vectors will pass continuously through the values 0°, 90°, 180°, 270°, 360°. Now, phase differences of 0° and 180° simply yield combined fields that are linearly polarized along directions at 45° to the original directions, while phase differences of 90° and 270° yield circularly polarized fields rotating in opposite senses about the beam axis. This pattern is repeated every half wavelength along the beam axis, as shown in Figure 15.5.

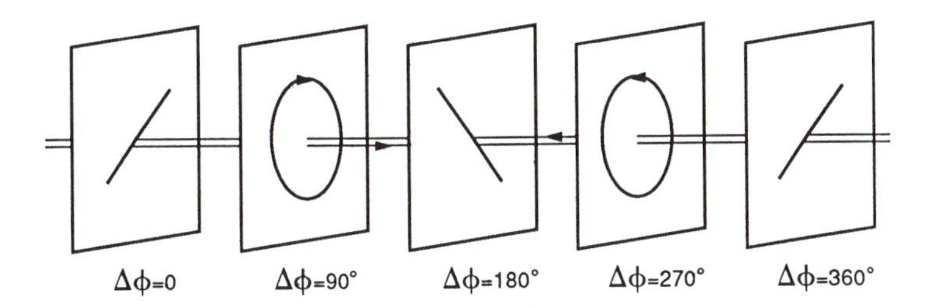

Figure 15.5 The polarization of the resultant of two counterpropagating beams with perpendicular polarizations

In order to bring out the consequences of having such variation in the light polarization on the optical pumping of an atom and the attendant light shifts, let us consider the specific example of sodium. Let us ignore its nuclear spin and assume that the optical pumping cycle occurs between the ground state having the total angular momentum quantum number J=1/2 and the excited state having J=3/2. We recall that space quantization of the angular momentum leads to a multiplicity of substates, so that the ground state is made up of substates in which the components of angular momentum along a given axis (chosen here to be in a fixed direction along the light axis) are m_J=+1/2 and −1/2, and the excited state m'_J=+3/2, +1/2, −1/2, and −3/2. Now, if the atom happens to be at a point where the polarization is circular, with the field rotating in the positive sense (the rotation of a right-hand screw advancing in the direction of the light beam), designated by σ^+, it can absorb a photon only if the excitation results in an *increase* in the m_J value by one unit. Of the two possible transitions, m_J=-1/2→m'_J=+1/2 and m_J=+1/2→m'_J=+3/2, it can be shown that the former will occur at one-third the rate of the latter, and consequently, the light shift for the latter m_J=+1/2 is greater. Moreover, it also happens that the optical pumping cycle tends to concentrate the atoms in this light shifted m_J=+1/2 state; this, after all, was the whole point of the Kastler optical pumping technique. Conversely, if the atom happens to be where the polarization state of the field is circular in the opposite direction, designated as σ^-, the m_J=−1/2 substate will suffer the greater light shift and become more populated than the other. At points where the polarization is linear, the two substates are equally shifted. Figure 15.6 illustrates the variation in the light shifts of atoms in the two substates as a function of their position within a range of one half wavelength along the axis of the laser beams. We see therefore that if we had a set of stationary atoms strung along the beam axis, they would alternate between being predominantly in one substate and predominantly in the other, with their energy varying in step. This picture is really a static one of an equilibrium situation, achieved after a sufficient time has allowed the pumping cycle to be repeated many times. As such, the spatial variation in energy we have found cannot, for the purposes of predicting the forces acting on a moving atom, be used as a static potential energy distribution analogous to hills and valleys on the earth's surface. The dynamical problem is considerably more complicated; we must take into account the time development of the pumping cycle itself, simultaneous with the varying polarization of the light seen by the moving atom. The static energy distribution found above would apply only if the pumping cycle took place so rapidly that the atomic substate population was in quasi-equilibrium at every point as the atom moved along the beam. The word "population" can refer to a single particle; it is used as the relative probability. Let us imagine the system set in motion: The pumping cycle repeats itself on the average every τ_p seconds, the mean pumping time, causing the populations to evolve in time, to alternate between one substate and the other, while the atom moves through the different polarization states of the light. Because of the delay in pumping the atoms into a particular substate, the population of that state reaches its maximum only after it has passed the point where the light shift is maximum, leading to an asymmetry in the population distribution relative to the periodic light shift, as shown in Figure 15.7. The result is that the atoms find themselves constantly being optically pumped into the substate whose energy is on the rising side of the energy curve, no matter in what direction they are moving. This brings to mind Sisyphus, the Greek mythological figure whose task in Hades was forever to push a rock uphill, only to have it roll back down; hence this method of atomic cooling has been dubbed *Sisyphus cooling*. Without entering into the difficult question as to how the photon–atom interaction leading to the light shifts can couple to the center-of-mass motion of the atom, we can use the law of conservation of energy merely to say that the rising light-shift energy leads to a corresponding loss of kinetic energy, that is, a cooling of the atom.

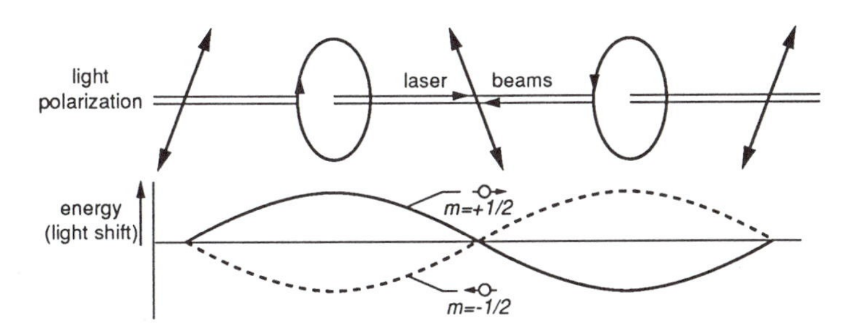

Figure 15.6 Light shifts in sodium atom in two opposed laser beams perpendicularly polarized

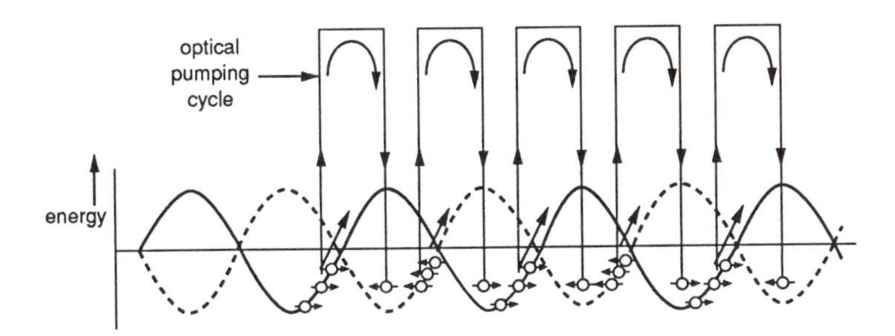

Figure 15.7 The asymmetry between light shift of energy and substate population distribution in sodium atoms (Cohen-Tannoudji, 1990)

The important question remains as to why Sisyphus cooling is effective at a much lower temperature than Doppler cooling. The answer must be sought in the essentially different physical processes involved in the cooling and the theoretical limits these processes set on the lowest achievable temperature. In the Sisyphus process it is the light shift in energy that plays a central role: If the potential well created by the light shift is designated as ΔE_m, then each optical pumping cycle results in the atom losing an amount of energy on the order of ΔE_m. After many such cycles, the atom will have lost so much energy that it no longer has enough energy to penetrate the adjacent regions, and it remains trapped. In this limit the thermal energy kT is on the order of ΔE_m. If the pumping light intensity is reduced, ΔE_m is reduced proportionately, and the limit can be far below the Doppler value of $h\Delta\nu_n$, where $\Delta\nu_n$ is the natural line width of the optically excited state.

Theory shows that the dependence of the friction coefficient α on the pumping light intensity is radically different for the two processes: In contrast with Doppler cooling, where α decreases linearly with the intensity as it approaches zero, in the Sisyphus cooling mechanism α remains essentially constant. To make this fact at least plausible, we note that although reducing the light intensity makes the light shift correspondingly smaller, this is offset by the lengthening of the pumping delay that enhances the population asymmetry, which, as explained above, is the root cause of the cooling effect. In Figure 15.8 are reproduced curves comparing the dependence of the average force experienced by the atom on the mean velocity for Doppler cooling and Sisyphus cooling. We see that in the latter, while the velocity capture range shrinks as the intensity of the pumping light is reduced, the slope of the linear portion of the curve where it crosses the origin (which is a measure of α) remains constant. It is apparent, then, that since the velocity capture range for Doppler cooling is greater, the initial cooling of an atom is most effectively done by that method; only when the cooling has progressed to a relatively low temperature would the Sisyphus mechanism become effective.

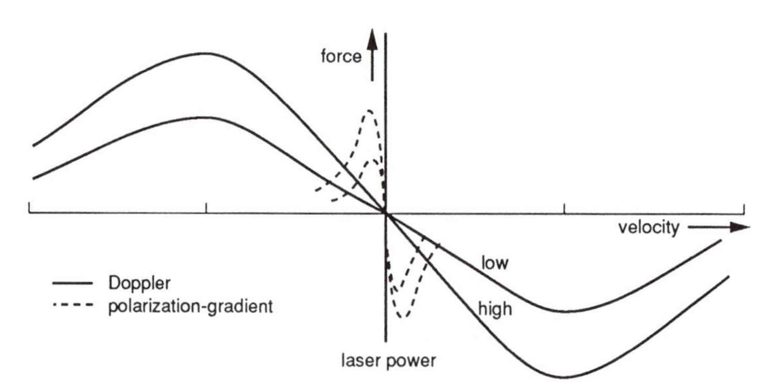

Figure 15.8 Comparison of forces due to Doppler cooling and polarization-gradient as a function of particle velocity (Cohen-Tannoudji, 1990)

Finally, we note that the optimum detuning of the laser to obtain the most effective cooling is determined in the Sisyphus cooling technique by the need to maximize the light shift, which can occur at much greater frequency departures from resonance than is allowed in the Doppler technique.

All these features have been brilliantly confirmed by experiments, initially on sodium but more especially on cesium, whose energy level structure permitted a wider confirmation of the dependence of the cooling on laser detuning. Furthermore, the unexpected sensitivity of the molasses to magnetic fields follows logically from the known displacement and "mixing" that such fields cause in the magnetic substates. Detailed studies of the effect of magnetic fields confirmed that their influence is diminished, as expected, at higher pumping light intensities.

The extent of the cooling was determined, as in the Doppler technique, by a time-of-flight method, which yields the velocity distribution of the atoms in the molasses. As expected, a nonequilibrium double-peaked distribution was observed, indicating distinct velocity capture ranges for the two types of cooling mechanisms. Velocities corresponding to absolute temperatures around 2.5 μK have been achieved; this amounts to but a few times the recoil of an atom in scattering a single optical photon. More recently, special techniques are being explored to attain subrecoil energies; temperatures as low as 3 nanodegrees Kelvin (along one dimension) have been reported by Cohen-Tannoudji et al.

15.10 Laser Cooling of Trapped Ions

Of particular interest to frequency standards is the cooling of isolated ions confined in the kinds of electric and magnetic field configurations described in an earlier chapter. We recall that the motion of ions in those fields, under conditions of stable confinement, is oscillatory with a discrete line spectrum. Thus for the Paul trap, the motion along each of the coordinate axes contains the following discrete frequencies:

$$\omega_n = (2n \pm \beta)\frac{\Omega}{2}, n = 0,1,2,\ldots, \quad 15.14$$

where Ω is the frequency of the rf trapping field, and β lies in the range $0<\beta<1$ and differs in general between the axial and radial motions. Under the conditions commonly chosen in practice, where the amplitude of the high-frequency ion jitter in response to the field is small, $\beta \ll 1$, and the higher frequencies with $n>1$ have a negligibly small amplitude. The case of the Penning trap is even simpler; there are only three frequencies: a simple frequency ω_z along the axis and two frequencies ω^+ and ω^- describing the radial motion, which in the limit of strong magnetic field correspond to motion in tight circles at the cyclotron frequency ω^+, precessing about the axis with the magnetron frequency ω^-.

The technique of laser cooling can be equally applied to an ion executing such an oscillatory motion as any neutral atom moving about randomly; in fact, it is somewhat simpler. First, because the motions along *all three dimensions* can be simultaneously slowed by using laser beams along a *single* axis, as long as it does *not* lie in the x-y symmetry plane of the coordinate system we have been using. In the absence of the cooling laser beams, our particular choice of axes, namely taking the z-axis to be along the axis of symmetry of the trapping field, ensures that the equations are "uncoupled"; that is, in the equation of motion for a particular coordinate, the other coordinates are not involved. However, the Doppler drag on the motion of an ion due to a beam directed at some angle to the mid-plane couples the motions, leading to 3-dimensional cooling. Also, in practice, departures from the assumed ideal field distribution, and to a lesser extent ion–ion collisions, could lead to exchange of energy between the different components of the ion motion. Second, the ions are thermally isolated in ultrahigh vacuum and therefore can gain heat from the environment (or through rf heating in the case of the Paul trap) only very slowly, making it easier to observe even a small rate of cooling.

The oscillatory ion motion does, however, have a distinctive effect on the optical absorption spectrum of the ions. If we imagine a plane monochromatic light wave of frequency ν falling on an ion executing simple harmonic motion of frequency ν_m along the direction in which the wave is traveling, the ion will see, because of the Doppler effect, a wave whose *frequency* is modulated, rising and falling as the ion swings back and forth, first counter to the wave direction, then with it. But we recall from Chapter 7 that there is a curious mathematical fact that runs counter to intuition here: Except in the limit of infinitely slow modulation, the frequency of the wave does not actually pass continuously through all the values in the Doppler range. In fact, the spectrum of such a frequency-modulated (FM) wave is discrete, containing the undisplaced frequency ν and a series of "sidebands" spaced at equal intervals of ν_m extending symmetrically to infinity above and below ν. The frequency spacing of the sidebands in this spectrum remains unchanged if the amplitude of oscillation, and therefore the Doppler frequency range of the ion, is changed; there is merely an increase in the amplitudes of sidebands at frequencies farther from the undisplaced center frequency. Quantitatively, the amplitude of the sideband at frequency $(\nu \pm n\nu_m)$ is proportional to $J_n(2\pi a/\lambda)$, where as usual, J_n represents a Bessel function of order n, and a/λ is the ratio of the amplitude of the ion oscillation to the wavelength of the light wave. The graph of the Bessel function $J_n(2\pi a/\lambda)$ of the first few orders n was given in Chapter 7 in Figure 7.4. If the particle is constrained to oscillate over a range below one wavelength, that is, if $a/\lambda<1$, then all the amplitudes rapidly approach zero for increasing n above zero. In this case the power resides principally in the undisplaced frequency, and the (first-order) Doppler effect is effectively absent. This is the Dicke effect we found to be so important in the hydrogen maser and the rubidium standard;

however, there the wavelengths involved were in the centimeter range, comparable to the amplitude, rather than the submicron range that concerns us here.

The extent to which the discreteness of the spectrum seen by an oscillating ion alters the circumstances of laser Doppler cooling depends on whether the natural line width for resonance absorption is greater or less than the spacing of the Doppler sidebands. If, for example, the spacing is *larger* than the natural line width, then for Doppler cooling it is clearly necessary that the laser be tuned to fall exactly on some harmonic of ν_m *below* the undisplaced resonance frequency; this ensures that the corresponding *upper* sideband seen by the ion will fall exactly on resonance. As the cooling proceeds and the amplitude of oscillation of the ion diminishes, the power in that upper sideband will also rapidly decrease, making further cooling less and less efficient.

The other extreme, in which the natural line width is much larger than the spacing of the Doppler sidebands, is expected to be more common, since ν_m is typically a fraction of a megahertz, whereas optical resonance line widths are perhaps a hundred times that. In this case we need the resonance line-shape function; for a free ion it is a Lorentzian function with a frequency width $\Delta\nu$ determined by the radiative lifetime of the states involved in the optical transition. Since the external fields used to confine the ions are expected to produce a negligible distortion of the resonance line shape, the cross section for resonance scattering can be written as follows:

$$\sigma(\nu) = \sigma_0 \left(\frac{\Delta\nu}{2}\right)^2 \frac{1}{(\nu_0 - \nu)^2 + \left(\frac{\Delta\nu}{2}\right)^2}. \qquad 15.15$$

Now, as already pointed out, the spectrum that the oscillating ion sees depends very much on the value of $2\pi a/\lambda$. At one extreme, $2\pi a/\lambda >> 1$, which is equivalent to the Doppler variation in the light frequency oscillating very slowly, that is, the oscillation frequency ν_m being very small compared to the maximum Doppler shift. In this case the spectrum consists of a large number of closely spaced lines of significant amplitude for all harmonic numbers n up to $(2\pi a/\lambda)$. For a given maximum Doppler shift, if the ion oscillation is assumed to be very slow (with a correspondingly large amplitude to keep the Doppler shift constant) the spectrum is so closely spaced as to form a nearly continuous band of width about equal to the Doppler shift caused by the ion velocity at its maximum. In this limit it would be a valid approximation simply to take the light spectrum seen by the oscillating ion as a single frequency displaced according to the instantaneous velocity of the ion. However, we would then have to allow for the fact that the ion does not spend the same length of time at different phases in its oscillation. In this approximation it is as if the ion were not confined at all, but moving freely, just as was assumed for the neutral particles discussed earlier. The same description of the Doppler cooling should then apply here also.

In general, however, $2\pi a/\lambda$ may assume any value, and in the *ion* frame of reference, it is as if it were being irradiated, not by a monochromatic beam of frequency ν, but rather by light having independently the several discrete frequencies $\nu \pm n\nu_m$. Furthermore, when an ion absorbs a photon, it does so at the sideband frequency that it sees, in its own frame of reference, as being at its resonance frequency ν_0. That is, the harmonic number n must satisfy $\nu_0=(\nu+n\nu_m)$; but in the *laboratory* frame this corresponds to a photon of energy $h\nu$ being absorbed. Now in this process of resonance absorption and re-emission, an atomic particle will, in its *own* frame of reference, re-emit radiation of very nearly the same energy as it absorbs. (There is a recoil energy of the ion that comes at the expense of the emitted photon, but for the present purposes this is negligible). However, this radiation will have, in the *laboratory* frame with respect to which the ion is oscillating, Doppler sidebands symmetrical about the center frequency, so that on the average, the energy of the photon emitted will be just that of the photon absorbed in the ion frame of reference, that is, $h(\nu+n\nu_m)$. The result is that in the laboratory frame of reference a photon of energy $h\nu$ has been absorbed and a photon of average energy $h(\nu+n\nu_m)$ has been emitted. If the laser frequency ν is set *below* the frequency of maximum resonance absorption ν_0, so that the absorption is more likely for the higher sidebands ($n>0$) than the for the lower ones ($n<0$), there will be a net loss of energy by the ion, which must come from its center-of-mass motion.

This way of describing the cooling process is physically equivalent to the one given earlier in terms of light pressure; a quantitative comparison is of course meaningful only in the range where both approximate treatments are valid. If we limit ourselves to the case where $2\pi a/\lambda>>1$, the instantaneous velocity of the ion is related to n by equating the Doppler shifts: $n\nu_m=(V/c)\nu$. Now, we had previously found that a photon beam of intensity I exerted an average force $F=I\sigma/c$ on an ion, and therefore the rate at which the ion energy is reduced (FV) is given by $I\sigma(V/c)$, that is, $I\sigma n\nu_m/\nu$, or $hn\nu_m$ per photon absorbed. This is just the result obtained by balancing the energy between absorbed photons Doppler-shifted in frequency and the emitted photons.

While the limiting case of $2\pi a/\lambda>>1$ is expected to be relevant in the initial stages of ion cooling, it is the opposite extreme, where this parameter approaches a value much less than unity, that will determine the ultimate level of cooling that can be reached. As already pointed out, in this limit the amplitudes of the Doppler sidebands above $n=1$ become negligible, so that the spectrum seen by the oscillating ion consists of the center frequency ν and only one sideband on each side of it, $(\nu+\nu_m)$ and $(\nu-\nu_m)$. In this case it is relatively easy to obtain an explicit expression for the rate of cooling, since the intensity of these sidebands, $J_{\pm 1}^2(2\pi a/\lambda)$, is approximately $(\pi a/\lambda)^2$. If we again assume a Lorentzian function of width $\Delta\nu$ for the resonance line shape with $\Delta\nu>>\nu_m$, we find that there is not just the loss of energy as a result of absorbing a photon of energy $h\nu$ followed by emission of one of average energy $h(\nu+\nu_m)$, but also a *gain* resulting from the absorption/emission cycle at the other ($\nu-\nu_m$) sideband. The overall rate of cooling, after some simplification, is given by the following:

$$\frac{dE}{dt} = -2\left(\frac{\pi a}{\lambda}\right)^2 I\sigma_0 \frac{\nu_m}{\nu}\frac{\nu_m}{\Delta\nu}. \qquad 15.16$$

In arriving at this, we have taken into account the effect on an existing motion as manifested by the Doppler shift and based the derivation of the laser cooling on the assumption that in the frame of reference in which an ion is initially *at rest*, it emits photons of the same frequency/energy as it absorbs. This is a valid approximation as long as the energy of *recoil* of an ion in the act of absorbing or emitting a photon is negligibly small compared to the energy of any prior motion. This may give rise to some confusion, since the Doppler cooling mechanism is also in a sense due to "recoil"; it is clear that a distinction must be made: They represent two levels of approximation in terms of the ratio between the photon and ion momenta. As the ion momentum is brought near zero by cooling, the higher-order approximation, which applies even to ions at rest, becomes necessary. It is precisely this ultimate recoil energy that will set the minimum ion energy attainable. It is the same recoil energy which accounts for the residual Brownian motion described earlier. Since the conservation of linear momentum requires that when a photon is absorbed, the ion recoil with a momentum $h\nu_0/c$, it follows that from a zero initial velocity the ion gains kinetic energy amounting to $(h\nu_0/c)^2/2M$. Now, under the present conditions $J_0^2(2\pi a/\lambda)\approx 1$, and thus nearly all the radiation the ion sees is at the center frequency ν, which has been assumed to be set at $\nu_0-\Delta\nu/2$. Since at this frequency $\sigma=\sigma_0/2$ (by definition), it follows that the rate of energy gain by the ion is given by the following:

$$\frac{dE}{dt} = \frac{I}{h\nu}\frac{\sigma_0}{2}\left(\frac{h\nu_0}{c}\right)^2 \frac{1}{2M}. \qquad 15.17$$

Finally, in order to obtain the ultimate equilibrium value of the ion energy, we equate the rate of energy loss to energy gain and obtain, after some simplification,

$$E_{\min} = \frac{1}{2}M(2\pi\nu_m a)^2 = \frac{h\Delta\nu}{4}. \qquad 15.18$$

While this analysis does not pretend to be anything more than a rough quantitative sketch of the basic physical processes involved, it nevertheless leads to the correct dependence of the cooling on the physical parameters. However, it is flawed in one serious physical respect: At or near the minimum energy, the conditions may be approached where the ion motion can no longer be treated using classical mechanics; it becomes a problem in quantum mechanics. This is seen from the fact that it is possible that E_{min} may not be much larger than $h\nu_m$, the quantum of energy for a harmonic oscillator of frequency ν_m. The energy of such an oscillator can assume only the following discrete values: $h\nu_m(n+½)$, where n is a whole number, and $½h\nu_m$ is the so-called *zero point energy*. The fact that the energy cannot reach precisely zero is, of course, a quantum phenomenon; the classical description is an approximation valid only for very large values of the quantum number n. To pursue the matter further would take us too far afield; suffice it to say that it is truly remarkable that the experimental implementation of the ideas we have described has been so successful as to reach the quantum energy level. An ion harmonically bound in a trap with a frequency $\nu_m=10^5$ Hz and cooled down to the point where there is a 95% chance that it is in the state of zero point energy would have a temperature around 0.8 μK. In the case of a Hg^+ ion it would have an amplitude of oscillation no more than 0.02 μm, that is, on the order of $\lambda/10$, and a second-order Doppler shift of no more than one part in 10^{21}!

Chapter 16
Application of Lasers to Microwave Standards

The advent of the laser changed the whole face of atomic standards research in a number of fundamental respects: from the cooling and confinement of large numbers of neutral atoms, directly applicable to the radical improvement of the cesium standard, to the cooling of single ions. This latter made possible, in one revolutionary advance, the realization of the ideal goal that motivated the original prelaser forays into field suspension of ions, culminating in the Hg^+ ion standard. That goal was to completely isolate the reference particle from its environment, and in so doing to observe its true resonance, free from Doppler shifts and uncontrolled random perturbations. In this chapter we will touch on the various new applications of lasers that have transformed the cesium and Hg^+ ion standards and taken them to new levels of performance.

16.1 Observation of Individual Ions

16.1 Introduction

One of the most significant developments made possible by the laser revolution is the direct optical observation of *individual* atomic ions. We can, by focusing a resonant laser beam onto the region of confinement of the ions, actually scatter enough photons to literally see the discrete particle nature of whatever is doing the scattering. It is true that even in the very early days of the Paul trap, the sensitivity of ion detection by electronic means was pushed to the point of resolving individual ions in a cluster, but now with the laser, the signal-to-noise ratio is incomparably higher, and the effect is far more striking.

This remarkable feat of directly observing individual cold ions as spots of light "crystallized" in a Paul trap, was of course the culmination of the work of many pursuing the many aspects of the interaction of radiation with matter, which the special properties of lasers made possible. At first, estimates suggested that the enormous signal-to-noise advantages of using resonance scattering of laser radiation held out the hope that the observation of individual atomic particles was within reach. Just how close one is to resolving individual particles starts with a theoretical estimate of the number of ions implied by the observed intensity of the scattering, based on what is known about the ion cross section for this process. This is useful in assessing progress toward the ultimate goal; however, the approach to resolving individual ions is signaled by the appearance of stepwise variation in the scattered intensity, jumping between a finite number of discrete levels, when ions are created or removed from the trap.

The ions are created by random ionizing collisions between electrons and the parent atoms, and a statistically fluctuating number of ions are accumulated in the trap during any given filling period. However, the frequency of occurrence, that is, the probability of different values of that number, follows a determinate law. If, as is generally the case, the number of available parent atoms is infinitely large compared to the number actually ionized, we can assume that any given ion had an equal probability of becoming part of the ion population at any time during the filling period. In that case it can be shown that the frequency of a given ion number follows what is called the *Poisson distribution.* If we represent the *average* number of ions trapped by $\langle n \rangle$, then the probability $p(n)$ that precisely n ions are trapped is given by the following:

$$p(n) = \frac{\langle n \rangle^n}{n!} e^{-\langle n \rangle}. \tag{16.1}$$

Figure 16.1 shows a plot of the Poisson distribution, giving the relative probability of a cluster being formed with different total numbers of ions, for a given average number of ions trapped. When this number is extremely large, as in the original Hg^+ ion standard, the discreteness in the fluorescent signal measuring the ion number would manifest itself only as a small fluctuation, namely *shot* noise. It is only when the signal from only a few ions is observed with a large signal-to-noise ratio that individual ions are clearly discernible.

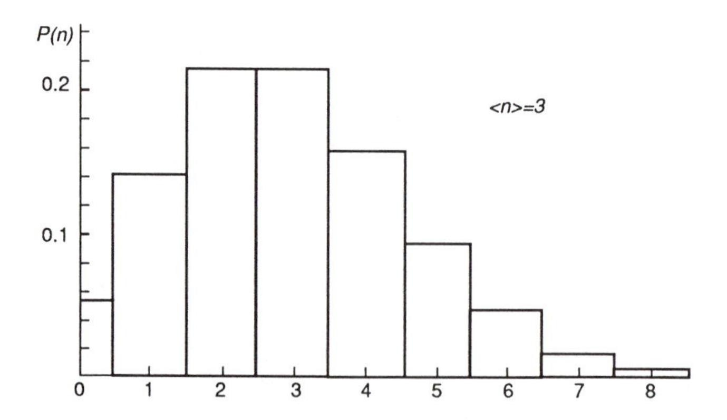

Figure 16.1 The Poisson distribution in the number of stored ions

The ability to observe a single ion suspended in a Paul trap in ultrahigh vacuum is extremely important; its isolation means that it is free from collisions that cannot only perturb its internal quantum states, but also limit how far its temperature can be lowered, and the (second-order) Doppler effect thereby reduced. As explained in an earlier chapter, collisions involving ions in the high-frequency electric field of a Paul trap result in abrupt changes in ion energy. We recall that in the absence of collisions, an ion pursuing a stable orbit will continue to do so indefinitely; however, if the motion of the ion is interrupted by a collision with another particle, the phase of its high-frequency motion is abruptly changed, and it may start on a higher-energy orbit. Thus the ability to observe an isolated single ion means that ion–ion collisions are eliminated; all that would remain is the possibility of collisions with background gas particles in the vacuum system. This means that now the experimental burden is on the ultrahigh vacuum system to maintain a sufficiently low particle density in the ion trap. Fortunately, another revolution, this one in vacuum technology, which, as we have seen, came with the introduction of the titanium ion pumps, makes it possible to reach pressures below 10^{-10} Pa, if extreme care is taken in the design and construction of the system. Although this is far from a "perfect" vacuum (by comparison, interplanetary space has an average particle density more than 10,000 times smaller), nevertheless, the particle density is so low that the average time between collisions involving the ion and a background particle can be several days! Thus although a collision may result in a substantial increase in energy, so much time elapses between collisions that the average rate of energy increase is very small. To really benefit from this low average collision rate, the trap must obviously be designed so that *several* collisions are required before the ion acquires sufficient energy to escape.

Unlike the Paul trap, the Penning configuration involves *static* electric and magnetic fields, and collisions in this case cause the ion to diffuse by small increments across the magnetic field to points of lower electrostatic potential, resulting in a spread in the ion kinetic energy. The size of the increment produced by a collision is on the order of the *cyclotron radius* (p/eB), where p is the momentum (mV), e is the ionic charge, and B is the intensity of the magnetic field. By using a sufficiently powerful magnet, the increment, and therefore the diffusion rate (for a given collision rate), can be made extremely small. However, for the purposes of atomic frequency standards, the use of an intense magnetic field introduces a complicating factor that can lead not only to systematic error in the reference frequency, but also to a loss in its stability. Therefore, we will limit ourselves for the present to experiments using ion confinement in a Paul trap.

16.1.2 Paul Trap Design

The design of the physical setup for laser cooling of ions in a Paul trap must begin with the design of the trap itself. A particularly important physical parameter at our disposal is the slow *secular* oscillation frequency of the ion. This, it will be recalled, is given by $\beta\Omega/2$, where β is approximately given by $\beta^2 \approx a+q^2/2$ under the so-called *adiabatic* condition, where ion oscillates at the field frequency with a small amplitude as it executes its slower oscillation about the trap center. In this case the ion oscillation spectrum is dominated by the frequency $\beta\Omega/2$ with components at the frequencies $(2+\beta)\Omega/2$ and $(2-\beta)\Omega/2$, which are very much weaker. There are two other important parameters: the inherent *optical* resonance line width of the ion and the spectral line width of the laser light. It simplifies the description of the Doppler cooling process considerably and can render it more effective if the laser line width is smaller than the optical resonance line width of the ion, and if in turn, the latter is smaller than the frequency separation of the Doppler sidebands, so that they are well resolved.

Since optical transitions suitable for laser cooling must occur with the ions presenting a relatively large cross section to the beam, their spectral line widths (due to the high transition probability) cannot be chosen far below the megahertz range; this implies that the ion oscillation frequency $\beta\Omega/2$ should be at least on the order of a megahertz, and thus for $\beta \approx 0.1$ (say) the field frequency Ω must be in the tens of megahertz range. Based on these considerations we can now arrive at the specifics of a typical trap design suitable for laser cooling of ions. For example, if we assume a field frequency of $\Omega/2\pi=10$ MHz, then for the Hg^+ ion we find $q_r=2.4\times10^{-10}V_0/r_0^2$. We are at liberty to choose the exact value of q_r, provided that it falls within the stable region of the a–q diagram; if we choose $q_r=0.05$ and $r_r=1$ mm, then the required radio-frequency amplitude would be $V_0=210$ volts. We note that the assumption of a relatively high oscillation frequency for such a massive ion has led us to trap dimensions on a miniature scale compared to those used in the usual mercury ion standard. The maximum kinetic energy an ion may have while confined in such a trap can be readily computed if we recall that under the assumed conditions, the average kinetic energy of the high-frequency oscillation at frequency Ω is equivalent to an electrostatic potential that simply adds to the applied static potential in determining the slow oscillation at frequency $\beta\Omega/2$. Using the formula $\beta^2 \approx a+q^2/2$, we can show that the maximum energy is $E_{max}=(eU_0/2+eqV_0/8)$. In the example we have chosen, this has the numerical value on the order of 1.25 electron volts, which, to give an idea of relative magnitudes, is the mean energy of a particle at a temperature of about 14,500°K.

The initial filling of the trap with ions is achieved through ionizing collisions between electrons in a beam intersecting a beam of parent atoms in the trapping region. Once the trap is filled, a process that takes only a fraction of a second, the beams must be interrupted in order on the one hand to minimize the density of background particles and on the other to remove any distortion of the electric field the charged electrons may produce. Since the intention is to trap only a few, and ultimately only one, supercooled ion, the large initial ion population having an energy spread reaching the maximum 1.25 electron volts must be reduced to only those near zero energy. This can be done, for example, by lowering the amplitude of the high-frequency field for a short time, thereby reducing the maximum energy and allowing the more energetic ions to escape. It might be thought that we would do as well by starting with the shallower trap, since the number of low-energy ions produced should be independent of the high-frequency field. While this may be true, the number of low-energy ions finally *trapped* may be considerably lower than would be achieved by having a deeper trap during the filling phase. There are at least two reasons why this might be so: First, the field distortion produced by the electron beam is relatively more serious when the trap is shallow and could reduce the efficiency of trapping. Second, a higher background pressure may accompany the presence of the electron and atomic beams during the filling period; if this happens and the background particles are lighter than the ions, the larger number of ions filling a deeper trap are cooled, and their energy distribution moves towards lower energy, increasing the number of low-energy ions.

16.1.3 Ion Refrigeration Cycle

In fact, this latter phenomenon could be exploited as a cooling cycle by purposely injecting a light gas such as helium, raising the amplitude of the high-frequency field for a time and then returning to the normal (lower) operating amplitude. If the amplitude of the field, and hence the oscillation frequency $\omega=\beta\Omega/2$, is changed so slowly that the ion makes very many oscillations across the trap in the time it takes for ω to change appreciably, it can be shown that for a given ion the ratio E/ω is constant, called an *adiabatic invariant* of the oscillatory motion. The amplitude a varies as follows: $a^2\omega$=constant. Thus by raising the field amplitude V_0 in the trap, ω is raised, which leads to a compression of the amplitude, with an *increase* in the energy of the ions. This increase in energy would then be transferred by collisions to the background gas, so that when the field amplitude is reduced to its original value, the ion energy falls below what it had been; the ions have been cooled. This would be analogous to the action of the compressor in a refrigerator, which raises the temperature of the working fluid as it compresses it to a higher pressure, allowing heat to be transferred through the heat exchanger to the now colder environment.

16.2 The Cooling Laser System

16.2.1 Laser System Design Considerations

The design of the laser system is determined primarily by the wavelength of the optical resonance transition in the particular ion species being used. Experimental work has been published on a number ions in addition to the Hg^+ ion; for example, the alkali-like magnesium ion Mg^+, the barium ion Ba^+, as well as the ytterbium ion Yb^+. From the point of view of constructing a suitable laser system, the ion of the heavy alkaline earth element barium, with a resonance in the green region of the spectrum at a wavelength of λ=493.4 nm, is the least demanding. Unlike the other ions, whose resonance occurs in the ultraviolet region of the spectrum, the Ba^+ ion resonance wavelength falls in a range for which there exists an adequate dye (coumarin 102) to use in a tunable dye laser to generate the desired wavelength directly.

16.2.2 The Ring Dye Laser

The main experimental challenge is to achieve a finely tunable laser output of great stability and spectral purity. We recall that dye lasers are based on optically pumped dye solutions that have relatively broad optical gain curves, each dye covering a band of wavelengths in different parts of the visible spectrum. To achieve a narrow spectral line width in the output requires that the laser oscillate in a single mode, whose frequency may require further active stabilization. In practice, a combination of interferometric narrow-band tuning elements, called *etalons*, with different *finesse* may be incorporated in the laser cavity to select a particular oscillation mode and sharpen the spectral width of the output. In more sophisticated systems, where stability of the output frequency is paramount, active stabilization is provided using a servo system to control the optical length of the laser cavity against a stable, external reference optical cavity. The output power of a CW single-mode dye laser using a standard standing-wave cavity is limited by a phenomenon called spatial *hole-burning*, in which the population inversion, and hence the optical gain, of the dye saturates at the points where the stationary optical field pattern has strong maxima. At intermediate points where the field is zero, the dye is effectively unused, retaining its high gain at those points. The laser therefore tends also to oscillate in modes making use of that unused gain, making it necessary to use high-loss frequency-selective elements to maintain single-mode operation. Efforts to remove these limitations culminated in the development of the now widely used traveling-wave ring laser.

In Figure 16.2 is shown schematically the layout of a commercially available ring dye laser using four cavity mirrors in a folded configuration. As with most CW dye lasers, the dye stream is optically pumped with a high-power Ar^+ ion laser and is equipped with a unidirectional filter to suppress the counterpropagating wave that would naturally be present in such a configuration of mirrors. Since the light waves propagate continuously around the cavity, all the dye molecules are subject to the same average optical field; they all contribute equally to the buildup of the laser field, and the phenomenon of hole-burning does not occur. This in combination with the natural tendency for the buildup of one mode to suppress other modes relaxes the need to sacrifice power in highly selective filters to maintain single-mode operation. The result is an order of magnitude increase in single-mode output power for the same pump power. Depending on the desired wavelength, and therefore the particular dye that is available for that wavelength, commercially available ring dye lasers are capable of producing CW output powers of several watts with spectral line widths less than 1 MHz. While special, stabilized long-cavity gas lasers, such as the He–Ne laser, have demonstrated the possibility of attaining far narrower line widths, the tunability and power of a stabilized ring dye laser makes it unparalleled as a laser source, with far-reaching applications.

The availability of high-power tunable dye lasers, which cover the visible region of the spectrum, provides a means for reaching ultraviolet wavelengths by *frequency doubling*, and other optical frequency synthesis techniques. This is of critical importance, since the resonance wavelengths of the interesting ions we mentioned above are in the ultraviolet (Hg^+ at λ=194.2 nm, Yb^+ at λ=370 nm, and Mg^+ at λ=279.7 nm) and are far beyond the direct range of dye lasers. We note that while the wavelengths for Mg^+ and Yb^+ can be reached by dividing by 2 the wavelength of light at λ=559.4 nm and λ=740 nm respectively, wavelengths well within the range of available dyes, the wavelength of the important Hg^+ ion is not so easily accessible. Nevertheless, it is achievable by a number of frequency synthesis methods involving nonlinear crystal optics to produce optical harmonic frequencies as well as mixing two or more optical frequencies to produce light at the sum or difference frequency. When the system is designed to produce the sum frequency, the process is called *up-conversion* of frequency and is one approach to synthesizing the Hg^+ ion wavelength.

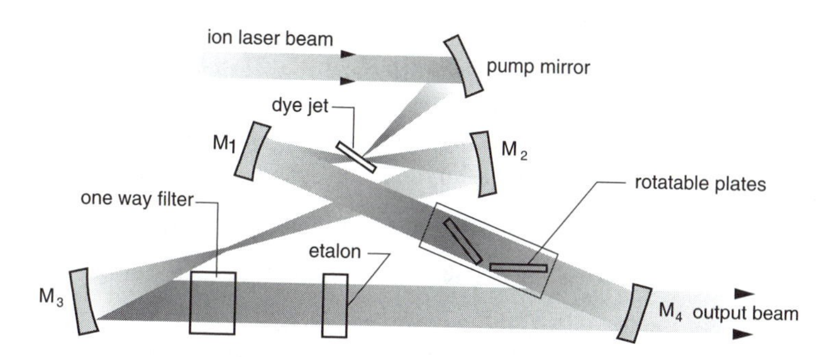

Figure 16.2 Typical layout of a traveling wave ring dye laser

16.3 Laser Detection of Hyperfine Resonance

16.3.1 Pumping of Hyperfine Populations

Once we have successfully synthesized the desired optical frequency for laser cooling for example the Hg^+ ion, we have in effect solved another problem we faced in constructing a microwave standard using the hyperfine resonance, namely pumping the ions preferentially into one of the hyperfine levels. This, it will be recalled, was first accomplished—and is still used in a number of portable embodiments of the Hg^+ ion standard—by using the fortuitous overlap of the ion resonance line from a mass 202 mercury lamp with just one of the hyperfine transitions in mass 199 ions. With a tunable laser source having a line width as small as 1 MHz, the pumping problem is solved. In fact, if the intent is only to cool the ions, the hyperfine pumping that will inevitably occur is a problem that must be circumvented, since the ions tend to end up in the nonabsorbing hyperfine sublevel.

16.3.2 The Signal-to-Noise Ratio

While the suspension of an individual cold ion provides an ideal reference, free from perturbations and Doppler shifts, it is limited in the signal-to-noise ratio that can be achieved using photon scattering to detect resonance; this is not the same thing as merely detecting the presence of ions. We recall that under ideal conditions, even if there were no other sources of noise, there will be *shot* noise due to the quantum nature of photons. If an average count of $\langle n\rangle$ photons is recorded in a given interval of time τ, it can be shown that there is a statistical uncertainty in n analogous to the uncertainty in the precise number of ions accumulated in a given period of time. In the latter case we saw that the ion number is expected to follow a Poisson distribution. If the same can be said of the distribution in the precise number of photons registered during the counting period, then we can show that σ, the *standard deviation* in n, defined as $\langle(n-\langle n\rangle)^2\rangle^{1/2}$, which is a measure of the unpredictable fluctuation in n, or noise, is $\sqrt{\langle n\rangle}$; and since the signal is proportional to n, therefore the signal-to-noise ratio is also $\sqrt{\langle n\rangle}$. It follows that if the average counting rate is $\langle\alpha\rangle$, then $\langle n\rangle=\langle\alpha\rangle\tau$, and the signal-to-noise is $[\langle\alpha\rangle\tau]^{1/2}$. As we saw in an earlier chapter, this determines the accuracy in locating the center of the resonance and is the basis for the definition given there of the figure of merit F of a resonator, namely

$$F=\left(\frac{\nu}{\Delta\nu}\right)\sqrt{\langle\alpha\rangle\tau}\,, \qquad 16.2$$

where $\nu/\Delta\nu$ is a measure of the resonator Q-factor and $[\langle\alpha\rangle\tau]^{1/2}$ is the signal-to-noise ratio; F and ultimately the frequency stability of the standard increase as $\sqrt{\tau}$. For a given sampling time τ, however, we gain stability by increasing the average counting rate $\langle\alpha\rangle$ of scattered photons monitoring the resonance.

16.3.3 A Proposed Trap with Multiple Reference Ions

What is needed is a large number of individual cold ions contributing to the photon detection signal in parallel, rather than having to accumulate over a long period of time the photon count from a single ion. The confinement of a large number of ions laser cooled to energies approaching absolute zero is incompatible with a conventional Paul trap, since it implies that all the ions must converge on the center as the temperature is lowered. This would imply extremely high ion density and consequent strong interaction, both as mutual electrical repulsion and an increased collision rate between them. Clearly, what is required is a field configuration in which many ions at rest can occupy separate points. One approach would be simply to have a large array of microtraps, each containing a single ion at its center. Such an array might be constructed as illustrated in Figure 16.3. A number of plane parallel conducting sheets, all identically perforated in a regular pattern, are stacked with the holes aligned. If we join alternate sheets electrically to a common terminal, so that all the sheets are connected to one or the other of two terminals, and then apply the high-frequency voltage between those terminals, we would create the desired quadrupole field at the center of each hole in the sheets. Such an arrangement would permit the observation of 10,000 individual ions in a fairly compact space.

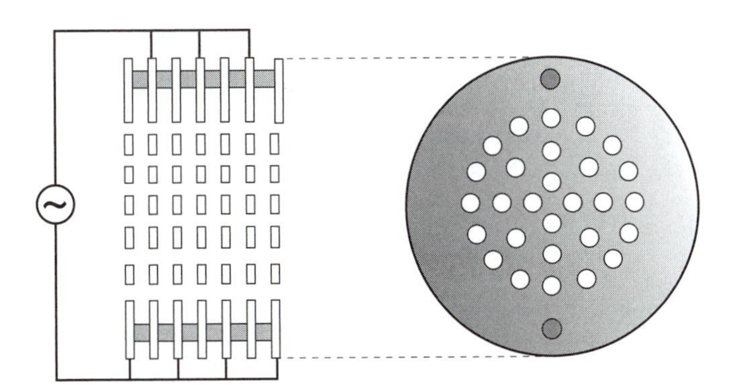

Figure 16.3 An array of microtraps each containing an individual ion

16.3.4 The Linear Paul Trap

A highly promising variant of the Hg^+ ion frequency standard, which is designed for multiple ion trapping, replaces the common hyperbolic cylinder and end cap geometry of the usual Paul trap with one whose principal section closely resembles the precursor of the Paul trap, namely the original Paul *mass filter* for separating ions in a beam according to their mass. It is a linear device, shown in Figure 16.4, consisting of four parallel conducting rods with electrodes at the two ends to prevent ions from escaping. Opposite rods are electrically joined, and a high-frequency voltage is applied between opposite pairs to produce a field in the neighborhood of the central axis, far from the ends, distributed as follows:

$$V(x,y,z,t) = \frac{V_0}{r_0^2}\left(x^2 - y^2\right)\cos(\Omega t), \tag{16.3}$$

where we have chosen coordinate axes such that the z-axis is along the central axis of the trap and the x- and y-axes pass through the centers of opposite conductors. The motion of an ion in such a field is governed by an equation of the same type (Mathieu) as in the usual Paul trap, with similarly defined coefficients a and q; however, here q_x=q_y and a_x=a_y=0. Again, for confined (oscillatory) ion motion in the radial direction we must choose q to lie in a stable region of the Mathieu equation.

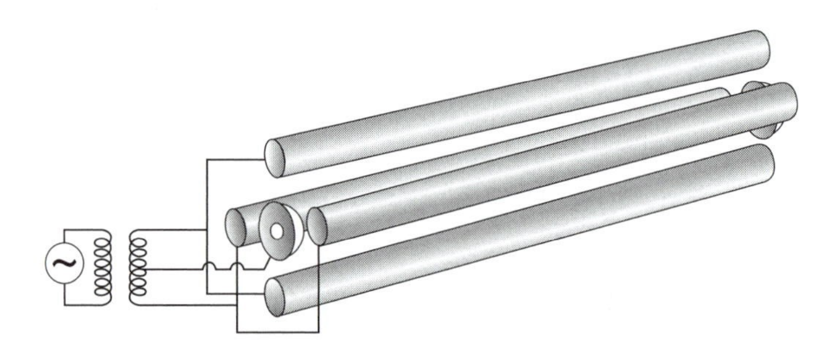

Figure 16.4 A linear Paul high-frequency ion trap

Confinement in the axial direction is provided for by applying a positive DC voltage to the end caps relative to the rods. The actual high-frequency and DC field distributions in the neighborhood of the end caps is expected to be intractable analytically and can at most be numerically approximated. Therefore, the exact ion motion cannot be predicted; however, in the adiabatic limit when

$q<<1$, we can draw on the treatment of the problem discussed in an earlier chapter: The motion can be separated into a small-amplitude oscillation at the field frequency, superimposed on a slow secular motion that is pursued as though it were in a static potential field U_{eq} that varies in space with the field amplitude E as follows:

$$U_{eq} = \frac{eE^2}{4M\Omega^2}. \qquad 16.4$$

On the basis of this we can anticipate broadly the necessary conditions at the ends of the linear trap for effective confinement of ions. In particular, to maintain radial confinement at the ends will require that the radial component of the high-frequency field there be strong enough to overcome the radially diverging effect of the static field applied to the end caps.

Since the linear portion of the trap is very much longer than the spacing of the rods, we may assume to a good approximation that at points far from the ends, the force on an ion is purely radial; and therefore, if by laser cooling the radial motion is brought nearly to rest, we would expect the ion to remain very near a fixed point on the axis. Thus through laser cooling it is possible to have a large number of ions "crystallized" along the axis contributing to the resonance signal.

16.3.5 "Time Bridge" Method of Resonance Detection

The extraordinary degree of freedom from perturbations that a single trapped ion enjoys means that we can realize the extremely narrow natural line widths expected of resonant transitions between very long-lived states. There are transitions, such as those between magnetic hyperfine substates, that have extremely long lifetimes for spontaneous emission. In the Hg^+ ion case, resonant transitions involving such states will have an unprecedented Q-factor, in excess of 10^{12}! Unfortunately, the very fact of the resonance transition having a small spontaneous probability also means that it will have a small absorption cross section, making it unsuitable for observing the resonance directly through the resonant photons. Even where the small transition probability can be overcome by using a sufficiently strong resonant field, as in the microwave reference transitions, the detection of resonance by the absorption or emission of the resonant radiation itself by a small number of atoms or ions is impractical. As we saw in this case, the resonance is observed by an indirect "trigger" method, in which the rather weak exchange of energy involved in the microwave resonant transitions caused a much larger observable effect: In the cesium standard it changed the trajectory of the cesium atoms, and in the rubidium and Hg^+ ion standards the absorption of resonant microwave photons affected the number of very much more energetic optical pumping photons absorbed in "allowed" (electric dipole) transitions between the reference magnetic substates and other states. Even for reference transitions in the optical frequency range, which must be selected to be between long-lived *metastable* states to yield an ultrasharp resonance, the resonant photon energy may be high enough that the detection of scattering from a small number of particles is possible, but it is still necessary to separate the act of applying the resonant optical field from the detection of the resulting transitions. Otherwise, the high rate of optical transitions necessary for good signal-to-noise in the detection of resonance will *power broaden* the reference transition frequency in the process. If we were to attempt to detect resonance directly using the scattering of resonant photons, in order to yield a good signal-to-noise ratio, we would need to use a beam of such power that the induced transitions would shorten the radiative lifetime. A shortened radiative lifetime translates, of course, into a broadened resonance.

This is avoided if the resonance is monitored using a strong transition linking one of the reference states, but with a wavelength far removed from the reference transition. The monitoring beam can then simply be turned off while the reference transitions are being induced by a weak probing field, and then turned back on to establish whether transitions were indeed induced; that is, we use a "time bridge."

In order to "interrogate" the ion, that is, determine what frequency of the field resonates with the reference transition, a version of the Ramsey separated fields method is adopted. We recall that in the cesium standard the resonance is observed by making the cesium atoms pass through two spatially separated phase-coherent microwave field regions, produced in the *Ramsey cavity*. As Ramsey originally pointed out, the method can be applied equally using phase-coherent fields separated in *time* rather than space; this is not surprising, since transitions obviously have to do with the evolution of quantum states in time. The Ramsey technique is important in realizing experimentally the sharpness of resonance implied by the long radiative lifetime of a suspended ion and its relative freedom from randomly induced transitions caused by unwanted perturbations. The ions are subjected to two coherent microwave bursts of precisely controlled intensity and duration, separated by a precise time interval. (Coherent here simply means that it is as though the two bursts are parts of the same continuous wave.)

We recall that in the discussion of the Ramsey field in the cesium standard, for the purposes of providing a concrete picture of the process, we spoke in terms of the motion of a magnetic dipole in a magnetic resonance experiment, in which a transition can be visualized as a rotation of the dipole axis with respect to a static magnetic field direction. The same analogy applies here. In quantum terms, the effect on an ion of a field inducing hyperfine transitions is to put it in a nonstationary quantum "mixed" state involving both initial and final states, with a magnetic moment varying periodically at the transition frequency. The time development of the transition is characterized by the *Rabi nutation frequency*, which is proportional to the amplitude of the field inducing the transition. Now, we recall that in quantum mechanics the absorption and stimulated emission of radiation are treated on an equal footing, and that if an otherwise free ion is kept in the microwave field, it will keep going back and forth between the two hyperfine states. An ion acted on by a resonant field characterized by the Rabi frequency ω_R for a period τ will, starting in one state, make a complete transition to the other state when τ reaches the value given by $\omega_R\tau=\pi$ and will return to the original state if it continues until $\omega_R\tau=2\pi$. In general, it can be shown that the probability that an ion starting in one state will after some time t have made a transition to the other state is given by the following:

$$P = \frac{\omega_R^2}{(\omega-\omega_0)^2+\omega_R^2}\sin^2\frac{1}{2}\sqrt{(\omega-\omega_0)^2+\omega_R^2}\,t \qquad 16.5$$

where ω and ω_0 are the (angular) frequencies of the field and ion transitions respectively. We easily verify that at resonance, when $(\omega-\omega_0)=0$, the probability that a transition has occurred, P, has the value one when $\omega_R\tau=\pi$, as stated above. The result is a good deal more complex when there are two separate periods during which the ion is subjected to the resonant field. In a common design, the amplitude of the field (and therefore the Rabi frequency ω_R) and duration τ of each of the two time-separated bursts are usually chosen such that $\omega_R\tau=\pi/2$. Now, if the field frequency is tuned to exact resonance with the ion transition frequency, then during the free time the ion spends between burst, the oscillating moment of the ion remains exactly in phase with the field, so that when the second burst begins, the situation is indistinguishable from the end of the first burst. In this case, at the end of the second burst $\omega_R\tau=\pi$, and the ion has completed its transition to the other state. The important point to make here is that if during the relatively long interval between bursts the ion transition frequency varies slightly, but randomly, the phase of the ion moment at the end of that interval tends to average out such frequency fluctuations. If the field is mistuned so that $(\omega-\omega_0)$ is not zero, then a phase difference $\Delta\phi=(\omega-\omega_0)T$ will develop between the ion moment and the field during the period T between bursts, and the probability of a transition by the end of the second burst will be less. Since only the *relative* phase between the ion moment and the field at the end of the free period T is physically significant, it follows that a mistuning leading to a phase difference of $\Delta\phi$ is indistinguishable from one giving a phase difference of $\Delta\phi+2n\pi$, where n is a whole number. Figure 16.5 illustrates the periodic nature of the transition probability as indicated by the scattered photon counts plotted against the frequency mistuning of the field.

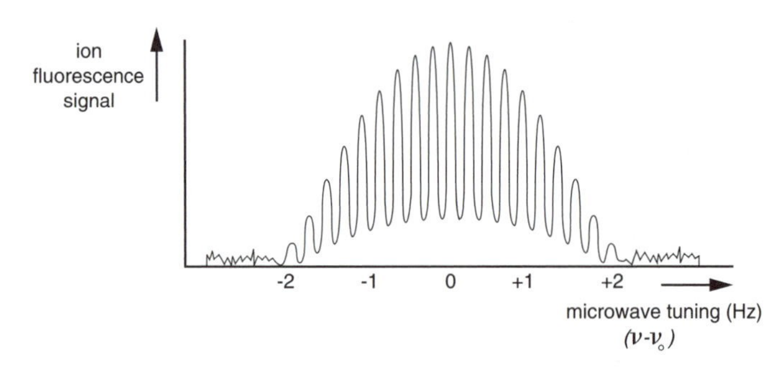

Figure 16.5 The periodic resonance signal seen in a time-separated Ramsey field

16.4 Laser-Based Mercury Ion Standards

While the prelaser Hg^+ ion standard achieves high performance in a compact portable unit, the use of laser cooling takes it to an altogether new level of sophistication and accuracy; but it is no longer portable. This is not, of course, a consideration for a *primary standard*, and as such its accuracy can be enhanced not only by the near elimination of the second-order Doppler effect, but also by reaching new limits on the other factors known to affect the hyperfine frequency: collisions of background atomic particles with the ions, and ambient electric and magnetic fields. Fortunately, the act of cooling the ions limits their motion to the immediate vicinity of the axis, where the high frequency field amplitude approaches zero. Therefore, the already minor effect the electric field has on the magnetic hyperfine splitting becomes entirely negligible. However, the magnetic field is another story.

The extraordinary sharpness of the resonance made possible by laser cooling has meant that the more common methods of achieving ultrahigh vacuum and magnetic shielding are no longer adequate. Wineland and the group at the Time and Frequency Division of the (U.S.) National Institute of Standards and Technology (Poitzsch, 1994) have proposed that both concerns can be effectively met by operating the linear ion trap system at the temperature of liquid helium (around 4.2°K). Since all forms of matter (except of course helium itself) are frozen at this temperature, with vapor pressure so small that it would be a challenge to measure it, this is the ultimate method of reducing background pressure, called *cryogenic pumping*. It of course introduces a whole new level of complexity into the picture, involving the technology of cryogenics. The ion trap system must be immersed in a liquid helium bath in what is called a *cryostat*, consisting of a double Dewar (vacuum flask): an inner one containing the liquid helium, and an outer one filled with liquid nitrogen. Since the liquid helium has to be continuously replenished, a constant source is obviously required. Apart from residual helium in the ion trap vacuum system, whose pressure can be maintained well below 10^{-7} Pa with ion pumps, cryogenic pumping then would put an end to the vacuum problem.

Furthermore, such a cryostat would also enable the ultimate in magnetic shielding: a superconducting metal enclosure. A superconductor is a material that below a certain critical temperature T_c, conducts electricity with zero resistance. The phenomenon was first observed in mercury at the temperature of liquid helium and has since been found to occur in many elements and a large number of alloys and compounds, including the high-T_c oxides discovered in 1985. The most common alloys are those of niobium with titanium or tin (Nb_3Sn), which has T_c=18°K. However, a superconductor is not just a body with zero electrical resistance. It was shown by Meissner in 1933 that if a material is placed in a magnetic field and then cooled below its transition temperature, it assumes a unique state in which the magnetic field is "expelled" from its interior, a phenomenon known as the *Meissner effect.* This effect, plus the fact that its zero resistance makes it perfectly *diamagnetic,* makes a superconducting enclosure an ideal magnetic shield. The diamagnetic property refers to its ability to exclude from its interior any external magnetic field by the free flow of currents whose magnetic field exactly cancels throughout its interior the externally applied field.

Thus using known technology, there is great promise that systematic uncertainties in a Hg^+ ion frequency standard can be pushed to levels inconceivable only a few years ago. Resonance frequency shifts due to the second-order Doppler effect, collisions, and magnetic field can be plausibly kept below parts in 10^{16}! Even frequency instability, which is a manifestation of the statistical error in fixing on the center of the resonance, can by the use of multiple ion traps reach parts in $10^{13}\tau^{½}$, where τ is the averaging time. In the long term, for averaging times exceeding, say, one day, the so-called *Allen variance*, the conventional measure of instability in frequency, would be less than $4{\times}10^{-16}$, surpassing all other standards.

16.5 The Proposed Ytterbium Ion Standard

The development of a *portable* laser-cooled Hg^+ ion standard is handicapped by the difficulty in synthesizing its 194.2 nm ultraviolet resonance wavelength with an all solid-state laser optical system. In this regard there is a great deal of interest in the development of an ion standard based on another species, namely the ion of the mass 171 isotope of the rare earth element ytterbium (Yb^+). In Figure 16.6 the relevant energy levels for this isotope of Yb^+ are shown, including (on a different scale) the 12.6 GHz magnetic hyperfine splitting of the ground state, which is used as the reference frequency, and the λ=369.5 nm ultraviolet transition used for hyperfine pumping and possible laser cooling.

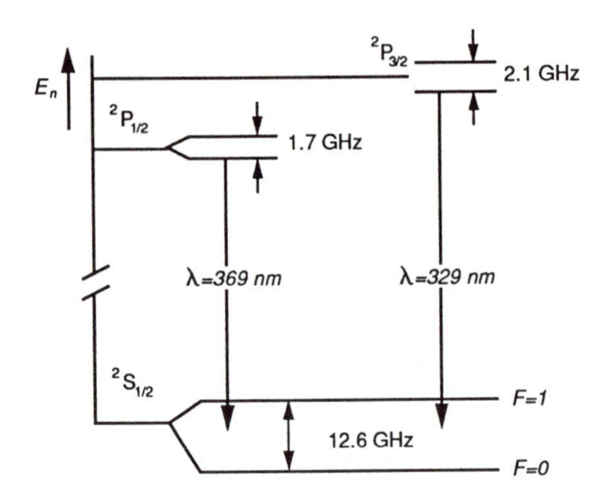

Figure 16.6 The relevant low-lying energy levels for a proposed Yb^+ ion standard

There are two essential differences between Yb^+ as a reference ion and the Hg^+ ion: First, the former has an ultraviolet resonance wavelength that is accessible by a simple doubling of the frequency of a diode laser output; second, the Hg^+ ion reference frequency at 40.5 GHz is over three times larger than that for Yb^+, which under equal conditions makes the Hg^+ ion resonance have a Q-factor three times greater and the standard based on it three times more stable. Nevertheless, until laser technology advances to the point where the synthesis of the ultraviolet frequency for the Hg^+ ion becomes possible using small solid-state components, the Yb^+ ion is a promising contender for field applications.

16.6 Beating Liouville's Theorem

The impact of the laser is no less significant in the development of standards based on neutral atoms, in particular, the all-important primary cesium standard, on which the definition of the unit of time is based. Again, by laser cooling, the thermal motion of the atoms can be all but frozen, drastically reducing the second-order Doppler shift and vastly increasing the "transit time," that is, the interaction time with the resonant field, thereby sharpening the resonance.

But in addition to these fundamental improvements, there is another remarkable laser application of particular relevance to atomic beam standards. We recall that in such standards, the signal-to-noise ratio of the resonance signal increases with the atomic beam intensity reaching the detector. Because the length of the resonance interaction region (the Ramsey cavity) must be as large as possible to achieve a sharp resonance, a definite limit is placed on the usable angular divergence of the atomic beam, while at the same time the beam aperture must also be kept small. Increasing the intensity of the beam under these constraints can be achieved only by increasing the *brightness* of the atomic beam source itself. We have already met the concept of brightness (radiance) in the context of light sources, particularly when comparing laser sources with conventional lamps. The term refers roughly to the number of particles (photons, atoms, etc.) emitted per unit cross-sectional area in unit time, per unit solid angle of divergence. This is a useful quantity, since in nearly all physical systems in which particle loss from the beam is negligible, it remains constant at all points in an optical system, no matter what combinations of focusing or deflecting elements are used. This result is based on a fundamental theorem in mechanics called *Liouville's theorem.* It is restricted to particles acted on by *conservative* forces, that is, forces that do not dissipate the energy in the form of heat in the way that frictional forces do. But this restriction leaves it with a very broad range of applications, including the usual (photon) optical systems as well as atomic beam systems. It relates, in a fundamental way, the space a group of neighboring particles can occupy to what they occupy in "velocity space," which for a beam translates into angular divergence. If we plot the three components of velocity as coordinates of a point with respect to a Cartesian system of perpendicular axes, then to each particle velocity corresponds a unique point in space, and velocities whose representative points lie within a certain region of space can be thought of as occupying a certain volume in velocity space. The theorem states that as the particles move under the action of the applied forces, the hypervolume in *phase space* (the product of volume in ordinary space with the corresponding "volume" in velocity space) spanned by a group of neighboring particles remains constant.

The conventional approach to increasing the brightness of an atomic beam source is to raise the atomic vapor density in the source combined with microchannel beam-forming openings to reach the highest possible intensity in a narrow cone. Recently, however, in a remarkable series of experiments it has been demonstrated that the brightness theorem can be "beaten" by the use of laser cooling. In other words, the brightness of an atomic beam can in fact be increased through the *nonconservative* interaction between laser beams and the atoms issuing from a source. The work was published by a group at the Eindhoven University of Technology, in the Netherlands (Hoogerland, 1994). They report a 1600-fold increase in the intensity of a beam of neutral atoms by the application of laser cooling! Although the work was on (metastable) neutral neon atoms and motivated originally by its application to atomic scattering studies, the technique clearly offers enormous potential improvement in the cesium beam standard.

The technique is essentially based on the use of radiation pressure (Doppler) cooling, as described in the last chapter, to reduce the velocity components of the atoms perpendicular to the beam direction, and to do it in a nonconservative way, as though by a frictional force. The result is that the angular divergence of the beam is reduced, making it more nearly parallel; that is, it is *collimated,* without the increase in beam cross section that would accompany collimation in a conventional optical system and that the theorem would demand of a conservative system. To take full advantage of this capability, the beam must pass through three stages: First, it is collimated, then focused down to a smaller diameter, and finally recollimated. The arrangement is depicted schematically in Figure 16.7. The first and last collimation stages are identical and incorporate two pairs of perpendicular laser beams in a plane across the atomic beam, forming a 2-dimensional optical molasses to resist the radial component of the particle motion. As this component tends to zero, so the direction of the particle motion becomes parallel to the beam axis. To be useful, it must be possible to achieve this with a reasonable amount of laser power, and to do so in a reasonable amount of space to a beam with a significant initial divergence as it enters the first stage. The maximum initial divergence angle accepted by the system, the so-called *capture angle*, is extremely important, as it determines the fraction of the atoms issuing from the source that contribute to the final beam. To achieve a large capture angle while keeping the cooling time as short as possible, the Eindhoven group extended the interaction region of each pair of collimating laser beams simply by multiple reflection between two plane mirrors slightly inclined to each other, as shown in Figure 16.7. The beams are injected between the mirrors from opposite sides in a direction slightly inclined in such a way that at each reflection the laser beam direction becomes more nearly perpendicular to the atomic beam axis. The geometry of the mirrors is chosen so that the laser beams are able to leave before they reach the point where they would begin to retrace their path. With this arrangement, if the laser frequency is tuned to the center of the atomic resonance, atoms having trajectories with a large initial angle to the axis are captured at the entrance to the mirrors, while the ones with a smaller angle are collimated farther downstream. This is clear if we note that by zero detuning of the laser, atoms moving perpendicular to a laser beam suffer the greatest radiation pressure, since only for them is the (first-order) Doppler shift zero.

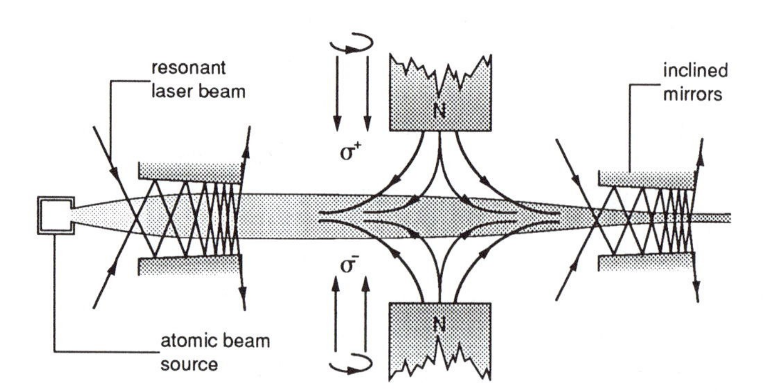

Figure 16.7 "Beating" Liouville's theorem: increasing the brightness of an atomic beam source

The middle stage, where the atomic beam is brought to a narrow focus, is a little more complicated. It consists of a so-called *magneto-optical trap*. It relies on the shift produced by an external magnetic field in the quantum energy states of atoms, and hence their optical resonance frequencies, that is, the Zeeman effect. Since the radiation pressure depends on the rate of photon absorption and reemission, and hence the frequency detuning of the atomic resonance with respect to the laser frequency, it follows that the light pressure acting on an atom is a function not only of its velocity because of the Doppler effect, but also of the magnetic field acting on it. If this field varies from point to point in space, then the light pressure the atom experiences will vary as a function not only of its velocity, but also of its position in space. Now, we recall that with pure Doppler cooling we were able to produce a net radiation pressure that changed sign with the velocity vector by having two oppositely directed laser beams tuned below the peak resonance frequency. To achieve a similar reversal with respect to magnetic field and hence spatial position, we use two oppositely directed laser beams whose polarization is such that the Zeeman shift in their frequencies of resonance changes in opposite directions with respect to a change in field intensity. We recall that in the Zeeman effect, the selection rules governing optical transitions between quantum states impose an essential connection between the polarization state of the radiation and the "permitted" change in the (quantized) orientation of the atomic angular momentum. This comes ultimately from the requirement that angular momentum of the photon–atom system be conserved in the process of emission or absorption.

Thus, for example, in Figure 16.8 we show the Zeeman effect in the so-called D_2-line of the principal resonance in the Cs atom, arising from the transition between the ground state, with its electron spin of ½ and zero orbital angular momentum, designated as $^2S_{1/2}$, and the first excited state with a combined (orbital + spin) angular momentum of 3/2, designated as $^2P_{3/2}$. In the presence of an external magnetic field, the substates with different orientations of the angular momentum with respect to the direction of the magnetic field, designated by the quantum numbers m, will be separated in energy by an amount dependent on the strength of the field. Now, a peculiar property of the photon, we recall, is that it carries one unit ($h/2\pi$) of angular momentum but has only *two* possible orientations with respect to a given axis, corresponding to the electric field in the light wave rotating clockwise or counterclockwise. (A linear polarization state is a mixture of those two.) Thus by absorbing a photon having the electric field rotating in a clockwise direction (looking along a fixed axis), designated as σ^+ polarization, an atom gains precisely one unit of angular momentum along that axis. Symbolically, we can write:

$$m' - m = +1 \quad \left(\sigma^+ - radiation\right), \qquad 16.6$$

where m' and m are the components (in units of $h/2\pi$) of the atomic angular momentum along the fixed axis, before and after the transition respectively. Similarly, for the oppositely polarized σ^--radiation we have

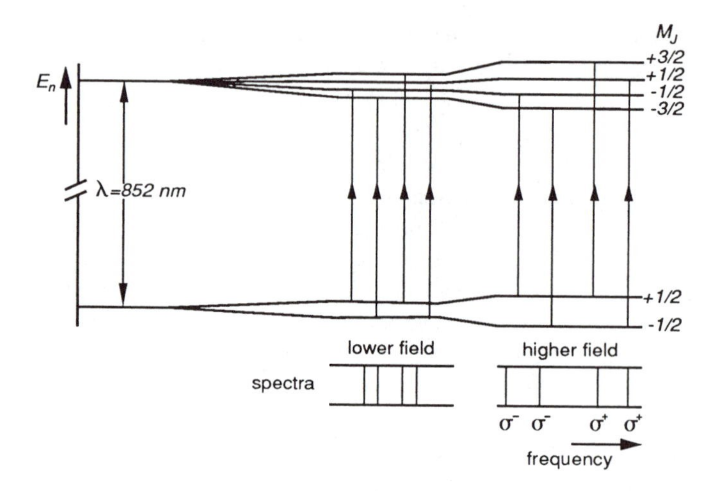

Figure 16.8 The Zeeman splitting of the D_2 line in Cs, neglecting the hyperfine structure

$$m' - m = -1 \quad (\sigma^- - radiation). \tag{16.7}$$

From the figure we see that as the magnetic field strength increases and the *m*-substates are more widely separated in energy, the resonance frequency for excitation with σ^+ radiation is shifted to a higher value, whereas excitation with σ^- radiation is resonant at a frequency shifted downward. Let us now limit our attention to how the resonance frequency of excitation by only one laser beam, having, say, σ^+ polarization, changes when the direction of the magnetic field is reversed with respect to our fixed axis. We note that the energy of the *m*-substates is proportional to the magnetic field, and if the latter is reversed in direction, we must reverse the sign of the energy so that the order of the *m*-values on an energy level diagram is reversed, as shown in Figure 16.9.

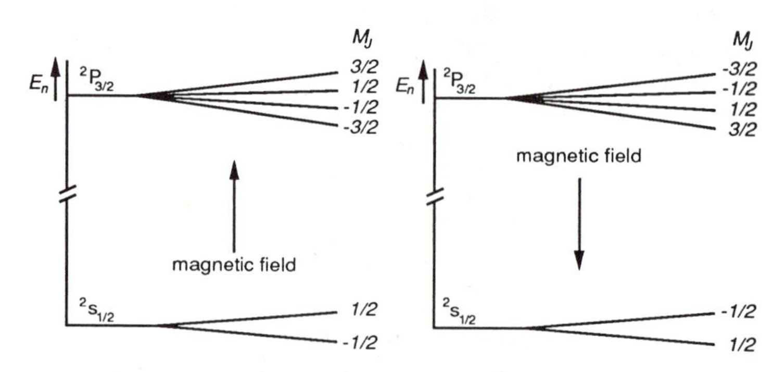

Figure 16.9 The effect of reversal of the magnetic field on the energy of m-sublevels

It is clear now that the excitation with the same σ^+ radiation has a resonance frequency that is shifted downward rather than upward as the magnetic field strength increases. By the same argument, excitation with σ^- radiation will have an upward shift on reversal of the field. Thus we see that having two oppositely directed laser beams, *circularly* polarized in opposite directions, with a parallel magnetic field of uniform gradient in space, passing through zero and reversing at some point, can result in an atom experiencing a corresponding reversal in the radiation pressure acting on it as it passes through that point. Such an arrangement, consisting of three mutually perpendicular pairs of lasers, is therefore called a σ^+–σ^- magneto-optical trap. The essential feature introduced by using σ-radiation in conjunction with a spatially varying magnetic field is that the simple "resistive" force $F_{rad}=\alpha kV$ is replaced by $F_{rad}=\alpha(kV+\beta B)$, where α and β are constants. It can be shown that for a sufficiently intense laser beam, the atoms will follow trajectories with a velocity whose Doppler shift compensates for the Zeeman shift in the resonance frequency. Stated mathematically, this amounts to the following:

$$kV = -\frac{2\pi g\mu_B}{h}B, \qquad 16.8$$

where $k=2\pi/\lambda$; g is the g-factor, the measure of the magnetic moment of an atom accompanying its mechanical angular momentum; μ_B is the Bohr magneton; and B is the magnetic field intensity acting on the atom. Suppose that the magnetic field distribution is chosen so that its (cylindrical) components vary in space according to the following:

$$B_r = -\frac{C}{a^2}r;\ B_\phi = 0;\ B_z = 2\frac{C}{a^2}z. \qquad 16.9$$

Such a field is produced, for example, in the neighborhood of the midpoint (taken as the origin) between two parallel coaxial circular coils separated by a distance equal to their radius, the so-called *Helmholtz* configuration, but having currents circulating in opposite directions. This is seen to correspond to the "pure" quadrupole field distribution familiar in the electrostatic case. To the extent that the light pressure acting on the atom can be accurately modeled as a velocity-dependent term due to the Doppler effect and a term having the spatial dependence of the magnetic field due to the Zeeman effect, its motion, for example in the radial direction, will be governed by an equation of the following form:

$$\frac{d^2r}{dt^2} + \alpha\frac{dr}{dt} = -\beta r, \qquad 16.10$$

where α and β are constants. This is recognized as the equation of motion for a damped harmonic oscillator, the form of whose solution for r, the radial distance of the atom from the axis (r=0), depends on the relative values of the "damping" coefficient α and the "restoring force" coefficient β. For $\beta>(\alpha/2)^2$ the radial motion, if allowed to continue, is a damped oscillation with a decaying amplitude; otherwise, the particle is drawn toward the axis without oscillation. The atoms are in this sense trapped in the radial direction, their distance from the axis falling exponentially; that is, they are focused.

It is just by using such a focusing arrangement between the two collimators as we have already described that it has been possible to "beat" Liouville's theorem and actually increase the brightness of an atomic beam source.

16.7 The Cesium Fountain Standard

16.7.1 Using Gravity to Return Atoms

Finally in this chapter we will briefly describe the immense new impetus that laser cooling techniques have given to realizing one of the earliest bold ideas for increasing the precision of atomic/molecular beam resonance spectroscopy: the molecular *fountain* experiment, proposed by Zacharias in 1953 (Zacharias, 1954, 1955). In this the resonance would be observed on atoms/molecules that are so slow that after passing vertically upwards through the resonance detection field they will fall back under the action of the earth's gravitational field, to pass again through the detection field in the other direction. For an apparatus of height h the maximum source velocity of a particle is given by $\frac{1}{2}MV^2=Mgh$, that is, $V=(2gh)^{\frac{1}{2}}$; if for example h=1.8 m then $V\approx 6\text{m}\cdot\text{sec}^{-1}$, and the average time of flight of an atom about 1.2 sec. Unfortunately, for an atom such as Cs (mass number 133), this would be about the average velocity of particles in thermal equilibrium at a temperature of about 0.5° above absolute zero. At ordinary operating temperatures of a Cs beam source, say 50°C, the number of atoms having a velocity below 6 $\text{m}\cdot\text{sec}^{-1}$ would be so small as to render such an approach not only extremely difficult, but one in which the gain in spectral line narrowing would be offset by a loss in signal-to-noise ratio. But this is precisely the sort of problem that the laser-cooling and beam-forming techniques we have been discussing are so spectacularly successful in overcoming. In fact, at the low temperatures attainable with laser cooling, an atomic beam cannot be formed with the usual linear geometry; the ballistic trajectories the atoms follow, which are parabolic arcs characteristic of bodies falling under gravity, will have a curvature that cannot be ignored.

The narrow Ramsey resonance patterns expected of such a long, free observation time have been demonstrated by a number of groups on Cs atomic fountains using laser-cooling; the development of this type of Cs standard is actively being pursued in several national standards laboratories.

16.7.2 The Prototype at the Paris Observatory

Preliminary results were reported in 1994 on the operation of a prototype Cs fountain standard at the Laboratoire Primaire du Temps et des Fréquences, of the Paris Observatory (Santarelli, 1994). Figure 16.10 shows schematically the essential elements of such a standard. As with the usual Cs beam standard, there are four principal functions that must be performed on the atoms: First, they must be formed into a well-defined beam; second, they must be put preferentially in one of the two hyperfine substates of the ground state; third, they must freely follow their vertical trajectories in a weak uniform magnetic field (the C-field), interacting with the resonant microwave field only at the beginning and end of their journey (the Ramsey field); and finally, their hyperfine substate population is monitored to detect resonance. In the present laser-based Cs fountain standard the first two functions are fulfilled in the laser-cooled source, while the Ramsey method of inducing transitions takes the form here of phase-coherent bursts of resonant microwaves applied sequentially in time rather than the space-separated regions in a Ramsey cavity.

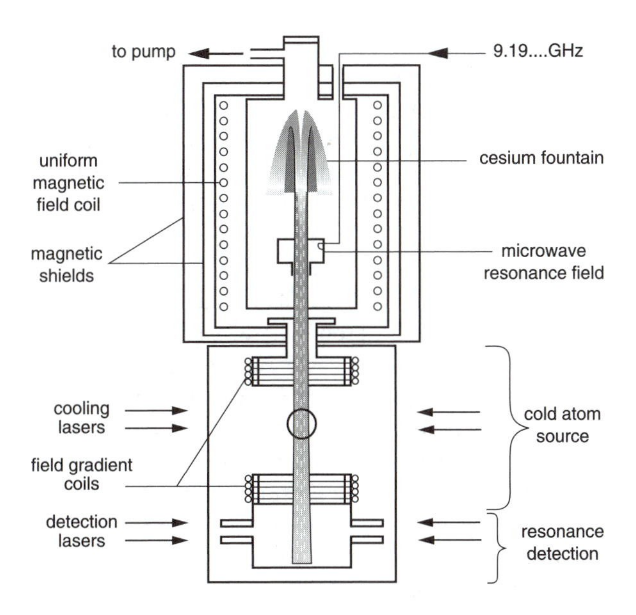

Figure 16.10 A schematic sketch of the Paris Observatory prototype cesium fountain standard

Finally, the detection of resonance is achieved optically using resonance fluorescence from a particular hyperfine substate, which the narrow spectral width of the laser light permits to be excited selectively. The Paris Observatory prototype has a cold atom source in which as many as 10^8 atoms are trapped and cooled down to around 5 microdegrees above absolute zero. It uses three pairs of mutually perpendicular laser beams tuned below resonance maximum, and it can be operated as a pure optical molasses or as a magneto-optical trap (MOT) with a magnetic field gradient (–8 gauss/cm) produced by two current-opposed Helmholtz coils.

16.7.3 Stabilization of the Laser Diodes

We recall that the Cs resonance wavelength has the advantage of being reached directly using compact solid-state diode lasers; in this case all the laser frequencies necessary for the operation of the fountain are derived from four diode lasers. These diode laser-based systems must, however, meet stringent requirements as to spectral purity and frequency stability in order to select a particular optical hyperfine transition and maintain a precise frequency offset from it. This requires active control of the laser output frequency, not with respect to an absolute reference optical cavity, as is often done, but with respect to the cesium resonance itself, using a special absorption cell containing cesium vapor. However, unless special techniques are used, the spectral width of the resonance in such a cell, due to the Doppler effect, would be too great for it to serve as a useful reference. In the days before the laser, efforts to reduce the Doppler width led to the development of atomic beams; now we have another approach, made practicable by the laser, called *saturated absorption* spectroscopy. It involves using two identical laser beams made to pass through the atomic vapor in opposite directions, in effect producing a *stationary* wave. If the laser beams cause a sufficiently high transition rate, there results a significant depletion in the number of those lower-state atoms whose Doppler-shifted frequency is resonant with the laser field, a depletion that indicates the beginning of saturation. If we imagine a third (weak) probing laser beam swept in frequency across the atomic resonance line, we would see successively absorption by different groups of atoms as their Doppler-shifted frequency comes into resonance with the probing beam, absorption that reflects the distribution of velocity among the atoms. However, two notches would appear in the absorption versus frequency curve, symmetrically placed about the maximum, as shown in Figure 16.11. The frequency width of these notches, which following Bennett are referred to as "holes," is the width the atoms would exhibit in the absence of the Doppler effect. These holes are due to the saturation of the absorption by those atoms whose Doppler-shifted frequency is resonant with one or the other of the two saturating laser beams. By varying the frequency of these two beams, the two holes can be made to coalesce at the maximum of the Doppler curve, corresponding to selecting atoms that are resonant with both beams at the same time, that is, those that have near zero velocity. This, then, is the basis of the *saturated absorption* technique for circumventing the Doppler broadening of spectral lines.

Using saturated absorption in a Cs vapor cell, the frequency of the output of one diode-based laser system is locked, by a servo-control loop, to a frequency precisely offset with respect to the resonance D_2 transition between the F=4 and F=5 hyperfine states of the ground ($^2S_{1/2}$) and first excited ($^2P_{3/2}$) states respectively. With a reported spectral linewidth of 100 kHz, this stabilized laser was used for the detection of microwave resonance as well in as stabilizing the frequency of two other free running (150 mW) diode lasers used in the entrapment and cooling of the atoms in the source. A fourth diode laser was stabilized on the F=3 to F=4 transition, shown in Figure 16.12, and used to restore atoms that are pumped out of the absorbing F=4 hyperfine state by the cooling lasers.

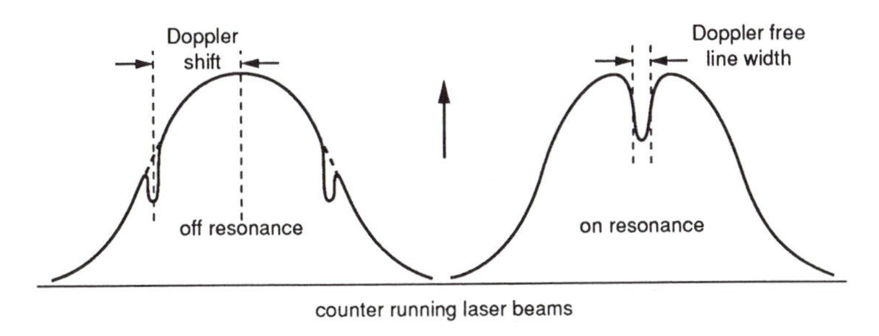

Figure 16.11 "Hole-burning" due to saturation in the Doppler broad-ended absorption profile

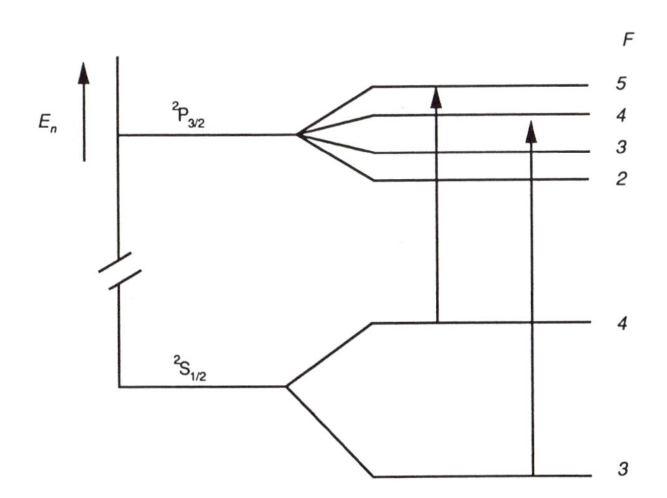

Figure 16.12 The transitions in Cs used to stabilize the diode lasers

16.7.4 Launching the Atoms

There remains one further critical requirement in order to fully achieve the potential of an atomic fountain standard: the "launching" of the atoms in a sufficiently narrow cone as to form a fountain in which most of the atoms traversing the microwave field in the upward direction will return through it in the downward direction. The effusion of atoms from an orifice as is usual in a conventional atomic beam source is unacceptably slow and divergent in the case of laser-cooled atoms in optical molasses. Ideally, the cooled atoms should be prepared in the nonequilibrium state of having their velocity components transverse to the fountain axis reduced to zero, retaining only the component along the axis. Initial experiments have used *moving optical molasses* to launch the atoms, that is, optical molasses in which a frequency offset is introduced between the oppositely directed lasers beams (whose frequency is offset with respect to the atomic resonance); this, through the Doppler effect, is equivalent to referring the motion of the atoms to a moving frame of reference. The properties of the molasses are therefore the same in this moving frame of reference as they would be in the rest frame in the absence of the frequency offset. Other techniques taking advantage of other mechanisms involved in the interaction of atoms with radiation have been studied as ways to manipulate the velocity distribution of atoms. The approach already described for increasing the brightness of an atomic beam source has an obvious application to atomic fountains.

Finally, we note that an atomic fountain standard not only has the potential of realizing a sharper resonance and therefore enhanced frequency stability, but can also reduce some important sources of systematic error. Errors arising from asymmetry in the microwave resonance field are reduced by the passage of the fountain in both directions through the same field, as is the (second-order) Doppler shift because of the low velocity of the atoms. These advantages clearly mark it as a prime candidate for a fixed installation primary standard of the highest accuracy.

Chapter 17
Measurement of Optical Frequency

17.1 Introduction

The early development of lasers was marked not only by the explosive proliferation of laser oscillation on different atomic and molecular transitions, but also by efforts to stabilize them and narrow their spectral line width. This was driven by the realization that the very attribute that makes the laser so remarkable is the one that still left room for spectacular improvement: spectral purity. The fundamental quantum limit on spectral purity far exceeds that of any common laser, subject to fluctuations in its optical cavity. Early success in stabilizing gaseous lasers made possible a number of important applications: from high-resolution interferometry to detect continental drift and seismic waves, to high-resolution spectroscopy.

With the development of a number of highly stabilized lasers having frequencies dotting the spectrum from the far infrared to the visible, and nonlinear devices effective in that frequency range to combine frequencies, it became a practical matter to actually measure the frequency of a light wave. This would have been an unimaginable prospect prior to the laser, and it is still far from a simple matter, involving multiple laser systems. Until recently, only the wavelength of such radiation could be measured; it was unthinkable that oscillations occurring at trillions of times per second could ever be followed by any detector! In fact, it is only since the advances we shall be considering in this chapter that it has become useful to describe radiation in terms of its frequency. One now speaks, for example, of the CO_2 laser (λ=10.6 μm) as having a frequency around 28.3 THz (1 terahertz=1000 GHz). This extension of the capability of measuring frequency to the optical regime requires a chain of fixed reference frequencies to be established, spanning the wide gulf between the microwave primary standard and optical frequencies. Fortunately, a number of stable lasers are available for this purpose, based on different active media, including HCN, H_2O, CO_2, Ne, and Ar^+.

17.2 Definition of the Meter in Terms of the Second

The effort to measure optical frequencies has been spurred by the adoption in 1983 of a new definition of the unit of length, the meter. It had been felt for some time that the relativistically invariant scale factor between time and space coordinates, the velocity of light, could be *defined* to have a particular value, thereby enabling the unit of distance to be defined in terms of the much more accurately kept unit of time. This came to pass at the 17th General Conference on Weights and Measures, in 1983, which adopted the definition of the meter as "...the length of the path traveled by light *in vacuo* in 1/299,792,458 of a second." This means that the velocity of light is now *by definition* exactly 299,792,458 m/s, the number being chosen, of course, so that the new meter is very nearly equal to the old standard. As a matter of principle, the new definition recognizes the relativistic point of view that space and time are not absolute and separate concepts; and as a practical matter, a standard of time interval (or frequency) can be maintained and measured with vastly greater accuracy than the distance between two points. Moreover, all distance measurements based on radar methods, and this includes all aerospace and much of interplanetary measurements, are in fact propagation times for electromagnetic waves of one wavelength or another. Previously, the unit of length was defined in terms of the wavelength of a certain line in the spectrum of krypton gas. With the new definition of the meter, the measurement of distance between two points is reduced ultimately to counting the number wavelengths of radiation of known frequency contained in that distance. It follows that for the greatest precision, the wavelength used for this purpose should be as short as possible; hence the need for the measurement of frequency of radiation extending to the optical region of the spectrum.

17.3 Theoretical Limit to Spectral Purity of Lasers

The development of stable optical frequency oscillators continues the progression toward oscillatory motion of higher and higher frequencies being used as the basis of time measurement; it is expected to lead to an *optical clock* whose level of performance is beyond what can be achieved in the microwave region. This is ultimately because probing a resonance requires a finite time, with its attendant uncertainty in frequency. It is only by raising the frequency that the *relative* sharpness of the resonance, or its Q-factor, can be raised, while keeping the probing time reasonably short in order to preserve short-term tracking stability of the atomic resonance. The stabilization of some lasers has already yielded performance competitive with the microwave standards, and optical frequency standards are ultimately expected to supersede those operating in the microwave region of the spectrum.

We are already familiar with the workings of laser oscillators: Unfortunately, unlike oscillators in the microwave region of the spectrum, such as the H-maser, the frequency of a laser is determined by the length and optical properties of the cavity and its associated optical elements, rather than the resonant quantum transition upon which the light amplification, and therefore self-oscillation, depends. As we have seen, the Q-factor of the longitudinal and the lowest one or two radial modes of oscillation of an optical cavity can be so large that cavity resonances are narrower than the atomic transition. This is because in the optical field of a gaseous laser, the atomic or molecular transition frequency is not immune from broadening by the (first-order) Doppler effect, as was the transition in the H-maser through the *Dicke effect.* In the case of liquid and solid laser media, there are strong perturbing intra-atomic forces that broaden complex quantum-level structures and the frequency of transitions between them. Laser oscillation therefore occurs primarily at the resonant frequencies of the cavity rather than at the peak of the atomic gain curve. But an optical cavity is clearly unsuitable as an absolute standard; the only fundamental unit that remains defined in terms of a particular object, namely a cylinder of platinum–iridium kept "under glass" at the International Bureau of Weights and Measures, is the kilogram.

From the point of view of an optical frequency standard, a laser oscillator serves to provide a strong, spectrally pure source of radiation, much as a klystron might do for a microwave standard. Long-term stability and reproducibility are achieved by locking the laser frequency on resonance with a suitable reference atomic or molecular quantum transition, free of Doppler and other sources of spectral line broadening. In this role the essential attribute of the laser is the spectral purity of its output. In practice, this is broadened by the fluctuations in the optical and mechanical properties of the cavity, particularly those due to environmental conditions, such as temperature and mechanical vibrations. We must distinguish, however, between these "technical" or "artificial" sources of phase/frequency fluctuations and those that are fundamental, that is, those that arise from the quantum properties of radiation and its interaction with atoms. These residual fluctuations would remain even if we had an ideal, perfectly stable cavity.

To understand the origin of this inherent, fundamental limit on the spectral purity of the laser output, and therefore the limit on the phase stability of the laser as a frequency standard, we must go back to the fundamental processes involved in its operation. There are two light-emission processes that atoms of the laser medium undergo: spontaneous and stimulated emission. In spontaneous emission, which occurs with a probability independent of the prior presence or absence of photons, the photons emitted by different atoms bear no phase relationship to each other, nor do photons emitted by the same atom at different times. In contrast, the stimulated emission of photons occurs with a probability proportional to the number of interacting photons already present, and the phases of photons emitted by different atoms, or by the same atom at different times, have a definite relationship; that is, they are coherent. It is the inevitable presence of spontaneously emitted, incoherent photons in the otherwise coherent stream of photons constituting the laser output beam that sets the limit on spectral purity mentioned above. Quantitatively, it can be shown that the mean square deviation in phase $\langle\Delta\phi^2\rangle$ is given by

$$\langle\Delta\phi^2\rangle = \frac{N_{spont}}{2N_{tot}}, \qquad 17.1$$

where the average $\langle\ \rangle$ is taken over a time during which N_{spont} photons are spontaneously emitted, and N_{tot} is the total number of photons in the given field mode. This result can be made plausible by noting that the ratio of photon numbers is proportional to $(E_{spont}/E_{tot})^2$, where E_{spont} and E_{tot} are the corresponding optical field amplitudes. That is, in terms of a field picture we have an oscillating optical field vector with its phase randomly fluctuating over a narrow range because of the addition of a (small) phase incoherent field component (the spontaneous photon). The size of the phase shift we can deduce if we recall further that the phase of a harmonically oscillating quantity $E\cos(\omega t+\phi)$ can be represented as the angle of rotation of a radius vector of length E turning with constant angular velocity ω, and that therefore the phase of the resultant of two fields oscillating at the same frequency can be obtained by vector addition of the two rotating vectors representing them, as shown in Figure 17.1. Now, under steady oscillating conditions where the population inversion is sustained by pumping at a constant rate, the mean optical field amplitude E_{tot} remains constant, and in cases of practical interest $E_{spont} \ll E_{tot}$. From the figure it is clear that the maximum change $\Delta\phi$ in the phase of the optical field due to the addition of a small vector increment occurs when the phase of the latter is at 90° to the main field vector. It follows that small fluctuations in the amplitude E_{spont} can produce a maximum phase change (in radians) given by $\Delta\phi = E_{spont}/E_{tot}$. Of course, the effect of the spontaneous component varies randomly, sometimes advancing the phase of the resultant, at other times retarding it. The situation will be recognized as reminiscent of a *random walk*, of which we have already given a simple model in an earlier chapter. As we saw there, while the average of the fluctuations is zero, being equally likely to be positive as negative, the average of the *square* of the fluctuations increases linearly with their number. In this case each spontaneously emitted photon corresponds to a new fluctuation in the phase, and since the emission occurs at a constant rate, we conclude that $\langle\Delta\phi^2\rangle$ increases linearly with time. It can be shown that this leads to a laser output with a Lorentzian spectral intensity distribution with a spectral line width $\Delta\nu$ given by

$$\Delta\nu = \frac{\pi h\nu(\Delta\nu_c)^2}{P}, \qquad 17.2$$

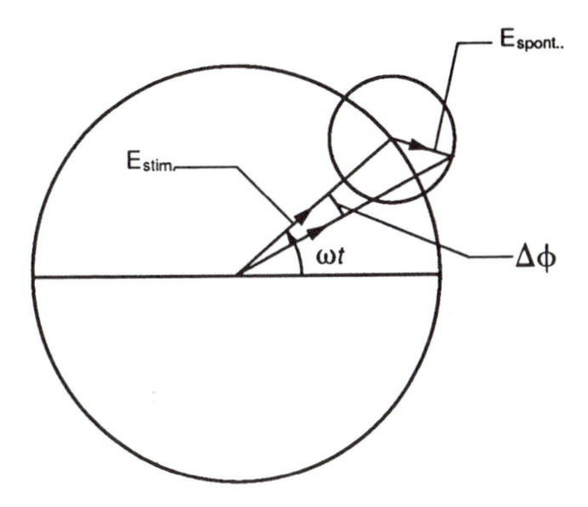

Figure 17.1 The rotating vector (phasor) representation of the stimulated and spontaneous optical fields

where $\Delta\nu_c$ is the *passive* cavity resonance linewidth (in the absence of the lasing medium) and P is the power in the cavity mode. It was on the basis of an expression of this form, first derived in 1958 by Townes and Schawlow, prior to the realization of a working laser, that the extraordinary potential spectral purity of lasers was predicted. Assume, for example, that we have a laser operating at 633 nanometers with an ideal cavity of length L=1 m and an output mirror with 1% transmission, so that 1% of the cavity intensity emerges as the output beam. The cavity resonance line width can be obtained from the average lifetime of a photon in the cavity. Thus a given photon has a 1% chance of leaving the cavity in the time required to traverse the cavity in both directions, and will therefore spend on the average 200 L/c before leaving the cavity. The corresponding (full) spectral width of the cavity resonance is then $\Delta\nu_c = c/\pi 200L$; that is, in this case $\approx$0.5 MHz. For a laser output power of 1 mW we find on substituting into the expression for the laser spectral line width $\Delta\nu = 2.5\times10^{-4}$ Hz! This quantum limit is so small that it was thought at the time it was first calculated that it was of no practical consequence; however, as we shall see, recent work on laser stabilization has led to claims of extraordinary spectral purity, approaching the quantum limit.

17.4 Stabilization of Lasers Using Atomic/Molecular Resonances

As with the microwave standards, the discovery of suitable reference atomic/molecular transitions and the methods of eliminating factors that broaden their frequency width parallel the study of high-resolution spectroscopy, and its drive to observe spectra under conditions of the utmost spectral resolution and accuracy. The reference transition should occur between quantum states whose (spontaneous) radiative lifetime is adequately long to allow a long interrogation time, and hence a sharp resonance. Also, the frequency of that resonance should be well separated from neighboring resonances and be relatively insensitive to environmental conditions such as ambient magnetic and electric fields. These are formidable conditions to fulfill; the best chance of satisfying them is in the rich spectra of molecules, spectra that characteristically consist of *bands*, each made up of closely spaced lines spanning a range of wavelengths. Moreover, there is almost a limitless variety of molecules, a far greater number than atomic species. To avoid spectral broadening due to intermolecular forces, the molecules should make up a low-pressure gas. The challenge is then to reach the resolution permitted by the natural line width of the absorber, in the presence of a much wider Doppler broadening due to thermal motion. There have in the past been many approaches to defeating the ever-present Doppler effect. Aside from actually reducing the Doppler effect itself by cooling the particles, the principal techniques that have been developed to achieve sub-Doppler line widths are saturated absorption (or fluorescence); two-photon absorption, about which more will be said in what follows; and Ramsey separated fields, with which we are familiar.

Great success has been achieved, as we have already seen in the case of stabilizing a diode laser against a resonance in atomic Cs as reference, using *saturated absorption* from two counterpropagating parallel laser beams, tuned to resonance with the desired atomic transition. Only atoms having a nearly *zero* velocity component along the beam axis are *simultaneously* resonant with both laser beams; any other molecules will see two field frequencies Doppler-shifted in opposite directions. The action of the combined beams can result in such a high rate of absorption that a significant number are in the excited state, leaving a depletion of the number in the absorbing ground state. This marks the onset of saturation in the amount of absorption: The gas becomes more transparent. It is important to distinguish between this process, in which each molecule absorbs only *one* photon at a time, from another *two-photon* technique for achieving sub-Doppler line widths, which we will encounter later.

17.5 Stabilization of the He-Ne Laser

One of the earliest attempts at stabilizing the output frequency of a laser with respect to a molecular resonance was on the helium–neon laser, because, as a gaseous laser, it has an inherently narrower spectral width than other types of lasers. However, because it can be tuned only over a very narrow range, it is necessary to find a species of atom or molecule that fortuitously has a transition that can be resonantly induced by the laser radiation. There are two molecular species that have been exploited with enormous success: one is the hydrocarbon methane (CH_4), otherwise known by the unglamorous name of "marsh gas," and the other is molecular iodine vapor (I_2). We recall that the He–Ne laser can generate not only the familiar red light at λ=633 nm, but also radiation in the near infrared at λ=3.39 μm; which wavelength dominates can be controlled by the design and operating conditions of the laser. It happens that certain lines in the spectra of CH_4 and I_2 fall within the narrow tuning range of the He–Ne laser operating at λ=3.39 μm and 633 nm respectively.

17.5.1 Stabilization on a Line in Methane

We begin with a general description of the methane case, which is representative of the essential features of both. At ordinary temperatures methane is a gas consisting of molecules each made up of one central carbon nucleus surrounded by four hydrogen nuclei (protons) at the corners of a regular tetrahedron, all immersed in a cloud of electrons that holds the molecule together through chemical bonds, as shown in Figure 17.2.

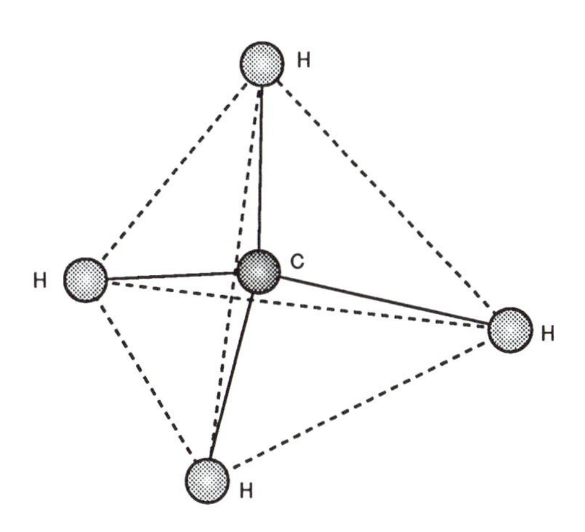

Figure 17.2 The symmetry of the methane (CH_4) molecule

The quantum energy states between which transitions resonate with the He–Ne output at 3.39 μm are ones that differ both with respect to vibration of the nuclei with respect to each other and rotation of the molecule as a whole. While for our present purposes it is fortunately not necessary to become entangled with the complexities of molecular spectra, there are some fundamental generalities that the transitions in methane illustrate. The bonds holding the hydrogen atoms to the carbon are like elastic springs, allowing the atoms to vibrate, not freely, but with each affecting the other. Such a system, like the simple coupled pendulums we met in an earlier chapter, will exhibit certain particular patterns of vibration: the so-called *normal modes*. In each of these modes all the particles vibrate with a common frequency, which differs in general from one mode to another. However, in the case of the methane molecule, since it has a high degree of symmetry, such as the three-fold axes of rotation through each of the hydrogen atoms and the carbon atom at the center, some of these modes have the same frequency. Such modes correspond in the quantum description to *degenerate* states, that is, ones of equal energy. It happens that methane, and other molecules having the same symmetry, have one nondegenerate mode, in which the atoms vibrate symmetrically with respect to the carbon; one doubly degenerate mode; and two triply degenerate modes. We will not bother with the nomenclature used by spectroscopists to classify these modes according to their symmetry types, etc. There is, however, an interesting physical consequence of having vibrational modes with the same frequency. We recall that if a particle executes two perpendicular oscillations simultaneously with the same frequency, the resultant motion will be uniform motion around a *circle* if their phases differ by 90°. It follows that it is possible for a system of particles to have *angular momentum* by virtue of excitation of degenerate modes of vibration. Naturally, such an angular momentum must be included among the other more familiar sources of angular momentum in arriving at the possible quantum states. Moreover, the vibrational motion and its associated angular momentum are not independent of the rotation of the molecule in space because of the *Coriolis force*, which is an inertial force arising in a rotating frame of reference according to Newton's laws. A more familiar example of a rotating frame of reference is, of course, one anchored in our Earth. A freely moving projectile does not simply arc to Earth in a vertical plane, as it would on a spinless Earth; rather it will deviate slightly in a horizontal plane. To preserve Newton's laws this deflection is attributed to a "force" perpendicular to the particle velocity and proportional to Earth's angular velocity.

A look at the rotational properties of the methane molecule brings out another rather surprising consequence of its symmetry. To appreciate this we recall that in computing the angular momentum associated with a given rate of rotation of an object, we require the *moment of inertia* for the given axis of rotation. If we consider the moment of inertia about an axis passing through the carbon atom, perpendicular to one of its axes of symmetry, we find that because of its three-fold axes of symmetry, the molecule behaves like a symmetric top with *complete* axial symmetry. This means that it behaves as though it were shaped like a symmetric top; the moment of inertia is the same for *any* rotation axis passing through the center and lying in the plane perpendicular to one of the hydrogen–carbon bond directions.

The transitions of interest are in the infrared among those between vibration–rotational states shown schematically in Figure 17.3. The states are grouped according to vibrational state but are differentiated within each group according to their angular momentum. We see that the angular momentum quantum number (J) can reach fairly large values. Transitions in which $J'-J=-1$ are conventionally classified as forming the P-branch of the spectrum; similarly, those in which $J'-J=+1$ form the R-branch. The specific transition that can be made resonant with the $\lambda=3.39$ μm output of the laser is in the fine structure of what is designated as the P(7) component in the $\nu_3=0\rightarrow\nu_3=1$ characteristic vibration band, in which $J=7\rightarrow J=6$.

The physical setup of the saturated absorption system for stabilizing a He–Ne laser consists essentially of a chamber several meters long containing methane gas at low pressure, provided with infrared transmitting windows. Beam-expanding optics and a retroreflector are provided to ensure precisely parallel counterpropagating laser beams passing through the gas. The return beam intensity is monitored with an infrared detector, such as a cooled In–Sb photoconductive detector. If the laser frequency is swept across the resonant frequency, there is a sharp peak in the intensity of the return beam, which can be used to derive an error signal for the servo control of the laser. The apparatus is shown schematically in Figure 17.4.

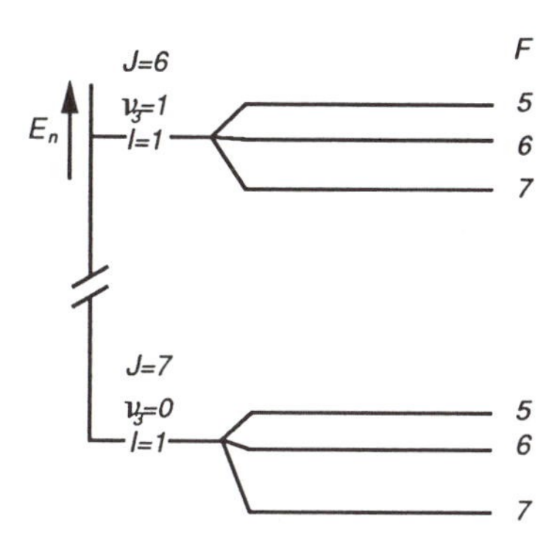

Figure 17.3 The vibration–rotational levels in CH_4 relevant to the stabilization of the He-Ne laser

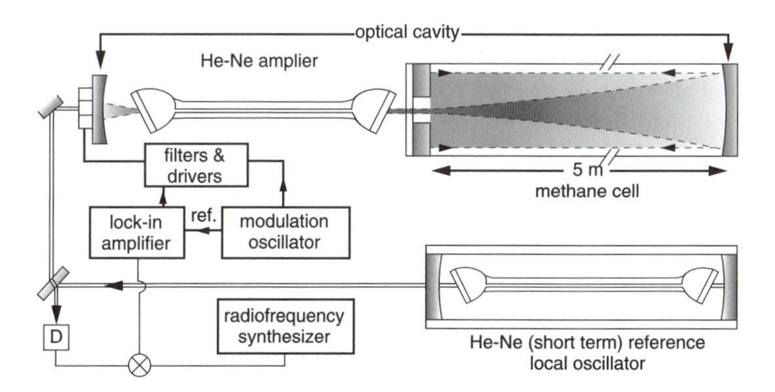

Figure 17.4 Experimental arrangement for laser stabilization using saturated absorption in methane gas

There are several factors that may limit the attainable sharpness of the resonance, and hence the tightness of control of the laser frequency. The most serious is the finite interaction time as the molecules cross the laser beams with their thermal velocity: A CH_4 molecule at 300°K will cross a 10 mm diameter laser beam in about 15 microseconds. This transit time can be doubled by operating at the temperature of liquid nitrogen (77°K). It is on this account that beam-expanding optics are used with larger apertures to increase the diameter of the beams. The other factors have to do with the parallelism of the reflected beam with the forward beam, and their angular divergence; these lead to (first-order) Doppler shifts in the resonance of a molecule moving across the beams. There is a fundamental lower limit on the divergence set by diffraction, which can be lowered only by widening the aperture. At the level of spectral resolution involved here, amounting to only a few hundred hertz in a frequency of 10^{14}, many other subtle sources of spectral broadening and frequency shifts become significant. These include the second-order Doppler effect; the energy of recoil accompanying photon absorption, which is on the order of $h\nu/Mc^2$; broadening and shift due to collisions between molecules; Zeeman broadening in Earth's magnetic field; etc.

17.5.2 Stabilization on a Line in Iodine

Another example of a widely used stabilized laser frequency using saturated absorption is in the visible at λ=633 nm, the familiar red beam generated by a He–Ne laser. The reference transition is a hyperfine component in the visible part of the band spectrum of molecular iodine I_2. At room temperature, iodine is a solid with a relatively high vapor pressure, which reaches around 100 Pa at 38°C. The stable isotope has mass number 127 and a nuclear spin of 5/2, giving rise to a hyperfine structure in a rich spectrum containing a high density of lines in the λ=500 nm–670 nm range.

As a diatomic molecule, the classification of its energy levels is a good deal simpler than that for methane. It has, of course, only one mode of vibration and two equal principal moments of inertia perpendicular to the molecular I–I axis. The curves showing the mutual potential energy of the two iodine nuclei as a function of their distance apart, an amount of energy arrived at approximately by solving for the electron energy states corresponding to fixed nuclei, are shown in Figure 17.5 for the ground state and the first excited electronic state. In the neighborhood of the points where the potential is a minimum, the shape of the curves is parabolic as for a simple harmonic oscillator. Thus within each electronic state there is an array of vibrational energy levels, represented by the horizontal lines, and each of these is further split into closely spaced levels of rotational energy. According to the *Franck–Condon principle*, in a quantum transition from one electronic state to another, the effect on the nuclear motion is negligibly small; that is, classically, the kinetic energy and separation of the nuclei would be the same before as after the transition. The kinetic energy of the nuclei at any given separation is simply represented in Figure 17.5 by the difference in energy between the appropriate potential curve and a horizontal line representing the total energy. Thus on the classical picture a quantum transition from a given electronic state to another is represented graphically by a vertical line drawn between points where the kinetic energy is the same. Since the two potential energy curves for the iodine molecule are not only dissimilar but are displaced horizontally with respect to each other, the final state classically predicted would vary according to the nuclear separation at the time of the transition. In fact, we see that it is possible for transitions to lead to states in which the kinetic energy remains finite even at great distances, that is, complete dissociation of the molecule into two separate atoms. As the upper rim of the potential well is approached, the vibrational levels are expected to become more and more closely spaced, providing a wealth of energy levels for laser stabilization.

As with methane, saturated absorption is used to obtain sub-Doppler resolution in stabilizing the He–Ne laser line at λ=633 nm using an electronic transition in I_2 vapor. In principle, the same basic design considerations, such as beam expansion to increase transit time across the beam, wave front curvature, and beam alignment are equally important here as in the methane case. In practice, however, because the I_2 line resonant with the He–Ne λ=633 nm light is weak, the absorption cell containing the iodine vapor is placed *inside* the laser cavity to interact with the intense optical field there. This circumstance has detracted from the accuracy and resetability that can be achieved in practice. Nevertheless, it has been widely used as a laboratory standard of wavelength, since it surpasses the standard krypton lamp.

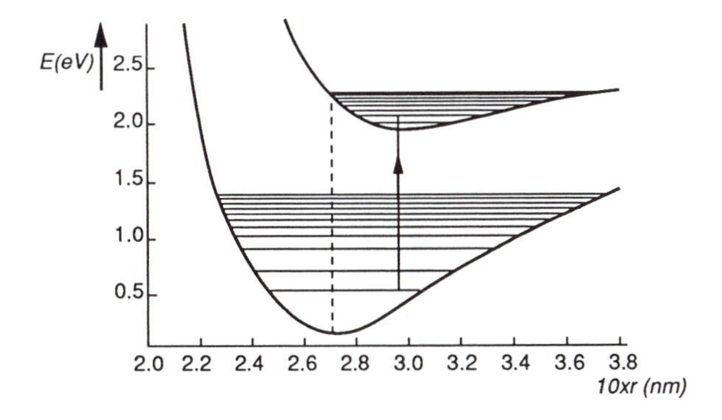

Figure 17.5 The molecular potential curves for the iodine molecule I_2 relevant to stabilization of He-Ne laser

17.6 Stabilization of the CO_2 Laser

Another type of laser, which, suitably stabilized, plays an important role in providing a reference frequency in the frequency chain, is the carbon dioxide (CO_2) laser, which operates in the infrared around 10.6 μm (28.3 THz). The CO_2 molecule consists of a carbon atom with an oxygen atom symmetrically placed on either side in a straight line. It has three normal modes of vibration; one, the so-called *bending mode*, is doubly degenerate, as illustrated in Figure 17.6.

The laser transition is between hyperfine components of rotational states in an upper, asymmetric stretching mode of vibration and a lower, symmetric stretching mode. It is effectively a 3-level laser system, with the excitation of the upper level enhanced by the presence of nitrogen and helium in the electrical discharge tube containing the carbon dioxide. Just as the He gas is added to transfer electronic excitation to the Ne in the He–Ne laser, the N_2 molecules in excited vibrational states transfer that vibrational energy to the CO_2 molecules. The laser is characterized by unmatched efficiency and output power, giving it an early advantage in industrial applications, where raw power was required. Now, by providing the conditions for high-resolution stabilization of its frequency, it has become important in providing a bridge in the frequency spectrum.

It is unique in allowing a useful saturation *fluorescence* signal from CO_2 itself to be used for stabilization. The frequencies of lasing transitions in several isotopes, such as $^{12}C^{16}O_2$ and $^{13}C^{18}O_2$, have been measured to almost 1 part in 10^{10}. There are altogether some 600 CO_2 frequencies that can serve as secondary standards in the infrared. Unfortunately, the reproducibility is ultimately limited by pressure shifts in the frequency of the CO_2 cell used for stabilization. The spectra of other molecular candidates for stabilization of the CO_2 laser have been studied; the most promising is osmium tetroxide, OsO_4, with which accuracies in the neighborhood of parts in 10^{12} have been achieved.

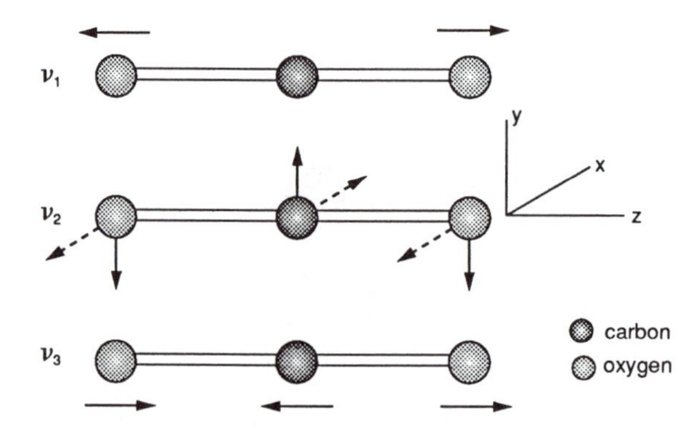

Figure 17.6 The normal modes of vibration of the carbon dioxide (CO_2) molecule

An absolute measurement of the frequency (473 THz) of the I_2-stabilized He–Ne laser, based on a comparison with the synthesized 16th harmonic of a stabilized CO_2 laser at 29.5 THz, was reported in 1993 by the French Laboratoire Primaire du Temps et des Fréquences. The accuracy given was about one part in 10^{11}, an accuracy limited by the poor reproducibility of the stabilized He–Ne laser itself.

17.7 Stabilization Using Two-Photon Transitions

In the stabilized laser systems so far discussed, sub-Doppler line widths were achieved in the reference molecular resonance by the saturated absorption (or fluorescence) method, in which each molecule absorbs *one* photon from counterpropagating laser beams. (Simultaneous resonance with both beams simply increases the probability of absorbing that one photon, in a given time.) But another important technique for achieving Doppler-free line widths, which has recently been exploited in the context of optical frequency standards, involves the simultaneous absorption by each atom or molecule of *two photons* from counterpropagating beams. Although the probability of such a two-photon process is vastly smaller than that for the one photon process, nevertheless, it can be significant with sufficient laser power and a favorable disposition of the quantum levels of the absorber. It had long been observed in magnetic resonance experiments at microwave and radio frequencies, where strong coherent fields have been readily available; however, it is only after the advent of the laser that it has become feasible to observe them at infrared and optical frequencies. To understand the way in which the Doppler effect is rendered ineffective in broadening the resonant absorption, suppose ν_0 is the frequency of the reference transition in a molecule placed in two counterpropagating laser beams of frequency ν_L. If the molecule has a component of velocity V in the direction of the common beam axis, then it will see the frequencies of the two beams Doppler shifted to $\nu_L(1+V/c)$ and $\nu_L(1-V/c)$. Now, the simultaneous absorption of two photons, one photon from each of the two beams, results in the molecule gaining an amount of energy $2h\nu_L$, no matter what its velocity happens to be; and therefore resonance will occur at $2\nu_L=\nu_0$ for *all* molecules. The quantum selection rules that govern two-photon transitions are naturally different from those of the usual one-photon process, in particular, the angular momentum of the initial and final states and their dependence on the polarization of the photons. If the two beams have the same polarization, then it is possible that two-photon absorptions occur from the *same* beam, in which case absorption can occur at Doppler-shifted frequencies by groups of molecules having different corresponding velocities. This would lead to a resonance line shape with a Doppler-broadened base on which is superposed a sharp Doppler-free peak. It is possible to eliminate the Doppler base if the two counter-propagating beams have different polarizations (for example, right and left circular polarizations) and the quantum selection rules applied to the given transition allow only one photon from each beam.

As an example of the application of this technique in the context of optical frequency standards we describe briefly the work reported in 1994 by the group at the Laboratoire Primaire du Temps et des Fréquences, in collaboration with other groups in France (Touahri et al., 1994), on the stabilization of a GaAlAs diode laser using two-photon transitions in the rubidium atom. In Figure 17.7 is shown a partial energy level diagram of the Rb atom, for which the two lower energy states are already familiar from the optical pumping of the Rb standard. The two-photon transition is between the (L=0) $5^2S_{½}$ ground state and the (L=2)$5^2D_{½}$ excited state. Such a transition is *forbidden* in a one-photon process; however, a two-photon process, although generally expected to be improbable, is in this particular case greatly enhanced by the presence of a level $5^2P_{½}$ approximately midway between the initial and final levels. This intermediate level is only 1.05 THz from the exact middle position. The sequence of events following the simultaneous absorption of two photons is that the atom is raised to the upper 5D state from which it quickly cascades down by allowed one-photon transitions, first to the 6P state and finally to the 5S ground state by the emission of a photon in the blue region of the spectrum at λ=420.2 nm. The emission of the blue line is eminently suitable for detecting the occurrence of two-photon transitions. Not only does it appear if and only if the desired transition has been induced, but it lies in a region of the spectrum where photomultipliers have low dark current, and so far removed from the laser frequency that spurious background scattering of the laser light is easily filtered out. We recall that the Rb spectrum contains many hyperfine components, and with the high spectral resolution implicit in all the techniques we are concerned with, the laser is stabilized on a particular hyperfine resonance. The general layout is shown schematically in Figure 17.8.

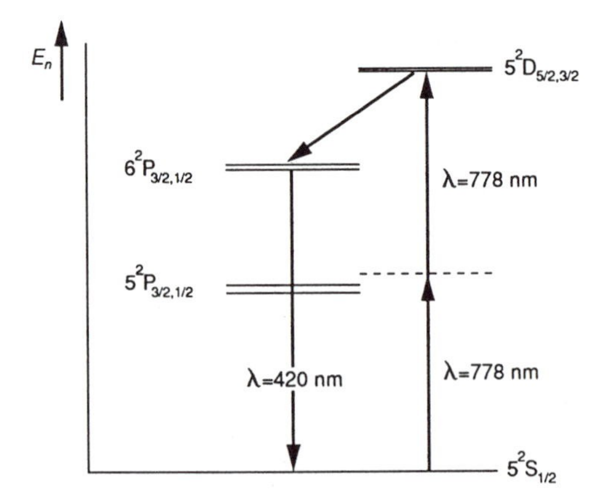

Figure 17.7 Partial energy diagram of Rb showing the 2-photon 5S–5D transition and the intermediate 5P level

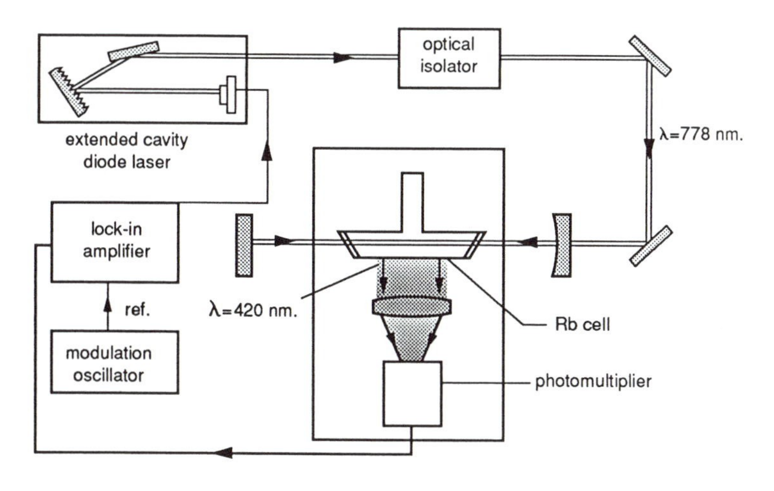

Figure 17.8 Schematic diagram of apparatus for 2-photon resonance in Rb to stabilize GaAlAs diode laser

The diode laser is in an *extended cavity configuration,* in which spectral narrowing is achieved by having a diffraction grating forming one of the reflectors, which is servo-controlled along with the diode current. The rubidium vapor absorption cell is placed in an external cavity, allowing for the control of the power and alignment of the counterpropagating beams. The laser frequency is modulated at 70 kHz, and the blue fluorescence is detected with a photomultiplier whose output signal goes to a lock-in amplifier in which the modulation frequency serves as reference. The output of the lock-in amplifier is the error signal used to servo-control the laser system. The frequency stability reported of a few parts in 10^{14} for an integration time of 300 seconds is an order of magnitude better than has been achieved with a He–Ne laser stabilized using I_2 as a reference. The largest correction is the light shift, which amounts to about one part in 10^{11}. However, it is relatively well understood, and extrapolation to zero light intensity is good enough to make this system competitive with the stabilized He–Ne/I_2 system.

17.8 Frequency Comparisons in the Optical Range

Even given a series of highly stable and reproducible reference frequencies extending to the visible range, there remains the equally challenging problem of interrelating their frequencies and referring them back to the primary microwave standard. The standard procedure in making frequency comparisons, one that is the basis of frequency synthesis techniques at radio frequencies, is to mix the optical signals in a nonlinear device and measure the beat frequency between the higher frequency ν_2 and a harmonic of the lower frequency ν_1 according to the following:

$$\nu_2 - n\nu_1 = \nu_{beat}. \tag{17.3}$$

This is practical, provided that ν_2 and ν_1 are sufficiently stable and are such that a coherent low-frequency beat is observable, preferably within the microwave range. Optical frequencies are so high that even a small fractional deviation will lead to very large excursions in the beat frequency. Completing the chain of comparisons from the microwave standard to the optical range is made difficult by the wide gaps between the limited number of reference lasers, and their narrow tuning range. This is aggravated by the unavailability of devices that can generate harmonics of sufficiently high order to bridge those gaps; in the microwave region it is possible to generate as high as the 1000th harmonic, whereas at infrared frequencies perhaps the 12th harmonic is possible. Even where a multiplicity of neighboring stabilized optical frequencies exist, the separations in frequency are still orders of magnitude larger than their tuning ranges. This is demonstrated in Figure 17.9, where we have plotted on a *logarithmic* frequency scale (each unit is a *tenfold* increase) some of the established laser frequencies spanning the spectrum from the microwave to the visible.

We recall that the generation of harmonics as well as the mixing of different frequencies to produce a sum or difference frequency is achieved using the nonlinear response of certain devices. At microwave and lower frequencies, where the wavelength can be much larger than ordinary electronic components, the nonlinear device is a diode that can be treated as a *lumped* circuit element of negligible size. This is no longer possible when we reach the infrared region of the spectrum and beyond. Either we must find a device whose interaction with the light wave extends only over a microscopic region or failing that, take into account the oscillatory variation of the light wave across the device, that is, treat it as having a *distributed* interaction.

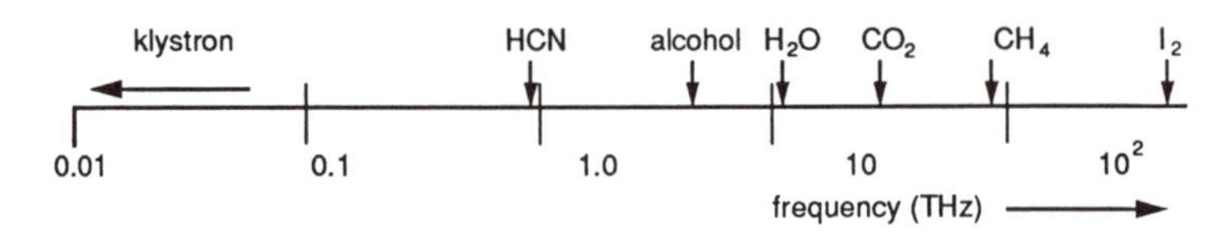

Figure 17.9 Available stabilized lasers plotted on a frequency scale extending from the microwave to the visible

17.8.1 The Point Contact MIM Diode

The most important nonlinear device responding to the first condition is the point contact metal–insulator–metal (MIM) diode. This is simply a finely etched tip of a tungsten wire in contact with the oxidized end surface of a nickel post, properly incorporated into a microwave circuit. These diodes respond to extremely high frequencies because the contact region has a diameter on the order of only 1/600 the wavelength of the CO_2 laser. Moreover, the resistance and capacitance across the junction are relatively small, since electrons can tunnel through the nickel oxide film, which is only on the order of a few molecules thick. These MIM point contact diodes can mix and generate sum and difference frequencies of input light waves up to 200 THz (λ=1.5 μm) in the near infrared. Given two infrared frequencies whose *beat* (difference) falls in the microwave region of the spectrum, focusing them onto such a device mounted in a microwave circuit will produce a microwave output for further frequency measurement.

17.8.2 Nonlinear Crystals

The second class of nonlinear devices useful beyond 30 THz (λ=10 μm) includes certain optical crystals that lack a center of symmetry. In these crystals the electric field component of the light wave induces an oscillating electric dipole moment that has a small quadratic dependence on the field amplitude. Moreover, the dipole moment in a given direction depends not only on the field component in that direction, but also on the components in other directions, reflecting the anisotropy of a crystalline medium. The nonlinear behavior is expected to be weak in general, since even in a 1 megawatt laser beam the electric field amplitude is on the order of $E \approx 10^6$ volts/m, whereas the interatomic fields in the crystal are on the order of 10,000 times that. This means that in order to gain a large cumulative effect, the light wave must interact with the crystal over large distances, containing many wavelengths, making the crystal fall in the class of distributed devices. This introduces a requirement on the velocities of light waves of different frequencies in the crystal medium, since any sum or difference frequency wave generated at points along the path of the input waves must reinforce waves generated at subsequent points, in order that there be overall buildup of the mixed frequencies. This means ideally that all frequencies must travel at the same velocity. In practice, it is only necessary that any phase difference that develops between waves generated at the beginning and those at the end of the finite path in the crystal be less than π radians. For example, let us consider what this means in the case of second harmonic generation: Let n_ν and $n_{2\nu}$ be the refractive indices of the crystal for the fundamental and second harmonic frequencies. Now, the second harmonic component of the electric polarization of the crystal, which acts as the source of the second harmonic wave, travels at the velocity of the fundamental wave, and hence the condition on the phase difference that can be allowed to develop over a distance L in the crystal, $\Delta\phi$, is the following:

$$\Delta\phi = 2\pi(2\nu)\left[\frac{n_{\nu}L}{c} - \frac{n_{2\nu}L}{c}\right] \leq \pi. \qquad 17.4$$

This is called the *phase matching condition.* If it is violated, then the contributions to the second harmonic wave originating from different points along the path of the fundamental wave will combine in opposing phases, resulting in destructive interference.

Regarded from the photon point of view, the process amounts to the conversion of two photons at the fundamental frequency into a single photon of twice the frequency. Looked at this way, the phase matching condition becomes a statement of the conservation of photon linear momentum h/λ, which in quantum theory allows some discrepancy, provided that it is within the Heisenberg uncertainty in momentum. We recall that this uncertainty is related to the uncertainty in distance in the same way that the uncertainty in frequency involves the time measurement. If L is in effect the uncertainty in the positions of the photons, then the Heisenberg uncertainty in momentum is on the order of $h/2L$; it follows that for momentum conservation we require

$$2\frac{hn_{\nu}\nu}{c} - \frac{hn_{2\nu}2\nu}{c} \leq \frac{h}{2L}, \qquad 17.5$$

which is clearly the same as the phase matching condition.

Unfortunately, in practice, all transparent media are to some extent dispersive; that is, the wave velocity, and therefore the refractive index, varies with the frequency. It can happen that a specific material may be found whose refractive index takes the same value at widely different frequencies, allowing phase matching at those frequencies. But this is rarely the case. A far more practical approach is to exploit the birefringence (double refraction) of certain crystals. Without getting too deeply involved in crystal optics, we will assert that there are crystals of certain classes of symmetry whose optical properties have symmetry about a single axis, the *optic axis*, and are called *uniaxial.* These crystals have a 3-fold, 4-fold, or 6-fold axis of symmetry, and their optical behavior is symmetric about these axes, reminiscent of the axial symmetry of the moment of inertia of the CH_4 molecule, which has a 3-fold axis of symmetry. Other crystals of a lower degree of symmetry have *two* preferred axes and are *biaxial.* In a uniaxial crystal, a light wave propagates along two wavefronts with the velocity of one, the *extraordinary* wave, depending on the direction relative to the optic axis. The other, the *ordinary* wavefront, advances with equal velocity in all directions. Along the optic axis both wavefronts advance with the same velocity. The directions of the electric (and magnetic) components of the two types of waves, that is, their polarization vectors, are always at right angles to each other. The velocity of the extraordinary wavefront may increase or decrease as a function of the angle with respect to the optic axis, depending on the specific material of the crystal, as shown in Figure 17.10. In such a crystal, phase matching is achieved when the phase velocity of the ordinary wavefront at one frequency matches the phase velocity of the extraordinary wavefront at the other frequency. Since the difference in phase velocity between the two varies continuously with the angle the laser beam makes with the optic axis, it is possible in principle to match phases with a suitable crystal. In practice, this requires a beam with extremely small divergence, critically adjusted to the correct angle with respect to the crystal axes. An approach less critically sensitive to beam adjustment is to find a suitable crystal in which the wavefront velocities can be adjusted by varying the temperature, while fixing the beam direction at right angles to the axis, where the velocity difference does not change (to first order) with angle. Just how critical the matching of phase velocities (or equivalently, refractive indices) is in practice can be estimated from the phase matching equation $(n_{2V}-n_V)\leq\lambda/4L$, where λ is the wavelength of the fundamental frequency wave. Since the degree of nonlinearity is relatively small, L is on the order of a centimeter; hence for a λ in the middle of the optical range, the difference in refractive index must not exceed one part in 10^5.

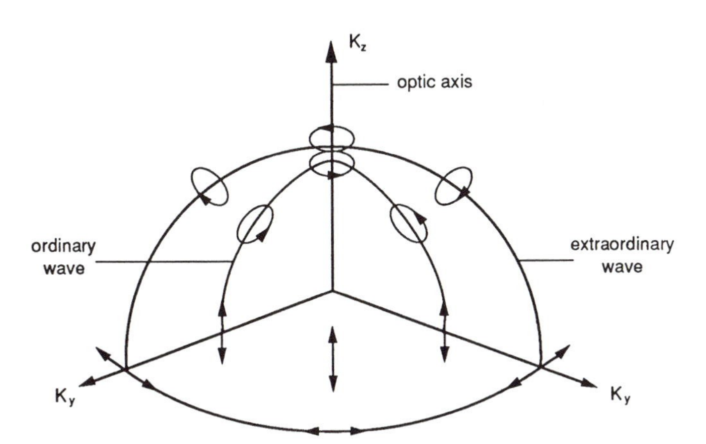

Figure 17.10 The ordinary and extraordinary wave vectors and polarizations in a uniaxial crystal such as quartz

Of the many birefringent crystals that show nonlinearity, those few that have an adequate nonlinear coefficient, are transparent in the desired wavelength range, and are resistant to surface damage will be useful as frequency mixers. Quartz has already been mentioned as the crystal in which the doubling of an optical frequency was first observed. Other crystals with greater nonlinearity that have been used include potassium dihydrogen phosphate (KDP) and ammonium dihydrogen phosphate (ADP), which are adequate for harmonic generation of 1μm infrared radiation. The widely used lithium niobate ($LiNbO_3$) crystal has a nonlinearity ten times that of KDP; potassium niobate ($KNbO_3$) has an even larger nonlinearity, providing phase matching into the blue part of the spectrum; and lithium iodate ($LiIO_3$) provides a further extension into the ultraviolet.

17.9 Measuring Optical Frequencies Relative to a Microwave Standard

Finally, we touch on the various ways the realization of frequency/phase transfer between an optical frequency standard and the primary microwave Cs standard have been pursued. The problem is as fundamental as that of gear trains in an analogous mechanical system. It has evoked increasing effort, with a number of successful determinations of optical frequencies based on different frequency chains having been accomplished. Although these differ in the detailed way in which phase coherence and stability are transferred from one member of the chain to the next, they are for the most part based on the harmonic generation principle already mentioned. Each link in the chain involves a *phase-lock loop* of some sort, that is, a servo loop in which the phase difference between the beat signal derived from a mixer and a synthesized (relatively) low frequency constitutes the error signal, properly conditioned, so that apart from the synthesized frequency offset, the phase of one laser is locked to a harmonic of the other. While this approach has been successfully applied to linking a specific optical frequency, such as that of a stabilized He–Ne laser to the microwave standard using a particular frequency chain, the narrow tuning range of lasers makes it difficult to apply to the measurement of arbitrary optical frequencies. On the other hand, for a clock based on an optical frequency standard, it is only necessary to determine the frequency of the *one* chosen optical standard in terms of the primary microwave standard. One such frequency chain, developed at the Physikalisch-Technische Bundesanstalt, is illustrated in Figure 17.11.

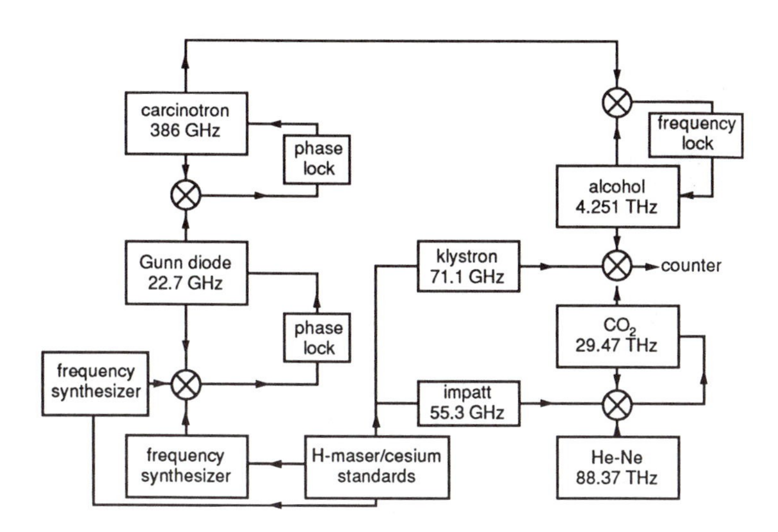

Figure 17.11 A frequency chain of the Physikalisch–Technische Bundesanstalt (Kramer, 1992)

17.9.1 A Frequency Division Chain

There are more recent approaches, based on *frequency division*, worth mentioning. In the first, it is not actually the frequency itself that is divided, but rather at each stage in the chain, there is a division by *two* of the frequency difference between two laser signals. If at some point in the chain the two frequencies whose difference is to be divided by two are ν_1 and ν_2, then the idea is to generate using nonlinear elements the frequency $\frac{1}{2}(\nu_1+\nu_2)$. This and the frequency ν_2 differ by $\frac{1}{2}(\nu_1-\nu_2)$; that is, we now have two frequencies that differ by half the original difference. If the process is repeated, the frequencies will differ less and less until the their beat frequency falls in the microwave range and can be compared directly with the primary standard. The original difference in frequency at the top of the chain can then be calculated; that difference may in fact be the frequency of the optical standard itself, simply by taking the other frequency as its second harmonic.

17.9.2 Frequency Division Using Parametric Oscillators

Another noteworthy approach to linking optical to microwave frequencies exploits the frequency-dividing property of optical *parametric* oscillators. The concept of a parametric excitation of an oscillator has already been encountered:

We recall the mechanical example of the "pumping" of a swing. It occurs whenever a parameter, such as the "spring constant," controlling the resonant frequency of an oscillatory system is modulated at certain frequencies. In the optical case, a modulation of an optical parameter in a medium, such as the refractive index, can occur if the medium is nonlinear, and an intense optical wave, called the *pump,* interacts with it. This can lead to the excitation of optical fields at two other frequencies called the *signal* and *idler* waves, which must satisfy the following conditions on their frequencies and wave vectors:

$$\nu_3 = \nu_1 + \nu_2 \; ; \mathbf{k}_3 = \mathbf{k}_1 + \mathbf{k}_2 , \qquad 17.6$$

where the indices 1, 2, 3 designate the signal, idler, and pump waves, respectively. These are a general form of the conditions we have already encountered as the *phase matching* conditions in the case of second-harmonic generation, which we would get in the so-called *degenerate* case when the signal and idler waves are one and the same. There is an important difference, however. Here the applied pump beam has twice the frequency of the excited signal/idler beam; that is, a wave has been produced at *half* the frequency of the pump beam, and not one at the second harmonic. This is the basis, then, for optical frequency division, provided that high quality nonlinear crystals and stable high-power laser sources are available to act as pumps.

Optical parametric division was demonstrated in 1992 by N.C. Wong and D. Lee at MIT (Wong, 1992) using a biaxial KTP crystal in a doubly resonant cavity configuration in which both the signal and idler waves are resonant at slightly different frequencies. The signal at the beat frequency between the signal and idler waves ($\nu_1 - \nu_2$) is phase-locked to a microwave reference frequency (ν_μ), so that from the known pump frequency ν_3 are produced two precisely known frequencies given by

$$\nu_{1,2} = \frac{\nu_3}{2} \pm \frac{\nu_\mu}{2} . \qquad 17.7$$

Since ν_μ is much smaller than the pump frequency, this constitutes a 2:1 frequency divider stage. It is important to note that the signal and idler wave *phases* are constrained to obey the condition $\phi_3 = \phi_1 + \phi_2$, and that since the phase of the beat signal is phase-locked to the microwave reference source, the output waves are phase coherent with the pump wave. In the actual experiment, illustrated in Figure 17.12, Lee and Wong used the output of a Kr^+ ion laser at λ=531 nm as the pump to generate infrared signal and idler beams at approximately the degenerate value of λ=1.06 μm. The corresponding threshold power for parametric oscillation was around 40 mW. The KTP crystal, 8 mm in length, was placed in an optical cavity made up as follows: At one end the surface of the crystal itself had a radius of curvature of 40 mm and was coated to provide maximum reflectance at the infrared wavelengths; at the other end was the output mirror, coated to have a 0.5% transmittance. On the other hand, the optical coatings of the reflecting surfaces were chosen so that the green pumping beam enters the cavity through the crystal and is reflected back by the output mirror at the other end through the cavity. The frequency separation of the two output infrared beams could be varied by about 1 THz by changing the angle of incidence of the pump beam with respect to the crystal. A piezoelectric crystal drive attached to the output mirror controlled the cavity length, and different mode pairs of signal and idler could be brought into resonance.

The application of this technique in a serial fashion, in which an optical frequency is sequentially divided by two until a microwave frequency is reached, suffers from loss power on each conversion and the need for high efficiency through the far infrared region of the spectrum. A clever alternative *parallel* scheme of conversion stages has been proposed by Lee and Wong, in which the same optical pump source is used to drive a parallel set of parametric oscillators whose signal–idler frequency separation increases in a regular progression. In this way a very wideband frequency comb spanning the spectrum between chosen (rational) fractions of the pump frequency is generated. Each frequency in this comb is further split into finer "secondary" combs by phase modulation driven by a reference microwave source. The scheme promises to provide the means of synthesizing the full range of coherent optical frequencies derived from an optical reference standard and to compare them to a microwave source. That is, it provides the "gear train" for an optical clock.

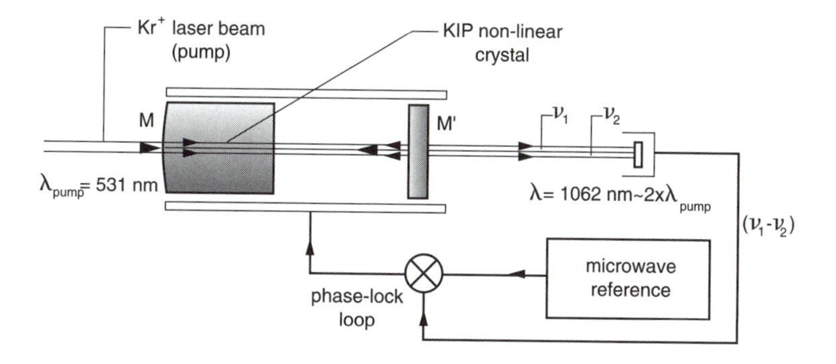

Figure 17.12 Optical frequency division using parametric oscillators (Wong, 1992)

Chapter 18
Applications: Time-Based Navigation

18.1 Introduction

Among the areas of application made practicable by the advent of atomic clocks we list the following:

Space Science: Long-distance tracking and data acquisition from "deep" space probes such as Voyager.

Radio Astronomy: Very long baseline interferometry (VLBI), made possible through a common phase reference.

Planetary Motion: The dynamics of the Earth as a planet and the variability of the length of the day.

Radio Navigation: Perhaps the most useful application. Land-based networks Loran-C and Omega, and the satellite-based systems the TRANSIT system, culminating in the NAVSTAR Global Positioning System (GPS) and GLONASS.

We will limit our discussion to the fundamental way in which precision timing is critical to the success of these applications—so fundamental, in fact, that they are unthinkable without atomic standards. We will devote only a brief discussion to the first three and reserve our attention to a more detailed description of GPS.

18.2 "Deep" Space Probes

In the tracking of interplanetary probes such as Mariner and Voyager, the great distances that must be bridged, reaching out over hundreds of millions of miles into space, and the limited electrical power available aboard such "deep" space probes clearly put a great burden on the system used to track and communicate with them. Overcoming the weakness of the return signals received from the probe at the ground stations is further compounded by the varying Doppler shift in frequency as the probe pursues its trajectory. Large, high-gain (directional) antennas must be used with tens of kilowatts of power transmitted in the uplink and every effort made to enhance the signal-to-noise ratio of the return signal.

This has led to the development of correlation techniques for the handling of digital codes imposed as phase modulation on a stable carrier wave and the use of phase-locked receivers. In such a receiver a servo loop forces the phase of a locally generated stable signal to track the phase of the incoming weak signal. The communication link with the spacecraft is made by way of a *transponder* on board the spacecraft; that is, after the microwave signal "beamed up" from the ground tracking station is received by the spacecraft and demodulated to retrieve the commands and data impressed on it, it is converted (coherently) to a *different* frequency, modulated with the desired data, and beamed back to the ground station. Since the propagation time for the microwave (S-band, $\nu \approx 3$GHz) signal to complete the round-trip journey may be tens of minutes, it means that noise and phase instability of the ground station oscillator over this relatively long time will limit the ability to communicate with the spacecraft. Phase noise on the order of less than tenths of a radian are required.

Of the atomic standards, the incomparable phase stability of the hydrogen maser in the time intervals involved in this application was early recognized; in fact, the earliest application of the maser was as a reference in the Deep Space Network operated by the Jet Propulsion Laboratory for NASA.

18.3 Very Long Baseline Interferometry

The second natural application is in radio astronomy, namely what is called very long baseline interferometry (VLBI). As already mentioned in connection with the Ramsey separated fields in magnetic resonance detection, the use of two separated antennas effectively increases the aperture of a radio telescope (along one dimension) to equal the baseline they define. We recall that the ability to resolve the detailed features of distant radio-emitting objects, that is, the resolving power, is determined by λ/D, where D is the diameter of the effective antenna and λ the wavelength. Since ground-based radio astronomy deals with radiation having wavelengths in the range from about 2 cm to 30 m, even for the short wavelength limit a resolving power comparable to the human eye for visible radiation implies an antenna diameter in excess of the world's largest steerable dish, the 100 m diameter one of the Effelsberg radio telescope in Germany. The largest fixed antenna, with a diameter of 305 m, is the one near Arecibo in Puerto Rico; it is a huge bowl of wire netting set in a natural valley with the focal point receiving antenna mounted on a girder 130 m above the dish. Although capable of detecting the faintest radio sources, its resolving power is not better than the human eye at visible wavelengths.

To understand how to get around this practical limit on antenna size, we must look at the way a radio "image" is constructed in a radio telescope—or an optical telescope, for that matter. In an ideal model, the radiation from each point in the distant object arrives at the antenna as a plane wave traveling in a slightly different direction and reaches different points on the antenna at different times and therefore with a different phase. From each of these points, then, the wave is reflected coherently to converge toward the focus, where it combines with the wave from other points on the antenna to produce a resultant with a certain amplitude and phase.

In an actual radio telescope using one dish antenna, only the intensity at the focus is detected, and a plot of the intensity distribution of the distant radio source is made by scanning the direction of the antenna. If instead of allowing the radio wave to converge to the focus, we imagine the amplitude and phase at the surface of the antenna to be measured at different points, then the resultant that *would* be obtained at the focus can be calculated theoretically. That is, an "image" can be theoretically synthesized using amplitude and phase information obtained over a finite area; it happens that it is not even necessary to have a very high density of points to gain in resolution. That is, a finite array of widely spaced phase-tracking receivers with smaller antennas can be used. This is not feasible at optical frequencies, but phase-lock techniques at radio and microwave frequencies allow phase information even for the weakest signals to be recoverable. Thanks to the existence of atomic standards to supply a constant, common phase reference for distant receivers, it is possible to have an effective antenna "the size of the Earth." Given such a phase reference, the relative phase of the radio waves reaching antennas that may be thousands of kilometers apart can be detected. Depending on the frequency range of the radio waves under observation, this implies synchronization of reference oscillators at the different antennas to within the order of nanoseconds. With present-day atomic frequency standards, it is possible to meet this requirement for antennas literally continents apart, with a proportionate increase in resolving power. The transfer of phase information, which is equivalent to time transfer, between such distant locations has been an important challenge for a long time because it is so critical to the establishment of a radio navigational system, about which more will be said later in this chapter.

18.4 The Motion of the Earth

The next application we shall briefly mention is in the detailed study of the Earth's motion. As we noted in an earlier chapter on time scales based on astronomical observations, the detailed motion of the Earth is complex. Superposed on the basic motions of spin about its axis and revolution around the sun we have the precession of the spin axis relative to the figure axis (movement of the poles), the precession of the axis of spin in space (precession of the equinoxes), and the slowing down of the spin rate, among other things. The precession of the spin axis relative to the axis of symmetry of the Earth is possible because of the slight oblateness of the Earth's shape, and it leads to a very small circular movement of the poles a few meters in radius with a period of 430 days (the *Chandler period*). The precession of the spin axis in space, we recall, is due to external torques exerted on the nonspherical Earth by the sun and moon, and it leads to the much slower precession of the equinoxes, requiring 26,000 years to complete one cycle. This manifests itself as the slipping of the seasons with respect to the months of the year. Since aside from the Chandler period, these phenomena occur on a relatively long time scale, their precise measurement requires clocks that establish a precisely *uniform* time scale extending over these long intervals. By uniform we mean a scale in which unit intervals are identical no matter at what point along the scale they happen to be. Faith in the atomic time scale being more uniform than the astronomical one stems from the belief that to the present accuracy there are no subtle long-term systematic effects on a quantum system to cause a departure from uniformity. On the other hand, the dynamical behavior of the Earth is expected to show those kinds of complex behavior on the basis of well-established theory. The keeping of precise atomic time over long periods enables data to be obtained that are useful in checking computational models based on that theory.

18.5 Radio Navigation

We take up now the main subject of this chapter: radio navigation and the Global Positioning System. Almost from the beginning of radio it was recognized that communication was not its only application. First came radio direction finding and radio beacons, then came the rapid development of radar during the second world war, followed by Loran (from the first letters of *lo*ng *ra*nge *n*avigation) and Omega, which are radio-navigational networks of fixed radio stations of known location. Finally, after Sputnik came space-borne stations using satellites, which provide global coverage: TRANSIT and NAVSTAR/GPS and finally GLONASS.

18.5.1 Radio Direction Finder

To provide some historical background, though atomic standards are not involved, we mention first the radio direction finder as the first application of radio to navigation. It enables a ship to obtain bearings on any broadcast station within range and frequency tuning band of the receiver. This may be achieved in one of two ways: If the broadcast is "all round," being equal in all directions, then the bearing of its signal must be determined by a receiver having sharp directionality in its reception; alternatively, as a result of prior information, it may be known that at a certain time the transmitter is radiating energy on a particular bearing, in which case an ordinary receiver is able to find the bearing of the transmitter. The latter method is the basis of the rotating navigational beacons used for the assistance of ships and aircraft.

A radio direction finder is based on the directional property of a loop antenna, shown schematically in Figure 18.1. Consider first the case when the plane of the loop is parallel to the direction of the radio wave, and assume it is a rectangular loop with the electric field component of the wave along the vertical sides of the loop. We see that since these two sides are at different points along the wave, the electric field does not have the same phase at the two sides, and there will be a net electromotive force around the loop. On the other hand, if the plane of the loop is perpendicular to the direction of the wave, the electric field component will be equal on the two sides, and the net electromotive force around the loop will be zero. In other words, if the plane of the loop is rotated about a vertical axis, the radio signal it picks up will be maximum when it lies in the direction of the wave and will pass through zero when it is perpendicular to the wave. Assuming that the radio wave propagates around the Earth along a great circle, the angular setting of the loop antenna to produce a maximum or minimum can be used to deduce the bearing of the transmitter. Radio beacons are provided in many parts of the world; their precise locations, frequencies, and identifying signals are published in navigational charts and publications. Radio bearings are subject to a number of different sources of error. Among the more serious for a shipboard direction finder is the perturbing and refracting effects of the metal in the ship's structure, necessitating a careful calibration under known conditions. Radio bearings may also be less accurate during sunrise and sunset, and to a smaller extent throughout the night, or when the radio wave crosses a shoreline obliquely.

Since the radio signal from a simple loop antenna clearly cannot distinguish between one position and another turned through 180°, it means that in the absence of additional equipment there is an ambiguity between the true bearing and its reciprocal. In the absence of prior navigational data, the reciprocal bearing could be mistaken for the true one, with disastrous consequences, as occurred in 1923 when seven U.S. destroyers ran aground.

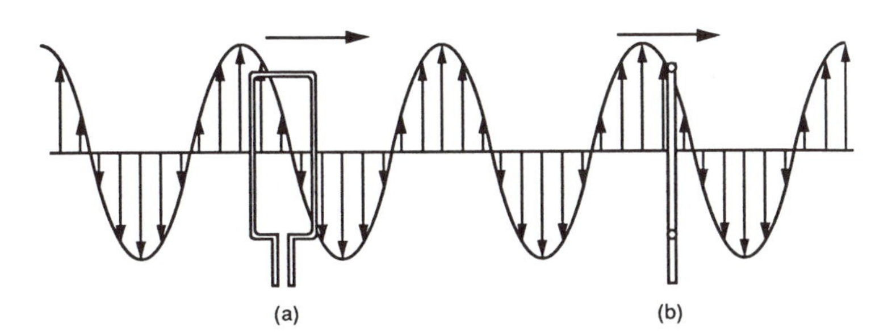

Figure 18.1 A loop antenna oriented (a) parallel and (b) perpendicular to the direction of an incoming plane radio wave

18.5.2 Radar

The detection of radio echoes from distant targets, which is the basis of what came to be called radar (the first letters of radio detection and ranging), was first achieved around 1937, two years before the beginning of the second world war. However, its development, a tightly held secret at the time, saw an enormous impetus during the war, since it proved to be of critical importance in providing early warning of the approach of enemy aircraft. Since then, of course, there have been developed many types of radars with varied characteristics to fit the many military and civilian applications. Of particular note is the Doppler radar, in which not only is the range determined, but also the relative velocity of the target.

In a typical radar system, a short burst of electromagnetic radiation is transmitted using a dish antenna, and the radiation back-scattered by illuminated objects is focused by the same antenna, detected, and suitably displayed. The range is determined from the time *delay* between the instants of transmission of the pulse and reception of its echo. Since the velocity of the radiation in free space is 3×10^8 m/s, to each microsecond increment in the delay corresponds an increment of 150 meters in the range of the target. The accuracy of the range determination then clearly depends on the accuracy with which the relative time/phase delay between the transmitted and received signals can be measured. To focus the return wave and thereby achieve adequate angular resolution in locating the target requires that the diffraction of the reflected wave at the antenna be kept small. This dictates the use of relatively short wavelength radiation in order that a dish antenna of reasonable size can be used. For that reason microwave radiation is generally used: an early common choice was 3 cm microwaves at a frequency around 10 GHz (X-band). Since the same antenna is commonly used for both transmission and reception, a fast-acting electronic switch must protect the receiver during transmission. As with radio receivers, the radar receiver has a local oscillator to heterodyne with the incoming signal to produce an intermediate frequency signal, which is then amplified and possibly taken through further lower-frequency amplifying stages before the signal is finally displayed on the screen of an oscilloscope. In the PPI (plan position indicator) display, the dot on the oscilloscope screen moves out from the center on a radial line, and the returning echo causes the intensity of the dot to brighten. The radial line rotates in synchronism with the antenna and therefore gives directly the relative bearing of the target.

Radars in which the phase of the transmitted wave is available to be used as reference, that is, coherent radars, are widely used to enable Doppler information to be extracted. An important application of Doppler radar is as a *moving target indicator* (MTI), which discriminates between a moving target, which reflects a Doppler shifted frequency, and the ground, sea, or cloud clutter that appear at the same range. The phase stability of the oscillator in such radars will clearly determine the velocity resolution; however, it is only very short term stability that counts, since even for a range of 1000 km the propagation delay is less than 10 milliseconds. This means that apart from a high-power maser or some exotic superconducting cavity oscillator with extraordinarily low phase noise, a high-quality quartz oscillator is the appropriate choice for this application.

18.5.3 Loran-C

Unlike the operating principle of radar, in which the user must actively transmit radiation in order to receive echoes, Loran involves only the reception of coded time signals broadcast from a network of fixed stations of known location. A radio-frequency carrier in the 1,750 kHz–1,950 kHz range was used in the original Loran-A chain, now superseded by the more accurate multichain Loran-C system operating at the even lower frequency of 100 kHz. This lower frequency gives the Loran-C system a much greater useful range, since the lower the frequency, the smaller the attenuation rate of the propagation mode in which the wave travels along the surface of the Earth, that is, the *ground wave*. Other modes of propagation involve waves reflected from electrically conducting layers of the atmosphere called the ionosphere; these are the *sky waves* responsible for global short-wave communication, shown in Figure 18.2. The system depends on the radio propagation time being an accurate measure of the distance between transmitter and receiver. Although the sky wave range is much greater than the ground wave, there is much greater uncertainty in the propagation time, since the altitude of the ionospheric layers is subject to change between daylight and nighttime hours, and depends on sunspot activity, etc. They arrive at the receiver later than the desired ground wave and must be carefully discriminated against in the timing signal circuitry. The propagation time of ground waves is to a first approximation proportional to distance; however, to achieve accuracies in the microsecond range requires making secondary corrections, depending mostly on the electrical conductivity of the surface and to a lesser extent the propagation properties of the atmosphere. These corrections can be computed accurately for propagation over seawater; however, transmission times over paths involving land with different types of terrain are much less predictable.

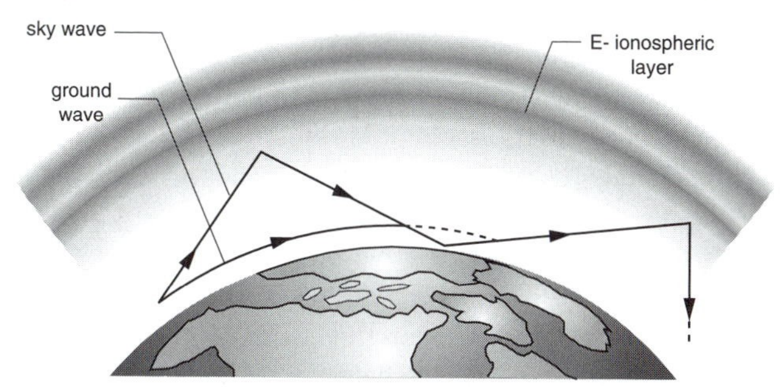

Figure 18.2 The ground wave and sky wave modes of radio propagation around Earth

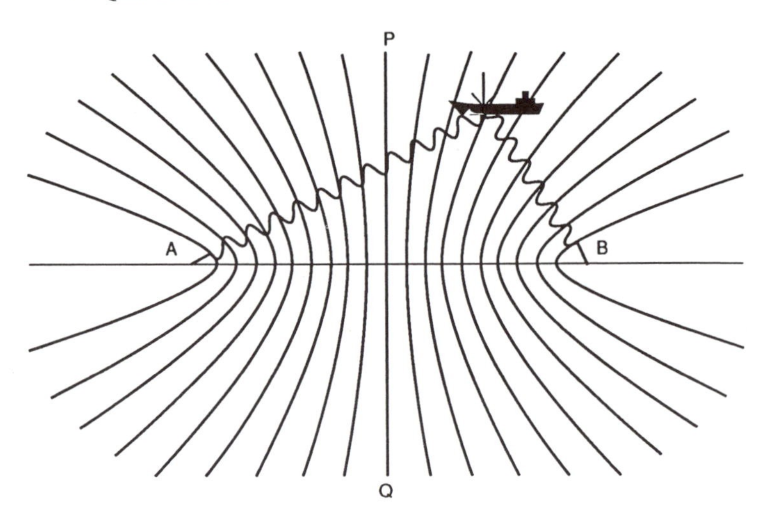

Figure 18.3 The hyperbolic lines of constant propagation delay between Loran stations at A and B

If the user ship or aircraft maintained precise time in synchronism with a transmitting station, then the one-way signal propagation time, assuming a known radio wave velocity, could be used to determine the range to that station. This, however, would require the user to carry a good atomic clock to maintain synchronism with sufficient accuracy to be useful, severely limiting the number of users who could avail themselves of the system. To overcome this limitation, the Loran system operates on the principle of the user determining the *differences* in the propagation delays of three or more precisely synchronized transmissions from widely separated stations, forming a network covering an extended geographical area. If we neglect at first the curvature of the Earth's surface, then surface navigation (no altitude information) requires a minimum of three stations. This can be seen graphically by plotting all the points that have a given constant difference of delay, and hence difference of range from two stations of known location. Figure 18.3 shows such a plot for two stations *A*, *B*; the locus of points having a constant *difference* in their distance from two fixed points is, in fact, a *hyperbola* with *A* and *B* as foci. For this reason this radio navigational system is sometimes called a hyperbolic system. The line *AB* is the *baseline*, and its perpendicular bisector, the *centerline PQ*, is the locus of points equidistant from *A* and *B* and therefore corresponds to zero difference. The *extensions* of the baseline to infinity away from the points *A* and *B* also correspond to a constant difference, namely the propagation time directly from *A* to *B*. To fix the position of the receiver, another set of hyperbolas is necessary, giving the lines of constant difference in time delay with respect to another pair of stations. From two observed delays between different pairs of stations, two hyperbolas are selected, one from each set associated with a pair of stations.

These may theoretically intersect at two points; however, in that event independent information or the delay difference from another pair may be required to resolve the ambiguity. Since the hyperbolas in the vicinity of the baseline extensions have arms that tend to close up into a hairpin shape, the likelihood of ambiguity is greatest there, and navigators avoid using station pairs whose baseline extensions are near enough that the chance of ambiguity is a source of concern. In fact, there is another important reason to avoid the baseline extension region: It has to do with the size of the error in range incurred by a given error in time delay. From Figure 18.3, where lines of constant time difference are plotted for *equal* increments in that time difference, we see that in the neighborhood of the baseline extensions the lines are spread out much more than, for example, along the baseline itself. This means that for a given error in the time delay measurement, the error in position is larger in this area than elsewhere. While on the subject of errors, we note that an important factor is the angle between the two hyperbolas at the point of intersection. If we admit to a certain error in the timing from each transmitter, then the hyperbolic lines should be replaced by hyperbolic bands, whose widths reflect the possible statistical spread in the time delay. The area of overlap of the two bands now defines the error in the position, an area that is clearly bigger if the bands are nearly parallel than if they are nearly perpendicular.

The Loran-C stations are grouped in *chains*, each covering a certain geographical region, with each chain having one station designated as *master* (M) and two or more other *secondary* stations (W, X, Y,...), any one of which may be paired with the master to form a master–slave pair, or *rate*.

Each station broadcasts the same 100 kHz carrier frequency, pulse-modulated in groups of pulses that are repeated at a rate unique to the chain, which helps to identify it. Pulse modulation is used with precisely defined pulse envelopes because of its advantages from the point of view of power utilization in communicating time with the best signal-to-noise ratio. The phase of the carrier is switched between successive pulses between 0° and 180° according to a binary code, the two phases corresponding to 1 and 0 in the binary system. These phase codes are chosen to distinguish between the master stations and the secondaries, and they alternate in time between codes designated as A and B, as shown in Figure 18.4. The phase modulation has the further benefit of helping to discriminate against the undesirable sky wave signal. The signal format for the secondary stations consists of groups of eight equally spaced phase-coded pulses 1 millisecond apart, while the master broadcasts groups containing an additional pulse 2 milliseconds after the first eight. The time interval between successive transmissions of the master's pulse groups is called the *group repetition interval* (GRI). Each chain is identified by a GRI *designator*, which is the group repetition intervals in units of 10 microseconds; for example, the northeast U.S. chain has the GRI designator 9960, and therefore the interval between groups of pulses is 99.6 milliseconds. This interval is chosen so that the signals have time to propagate throughout the chain well before the beginning of the next group of pulses. Possible confusion is avoided by having a well-ordered, precisely timed sequence of transmissions from members of the chain. The transmissions occur in a sequence started by the master followed usually in the order of the letter designations of the secondary stations: *W* followed by *X*, then *Y*, then *Z*, etc. First the master begins the cycle; the first secondary does not transmit until it has received the master's signal and added a further delay called the *coding delay*; then the next secondary transmits after similar delays; and so on. Since the period between pulse groups is nearly 0.1 second, and the propagation times are typically less than 10 milliseconds, there is no ambiguity as to the identity of pulses from master and slave—the slave signal is known to arrive later, except along the extension of the base-line away from the slave station, where the signals would coincide were it not for the coding delay. This delay not only facilitates reading the time difference in this case, but it also may be changed at will, and it can be used for security in times of war. The timing information is recovered from the phase-modulated pulse signals using phase-sensitive detection and fixing a fiduciary point in a pulse to mark the time, independently of the pulse amplitude. The waveform of each pulse in a group is shown in Figure 18.4. The timing of the pulse is defined by the third zero crossing of the signal, as indicated in the figure, to take advantage of the sharp rise in amplitude early in the pulse, where little sky wave "contamination" is expected. The instrumentation used in the system has a resolution of a few hundredths of 1 microsecond. By radio communication, the oscillator at the master station provides corrections to the other station oscillators in the network as slaves, keeping them within microsecond synchronism. To maintain this degree of synchronism independently, without corrections over a period (say) of one week, implies a long-term stability on the order of 1 part in 10^{12}, which is assured by using a cesium standard. In order to provide a precise time distribution service in addition to navigation, and to tie networks covering diverse regions of the globe, the master oscillator is tracked relative to the time standards at the U.S. Naval Observatory, and adjustments are made to maintain accuracy within tolerance.

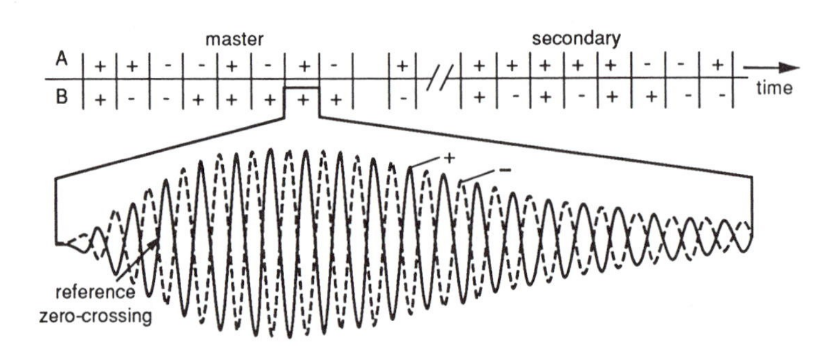

Figure 18.4 Waveform of a pulse in the groups broadcast by Loran-C stations (Maxim, 1992)

18.5.4 The Omega Network

The main drawback to Loran is the limited range of the ground wave, on the order of 1,500 km, necessitating a proliferation of chains and complex receivers to reach global coverage. An alternative radio navigational network called Omega achieves long-range coverage with a network of only eight stations by transmitting at much lower frequencies, in the *very low frequency* (VLF) band allocated for navigation, between 10 kHz and 14 kHz. The stations are scattered throughout the world, from Norway to Hawaii, and they transmit according to a precisely timed schedule; for example, station *A* (Norway) begins its transmission format on 10.2 kHz for 0.9 s, then is silent for 0.2 s, comes back on 13.6 kHz for 1 s, is silent again for 0.2 s, and finally transmits on 11.33 kHz for 1.1 s and falls silent for the balance of 10 s before repeating the sequence. Like Loran, Omega is a hyperbolic system; however, unlike Loran, where the leading edges of pulses received from pairs of stations are matched to obtain the *time* difference, in the Omega system it is the *phase* difference that is observed. Since the phase repeats periodically along the wave, there will be an ambiguity as to the whole number of cycles difference that may be present. To resolve this ambiguity, the Omega receiver must be "initialized" by setting the initial coordinates, after which the receiver will automatically track the phase relationship between the received signals.

As already pointed out, by choosing very low frequency transmissions, long-range propagation is assured—even some penetration into seawater is possible. However, the waveguide-like modes of propagation between the Earth's surface and the lower D-region of the ionosphere have a phase velocity that is sensitive to the behavior of the ionosphere, giving rise to a number of phenomena that must be taken into account when assessing the reliability of a fix. Thus there is a change in the effective height of the ionosphere from around 70 km during daylight hours and 90 km during the night. Furthermore, disturbances occur in the ionosphere during and after "sunspot activity," which has been observed to occur with some regularity, repeating on the average every 11.4 years. Other potential sources of error include interference between different waveguide modes of propagation, and *wrong way* propagation, where the actual signal has traveled the long way around the Earth, rather than directly from the transmitter.

18.6 Navigation by Satellite

The use of satellites for navigation, surveying, and time dissemination has the major advantage of line-of-sight radio communication to cover very large geographical regions. It avoids the problem in surface communication from ground stations of uncertainties in the propagation velocity of radio waves over varying surface conditions around the Earth's curvature, as well as sky wave contamination. It is interesting that it was not many years after the launching of the first satellite, Sputnik I, in 1957, that I. Smith filed a patent describing a satellite system from which time codes could be emitted that would be received on Earth delayed by the propagation times, setting up hyperbolic *lines of position*, a straightforward extrapolation from the Loran concept.

In the U.S., support for a space-based navigational system came through military funding first by the Navy, at the Johns Hopkins Applied Physics Laboratory, and then by the Air Force, at the Aerospace Corporation. This typifies the channeling of research funds in the U.S. through the military services. There are numerous examples of technological advances made possible by government funding, usually through the military, which otherwise would not have been realized. A prime example is the H-maser. Private industry would not have developed it because the market for it was too small to defray developmental costs and turn a profit in a reasonable time frame. As it was, the initial development was undertaken at NASA and under NASA contracts, first to Varian Associates, and then to Hewlett Packard Co. Fortunately, in instances such as atomic time standards and satellite technology the interests of the military happen also to accrue benefits to the public at large. This is proving to be particularly true of the Global Positioning System.

The Navy-sponsored effort led to the satellite navigational system TRANSIT, designed as an all-weather *surface* navigational system for Navy vessels, including strategic submarines. Its limited application to surface navigation and intermittent coverage made it unsuitable for high-speed aircraft and missiles, where continuous three-dimensional navigation is necessary. For this reason the Air Force opted in 1963 for a global navigational system first studied at the Aerospace Corp., a system that evolved into what came to be called the Global Positioning System (GPS). By 1965 the Air Force had decided to let out contracts for the development of user receivers. Other systems, such as the Army SECOR, were also being evaluated at that time, but by 1974 it was determined that a joint military project would be undertaken based on the Air Force GPS concept.

Such a system would have the advantage of *continuous* and total geographical coverage with line-of-sight communication. Its realization, however, clearly depends on the ability to sustain satisfactory operation of an accurate time standard on board the spacecraft, and to broadcast time signals and orbital data with sufficient power and accuracy from an orbiting spacecraft. The use of satellites also relies on the ability to compute accurate *ephemerides* (plural of ephemeris: orbital position as a function of time); this is essentially different from fixed land-based stations—a satellite occupies many different known positions (albeit at different times), almost like having several transmitters in a network stretched out along the orbit. However, to measure propagation delays from the same satellite at widely separated points along its orbit presumes that the receiver is able to maintain precise time over the relevant part of the orbit. In the absence of that, to obtain a fix in three dimensions, that is altitude in addition to latitude and longitude, requires signals to be received simultaneously from four satellites with synchronized clocks.

In the TRANSIT navigational satellite system developed by the U.S. Navy, first declared operational in 1964, there were four, and later six, satellites orbiting at an altitude of around 1100 km in nearly circular *polar* orbits. They completed their orbits in a little over 100 minutes, and as the Earth rotated under them, they provided good global coverage. The system was originally developed to determine coordinates of Navy vessels and aircraft, but eventually civilian use was authorized, and the system was used for surveying as well as navigation. Early experiments by the U.S. Defense Mapping Agency and the U.S. Coast and Geodetic Survey showed accuracies on the order of 1 m at a fixed point after several day's observation, using postprocessed precise ephemerides.

The desire to have the system accessible to users equipped only with a quartz oscillator with good short-term stability, rather than an atomic standard, is met, in effect, by measuring a *difference* in signal delay, but this time not necessarily between different transmitters as in Loran, but between successive incremental positions of the *same* satellite. But the continuously changing signal delay due to the motion of the source is nothing more than the Doppler effect. One of the important observables is the point at which the Doppler shift in the signal passes through zero and reverses sign: this occurs when a satellite passes through the point closest to the receiver. Noting the precise times when this occurs and having an accurately computed ephemeris, so that the positions of a satellite are accurately known for those times, ultimately fixes the position of the receiver.

Needless to say, the success of such a system depends on the stability of the clocks on board the spacecraft and the ability to communicate accurate orbital information. The system's reliance on Doppler suggests that it is the short-term phase/frequency noise that will limit precision, while long-term drifts should be small to avoid the need for frequent corrections to stay within the tolerance limits. The Transit satellites were equipped with ultrastable quartz oscillators with a drift rate of a few parts in 10^{11} per day—atomic standards for on-board spacecraft applications were still under development. The long-term drift of these quartz oscillators could be modeled mathematically, allowing time corrections to be extrapolated. They controlled the frequency of transmission at 150 MHz and 400 MHz for Doppler tracking and navigation. The satellites were tracked by widely separated fixed ground stations of known location (TRANET) using the same basic Doppler technique used in tracking Sputnik. While the Doppler frequency shift itself gives information on the relative velocity (range rate), the accumulated *phase* shift it causes (computed mathematically by integrating the Doppler frequency with respect to time) gives the *change* in satellite–receiver distance. To derive the actual distances as the satellite continues in its orbit requires independent knowledge of its distance at least at one point (mathematically, to determine the integration constant).

The TRANSIT system had two important shortcomings: First, although the six orbiting satellites were able to provide global coverage, it was not continuous. A satellite passed overhead (at the equator) every 100 minutes or so, and users had to interpolate their position using dead reckoning between passes; under worst-case conditions a user might require several hours between fixes. Second, the navigational accuracy was only slightly better than Loran-C, relying as it did on on-board quartz clocks rather than atomic clocks.

18.7 The Global Positioning System (GPS)

In view of these deficiencies, and with the intervening developments in portable atomic clock and satellite technology, the U.S. Department of Defense in 1973 directed its Joint Program Office to oversee the development, evaluation, and deployment of an accurate space-based *global positioning system* (GPS). The product of that effort is the present Navigation System with Timing and Ranging (NAVSTAR). Specifically, the functions of the system are to enable military, and now civilian, users to determine accurately, under all-weather conditions, their position and velocity, and to transfer precise time on a continuous basis anywhere on or near Earth. This was to be achieved by having coded time and ephemeris signals broadcast from a number of satellites, each carrying an atomic clock, in such orbits as to ensure that a sufficient number of them are in view at all times, anywhere on Earth. The ranging method is again based on the propagation time of radio waves from the satellites to the user: either using the propagation delay in receiving the coded time signal, or the accumulated phase difference between the broadcast carrier wave and the user's reference oscillator/clock. As with the other time-based navigation systems, this system is designed to require receivers equipped only with a relatively inexpensive quartz clock. Since in general, the receiver clock will not remain in exact synchronism with the satellite clocks, the range computed using the uncorrected propagation time observed is called the *pseudo-range*; the true range is obtained by correcting for the clock error. In order to fix the position of the receiver in *three* dimensions—longitude, latitude, and altitude—three *true* ranges are necessary. This can readily be seen if we imagine spherical surfaces drawn around the satellites as centers with radii equal to the true ranges to the receiver. If only two ranges are known, then the receiver may be at any point on the circle of intersection of the two corresponding spherical shells, whereas if the range to a third satellite is known, the position of the receiver is uniquely determined as being at the intersection of the third spherical shell with that circle. If the ranges to the satellites are only pseudo-ranges because of the clock error, then of course, the position so determined will be in error, and the spherical shell drawn around a *fourth* satellite with the pseudo-range as radius will not pass through the same position as the other three. However, if we assume that all the satellite clocks are in perfect synchronism, so that a single clock error exists because of a drift in the receiver clock, then it will be possible to compute the correction to the receiver clock, which will convert the pseudo-ranges to true ones, and make the *four* spherical surfaces with corrected radii pass through a unique point, the true position of the receiver. This system then requires that the signals from at least *four* satellites be in full view globally at all times.

A description of the system separates naturally into three *segments*: the satellite system, the ground monitor and control stations, and the users, including the many types of receivers. We will give a brief description of these in that order.

18.7.1 The Satellite Constellation

From what has been said, the number of satellites and their orbits must be chosen so that signals from at least four of them can be received simultaneously anywhere on Earth, 24 hours a day. This assumes that because of the unpredictable motion of the receiver four signals must be received at each *epoch*; however, in the event that the receiver is stationary or moving slowly, then signals received sequentially from satellites at several different epochs would suffice to fix the position of the receiver, since the satellites will occupy different known positions at these epochs. Nevertheless, even in this case, to be assured of an immediate and accurate fix, four satellites must simultaneously be in full view at all times.

The ultimate choice of the number of satellites in the *constellation* and their orbits evolved from a number of proposals early in the 1980s, ranging from a 24-satellite constellation in 3 orbital planes inclined 63° to the equator to 18 active satellites with three in each of six orbital planes. The present policy calls for a constellation consisting of four active satellites orbiting in each of six planes with an inclination of 55°, making a total of 24 satellites, plus four more spare satellites to replace mal-functioning operating satellites.

There are five categories of GPS satellites, designated as Block I, II, IIA, IIR, and IIF satellites. The typical appearance of a GPS satellite is illustrated in Figure 18.5. The eleven satellites making up Block I were launched in the period between 1978 and 1985. With the exception of one booster failure on the seventh in the series, all the launches were successful. The design life of these satellites was only 4.5 years, yet two of them were still operating satisfactorily after twice that period. In common with all the subsequent satellites, these carried atomic clocks in addition to sophisticated radio communication equipment, as well as a propulsion system for orbital corrections. They were powered by two 7 sq. meter solar panels, weighed 845 kg, and were placed in orbits inclined at 63°. The Block I satellite signals were not secured to prevent general accessibility to civilian users; this is in contrast to the Block II satellites, some of which broadcast inaccessible coded signals. By 1994 the remaining Block I satellites, still functioning at reduced power, had been made redundant and have been boosted out of orbit and left for scientific tests.

Figure 18.5 A typical GPS satellite

The first of the Block II satellites, weighing over 1,500 kg, was launched in 1989 using a Delta II rocket; subsequently, over the period 1989–90 eight more satellites in this series were launched and placed in four different orbital planes inclined at 55° to the equator, with an altitude of about 20,000 km. From 1990 to 1994 fifteen more satellites were launched, which are classed as Block IIA, the "A" denoting advanced, which have the capability of communicating with each other, and some of which carry *optical* corner cube reflectors, which reverse the direction of a tracking laser beam from a ground station, independently of the orientation of the satellite. This facilitates the tracking of a satellite using laser ranging, the optical analogue to radar, with an accuracy approaching the centimeter range. This capability is matched by the accuracy of frequency and time made possible by *four* on-board atomic standards: two rubidium and two cesium, with long-term frequency stability of a few parts in 10^{13} and 10^{14} per day, respectively. This corresponds to an average drift of about 3 nanoseconds

per day for the cesium standard, which for a radio wave traveling at 3×10^8 m/s implies an error of about 1 meter in ranging.

The Block IIR (the "R" for replacement) satellites were scheduled for launch using the Space Shuttle in 1996, however, due to a number of technical issues, including the production of the sophisticated on-board atomic clocks, the delivery and launch have been delayed about ten months. The first Block IIR satellite was delivered to Cape Canaveral in September, 1996, and after extensive tests to verify a smooth integration with the existing GPS, was scheduled to be launched in January 1997. Unfortunately, that launch was unsuccessful due to failure in the Delta II launch vehicle. The launch was rescheduled for a later date. This new generation of satellites has a design operational lifetime of ten years, two and a half years longer than the Block II/IIA satellites. The satellites are considerably heavier, at over 2,000 kg, and are able to accommodate hydrogen masers, especially adapted for spacecraft operation, as onboard frequency/time standards. As a class, the masers have long ago demonstrated superior stability, and the thought of flying them in satellites has long been cherished from the beginning. However, the massive ion vacuum pumps and large magnetic shields that characterize laboratory installations presented a formidable obstacle in meeting the size and weight constraints of a spacecraft. The frequency stability expected of the hydrogen space masers is less than one part in 10^{14}, a tenfold improvement over the Block IIA standards, with, it is hoped, a corresponding upgrade in system performance.

18.7.2 The Orbital Parameters

A contract worth approximately $1.3 billion was awarded Rockwell by the U.S. Air Force in April, 1996, for 33 fourth-generation GPS satellites (Block IIF), the first to be delivered ready for launch in 2001. These advanced satellites will be able to accommodate expanded missions, provide a significantly higher accuracy signal accessible to civilian users, and have an operational life of over 12 years.

A major premise in the use of satellites as platforms for radio transmitters forming a navigational network is that their precise positions are predictable at all times, and that this information can be communicated to the user. That this is the case hinges on the fact that the satellites follow orbits that to a very good approximation are Keplerian ellipses with Earth's center at one of the foci. To completely specify the motion of a particle in 3-dimensional space, acted on by known forces, requires in general six numbers, which may, for example, be the three coordinates and the three components of velocity at some point in time (epoch). It follows that the most general elliptical orbit in space requires five parameters to specify it completely, and one parameter to specify the position of the particle in the orbit. The number of orbital parameters results from the two angles required for the orientation of the plane of the orbit, another angle to specify the orientation of the ellipse in that plane, and two more to specify the

semi-major axis and ellipticity of the ellipse. These parameters are illustrated in Figure 18.6, where the position of the particle in the orbit, historically known as the *anomaly* (sic), is shown as the angle θ at the focus of the ellipse where the center of mass of the system is located. For GPS satellites the semi-major axis is nominally 26,560 km, and the orbital period is *half* a sidereal day, that is, half the time for a complete rotation of the Earth with respect to the stars (actually with respect to the vernal equinox, which is very nearly the same thing). Having the satellite complete its orbit a whole number of times each sidereal day ensures that its ground track repeats every sidereal day; that is, ideally it will pass overhead at a given point on Earth at the same time every sidereal day.

The simple reentrant (closed) elliptical Keplerian orbit is predicted theoretically according to Newton for a satellite attracted only to a rigid, homogeneous, spherical Earth, which acts on the satellite as would a point mass at the center of the Earth. An actual Earth satellite is subject to conditions that differ slightly from that, differences that are called *perturbations*. Fortunately, these are all small compared to the forces that give rise to the Keplerian orbits, a fact that is exploited theoretically in arriving at corrections by a process of successive approximation. The result is that the orbital parameters are subject to change in time and must therefore be corrected either by actually activating thrusters, or updating actual tracking data and computing the changes in the parameters they imply.

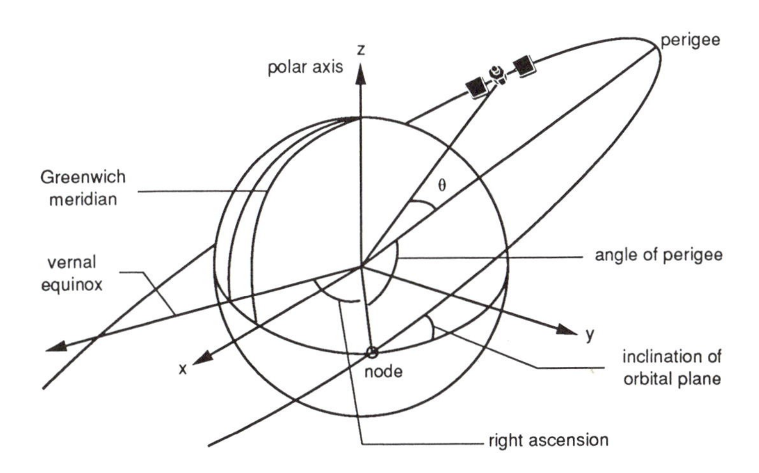

Figure 18.6 The definition of the satellite orbital parameters (Hofmann–Wellenhof, 1994)

18.7.3 Perturbations Affecting the Orbit

There are two types of perturbations: those that have a gravitational origin, such as those arising from the presence of the moon and sun, and the oblateness of the Earth; and those that are nongravitational, such as solar radiation pressure, solar wind, and air drag. Since there is potentially a bewildering number of different factors that may perturb the motion of the satellite, we need to stipulate what would be a tolerable error in the satellite position. If we set that tolerance at 1 meter deviation over one orbital period, we would find that a *constant* perturbing force must not cause an acceleration greater than 10^{-9} ms^{-2}. To appreciate the relative size of such a force, we note that the primary gravitational pull on a GPS satellite due to the Earth, which keeps it in orbit, is GM_E/r^2, where G is Newton's gravitational constant, M_E is the mass of the Earth, and r=26,560 km is the semi-major axis of the orbit. A quick calculation yields $\approx$0.57 ms^{-2}, almost a billion times greater than the tolerable perturbation!

The largest perturbation comes from the nonsphericity of the Earth. Because of the Earth's rotation about its axis, there is a centrifugal force that varies from a maximum at the equator, diminishing with latitude, becoming zero at the poles. In consequence of this, the net inward force at the surface, the observed weight, is greatest at the poles and diminishes toward the equator. The equilibrium figure of a plastic body is an oblate spheroid, a slightly flattened sphere with an elliptical cross section through its axis. The actual oblateness of the Earth is small—the diameter from pole to pole is only about 43 km shorter than through the equator, or about one part in 298. Of course, the detailed shape and structure of the Earth is far from a smooth homogeneous oblate spheroid; the degree of detail that is significant clearly depends on how far the satellite is from the surface. Indeed, if the object had been to study the topography of the Earth through satellites, the orbits would have been chosen to bring out the very perturbations that the GPS system must avoid. At an altitude of about 20,000 km, the GPS satellites are sufficiently far from the Earth's surface that the oblate spheroid model, which introduces less than one part in 10^4 correction, is considered adequate, and higher-order approximations are expected to yield negligible improvement in orbital accuracy. An analysis of the effect of the Earth's oblateness on a satellite orbit shows that there is a slow precession (a rotation of the perigee) at a rate proportional to $(5\cos^2\theta-1)$, where θ is the angle the orbital plane makes with the Earth's equatorial plane. This rate is nearly zero if $\theta\approx63°$, which provides a rationale for choosing that angle for the early satellites. Moreover, the analysis shows that the mean time to complete an orbit is also a function of θ, in this case proportional to $(3\cos^2\theta-1)$, which is nearly zero for $\theta\approx55°$, hence the choice of that angle for the later satellites.

The other gravitational type of perturbation comes from the presence of sun and moon, and is referred to as *tidal effects*, since the attractions of the same two bodies account for the tidal action on the Earth. A rough estimate of the variation of the sun's gravitational pull over the orbit of the satellite using Newton's inverse square law of gravitation yields for the sun's perturbation about 2×10^{-6} ms^{-2}, while for the moon the figure is about 5×10^{-6} ms^{-2}. Related, indirect perturbations due to the tidal deformation of the "solid" Earth, as well as the oceanic tides, are very much smaller, in the range of 10^{-9} ms^{-2}.

Of the nongravitational perturbations, the most important is the solar radiation pressure. We recall that radiation, whether a laser beam or radio wave, carries linear momentum, and therefore the absorption and scattering of sunlight by a satellite results in forces being exerted on it. Since the scattering in general is not the same in all directions, it follows that the force experienced by a satellite is not necessarily in the direction of the incoming rays of the sun—there will be a smaller transverse component. The actual perturbation produced obviously depends on the *solar constant*, a measure of the intensity of solar radiation falling on the satellite (S=1.4 kW/m^2), the cross section presented by the satellite to the sun's rays, the reflectivity of the surfaces, etc. There is the further complication that the satellite may pass through the shadow of the Earth; that is, it may experience periods of solar eclipse. The computed size of the perturbation is on the order of SA/cM_s, where A/M_s is the ratio of cross section to mass of the satellite, which yields $\approx10^{-7}$ ms^{-2}. This shows that radiation pressure produces a very significant perturbation, one that must be well modeled and taken into account if the desired accuracy is to be achieved.

Finally, in addition to solar radiation there is the *solar wind:* the sun continuously emits particles, mostly high-speed electrons and protons. Near the Earth's orbit the average speed of the protons is about 400 kms^{-1}, and its number density ranges from 2×10^6 to 10^7 particles m^3. Assuming that the particles are completely stopped on collision with the spacecraft, the resulting acceleration, for example on a spacecraft having A/M_s=0.03, is less than 10^{-10} ms^{-2}, and therefore negligible.

18.7.4 The Control Segment of GPS

The crucial functions of monitoring the orbits of the satellites and the frequency and phase of their on-board atomic clocks, and updating orbital parameters for ephemeris prediction, are the responsibility of the control segment of GPS. This comprises ground-based stations of known geodetic position, including a master control station and three other control stations, as well as a worldwide network consisting of five monitoring stations. The master control station is at the Consolidated Space Operations Center in Colorado Springs, Colorado. It collects the satellite tracking data from the worldwide monitoring stations, from which it computes the updated orbital and atomic clock parameters. This information, along with other operational commands, is sent to the three ground-based control stations to *upload* to the satellites. The monitor/tracking network stations are located at Colorado Springs, Ascension Island (South Atlantic), Diego Garcia (Indian Ocean), Kwajalein (North Pacific), and Hawaii. These stations are equipped with precise cesium clocks and receivers that continuously track all satellites in view. The signal propagation times, which yield the *pseudo-ranges*, are obtained every 1.5 seconds, are "smoothed" to allow for ionospheric and meteorological variables, and transmitted as 15 minute interval data to the master control station. The three ground control stations are positioned also at the three monitor station sites on Ascension, Diego Garcia, and Kwajalein.

The implementation of a global time-based navigational system clearly requires the definition of a suitable frame of reference, including time. Position and time of any element of the system must be referred to a common, invariant set of coordinate axes and clocks. To specify the position of a point over a finite region of the Earth's surface, as with Loran-C (or ordinary surveying, for that matter), requires only certain fiduciary reference points to establish a baseline and the measurement of appropriate angles, which are then in a sense the coordinates of any point in that surface region. Obviously, this will not work for a global system, where positions in *space* surrounding the Earth is included: For that we need ideally an *inertial* frame of reference fixed in space and not tied to the Earth and partaking of its complicated gyroscopic motion, etc. One such system that neglects only the residual variation in the gravitational field over the Earth–satellite system, takes the direction at a specified epoch of the Earth's angular momentum axis (which is constant apart from the slow precession of the equinoxes) as the coordinate z-axis, and the direction of the vernal equinox, which is perpendicular to the Earth's axis and lies in the orbital plane (the *ecliptic*), as the origin of the longitude coordinate. The angle of latitude with respect to the z-axis and the radial distance from the Earth's center complete the system. It is in terms of the coordinates in this *quasi-inertial* (nonaccelerating) geocentric system that the computation of the satellite orbits is carried out. However, for the practical purposes of the navigator, what is required are his coordinates and altitude with respect to an Earth-fixed system using the prime meridian through Greenwich as the origin of longitude. Such a system is the (Conventional) Terrestrial Reference Frame (TRF), in which the z-axis is taken to be the *mean* position of the Earth's rotational axis during the arbitrary period from A.D. 1900 to 1905. For a true prediction of altitude it is not sufficient to assume a spherical Earth—the oblateness must be taken into account—and therefore, since 1987 GPS has used the World Geodetic System WGS-84 (these things are regularly updated and therefore their designation includes the last two digits of the year), in which the Earth's figure is an ellipsoid with semi-major axis 6,378,137 meters, and the geometric flattening is 1 in 298.2572. This system is operationally established by a set of ground-based control stations serving as reference points, with particularly accurate position-fixing facilities including laser ranging and very long baseline interferometry (VLBI). Once the orbits of the satellites are computed using coordinates in the free quasi-inertial system, they must then be *transformed* into those in the practical Earth-fixed system, using the known motions of the one system relative to the other.

The operation of the GPS system presumes that all elements, including the navigator, maintain close time synchronism, and that when clocks drift apart, as they naturally will, it is possible to model their behavior mathematically to predict clock errors. The satellites and ground-based stations are all equipped with precise atomic clocks, which of course keep what is defined as atomic time. Navigators, on the other hand, are closely tied to what is called Universal Time (UT), which is defined in terms of the *mean* solar day, the basis of civil time. This is the average of the *apparent* solar (24 hour) day, which varies throughout the year, taken over that period. The time scale used by GPS is based on what is called Universal Time Coordinated (UTC); the unit in this system is the atomic second. However, because of the possibility of long-term drift of the universal time with respect to atomic time, they are kept in step to within one second by inserting a "leap" second as required. This results in a piecewise uniform time scale that tracks universal time; any fractional difference between the different time scales is monitored and published by national observatories charged with this service.

We should recall at this point the *Sagnac effect,* a relativistic effect on time measurement associated with the rotation of the Earth, which we introduced in Chapter 7. A coordinate system fixed in the Earth is noninertial, since its rotation with respect to "the fixed stars" constitutes an accelerated motion (not of speed, but changing direction). Consequently, as stated earlier, it is Einstein's theory of *general relativity* that is involved. According to the theory, if we imagine we have two identical, precise clocks at some point on the Earth's equator, and one remains *fixed* while the other is taken *slowly* (with respect to the Earth) along the equator all the way around until it reaches its starting point, then the time indicated on the two clocks will not agree. The difference $\Delta\tau$ was quoted as being given by the following:

$$\Delta\tau = \pm\frac{2\Omega}{c^2}S\,, \qquad 18.1$$

where Ω is the angular velocity of the Earth (7.3×10^{-5} rad/sec) and S is the area ($\pi R_E^{\,2}=1.3\times10^{14}$ m^2) enclosed by the path of the moving clock. The formula yields a significant time difference of about $\pm1/5$ microsecond. This is not insignificant in the present context and must be taken into account.

The determination of satellite orbital position as a function of time, from tracking data obtained by monitoring stations of known location, is the reciprocal problem to that of navigation using signals received from satellites having known orbital positions. We have seen that to completely predict a satellite position requires six numbers: These may be the three coordinates of position in 3-dimensional space and the three components of velocity at a given epoch, or the three coordinates at two distinct epochs. The first instance casts the problem as one of using the equations of motion to predict the motion subsequent to given *initial values*, and the second as fitting the general solution to given *boundary values*. The satellite orbital motion is solved using the quasi-inertial space coordinate system, whereas the positions of the tracking stations (as well as coordinates required by the users) are naturally with respect to the Earth-fixed conventional terrestrial system. The observational nexus between the two systems is through the precise tracking of GPS satellites afforded by laser ranging and VLBI from points coincident with some of the GPS monitoring stations. The need for mathematical transformations between the two systems makes the task of determining and updating the satellite ephemerides a complicated one, but nevertheless a manageable one thanks to integrated circuits and on-board computers.

18.7.5 Coding of GPS Satellite Signals

From the beginning, the idea of a precise global system of satellite navigation was inspired and brought into being by the U.S. Department of Defense, and its military importance makes the need for security pretty obvious. On the other hand, there was a desire to accommodate civilian users of the system, which has led to an elaborate coding scheme for controlling full accessibility to the system; unauthorized users were to have only a purposely degraded accuracy of positioning. Two carrier frequencies are actually broadcast in order to provide information on the dispersion of the radio waves as they traverse the ionosphere, that is, the frequency dependence of the propagation velocity of the radio waves through the ionized layers surrounding the Earth. This enables the ionospheric delay to be mathematically modeled in correcting the observed satellite pseudo-range. The two frequencies are the 154th and 120th harmonic of the fundamental frequency at 10.23 MHz based on the on-board atomic frequency standards, with stability on the order of parts in 10^{13}. The two frequencies, which fall in the so-called microwave L-band, are at L_1=1.57542 GHz and L_2=1.22760 GHz, with wavelengths about 19 cm and 24 cm respectively. In addition to the time-ranging codes already mentioned, these carrier frequencies are also modulated with data including the updated satellite orbital parameters, GPS time and satellite clock reading and drift, etc.

The type of code used is the so-called *pseudorandom number (PRN)* code, a particularly appropriate choice both for security and precision time comparison. It is a binary code generated by a shift register in such a way that within a certain group that repeats periodically, the binary bits are more or less randomly distributed. As an example, consider a 5-bit shift register in which at every clock pulse the bits advance to the right one place, the extreme right-hand bit becoming part of the output. The criterion for choosing the bit that replaces the one on the far left is the key to the code: For example, it may be chosen to be 0 if the bits in the third and fourth places are the same, otherwise a 1. A possible PRN code generated this way would be (011010111100010). A different choice of the criterion clearly would lead to a different output sequence. This binary code is impressed on the carrier wave as a biphase modulation; that is, the phase for a binary 1 is shifted by 180° with respect to a binary 0, as shown in Figure 18.7. The extent of "randomness" of the PRN code can now be seen by computing the *autocorrelation* function of the signal biphase modulated according to it.

To do this we simply form the product of the signal with its duplicate *shifted* a whole number of clock intervals, and then sum (integrate) over the whole period of the sequence. Since a phase shift of 180° is equivalent to a reversal of sign, we find that the autocorrelation has a maximum when there is *no* shift, since products of signals with the same sign are summed in the entire code. On the other hand, if we compute the correlation of the signal with its duplicate shifted even one space to the right or left, the result is zero, as can be verified for the PRN code we gave as an example. Moreover, the correlation will be small between the PRN coded signal and *any* other binary sequence that does not match it identically, since some negative products of signals of opposite signs are included in the sum. It follows that not only does the code provide a sharp time-matching function between a coded signal and its duplicate, but it also enables this match to be available only to those who can generate a duplicate code. It therefore fulfills the added function of controlling access to the information carried on the satellite signals. By assigning different individual PRN codes to the many satellites that make up the GPS constellation, the need to be able to identify each satellite is fulfilled without any ambiguity.

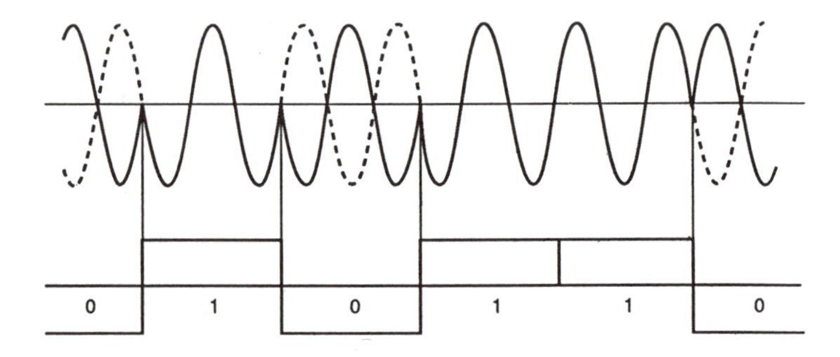

Figure 18.7 Binary phase modulation of satellite pseudorandom number signals

The original concern of the U.S. Department of Defense that the precise positioning capability of GPS be secure against compromise or use by an enemy in times of war, while allowing a degraded capability to the general public, led to a rather complicated coding scheme. The dual precision capability is achieved principally by having two distinct PNR codes for the satellite clock readings impressed on the carrier waves broadcast from the satellites: the so-called C/A (Coarse/Acquisition) code, and the P (Precision) code. A further security provision is A-S (Anti-Spoofing), by which is meant a countermeasure against anyone sending out false signals with the GPS signature, thereby confounding the system, or worse. The anti-spoofing feature consists in using a code (the W-code) to encrypt the P-code, yielding what is designated as the Y-code. The C/A-code is available to the general public, as is the P-code; but of course not the Y-code, although there is strong support to the idea of making the full capability of the system available to everyone.

The C/A-code is generated by combining the outputs of *two* 10-bit fed-back shift registers using binary addition modulo-2, that is, without "carrying." The clock rate for generating the C/A code is one-tenth the 10.23 MHz atomic-based reference, that is, 1.023 MHz, and it is repeated every millisecond. The interval between two bits in the code of just under 1 microsecond corresponds to the propagation delay for an increment in the range of 300 meters. It is impressed only on the L_1 carrier wave in phase quadrature (displaced 90°) to the Y-code, which is on both carrier waves L_1 and L_2.

To generate the P-code is somewhat more complicated: It is a certain combination of *two* PRN sequences, each generated by two registers, repeating about every 1.5 seconds, one containing over 15 million bits, while the other contains an *additional* 37 bits. Because of the difference in the number of bits in the two PRN sequences, the combined sequence will repeat only when a whole number of repetitions of the one sequence has an equal number of bits as another whole number of repetitions of the second sequence. Expressed symbolically, if n_1 and n_2 are the numbers of bits in the two PNR sequences, then the combined sequence will repeat after p repetitions of the first sequence, or q repetitions of the second, provided that p and q are the *smallest* whole numbers for which $pn_1=qn_2$. For example, if $n_1=6$ and $n_2=4$, then $p=2$ and $q=3$, and the combined sequence repeats every $pn_1=qn_2=12$ bits. Of course, in the language of elementary arithmetic, 12 is simply the least common multiple of 6 and 4. Getting back to the P-code, we find that the sequence that results from combining one having about 15 million bits with another of slightly greater number will not repeat until after about 200 trillion bits! At the clock rate of 10.23 MHz, the interval between successive bits is less than one-tenth of a microsecond (corresponding to a range interval of 30 meters), and the code repeats every 266.4 days. The total code length is divided into one-week segments, which are assigned to satellites defining their PRN identification number. To show explicitly the broadcast signal as a function of time, let $S_{C/A}(t)$, $S_Y(t)$, $S_D(t)$ be the sequence of +1/−1 constituting the binary codes C/A, Y, and the navigational and other data. Then we have

$$\begin{aligned} L_1 &= a_1 S_Y(t) S_D(t)\cos(2\pi\nu_1 t) + b_1 S_{C/A}(t) S_D(t)\sin(2\pi\nu_1 t), \\ L_2 &= a_2 S_Y(t) S_D(t)\cos(2\pi\nu_2 t), \end{aligned} \tag{18.2}$$

where ν_1 and ν_2 are the frequencies of the carrier waves L_1 and L_2. The total navigational data message consists of 1500 bits subdivided into 5 subframes, generated at a clock frequency of 50 Hz, so that it takes 30 seconds for the whole message. It contains the satellite orbital position (ephemerides) update, GPS and satellite clock time including various numbers to model the satellite clock correction, and other data of a "housekeeping" nature. The first subframe contains among other things the GPS week number, numerical coefficients to model the satellite clock correction, predictions of range accuracy, and age of data. The second and third subframes contain the satellite ephemerides. The contents of the fourth and fifth subframes change from one message to the next, repeating after 25 "pages"; the total information contained in all the different pages therefore takes 25×30 seconds, or 12.5 minutes, to broadcast. The pages of the fourth and fifth subframes are broadcast by all satellites; in addition to those reserved for military use, these pages contain data relating to the ionosphere, UTC, satellite health status, and low-precision orbital data on all GPS satellites.

18.7.6 Corrections to Signal Propagation Velocity

To reach ground-based or airborne receivers, the satellite transmission must penetrate the ionospheric layers of the atmosphere, as well as the troposphere. Although the velocity of propagation of the signal is little different from that in free space, nevertheless, the great distances involved lead to an accumulated effect on the arrival time of the signal, which because of *dispersion* differs for the two carrier frequencies L_1 and L_2. It can be shown on the basis of a simple model, in which the electric field component of the radio wave drives otherwise free electrons into oscillation, that the frequency dependence of the *phase velocity* is approximately as follows:

$$V_{\text{phase}} = \frac{c}{\sqrt{1 - \dfrac{80.6 N_e}{\nu^2}}}. \qquad 18.3$$

The electron concentration N_e is stratified horizontally, increasing stepwise with altitude from the E-layer with around 10^{11} electrons/m^3 at 100 km to the F_2-layer above 300km with around 10^{12} electrons/m^3. At the GPS signal carrier frequencies, the effect on the velocity is on the order of 3 parts in 10^5; this corresponds to a correction on the order of 10 m. We should note that for a dispersive medium, such as the ionosphere, it is necessary to specify exactly what the velocity refers to: Is it the crests of an infinite wave train, or the leading edge of a pulse, or what? We recall that a complex waveform can be analyzed into its Fourier frequency components, and these will travel with different velocities, so that the waveform will in general change while it advances, leaving no feature whose rate of advance would be useful to define. If the waveform is such that its spectrum is contained in a narrow range centered on one frequency, such as we have in each of the GPS broadcasts, then the modulation pattern on the carrier is preserved and travels with the so-called *group velocity*. In the case of propagation through the ionosphere the group velocity is smaller than the velocity of light in free space, while the phase velocity is greater. Hence pseudo-ranges derived from time-code matching are based on a different velocity from those based on the relative phase between the carrier wave and local reference oscillator. It is interesting to note in passing that there are instances where the computed group velocity is actually *greater* than the velocity of light in free space: The dispersion in such media is described as *anomalous*. In the days before Einstein's theory of relativity this would not have been considered particularly unsettling; as it was, in the early days of that theory, there was general relief when Sommerfeld and Brillouin showed that in fact the *beginning* of a radio transmission always travels with just the velocity of light, and that *signal* velocity is not the same as the group velocity in cases where the dispersion is anomalous. The signal velocity is always less than or equal to the velocity of light.

As already pointed out, the concentration of electrons in the ionosphere N_e and its distribution with respect to altitude vary according to exposure to the sun and sunspot activity. Furthermore, the radio waves must pass through the layers of the ionosphere along a path that obviously varies with the position of the satellite in its orbit. The ionospheric correction to the signal delay must therefore be continuously monitored. This is the justification for using the two broadcast frequencies, L_1 and L_2; the signal delays provide two numbers to solve for the two unknown quantities: the pseudo-range and the effective electron concentration.

Unlike the ionosphere, the neutral troposphere is nondispersive; that is, the velocity of a wave does not depend on its frequency, and there is no difference between the velocity of propagation of the phase of the carrier and the modulation impressed on it. The refractive index is a function of the atmospheric temperature, pressure, and water vapor concentration. Several semiempirical models of the refractive index as a function of altitude have been developed; based on one of these the tropospheric delay is estimated along the slant path from the satellite to the receiver. The correction for this amounts to only a few meters in the pseudo-range.

18.7.7 The User Segment

Finally, we come to the user segment of the system. This serves in addition to the military services a large and expanding body of civilian users: navigators ranging from those of high-speed aircraft, to pleasure boat operators, to hikers in the woods. With the ongoing drive to make the full capabilities of the system accessible to the general public it may not be long before GPS is incorporated into a multitude of technologies affecting the lifestyle of ordinary people.

The types of GPS receivers currently available are many, and they vary widely in sophistication and cost. They may have special features to enhance their performance in specific applications, such as high-speed navigation, or large-scale surveying, or precise synchronization of remote clocks. But basically, they are special radio receivers with precise phase/time tracking and navigational data processing capabilities.

As radio receivers, their first essential component is the antenna. For a GPS receiver, the design of the antenna and it physical environment are particularly important. Ideally, it should convert the oscillating electric (or magnetic) field component of an otherwise freely propagating wave into an oscillation of current in the receiver circuitry. The phase of that current oscillation must track exactly that of the wave, allowing at most a fixed phase offset irrespective of the orientation of the antenna. This is obviously crucial in applications where the receiver is subject to rapid movement. To receive signals from several different satellites, whether simultaneously or sequentially, requires an antenna whose response is not strongly dependent on the direction of the incoming wave; that is, it should be *omnidirectional*, although it is desirable to discriminate against low-elevation signals, which are likely to be contaminated by spurious reflected waves. The directional properties of a simple dipole antenna (a straight conductor) or loop antenna rule them out; most antennas in general use are microstrip antennas. The antenna section of the receiver may include a preamplification stage and a down-conversion of the frequency before transmission to the radio-frequency section. Some units are designed to receive only the primary L_1 frequency, whereas others receive both L_1 and L_2.

In the radio frequency section, the phase of each carrier frequency is tracked using a phase-lock loop, in which the phase of a controlled oscillator is locked to that of the received carrier using the output of a phase-comparator in a feedback loop. The separation of signals from the different satellites is achieved using correlation techniques on the C/A pseudorandom number codes. The locally generated code is automatically shifted in time to produce a maximum correlation with the incoming signal; the time shift gives, aside from clock errors etc., the wave propagation delay. An important indicator of the degree of sophistication and cost of a receiver is the number of satellites it can track simultaneously. Either the signal from each of four satellites is directed along separate parallel circuits, constituting four channels, or the same channel may be used to sequentially process the signal from different satellites. Lower-cost units are of the latter type.

In addition to the radio frequency section, a microprocessor is incorporated in a GPS unit with memory, keyboard and display. This is necessary, of course, to make all the necessary corrections to the observed time delays, to use the broadcast ephemerides of the satellites, and to solve the equations to obtain the coordinates and time of the user. Care must be exercised in the choice of location of the antenna; even the best geometric and electrical design of the antenna will be to no avail if the physical surroundings can reflect portions of the wavefront, causing them to arrive at the antenna along different paths, as shown in Figure 18.8. The differing delays thus produced in the arrival time of the signal are called *multipath* errors.

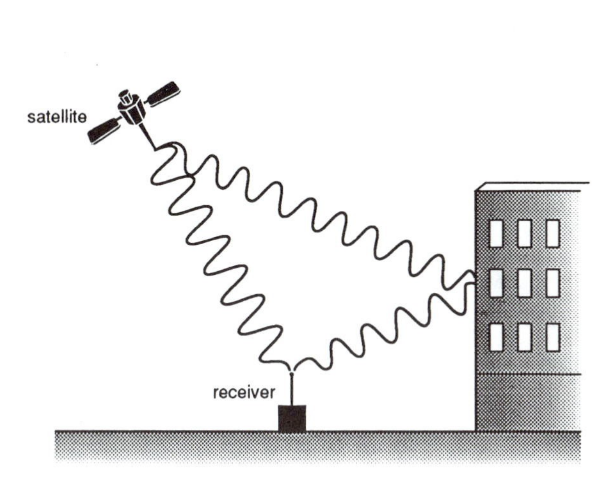

Figure 18.8 Multipath signals produced by reflections from the environment of the antenna

Chapter 19
Concluding Thoughts

When we look back over the course of evolution of timekeeping methods, we are struck again by the relentlessly accelerating pace of technological change: From centuries of sundials and water clocks, tower clocks, and pendulums, until within our own century we have electronic timekeeping, first with quartz crystals, then with quantum-based atomic clocks. Within the span of the last twenty-five years we have come close to the ultimate timekeeping mechanism: the inner quantum beat of a single atom frozen in space. This, at least, is the ideal for timing events within the realm of human experience.

It would be difficult to exaggerate the importance of the impact that such stable and *portable* time standards have already had on civil and military technology; this has come mainly from the realization of precise three-dimensional, high-speed navigation, which promises to bring about a wide variety of new applications affecting people's everyday lives. It does not require a lot of imagination to predict the many ways one might exploit the ability to pinpoint one's position anywhere on Earth using an inexpensive portable radio receiver. Already, luxury automobiles have as an option a built-in GPS navigational computer/receiver with a display of street maps and itinerary programming.

19.1 The Synchronization of Clocks

Precise timekeeping on a cosmological scale, however, would confront all the relativistic complexities of space–time distortions where gravitational masses exist, with such spectacularly singular examples as *black holes*, which have captured the popular imagination. As indicated in a previous chapter, the abandonment of an absolute and universal time in Einstein's general theory of relativity means that an operation as fundamental as the synchronization of remote clocks requires careful redefinition, which may not even be possible unambiguously in certain gravitational fields where there is a circulation of matter. This obviously brings into question the very idea of simultaneity of events occurring "at the same time" at separated points.

The concept of simultaneity clearly bears on the question as to whether an event, defined as something occurring at a given point in space at a particular time, precedes or follows another event; it is no less than a question about the future and the past. Now, the concepts of cause and effect, which are at the very foundation of empirical science, have meaning only when the relationships of *earlier* and *later* have a well-defined meaning, independent of the state of the observer. This, according to relativistic theory, imposes a restriction on events that are admissible as possibly connected as cause and effect. This restriction arises essentially from the fact that according to the theory, no interaction can propagate faster than the velocity of light. The relationship of events occurring at different points and at different times are best presented using graphs plotted with respect to a set of Cartesian coordinate axes. In the spirit of relativity, events are represented as *world points*, that is, points having four coordinates, the three space coordinates x, y, z and time t, as shown in Figure 19.1.

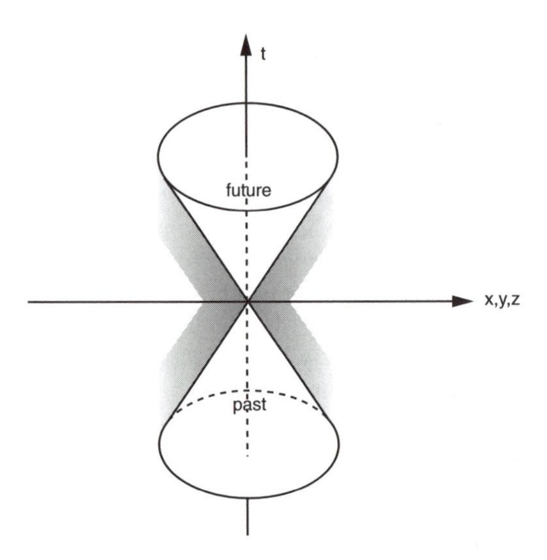

Figure 19.1 Graphical representation of the light cone separating "world points" that are in the absolute future and past relative to the origin

In terms of such a representation, the events whose world points lie in the *interior* of a certain geometric cone, called the *light cone*, are the only events that can be related to the event represented by the origin of the coordinate system as cause and effect. The light cone is defined as having its axis along the t-axis and its surface inclined to that axis at an angle whose tangent is numerically equal to the velocity of light, 2.99797×10^8 meters per second. Events represented by points in the upper half of the cone are in the absolute future and those in the lower half in the absolute past relative to an event represented by the origin. All events that are represented by points outside the cone can occur before or after the event at the origin, according to the state of relative motion between their coordinate systems and that of the observer. Of course, the very idea that one event can occur either before or after another event depending on the relative motion of the observer runs counter to our intuitive sense of absolute time and time ordering. We cannot quite escape the notion that the time ordering merely *appears* to depend on the state of relative motion, as some sort of illusion. This attitude presumes an underlying "reality" that forms a backdrop against which the observed universe can be compared, and thereby the "effect" of relative motion made manifest. The fact is that the instinct to look at time as an absolute is misdirected.

On a terrestrial, human scale, where these radical nonintuitive relativistic effects are extremely small, the precision which the new standards have made possible puts these radical, nonintuitive effects within range of study. The gravitational red shift and the Sagnac effect, described in Chapter 7, are examples. It is likely that any direct knowledge of such effects will, in the foreseeable future, come through the use of ultrastable atomic clocks and the ability to transfer ultraprecise time between distant points.

19.2 The Direction of Time

At the other, subatomic, end of the time scale, other interesting questions arise that again call into question the intuitive classical notion of time. These have to do with the symmetry (or lack thereof) of the laws that govern physical phenomena with respect to the operation of time reversal. This operation means physically reversing the direction of time, as we would if we ran a motion picture backwards. The fundamental question is whether the laws of physics are symmetrical with respect to the reversal of time; that is, whether all the laws of physics for the phenomena depicted in the time-reversed movie are indistinguishable from those in real life. In other words, can we tell, through any analysis of the action observed, whether a movie is being projected forwards or backwards.

The answer based solely on our everyday experience is an obvious yes! We must not, of course, use as criteria for distinguishability common human practices or expected patterns of behavior of living objects. It would not be difficult to conclude that a movie is being projected backwards if it shows all the cars racing down a highway backwards! Such a bizarre picture would not violate any physical law, but our knowledge of traffic laws makes it quite certain that the movie is being run backwards! In the realm of everyday experience there would seem to be little question about it: Time is an arrow, pointing in one direction, from a definite past to a definite future.

Quite apart from examples of the kind just cited, however, there are common experiences that are, for far more fundamental reasons, never observed in the reverse direction of time. For these so-called *irreversible processes* time indeed does seem to point one way. There are many obvious examples of this: the act of dissolving a spoonful of sugar in a cup of coffee, and the cooling of that cup; or on a far more complex level, the aging process in animals.

But the laws of physics, except possibly those operating in the realm of high-energy subatomic particle physics, are in fact symmetric with respect to time reversal, and the irreversibility of the processes mentioned above has to be reconciled with the physical laws involved. The time reversal invariance of Newton's laws of motion, for example, can readily be verified on a process sufficiently lacking in complexity, such as a two-particle collision, as exemplified by two ideal billiard balls colliding on a frictionless table top. In such a case there is absolutely no way of knowing the direction of time (assuming that the man with the billiard cue is out of the picture). However, in the examples of the coffee cup or the aging animal cited above, we are dealing with systems that comprise a vastly greater number of variables, with complex internal structures involving chemical and thermal processes. The reconciliation of the large-scale global irreversibility of the processes undergone by these complex systems with the microscopic particle-on-particle interactions that are reversible was a major challenge in the history of physics. Its solution, and the great advances made in the branch of physics called statistical mechanics in the 19th century, are due mainly to Boltzmann, a name with which we are already familiar.

In order to understand the basis of the arguments that provide a way out of what appears a logical dilemma, consider a much simpler example that nevertheless retains the essentials of the problem. Suppose a shotgun is discharged, resulting in a large number of lead shot from the cartridge being scattered at high velocity, usually, it is hoped, in the direction of some hapless game bird. It would indeed seem miraculous if the time-reversed sequence were ever witnessed, in which a large number of lead shot starting from widely scattered points converged on the muzzle of a shotgun and into a cartridge!

Now, because the classical equations of motion are symmetrical with respect to time reversal, if a particle starts from its final position with precisely the reverse of its final velocity, it will return along the same trajectory to precisely its original position, with the reverse of the original velocity. Thus if the motion of an individual shot could be followed by some form of high-speed cinematography, then a reversal of the motion picture would show a motion equally in conformity with the equations of motion. Again, we must exclude air resistance, since this introduces interactions with a vast number of particles (the air molecules), whose motion would then have to be included.

The question remains; Why is the process observed to proceed in only one direction in time? The answer essentially rests on the knowledge or lack of knowledge of the positions and velocities of all the particles involved. If the act of discharging a shotgun was one requiring that every single lead shot should hit a precisely prescribed point in the target area with a precisely prescribed velocity, then firing a shotgun would be just as rare an event as having the same number of lead shot converging from the target area back into the muzzle of the shotgun! If the discharge of a shotgun is depicted in an ordinary motion picture, and it is run backwards starting at a point where the lead shot have almost reached the target area, we have set up an initial state consisting of the scattered lead shot in their final positions and reversed final velocities. With these very special initial values, the lead shot would indeed go right back into the muzzle of the shotgun! Of course, in principle, it would be possible to set up an elaborate set of mechanisms that could project lead shot with precisely prescribed velocity vectors from precisely the right points in space so that they all converge more or less simultaneously onto the muzzle of a shotgun; and it is possible that the National Science Foundation would fund such a project! But that is not what we see when we view the backwards-run movie; the lead shot are all miraculously in the correct initial state.

In summary, when the motions of particles are followed in microscopically fine detail, time reversibility is observed in accordance with the invariance under time reversal of the equations of motion. That is, there is no preferred direction to time. However, on what is called a macroscopic scale, in which we deal with large, complex systems comprising a vast number of constituent particles in their broad behavior, there are certain processes that are irreversible. To observe these processes in reverse would require the presetting of such a large number of initial positions and velocities that it would be as unlikely as the proverbial monkey randomly hitting the keys of a typewriter and coming up with Shakespeare's *Hamlet*!

Without getting into the matter more deeply, we will simply state that such irreversible processes are always ones in which a state of "order" leads to one of higher "disorder." It was Boltzmann who gave definitions of what should be an appropriate measure of disorder, called *entropy*. Loosely speaking, it refers to the number of possible assignments of position and velocity to the particles making up the system occupying a given volume and energy. Time flows in the direction in which systems evolve into states of increasing disorder. In applying this law, care must be taken to include in the definition of the system all interacting entities.

19.3 Time-Reversal Symmetry in Subatomic Events

Finding an explanation that reconciles the unidirectional flow of time for macrosystems consistent with the microscopic reversibility of time does not prove or disprove the validity of that reversibility for physical processes involving particles at a fundamental microscopic level. This is a fundamental question, since it has been demonstrated in a number of important advances in physics that symmetry is often a fundamental principle in the structure of physical theory.

Symmetry takes many forms: We have already encountered several important examples of symmetry in molecules and crystals. The most familiar example is manifested approximately by the human figure, in which the left side is nearly a mirror image of the right side. More generally, any operation on a physical system, such as a reflection in a plane or rotation about an axis or displacement along a straight line, that results in the system being indistinguishable from the original is called a symmetry operation. Regular solids such as a cube or sphere may possess symmetry with respect to several operations. Thus a cube has 9 mirror planes, three 4-fold and four 3-fold rotation axes, that is, axes about which complete rotations take the cube through four or three indistinguishable positions. A sphere can be rotated through any angle about any axis passing through the center and the result is indistinguishable for any angle: This is spherical symmetry. More generally, whether it is an object such as a cube or a mathematical expression, we are dealing with a form or property returning, after some type of transformation of relevant variables, to what it was initially. For the cube it was its invariance under a transformation of the angles, variables defining its orientation, or reversal of signs of coordinates defining mirror images.

One of the most fundamental questions in the realm of elementary particle physics, where the fundamental building blocks of matter are studied, is whether particle interactions exist in which time-reversal symmetry is broken. When elementary particles of matter are broken up into their elementary constituents through high-energy collisions in accelerators, many of these fragment particles keep their identity for only very short intervals of time before other particles are created through their decay, which in turn decay into still other particles. This leads to a complex genealogy, which has been largely reduced to some order. The pillars on which elementary particle theory is built are the conservation laws. These go well beyond the familiar conservation of energy and momentum. They are associated with various kinds of symmetry in the (quantum) mathematical expressions describing the particle interactions, a fact of profound importance that cannot be discussed within the compass of this book. The conservation of momentum (both linear and angular) and energy is associated with the invariance of the system with respect to any (continuous) displacement of its linear and angular position, or the origin of the time coordinate. The existence of these symmetries in a system means that the choice of the position of the origin and orientation of the coordinate axes can be made arbitrarily; indeed, the sense that our space is uniform and isotropic is so ingrained that these choices are made arbitrarily without our being conscious of their having any possible physical implications. Other kinds of symmetry are invariance under such transformations as changing the sign of all the space coordinates (called reflection through the origin, or *inversion*) or reversing the sign of electric charge or, what is of particular interest to us, changing the sign of the time coordinate.

Symmetry under a change of sign of the spatial coordinates deserves further explanation to appreciate its significance. It may at first seem no more than a matter of convention as to the relative positive directions chosen for the x-, y-, and z-axes. Actually, unless the system is symmetric under inversion, it matters very much how the positive directions of the axes are chosen. This is best shown by an example: Consider the geometrical figure called the helix, as embodied, say, in the threads of a machine screw or a wire spring, as illustrated in Figure 19.2. This figure, as well as the coordinate axes themselves, can be *right-handed* or *left-handed*; this distinction must be made, since there is no *proper* rotation that will take one type into the other. A right-handed screw cannot be made into a left-handed screw simply by changing its orientation; nor can coordinate axes belonging to one type be brought into coincidence with the other type by any proper rotation. If we take the coordinates of points lying on a given helix with respect to some coordinate axes, change the sign of all the coordinates, and replot them on the same set of axes, the result will be another helix, but with the opposite helicity; that is, a right-handed helix becomes left-handed, and vice versa. Furthermore, we can verify that the result of inversion can also be achieved by a mirror reflection followed by a proper rotation, so that the mirror image of a right-handed screw is a left-handed one. The invariance of a physical interaction between particles under reflection through the origin means that the interaction and its consequences are indistinguishable from their mirror image, and cannot depend on the choice of "handedness" in the coordinate system used to describe them. This is more than a mathematical curiosity; where it exists, an important conservation law flows from it, namely the conservation of what is called *parity*. This term is used in this context to mean the attribute of a mathematical expression describing the quantum state being *odd* or *even*, in the sense of whether its sign simply changes or remains unaltered when the signs of the coordinates are changed. Thus finally, in an interaction that conserves parity, if the system starts in a state of definite parity it must retain it throughout the interaction: A state of even parity remains even, and similarly for an odd state.

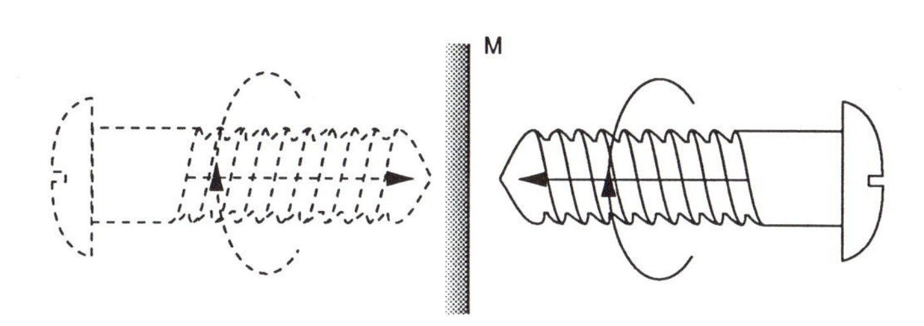

Figure 19.2 A right-handed screw is transformed into a left-handed one by reflection in a plane mirror

In quantum mechanics, parity of states plays an extremely important role, for example in the analysis of atomic and nuclear spectra. It had tacitly been assumed from the beginning that parity is always conserved, that the universe and its mirror image are indistinguishable. It happens, however, that there is an important example of a particle interaction in which this type of symmetry is "broken," namely the so-called "weak interaction," which is responsible for the radioactivity of some atomic nuclei in the form of high-energy electrons (beta particles). The violation of parity conservation was postulated by Lee and Yang in 1956 to explain what seemed bizarre decay patterns of certain fundamental particles and was brilliantly confirmed in the laboratory by Wu et al. using beta emission from oriented cobalt nuclei at temperatures approaching absolute zero.

Of more interest to us, however, is the discovery many years later by Fitch and Cronin (Fitch, 1980) that in the decay of a certain elementary particle, called K°, into two other elementary particles, π^+ and its antiparticle π^-, there is a break in the symmetry under the combined operations of charge and spatial coordinate sign reversal. This is referred to as *CP violation*: C for reversal of electric charge, and P for reversal of parity. This strongly suggests that time-reversal invariance is also violated, since it is expected on the basis of a theorem that goes to the very foundation of relativistic theory, known as the *CPT Theorem*, that all systems should be invariant under the *combined* operations of C, P, and time reversal, T. And in fact, the experimental evidence suggests that indeed time-reversal symmetry is broken in this particular decay process of K°. Of course, the significance of this result lies in its implications for physical laws as they apply at the most fundamental level, to processes involving the creation and decay of elementary particles. For us it can only be a source of wonderment as to how sophisticated physical theory has become and how the concept of time has evolved into an even more puzzling enigma.

References

Audoin, C. and Diener, W.A. (1992): Frequency, phase and amplitude changes of the hydrogen maser oscillation, *Proc 1992 IEEE Frequency Control Symposium*, 86–89.

Barnes, J.A., Allan, D.W., and Wainwright, A.E. (1972): The ammonia beam maser as a standard of frequency, *Precision Measurement and Calibration*, (5) (Frequency and Time), NBS Special Publication 300: 73–77.

Barrat, J.P. and Cohen-Tannoudji, C. (1961): Etude du pompage optique dans le formalisme de la matrice densité, *J. Phys. Radium*, (22): 329–336 and 443–450.

Bell, W.E. and Bloom, A.L. (1958): Optically detected field-independent transition in sodium vapor, *Phys. Rev.* (109): 219–220.

Bender, P.L., Beaty, E.C., and Chi, A.R. (1958): Optical detection of narrow Rb87 hyperfine absorption lines, *Phys. Rev. Letters* (1): 311–313.

Boyd, G.D. and Gordon, J.P. (1961): Confocal multimode resonator for millimeter through optical wavelength masers, *Bell Sys.Tech. J.* (40): 49.

Chantry, P.J., Liberman, I., Verbanets, W.R., Petronio, C.F., Cather, R.L., and Partlow, W.D. (1996): Laser pumped cesium cell miniature oscillator, *Proc. 52nd Annual Meeting, the Institute of Navigation*, Cambridge, U.S.A.: 731–739.

Cohen-Tannoudji, C. and Dalibard, J. (1989): *Journal of the Optical Society of America*, (B6): 2023.

Cohen-Tannoudji, C. (1962): *Annales de Physique* (Paris) (7): 423–460.

Cohen-Tannoudji, C. and Phillips, W.D. (1990): New mechanisms for laser cooling, *Physics Today*, Oct. issue, Am. Inst. of Phys.: 33.

Cohen-Tannoudji, C. (1962): *Annales de Physique* (Paris), (7): 423

de Camp, L. Sprague (1960): *The Ancient Engineers*, Dorset Press, New York.

Dehmelt, H.G. (1957): Slow spin relaxation of optically polarized sodium atoms, *Phys. Rev.* (05): 1487–1549.

Dehmelt, H. G. (1957): Modulation of a light beam by precessing atoms, *Phys. Rev.* (105): 1924.

Dehmelt, H.G. (1981): Invariant frequency ratios in electron and positron spectra..., in *Atomic Physics*, (7), D. Kleppner and F.M. Pipkin, eds. Plenum, New York.

Dehmelt, H.G. (1983): Stored ion spectroscopy, *Advances in Laser Spectroscopy*, F.T Arecchi, F. Strumia, and H. Walther, eds., Plenum, New York.

Dicke, R.H. (1953): The effect of collisions upon the Doppler width of spectral lines, *Phys. Rev.* (89): 472–73.

Drake, Stillman, (1967): *Galileo: a Dialogue Concerning Two Chief World Systems*, Univ. California Press.

Frazer, Sir James George (1958): *The Golden Bough*, Macmillan Co., New York, (Originally publ. 1922).

Gerber, E.A. and Sykes, R.A. (1966), State of the art quartz crystal units and oscillators, *Proc. IEEE* (54): 103–115.

Gordon, J.P., Zeiger, H.J., and Townes, C.H. (1955): *Phys. Rev.* (95): 282L.

Hall, John L. (1975): Sub-Doppler spectroscopy, methane hyperfine spectroscpy, and the ultimate resolution limits, *Physics of Quantum Electronics* (2) (Laser Applications to Optics and Spectroscopy), S.F. Jacobs, M.O. Scully, and M. Sargent, eds.: 401. Addison–Wesley, Reading.

Hoffmann-Wellenhof, B., Lichtenegger, H., and Collins, J. (1994) *Global Positioning System*, 3rd ed., Springer–Verlag, Wien.

Hoogerland, M.D., Driessen, J.P.J., Vredenbregt, E.J.D., Megens, H.J.L., Schuwer, M.P., Beijerinck, H.C.W., and van Leeuwe, K.A.H. (1994): Factor 1600 increase in neutral atomic beam intensity using laser cooling, *Proc. 1994 IEEE International Frequency Control Symposium*: 651.

Javan, A., Bennett, W.R. Jr., and Herriott, D.R., (1961): Population inversion and continuous optical maser oscillation in a gas discharge containing a He-Ne mixture, *Phys. Rev. Lett.* (6): 106.

Kastler, A. (1950): Quelques suggestions concernant la production optique et la détection optique d'une inégalité de population des niveaux de quantification spatiale des atomes, *J. Phys. Radium* (11): 255–265.

Kastler, A. (1957): Optical methods of atomic orientation and of magnetic resonance, *J. Opt. Soc. Amer.* (47): 460–465.

Kleppner, D., Berg, H.C., Crampton, S.B., Ramsey, N.F., Vessot, R.F.C., Peters, H.E., and Vanier, J. (1965): Hydrogen-maser principles and techniques, *Phys. Rev.* (A138): 972–983.

Kramer, G., Lipphardt, B., and Weiss, C.O. (1992): *Proc. 1992 IEEE Frequency Control Symposium*, Hershey: 39–43.

Kramer, Samuel N. (1963): *The Sumerians: Their History, Culture and Character*, Univ of Chicago Press, Chicago.

Lamb, W.E., Jr., (1964) Theory of an optical maser, *Phys. Rev.* (134): A1429.

Landau, L.D. and Lifshitz, E.M. (1969): *Mechanics* (2nd ed.), Pergamon Press, London, p. 82.

Maiman, (1960): Stimulated optical radiation in ruby masers, Nature (London) (187): 493–494; also "Optical and microwave optical experiments in ruby", *Phys. Rev. Lett.* (6): 564–565.

Major, F.G. (1977): Réseau à trois dimensions de quadripôles électriques élementaires pour le confinement d'ions, *J. de Physique-Lettres* (38): L-221–225.

Major, F.G. (1969): Microwave resonance of field confined mercury ions for atomic frequency standards, NASA Goddard Tech Report X-521-69-167.

Major, F.G. and Werth, G. (1973): High resolution magnetic hyperfine resonance in harmonically bound ground state mercury 199 ions, *Phys. Rev. Lett.* (30): 1155–1158.

Maxim, L.D. (1992): *Loran-C User Handbook*, U.S. Coast Guard Office of Radio Navigation Systems, Commandant Publication P16562.5.

Morley, Silvanus G. (1956): *The Ancient Maya*, Stanford Univ. Press (1st ed. 1946).

Norton, J.R. (1994): Performance of ultrastable quartz oscillators using BVA resonators, *Proc. 8th European Frequency and Time Forum*, Weihenstephan, (1), 457–465.

Panish, M.B., Hayashi, I., and Sumski, S., (1969): *IEEE J. Quantum Electron* (5): 210.

Paul, W., Osberghaus,O., and Fischer, E. (1958): Ein Ionenkäfig, Forschungsberichte des Wirtschafts- und Verkehrsministeriums Nordrhein-Westfalen, Nr 415, Westdeutscher Verlag, Cologne.

Peters, H.E., McGunigal, T.E., and Johnson, E.H. (1969): *Proc. 23rd Annual Frequency Control Symposium*, Atlantic City.

Peters, H.E., Owings, H.B., Koppang, P.A., and MacMillan, C.C. (1992): *Proc. 1992 IEEE Frequency Control Symposium*, 92–103.

Pierce, J.R. (1954): *Theory and Design of Electron Beams*, van Nostrand, Princeton, p. 41.

Poitzsch, M.E., Bergquist, J.C., Itano, W.M., and Wineland, D.J. (1994): Progress on cryogenic linear trap for Hg199 ions, *Proc. 1994 IEEE International Frequency Control Symposium*, p. 744.

Priestley, J.B. (1964): *Man and Time*, Aldus Books, London.

Ramsey, N.F. (1972): History of atomic and molecular standards of frequency and time, *IEEE Trans. Instrumentation and Measurement* (IM 21): 90–99.

Ramsey, N.F. (1956): *Molecular Beams*, Oxford University Press, Oxford, p. 124.

Ramsey, N.F. (1949): *Phys. Rev.* (76): 996.

Rochat, P., Schweda, H., Mileti, G., and Busca, G. (1994): *Proc. IEEE International Frequency Control Symposium*, Boston. 716–723.

Santarelli, G., Laurent, P., Lea, S.N., Nadir, A., and Clairon, A. (1994): Preliminary results on Cs atomic fountain frequency standard, *Proc. 8th European Frequency and Time Forum*, Technical University, Munich, p. 46.

Schove, D. Justin and Fletcher, Allen (1984): *Chronology of Eclipses and Comets AD 1–1000*, Boydell and Brewer Co. Suffolk.

Sorokin, P.P. and Lankard, J.R. (1966): Stimulated emission observed from an organic dye, chloro-aluminum phthalocyanine, *IBM J. Res. Dev.* (10) 2: 162–163.

Stephenson, F.R. and Morrison, L.V. (1982): *History of the Earth's Rotation, Tidal Friction and the Earth's Rotation II*, Brosche, P. and Sundermann, J., eds., Springer-Verlag Berlin, p. 29.

Tessier, M., Vanier, J. (1971) Théorie du maser au rubidium 87, *Canadian J. Phys.* (49): 2680–2689.

Touahri, D., Abed, M., Hilico, L., Clairon, A., Zondy, J.J., Millerioux, Y., Felder, R., Nez, F., Biraben, F., and Julien, L. (1994): Absolute frequency measurement in the visible and near infrared ranges, *Proc. 8th European Frequency and Time Forum*, Technical University, Munich, p. 19.

Townes, C.H. and Schawlow, A.L. (1955): *Microwave Spectroscopy*, McGraw Hill, New York.

Townes, C.H. (1962): *Topics on Radiofrequency Spectroscopy*, International School of Physics "Enrico Fermi": Course 17, p. 55.

Wong, N.C. and Lee, D. (1992): Optical parametric division, *Proc. 1992 IEEE Interna\tional Frequency Control Symposium*, p. 32.

Further Reading

General Reference

Vanier, J. and Audoin, C., *The Quantum Physics of Atomic Frequency Standards* Inst. of Phys. 1989.

Historical

Landes, David S., *Revolution in Time,*, Harvard Univ. Press, 1983.

Aveni, Anthony, *Empires of Time*, Kodansha America, Inc. 1995.

Darwin, George H., *The Tides*, W.H.Freeman 1962 (1st ed 1898).

Fredrick, L.W. and. Baker, R.H., *An Introduction to Astronomy* 8th ed. van Nostrand 1974.

Golino, Carlo, ed. *Galileo Reappraised*, Univ. of Cal. Press 1996.

Sobel, Dava, *Longitude, the true story of a lone genius*, Walker and Co. 1995.

Oscillations

Knudsen, J.M. and Hjorth, P.G., *Elements of Newtonian Mechanics*, Springer, 1995.

Pierce, John R., *Almost All about Waves*, MIT Press, 1974.

Nettel, S. *Wave Physics* 2nd ed., Springer Verlag, 1995.

Stuart, R.D., *An Introduction to Fourier Analysis*, Methuen, 1961.

Crystals

Wood, Elizabeth, *Crystals and Light*, van Nostrand Co., 1964.

Bottom, Virgil E., *Introduction to Quartz Crystal Unit Design*, van Nostrand Reinhold, New York, 1981.

Dmitriev, V.G., Gurzadyan, G.G., and Nikogosyan, D.N., *Handbook of Nonlinear Optical Crystals*, Springer, 1991.

Quantum Theory

de Broglie, Louis, *Matter and Light the New Physics*, W.W. Norton, New York, 1939.

Feynman, Richard P. *QED The Strange Theory of Light and Matter*, Princeton U., 1988.

Mott, N.F. *Elementary Quantum Mechanics*, Springer–Verlag, 1972.

Optics

Andrews, C.L. *Optics of the Electromagnetic Spectrum*, Prentice Hall, 1960.

Atomic Spectra

Hertzberg, Gerhard, *Atomic Spectra and Atomic Structure*, Dover, 1944.

Haken, H. and Wolf, H.C., *The Physics of Atoms and Quanta* 4th ed., Springer, 1994.

King, W.H., *Isotopic Shifts in Atomic Spectra*, Plenum Press, 1984.

Magnetic Resonance

Agarbiceanu, I.I. and. Popescu I.M., *Optical Methods of Radio-Frequency Spectroscopy*, Wiley, New York, 1975.

Relativity

Einstein, A., Lorentz, H.A., Minkowski, H., and Weyl, H., *The Principle of Relativity*, Dover, 1923.

Schwartz, Jacob, *Relativity in Illustrations*, Dover, 1989.

Schwinger, Julian, *Einstein's Legacy: Unity of Space Time*, Scientific American Library, No. 16, 1987.

Quantum Oscillators

Audoin, C., Schermann, J.P., Grivet, P., Physics of the Hydrogen Maser, *Advances in Atomic and Molecular Physics* (7), Academic Press, New York, 1971.

Bertolotti, M., *Masers and Lasers: an Historical Approach*, Adam Hilger, Bristol, 1983.

Hecht, Jeff, *Laser Pioneers*, Academic Press, Boston, New York, 1992.

Yariv, A., *Introduction to Optical Electronics*, Holt Rinehart and Winston, New York, 1976.

Demtröder, W., *Laser Spectroscopy*, 2nd ed., Springer, 1995.

Feynman, R.P., Leighton, R.B, and Sands, M., *The Feynman Lectures on Physics Vol III*, Addison Wesley, 1965 (for quantum mechanics and nitrogen maser).

Ion Traps

Ghosh, Pradip K., *Ion Traps,* Oxford Univ. Press, Oxford, 1996.

Laser cooling

Arimondo, E., Phillips, W.D., Strumia, F., eds., *Laser Manipulation of Atoms and Ions*, International School of Physics "Enrico Fermi", Course 118 Varrena, North-Holland, New York, 1992.

Minogin, V.G. and Litokhov, V.S., *Laser Light Pressure on Atoms.*, Gordon and Breach Science Publishers, 1987.

Metrology

Evans, A.J., Mullen J.D., and Smith D.H., *Basic Electronics Technology*, Texas Instrm., 1985.

Essen, L., *The Mesaurement of Frequency and Time Interval*, Her Majesty's Stationery Office, London, 1973.

Navigation

Hoffmann-Wellenhof, B., Lichtenegger, H., and Collins, J., *GPS Theory and Practice*, 3rd ed. Springer-Verlag, Wien, 1994.

USCG-Headquarters *Radionavigation Systems/United States Coast Guard*, Radionavigation Div., Doc. No. TD 5.8:R 11/4, 1984.

Maloney, Elbert S., *Dutton's Navigation and Piloting*, 14th ed. Naval Inst. Press, 1985.

Schmid, Helmut H., *Three-Dimensional Triangulation with Satellites.*, U.S. Dept. Commerce NOAA, US Govt. Printing Office, 1974.

Index

3-level laser, 325, 406

absorption, 33, 75, 89, 92, 113, 114, 120, 131, 135, 136, 140, 158, 205, 290, 314, 393, 403, 407
additive noise, 61, 300
adiabatic approximation, 372
ADP, 414
adsorption, 76, 184, 230
aeolian harp, 37
aging, 66, 75, 76, 236, 304, 452
Al_2O_3, 322
Allan variance, 81, 83, 177, 252
alumina, 298, 322, 324
A-magnet, 182, 186, 188, 206
ammonia maser, 207, 212, 215, 218, 220, 221, 224, 225, 228, 243, 245
amplification, 47, 55, 56, 89, 205, 211, 323, 397, 446
amplitude, 24, 28, 33, 35, 38, 40, 43, 152, 317
angular momentum, 93, 96, 97, 98, 99, 104, 107, 124, 257, 387, 402, 439
anomaly, 107, 436
antenna, 31, 63, 90, 121, 192, 350, 420, 421, 423, 424, 446, 447
antireflection, 320
antisymmetric, 102, 109
Ar^+ ion laser, 333, 338, 339, 348, 374
Arditi, M., 172
Arecibo, 420
Ashkin, A., 348
Astrolabe, 10, 11
atomic beam, 130, 131, 132, 133, 138, 185, 223, 242, 246, 384, 385, 390, 392, 394, 458
atomic fountain, 390, 394, 460
atomic hydrogen, 138, 225, 235, 236, 238
atomic magnetism, 117
atomic plane, 73, 74
atomic second, 440
atomic transitions, 267, 309

balance-wheel, 19
band theory, 109
beam, 14, 15, 34, 55, 57, 71, 74, 75, 180, 182, 183
Beaty, E.C., 168, 169, 171, 457
Bender, P.L., 168, 169, 171, 457
Bernoulli's principle, 12
binary counter, 86
blackbody radiation, 42, 90, 91, 92, 310
B-magnet, 183, 188, 196
Bohr magneton, 104, 118, 164, 226, 256, 389
Bohr, N., 93
Boltzmann constant, 149, 357
Boltzmann, L., 129, 452, 453
bond, 67, 68, 103, 112, 235, 248, 249, 402
Bragg reflection, 74
Bragg, W.H., 73
Bragg, W.L., 73
Breit, G., 163
Breit–Rabi formula, 163, 164, 187, 245, 296
Brewster angle, 321, 338
Brewster window, 321

buffer gas, 150, 153, 159, 161, 168, 170, 171, 178, 222, 255, 277

Cady, 65
calendar, 2, 3, 11
causality, 33
cavity, 54, 55, 56, 308, 309, 310
centrifugal force, 6, 437
cesium standard, 173, 176, 180, 369, 379, 380, 384, 428, 435
charge exchange, 266, 277
Chaucer, G., 11
Chi, A.R., 168, 169, 171, 457
Chilowsky, C., 64
circadian rhythm, 1
circular polarization, 124, 134, 135, 136, 139, 165, 407
Clairaut, 18
clepsydra, 12, 14
clock paradox, 154
CO_2 laser, 395, 406, 407, 411
Cohen-Tannoudji, C.N., vi, 172, 358, 359, 361, 362, 363, 457
coherence, 57, 224, 247, 249, 308, 318, 325, 414
collimator, 238, 242
collisions, 32, 90, 94, 126, 128, 142, 143, 152, 170, 230, 235, 247, 248, 249
communication, 64, 339, 341, 420, 422, 425, 428, 429, 430, 433
conductors, 54, 378
confocal optical cavity, 311
conservative system, 386
control stations, 433, 438, 439
cooling of atoms, 159
cooling of ions, 371, 372
coordinate system, 125, 145, 156, 309, 314, 364, 440, 441, 450, 455
Coriolis force, 402
coumarin, 334, 374
covalent bond, 67, 69, 103, 112
Cronin, J.W., 456
cross section, 230, 261, 291, 292, 355, 356
crown wheel, 13, 15
crystal symmetry, 69
Curie, P., 65
current density, 341, 343
cyclotron frequency, 262, 363
cyclotron radius, 266, 267, 371

dark current, 293, 300, 408
de Broglie, L., 93, 94, 462
Dehmelt, H.G., v, 107, 138, 166, 457, 458
detectors, 65
Dicke effect, 150, 153, 171, 233, 293, 364, 397
Dicke, R.H., 150, 152, 294, 458
dielectrics, 319
diffraction, 31, 58, 73, 192, 308, 310, 404, 409, 424
diffusion, 131, 138, 178, 180, 206, 255, 330, 333, 340, 343, 357, 371
Dirac, P.A.M., 99
direction finder, 422, 423
direction of time, 451, 452
dish antenna, 421, 424
dispersion, 28, 32, 33, 61, 89, 142, 175, 312, 441, 444, 445
Doppler cooling, 358, 359, 362, 365, 367, 372, 387
Doppler effect, v, 83, 146, 147, 148, 202, 203, 294, 392, 407, 431
Doppler side-bands, 294, 295, 365, 366, 372
Doppler, C., 146
Dunoyer, L., 130, 131
dye laser, 334, 337, 338, 339, 374, 375
dynode, 198, 300

Earnshaw's theorem, 259
eclipses, 3, 4, 5

ecliptic plane, 8
effusion, 185, 394
Einstein, A., 92, 136, 142, 144, 153, 155, 157, 462
elastic constant, 79
electric dipole moment, 207, 210, 212, 214, 215, 220, 312, 352, 411
electric field component, 32, 89, 352, 411, 423, 444
electromagnetic wave, 29, 34, 42, 53, 73, 89, 90, 124, 191, 202, 346, 348, 396
electron magnetic moment, 226
electron multiplier, 198, 283
electronic charge, 102, 120, 342
electrostatic fields, 299
electrostatic quadrupole, 212
emission, 57, 60, 90, 92, 113, 114, 120, 136, 137, 288, 290, 313, 314
energy bands, 113, 324, 340, 341
energy defect, 258
energy levels, 95, 97, 100, 101, 107, 108, 110, 111, 119, 287, 312, 324, 358, 383, 384, 404
entropy, 453
ephemerides, 430, 431, 441, 444, 446
equinox, 436, 439
escapement, 13, 15, 16, 18, 20, 21, 250
etalon, 334, 338
Euler, L., 346
excitation, 15, 36, 37, 39, 40, 57, 74, 90, 92, 208, 209, 238, 327, 329
extraordinary wave, 139, 412, 413

Fabry–Perot interferometer, 308
feedback, 47, 48, 50, 52, 78, 79, 85, 87, 175, 176, 218, 316, 317, 341, 343, 446
Fermi, E., 107, 226, 460
figure of merit, 27, 180, 215, 251, 304, 376
filter, 40, 55, 64, 167, 168, 169, 178, 197, 237, 251, 267, 268, 300, 301, 321, 334, 338, 343, 374, 378, 408
fine structure, 168, 225, 332, 403
Fischer, E., 268, 459
Fitch, V.L., 456
flicker noise, 60, 82, 177
flip-flop circuit, 86
flop-in, 133, 183, 188
flop-out, 133
fluorescence, 133, 136, 290, 292, 297, 299, 300, 302, 303, 306, 335, 336, 337, 338, 350, 392, 400, 406, 407, 409
flywheel oscillator, 198, 301
focusing magnet, 187, 188, 196, 238, 241
foliot, 15, 16, 18
fountain standard, 391, 394
Fourier analysis, 39, 49, 82, 150
Fourier, J., 39
Frazer, J.G., 1, 2, 458
frequency chain, 406, 414, 415
frequency dividers, 251
frequency pulling, 218, 219, 246, 317
frequency response, 49, 130, 141, 314
frequency stability, 28, 65, 80, 83, 223, 224, 246, 252, 334, 343, 377, 392, 394, 409, 434, 435
frequency standards, v, 177, 219, 223, 257, 259, 265, 272, 307, 339, 345, 363, 371, 396, 407, 408, 421, 441, 459
Fresnel formula, 319
Friedrich, 73
future and past, 450

GaAlAs diode laser, 140, 408, 409
GaAs diode laser, 341

gain factor, 317
Galileo, 16, 17, 26, 458, 461
Gauss's law, 265
Gaussian line shape, 149
Gaussian optics, 309
general relativity, 156, 158, 440
getter, 76, 184, 333
global positioning system, 432
gold black, 300
GPS, 157, 432, 433, 434, 435, 436, 437, 438, 439, 440, 441, 442, 444, 445, 446, 449, 464
Graham, G., 18
ground state, 57, 67, 101, 104, 105, 107, 108, 132, 133, 137, 138, 207, 225, 286, 324
ground wave, 425, 429
Gunn diode, 301
gyromagnetic ratio, 119, 122
gyroscopic motion, 8, 122, 128, 439

Halley, E., 20
Harrison, J., 20, 21
Heisenberg Uncertainty Principle, 37
Heisenberg, W., 37
helium, 76, 138, 170, 257, 258, 277, 285, 305, 322, 327, 328, 329, 330, 331, 332, 333, 373, 382, 383, 400, 406
He–Ne laser, 309, 327, 331, 375, 400, 403, 404, 405, 406, 407, 409, 414
hexapole, 187, 188, 238, 239, 240, 241
Hipparchus, 8
hole-burning, 374
homogeneous broadening, 142
Hooke, R., 20
hot wire detector, 197
Huygens, C., 17, 20
hydrogen atom, 93, 100, 107, 148, 207, 210, 224, 225, 228, 230, 231, 235, 237, 242, 247, 248, 253, 257, 287, 401
hyperfine interaction, 105, 161, 227, 230
hyperfine pumping, 167, 168, 169, 180, 286, 289, 292, 376, 383
hyperfine transition, 165, 170, 201, 221, 227, 244, 272, 287, 292, 296, 376, 380, 392

inelastic collision, 258, 264
inertial frame of reference, 439
inhomogeneous broadening, 142, 144, 335
intensity gradient force, 132
interference, 35, 57, 93, 114, 156, 167, 192, 301, 318, 319, 320, 343, 346, 412, 429
inversion spectrum, 205, 210, 220
iodine, 400, 404, 405
ion confinement, 260, 274, 285, 371
ion motion, 259, 261, 269, 271, 273, 295, 296, 297, 305, 364, 367, 378
ion pump, 184, 238, 256, 298, 371, 383
ion resonance detection, 280
ionization, 197, 333
isotope filter, 168, 178
isotope shift, 167
Ives, H.E., 148

Javan, A., 307, 327, 458
Johnson noise, 59, 60, 61, 251, 282

Kastler, A., 133, 136, 138, 172, 307, 358, 458
KDP, 414
Kepler, J., 346
klystron, 54, 55, 199, 200, 219, 301, 397
Knipping, 73

Lamb dip, 317, 318
Lamb, W.E., 236, 317, 459
Lambert's law, 314
Lamb-Retherford experiment, 236
Land, E.H., 139
Langevin, P., 64
Larmor theorem, 272, 273
lasers, 307, 312, 322, 327, 331, 332, 333, 334, 339
Laue pattern, 74
L-C circuit, 140, 279
light cone, 450
light polarization, 359, 360
light pressure, 346, 348, 366, 387, 389
light shift, 159, 172, 222, 352, 358, 360, 361, 362, 363, 409
linear medium, 34
linear Paul trap, 378
linewidth, 290, 393, 399
Liouville's theorem, 386, 390
liquid dye lasers, 334
longitudinal modes, 57, 308, 333, 343
loop antenna, 121, 423, 446
Loran-C, v, 419, 425, 427, 428, 432, 439, 459
Lorentz force, 256, 261, 262, 263, 346
Lorentz transformation, 145, 147
Lorentz, H.A., 32, 89, 120, 142, 143, 145, 312, 462
Lorentzian line shape, 141, 355

magnetic fields, 54, 129, 137, 161, 162, 182, 189, 195, 201, 223, 225, 227, 238, 245, 253, 255, 259, 274, 323, 346, 363, 371, 382
magnetic moment, 97, 99, 104, 105, 106, 107, 118, 119, 122, 127, 128
magnetic resonance, 117, 122, 123, 126, 128, 132, 133, 134, 135, 136, 138
magnetic shielding, 189, 255, 382, 383
magneton, 104, 118, 164, 226, 256, 389
magneto-optical trap, 387, 389, 392
Major, F.G., vi, 285, 286, 459
Majorana transitions, 188, 244
masers, 220, 222, 236, 251, 252, 307, 312, 322, 435, 457, 459
mass filter, 197, 267, 268, 378
mass spectrometry, 278, 285
Mathieu equation, 270, 378
Mathieu, E., 270
Maxwell, C., 29
Maxwell's theory, 29, 30, 89, 90, 144
Mayan calendar, 2
mean free path, 131, 148
mercury 199 ion, 289, 290, 459
metastable state, 134, 328, 330, 380
meter, v, 145
methane, 170, 400, 401, 402, 403, 404, 405, 458
methane-stabilized He-Ne laser, 400
microwave cavity, 54, 55, 178, 191, 192, 207, 212, 213, 214, 217, 221, 242, 243, 310
microwave source, 191, 203, 417
MIM, 411
Minkowski, H., 145, 462
mirrors, 56, 57, 156, 192, 308, 309, 310, 316, 318, 320, 321, 326, 331, 338, 339, 374, 386
mixer, 63, 302, 414
modes of oscillation, 44, 126, 295, 397
modulation, 38, 87, 150, 151, 152, 166, 173, 190, 199, 203, 218, 247, 280, 293, 294, 303, 339, 364, 409, 416, 417, 420, 427, 441, 442, 445
momentum conservation, 412
monitoring stations, 438, 440

multilayer dielectric mirror, 318, 319, 321
multipath error, 447

NAVSTAR, 419, 422, 432
Nd-YAG laser, 327
neodymium, 326
Newton, I., 18, 24, 26, 32, 42, 61, 345, 436
Nicholson, A.M., 66
noise figure, 251
Nonlinear Crystals, 411
Nonlinear Media, 33
normal modes, 36, 39, 44, 56, 57, 93, 95, 96, 97, 264, 401, 406
nuclear magnetic moment, 105, 257, 286
nuclear magneton, 104, 164
nuclear spin, 105, 107, 108, 161, 181, 209, 210, 220, 257, 286, 287, 288, 358, 360, 404

Omega, 419, 422, 429
optical cavity, 56, 57, 309, 310, 311, 320, 322, 326, 331, 333, 338, 343, 374, 392, 395, 397, 416
optical frequency, 56, 71, 312, 316, 322, 323, 330, 334, 346, 352, 375, 376, 380, 396, 397, 407, 408, 414, 416, 417
optical gain, 326, 331, 335, 336, 338, 341, 374
optical molasses, v, 357, 386, 392, 394
optical pumping, v, 136, 137, 138, 140, 165, 166, 172, 180, 257, 258, 286, 290, 305, 307, 313, 326, 350, 358, 360, 362, 379, 408
ordinary wave, 412
organic dye, 334, 336, 460
Osberghaus, O., 268, 459
oscillator, 25, 27, 28, 29, 47, 52, 53, 54, 55, 56, 58, 59, 62
oxazine, 339

pallet, 15, 19
paramagnetic atoms, 122
parametric oscillation, 37, 416
particle nature, 92, 369
particle suspension, v
Paul trap, 267, 272, 274, 275, 285, 289, 295, 296, 297, 363, 364, 369, 371, 377, 378
Paul, W., 133, 238, 267, 268, 271, 272, 278, 282, 459
Pauli exclusion principle, 101, 230
pendulum, 16, 17, 19, 20, 24, 25, 26, 34, 38, 43, 44, 117, 208, 277
Penning trap, 259, 263, 272, 276, 296, 363
Penning vacuum gauge, 259, 260
Penning, F.M., 259
perigee, 437
perturbation, 119, 141, 172, 195, 223, 224, 234, 255, 285, 437, 438
Peters, H.E., 229, 247, 251, 252, 458, 459
phase, 2, 24, 27, 28, 33, 49, 50, 147, 199, 200, 201, 250, 316, 385, 414, 427
phase matching, 412, 414, 416
phase shifts, 37, 79, 249, 250
Phillips, W., 357, 457, 463
photomultiplier, 293, 298, 300, 302, 304, 409
photon, 92, 113, 121, 124, 134, 135, 136
photon counting, 303, 304
Pierce, G.W., 66
Pierce, J.R., 260
piezoelectricity, 65, 70
Planck, M., 90, 91, 92, 93
Planck's constant, 91, 94
p–n junction, 340, 341
point contact, 87, 411
Poisson distribution, 283, 370, 376
Polarization Gradient Cooling, 358

polymer, 223
population inversion, 211, 313, 314, 323, 324, 325, 326, 327, 330, 331, 332, 334, 335, 336, 337, 338, 341, 374, 398
power, 42, 59, 60, 317, 326, 339, 374, 375
precession of the equinoxes, 8, 421, 439
principle of superposition, 34
PRN code, 441, 442
pseudorandom number code, 446

Q-factor, 27
quadrupole field, 212, 214, 269, 273, 274, 276, 279, 297, 377, 389
quantum, v, 89, 90, 92, 93, 94, 95, 96, 97, 98
quantum number, 96, 97, 98, 99, 100, 101, 104, 105, 108
quantum state, 94, 99, 101, 103, 105, 109, 114
quarter-wave plate, 139
quartz crystals, 65, 79, 449
quartz movement, 88
quartz plate, 71
quartz resonator, 63, 64, 65, 67, 74, 75, 76, 77, 78, 117
quartz watch, 88

Rabi, I., 132, 163, 193
radar, 54, 64, 83, 205, 263, 396, 422, 424, 425, 434
radio direction finder, 422, 423
radio receiver, 63, 159, 424, 446, 449
radio telescope, 31, 420, 421
Raether, M., 267
Ramsey cavity, 195, 203, 380, 384, 391
Ramsey, N.F., vi, 132, 133, 135, 182, 189, 190, 194, 195, 224, 229, 231, 458, 460
random walk, 61, 62, 82, 153, 266, 356, 398
rate equations, 325
refraction, 31, 139, 319, 348, 412
refractive index, 31, 32, 33, 318, 320, 334, 341, 348, 412, 416, 445
relativistic Doppler effect, v, 147
resolving power, 192, 420, 421
resonance, v, 27, 28, 36, 37, 38, 39, 52, 56, 76, 77, 79, 80, 84, 117, 133, 134, 135, 136, 138, 139, 141, 142, 143
resonators, 41, 54, 63, 65, 74, 75, 76, 91, 117, 254, 459
rf heating, 277, 364
rhodamine 6G, 334, 335, 339
ring laser, 374
Robinson, H., vi, 172
rubidium, v, 104, 107, 108, 119, 120, 136, 140, 159, 161, 163, 167, 171, 175, 176, 178, 221, 222, 321, 350, 358, 364, 379, 408, 409, 434, 460
ruby laser, 71, 322, 323, 324, 326
Rutherford, E., 92

Sagnac effect, 156, 440, 451
satellite orbit, 157, 436, 437, 439, 440, 441, 444
saturated absorption, 392, 393, 400, 403, 404, 405, 407
saturation, 217, 220, 231, 233, 240, 298, 317, 392, 393, 400, 406
scattering of light, 348, 350
Schawlow, A.L., 307, 322, 399, 460
Schrödinger equation, 94, 95, 97, 100
second harmonic generation, 411
seconds, 9, 10, 81, 86, 87, 88, 177, 178, 267, 285, 324, 326, 327, 332, 336, 360, 404, 409, 421, 425, 427, 434, 439, 443, 444

secular motion, 379
selection rules, 113, 120, 121, 136, 165, 166, 167, 257, 328, 358, 359, 387, 407
semiconductors, 110, 112, 114, 339
servo control, 78, 244, 403
shell structure of atoms, 99
shot noise, 51, 59, 60, 61, 180, 196, 279, 282, 292, 293, 300, 304, 370, 376
sidereal day, 7, 8, 436
signal-to-noise ratio, 180, 183, 196, 197, 222, 224, 276, 292, 293, 295, 369, 370, 376, 377, 380, 384, 390, 419, 427
simple harmonic motion, 17, 19, 24, 33, 135, 139, 150, 240, 261, 364
simultaneity, 153, 449
singing condenser, 64
singlet states in molecules, 336
Sisyphus Effect, 358
sky wave, 425, 427, 429
solar eclipse, 4, 5, 438
solar wind, 437, 438
solar-blind photomultiplier, 300
sonar, 64
space charge, 276
space quantization, 97, 132, 181, 225, 360
spectral line, 120, 130, 131, 135, 140, 142, 143, 148, 167, 168, 170, 220, 243, 288, 290, 291, 307, 324, 326, 372, 374, 375, 390, 393, 395, 397, 398, 399, 458
spectral purity, 58, 83, 87, 176, 203, 301, 303, 306, 311, 332, 374, 392, 395, 397, 399
spectroscopic notation, 324, 327
spectrum, 35, 36, 37, 40, 41, 42, 56, 60, 82, 90, 119, 152, 203, 205, 210, 271, 294, 396
spin exchange, 230, 234, 257, 291, 302
spontaneous emission, 57, 92, 136, 137, 172, 223, 290, 313, 314, 326, 329, 350, 352, 355, 356, 379
stability of oscillators, 65
stabilized CO_2 laser, 407
Stark effect, 172, 210, 211, 220, 353
stationary state, 93, 95, 96, 98, 126, 172, 208, 209, 353
stellar interferometer, 192, 193
Stephenson, F.R., 460
Stern–Gerlach experiment, 131, 180
stimulated emission, 57, 92, 134, 135, 136, 164, 205, 290, 311, 312, 313, 316, 322, 326, 334, 350, 352, 381, 397
strong focusing principle, 269
sub-Doppler line widths, 400
sundials, 9, 449
sunspot activity, 425, 429, 445
superheterodyne, 63
surface ionization, 197
symmetry, v, 65, 67, 68, 69, 70, 74, 96, 102, 105, 110, 121, 127, 139, 154, 187, 207, 209, 210, 212, 213, 230, 239, 260, 271, 297, 312, 323, 350, 364, 401, 402, 411, 412, 421, 451, 453, 454, 455, 456
synchronization of clocks, 155
synthesizers, 87, 175

Teflon, 223, 224, 230, 242, 247, 248, 288
temperature compensation, 77, 78, 79
temperature stabilization, 244
thermal noise, 59, 224, 225, 282
threshold for oscillation, 221, 253, 326, 331
tides, 438
time, v, 1, 2, 3, 9, 37, 84, 86, 153, 154, 155, 156, 157, 158, 179

time bridge, 303, 380
time reversal, v, 451, 452, 453, 456
time-based navigation, 419
titanium, 184, 322, 371, 383
tower clocks, 15, 16, 449
Townes, C.H., 205, 206, 215, 217, 224, 228, 322, 399, 458, 460
TRANSIT, 419, 422, 430, 431
transition probability, 127, 195, 325, 372, 379, 381
triplet state, 337
two-photon transition, 407, 408

ultrasonic transducer, 64

vacuum pump, 131, 183, 206, 238, 253, 255, 264, 435
vacuum shell, 183, 237, 253, 293, 297
van der Waals force, 224, 249
vapor pressure, 178, 184, 298, 299, 382, 404
VCXO, 302
vector model, 98, 104, 106, 122, 123, 128, 227
verge, 15, 16, 18
very long baseline interferometry, 420, 439
VLBI, 419, 420, 439, 441
von Laue, M., 73

wall shift, 159, 247, 248, 250, 251, 252, 253, 254, 255
water clock, 12, 13, 14, 117, 449
wave function, 59, 94, 95, 96, 99, 100, 101, 102, 103, 109, 126, 208, 209, 213, 230
wave mechanics, 94
wave motion, 39, 147
wave propagation, 446
wave propagation in ionosphere, 425
wavefront, 31, 412
waveguide, 214, 429
wavelength, 31, 53, 56, 64, 73, 74, 89, 91, 94
Werth, G., 286, 459
Wood's horn, 300
Wood's tube, 236, 237
work function, 197, 301
wrist watch, 21

X-ray diffraction, 73

Young, T., 57, 346

Zeeman effect, 97, 120, 121, 129, 218, 220, 333, 387, 389
zeolite, 184
zero point oscillation, 134